Foundations of Engineering Mechanics

Series Editors: V. I. Babitsky, J. Wittenburg

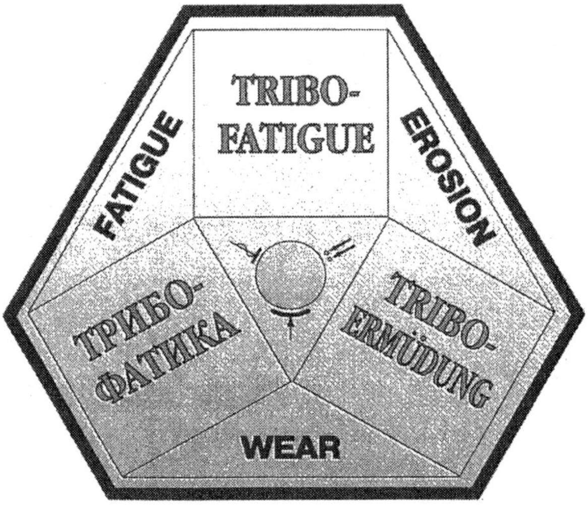

L. A. Sosnovskiy

TRIBO-FATIGUE

Wear-Fatigue Damage and its prediction

With a Preface by
Professor *K. V. Frolov*, DSc,
Academician of the Russian Academy of Sciences
and National Academy of Sciences of Belarus, and
Professor *N. A. Makhutov*, DSc, Corresponding Member
of the Russian Academy of Sciences

Translator *L. F. Burtsev*
Editor of the translation *R. S. Sosnovskaya*

With 209 Figures

Springer

Series Editors:
Vladimir I. Babitsky
Mechanical and Manufacturing Engineering
Loughborough University
Loughborough LE11 3TU, Leicestershire
United Kingdom

Jens Wittenburg
Institut für Technische Mechanik
Universität Karlsruhe (TH)
Kaiserstraße 12
76128 Karlsruhe
Germany

Author:
Professor Leonid A. Sosnovskiy
Belarusian State University of Transport
Volotovskaya St. 4
246050 Gomel-50
Belarus

Translator:
L. F. Burtsev
Editor of the translation
R. S. Sosnovskaya

ISBN 3-540-23153-6 Springer-Verlag Berlin Heidelberg New York

Springer is a part of Springer Science+Business Media
springeronline.com

Printed in Germany

Typesetting: data delivered by editor
Cover design: deblik, Berlin
Printed on acid free paper 62/3020/M - 5 4 3 2 1 0

PREFACE TO THE ENGLISH EDITION

The author read several lectures on the concepts of tribo-fatigue first in 1985/86 at the Belarusian State University of Transport and they were published in 1988 under the title 'A Complex Assessment of Reliability of Active Systems Based on the Criteria of Fatigue and Wear Resistance (the Fundamentals of Tribo-fatigue)'.

It is a challenging task to write a textbook on any discipline intended for university students. It is still harder to write a textbook on a new discipline first introduced into the curriculum of the university. It explains the gap of 14 years between the publication of the first textbook and the present publication. They were the years of continuous intensive research. It is suffice to say that during this time over 200 scientific works were published, including several monographs; four international symposia on tribo-fatigue were held (Gomel, Belarus, 1993; Moscow, Russia, 1996; Beijing, China, 2000; Ternopil, Ukraine, 2002).

No information is available if such a manual exists in English at all. Now the Springer Publishers fill up the gap providing specialists from various countries with firsthand information about the ideas of tribo-fatigue and the results of research in this domain.

Tribo-fatigue is a new vigorously developing branch of mechanics that has emerged in response to practical challenges of machine building. It was impossible, from the standpoint of both tribology exclusively or mechanical fatigue solely, to make any valid assessment (theoretical or experimental) of damage, durability or limiting states of such specific mechanical systems that take up and transmit cyclic loads while operating under friction (be it sliding, rolling, impact and others). It is explained by the fact that in operation complex wear-fatigue damage emerges in such systems (termed 'active systems' in tribo-fatigue). It is exactly due to this fact that these active systems are, as a rule, most essential in any machine. Any failure of these systems leads both to significant material losses and breach of guarantees of people's safeguarding.

Now that the main ideas of tribo-fatigue have been formulated, described analytically and validated experimentally, they can be systematized in the following manner.

A. It was found that damages due to contact (friction) and off-contact (cyclic) loading do not add up, they interact in an intricate manner. The traditional theory of damage summation has acquired a new and unexpected furtherance in the nonlinear statement. Yet, two problems have emerged immediately: how a variety of multiform and numerous damages should be assessed quantitatively? What is the possible result of their interaction?

W. Weibull and V. Bolotin were first to develop the statistical theory of size effect in linear stressed state. The dependence of the limiting stress on the volume of the loaded specimen was determined according to this theory. Using the works of W. Weibull and V. Bolotin L. Sosnovskiy formulated the statistical model of the deformable solid with a dangerous volume for any combined stress states. It provided explicit answers to the first question: a procedure of calculating the new measure of damage has been developed – the relative dangerous volume during cyclic deformation, friction, complex loading. The idea of the damage tensor generalized this research. In fact, it is justifiable to treat the dangerous volume as a phenomenological equivalent of damage of a deformable solid under any loading conditions.

In order to answer the second question, a (phenomenological) theory of interaction of damages had to be developed. It is based on the idea about Λ-function (or R-parameter) of interaction, that can acquire three groups of values. If $\Lambda >> 1$, the processes of damaging due to various loads strongly intensify. On the opposite, if $\Lambda << 1$, the processes of damaging strongly slow down. While at $\Lambda = 1$ there is a usual damage summation studied traditionally. It is easy to realize that at $\Lambda >> 1$ they are the systems that soften spontaneously in the process of operation, at $\Lambda << 1$ they are the systems that harden spontaneously and at $\Lambda = 1$ the system is stable.

B. It would seem clear that various signs indicate when active systems can reach their limiting states: whether it is the criterion of mechanical fatigue (volume fracture), or just the criterion of wear (critical surface damage), or both these criteria simultaneously. In reality a complex interaction of damages due to contact and off-contact loads corrects these ideas radically. It turns out that the characteristics of resistance to fatigue strongly depend on the conditions and processes of friction. Moreover, it is established theoretically and experimentally that friction with wear can both strongly weaken and reinforce the resistance of a system to fatigue. Whence the idea of the direct effect emerged.

While the direct effect was seemingly expected and perceived by specialists naturally, the idea of the back effect was unexpected; on the contrary, it was initially categorized as impossible. Nevertheless, the revolution in the thinking has already occurred. While tribologists had attributed the processes of wear and friction for over 150 years only and exclusively to contact loading, we are now definitely aware that off-contact cyclic loading can strongly intensify (or reduce) wear and correspondingly alter the friction coefficient.

C. Since the direct and back effects were established, it necessitated to develop the theory of limiting states of active systems with allowance for these effects, i. e. to apply non-traditional methods. For the time being the terms of stresses failed to contribute to such method. It was developed on more general, energy concepts.

Like in the case of development of the theory of damage interaction, two problems emerged immediately: how to identify in the total input energy that part which is expended exceptionally for appearance and development of damages? What is the critical energy beyond which the state of the system becomes limiting?

The first question was easily, though formally answered from the boundary conditions for the equation of limiting states. Though valid mathematically this answer does not seem fully validated from the point of view of the mechanics of damage; the parameter that identifies the effective part of the total energy should apparently depend on the loading conditions. The procedure of determination of this parameter given in the book does not make this condition binding. This possible inaccuracy is compensated, in fact, that the Λ-function of damage interaction is introduced into the criterion of the limiting state strongly responsive to any variation in the loading conditions.

A specific situation emerges in relation to the critical energy that transforms the active system into a limiting state. This energy, on the opposite, should not depend on the manner how the limiting state is reached and what mechano-physical mechanisms of damage come into effect. Therefore, this energy should be a fundamental characteristic of matter. The author realized that this characteristic could and should be the energy of breaking of atomic bonds (or the energy of activation of the fracture process).

The traditional criteria of the limiting state assert that intensification of effective stresses is equivalent to a loss of reliability and durability by an object. The energy criterion of limiting states of active systems given in the book 'permits' different situations in reality: contact and off-contact loads grow under definite conditions and the system's reliability increases concurrently; meanwhile such loads are light in other conditions, yet opposite to expectations, they accelerate degradation of a system. Such unusual statements have been proved experimentally. For example, during contact-mechanical fatigue (when contact and off-contact loads are effective concurrently) the endurance limit can increase more than 1.5 times versus the limit unaffected by any contact load (the direct effect). The ultimate contact pressure the system can tolerate increases 1.25 times if an extra cyclic load is excited in it (the back effect). It is explained convincingly and easily: it is the matter of real processes of damage interaction that occur under given conditions.

D. It is hard to overestimate the significance of experimental determination of new characteristics of resistance of materials to wear-fatigue damage. A number of inventions and ingenious designing solutions have led to new methods and high technologies of wear-fatigue tests and at present these characteristics are determined for a variety of conditions. In fact, a new class of testing equipment has been developed represented by the SI-series machines for wear-fatigue tests. Their unique potential is briefly described in the book.

E. The main task of tribologists is apparently to combat wear. Huge effort and means are spent for the purpose all over the world. Prevention of fatigue breakdowns is presumably the main task of specialists in mechanical fatigue. Again huge effort and means are spent for the purpose all over the world. From the viewpoint of tribo-fatigue, it is time, at least in some situations, to control reasonably the damaging phenomena rather than to try to suppress them, because it is stated above that wear, similarly to cyclic stresses, can be useful in the sense of performance of active systems. The book shows an algorithm how to solve the problem of optimum control over complex wear-fatigue damage of active systems of machines and equipment.

G. The method of strength calculation has presently reached certain perfection. It is impossible to state that similar achievements have been made in the calculations of friction and wear. In our view, it is in part explained by the fact that they are based on the mechanics of a discrete contact between bodies with rough surfaces rather than on the mechanics of a deformable solid. If the linear wear to the depth corresponding to the height of projections on a rough surface is accepted zero, then the interaction between bodies in friction and wear should be evidently described with allowance for deformation and surface damage of nonrough contacting surfaces.

The present book describes the next essential step in developing the methods of strength calculation: an engineering procedure is developed to perform calculations using the criteria of surface strength, i.e. wear resistance. It is based on the fact experimentally established that full fatigue curves during cyclic deformation and friction have similar patterns and comprise four main regions that describe quasistatic, low-cycle, multicycle and gigacycle damage and fracture. Then, irrespective of the mechanisms of damage and fracture, it is possible to formulate a unified and orderly procedure of calculating strength of structural elements, friction pairs and active systems. The method of determination of cross sectional dimensions of a loaded object described in the strength of materials is modified: the effect of friction processes on the change of mechanical fatigue characteristics are allowed for. In addition, similar procedures of calculating the required size of the contact area in friction and what the friction coefficient should be to ensure the normative reliability of a system, nave been worked out.

It is a prime feature of the methods of calculation of volume and surface strength of components of an active system is that both direct and back effects are duly taken into account.

So, it is time now to switch over from calculation of strength of individual pieces to designing of active (mechanical) systems of machines using the tribo-fatigue criteria. This switch-over is supported by the methods of design and experimental assessment of damage and limiting states of active system provided by tribo-fatigue.

The book contains some other new results that specialists will definitely appreciate. For example, they will learn the concept of assessment of risk and safety which is free of any subjective rating of material damage; the method of quantitative analysis of quality of materials based on mechanical characteristics; the concept of the friction coefficient in an active system that is determined with allowance for the effect of an off-contact cyclic load, etc. We share the common view that tribo-fatigue is a specific way of setting up and solving practical problems leading to development and implementation of methods of improving reliability of active systems of modern machinery together with saving labor and material cost in production and operation. We would like to draw attention of specialists that it is necessary to develop the theory of translimiting states of systems. The initial foundation for this theory is briefly outlined in the present book.

...When we mention here some works of Professor L. Sosnovskiy disclosed in the present book, the reader will certainly understand that his colleagues and associates have contributed and continue to contribute largely to the results of research in this new and promising domain of knowledge. Specialists of the S&P Group TRIBOFATIGUE Ltd (Gomel) have been cooperating highly fruitfully for many years with researchers of the IMASH of the Russian Academy of Sciences (Moscow), the Institute of Problems of Strength of the National Academy of Sciences of Ukraine (Kiev), the IMINMASH of the Belarusian National Academy of Sciences (Minsk) and others. The fifth international symposium on tribo-fatigue will take place in 2005. It will provide an opportunity for researches from many countries to evaluate the past stage of progress of tribo-fatigue research. Publication of the present book in English will certainly favor mutual understanding between scientists from different countries in a new and interdisciplinary field of knowledge.

It is believed that the book will be useful both for university instructors and students, for scientific workers, post-graduate studies, engineers and for all those who are eager to know about the problems how to rate and improve the service life of mechanical systems operating under complex loading.

Professor *K V Frolov*, DSc,
Academician of the Russian
Academy of Sciences and
National Academy of Sciences
of Belarus

Professor *N A Makhutov*, DSc,
Corresponding Member of the
Russian Academy of Sciences

Moscow, May 2004

PREFACE TO THE RUSSIAN EDITION

It is the first effort to write a textbook on fundamentals of tribo-fatigue, therefore I believe that its content should be explained somehow.

(1) The first chapter introduces tribo-fatigue and deals with a general analysis of the problems of volume fracture and surface damage of materials. It discloses basic information of the disciplines that all future engineers and designers study in some way and serves as a foundation of tribo-fatigue. This information is systematized to help understand what tribo-fatigue is, on the one hand, and it reflects my main concern that it can be directly used to convey the essential sense of the textbook, on the other hand.

As a rule, future engineers learn the strength of materials profoundly, therefore the key problems of static strength are disclosed just briefly. Yet, mechanical fatigue is described exhaustively. There are two reasons why: first, the traditional course of strength of materials treats it in an utterly sketchy way; second, tribo-fatigue is based on modern ideas about fatigue damage and fracture of materials and structures. When writing about friction and wear, I was keeping in mind that future engineers study this discipline, so a variety of common, usually taught methods of calculations are not repeated here.

The end of the chapter introduces the theory of reliability of mechanical systems, the criteria of fatigue and wear resistance, in particular.

The first chapter thus systematizes and covers briefly the initial data the student should know and useful for instructors to gain experience.

(2) The basic sense of tribo-fatigue is disclosed in chapters 2–5.

According to Interstate Standard, GOST 30638–99, tribo-fatigue is a "science of wear-fatigue damage and fracture of active systems of machines and equipment". The active system is any mechanical system that bears and transmits alternating working loading and in which the process of friction appears in any its manifestation simultaneously whether it is sliding, rolling, slip, impact, etc (Chapter 2). Complex wear-fatigue damage is typical for active systems due to kinetic interactions between the phenomena of fatigue, friction and wear, erosion or corrosion. Exceptionally basic methods of analyzing and predicting such damage are disclosed (Chapters 2, 4 and 5) and they are based on the following:

(a) a statistical model of the deformable solid body with a dangerous (damaged) volume that enables to assess real damage of the object under the effect of a given system of loads;

(b) a phenomenological concept of interaction between dangerous volumes due to contact and off-contact loads that serves to describe integrally and to reflect equivalently the statistics and the direction (hardening – softening) of real interactions of damages in the loading of the object;

(c) an experimentally established similarity between the full fatigue curves during cyclic deformation and during friction enables to describe the types of fracture (damage) in a single manner and at the same time to discriminate the types without ambiguity typical for given conditions of operation.

Tribo-fatigue establishes (and describes Chapters 4 and 5) two effects: direct (the effect of friction processes on the change of characteristics of resistance to fatigue) and back (the effect of cyclic stresses on the changes of characteristics of resistance to wear). Knowledge of basic mechanisms of wear-fatigue damage when these effects occur leads from designing individual components of machines and equipment to life designing of active systems of machines and equipment allowing for interactions between their components. Chapter 5 describes the principles of calculation and design of active systems. It requires to develop and introduce a complex of methods and means of control over the processes of wear-fatigue damage of specific systems to achieve savings of labor, means and materials making production and operation cheaper and at the same time to achieve gains of reliability and durability.

A new class of testing machines of SI-series (based on a number of inventions) has been developed for experimental studies of behaviour and determination of wear-fatigue damage characteristics. Chapter 3 describes machines and methods of wear-fatigue tests.

A modern trend of designing special purpose objects is to assess quality, risk and safety, the textbook describes basics of these problems. It contains a brief survey of traditional concepts of risk and safety (Chapter 1) but the emphasis is on the concept of risk as expectation of unfavorable events (situations); this interpretation relates the risk indicator both to the damage and to safety of an object (Chapter 5).

Each chapter (but the first) finishes with self-test questions that can be helpful for both better comprehension of the information and more comprehensive digestion of basic knowledge.

(3) The textbook contains one normative document and two scientific presentations. The document is an Interstate standard of tribo-fatigue terms containing strict (concise) definitions of basic notions with the English translation attached that I believe useful.

The paper "On Methodology of Tribo-fatigue" was prepared by a group of professors for the 3rd International Symposium on Tribo-Fatigue (Beijing, October 2000). It characterizes briefly the sphere of tasks and interests of tribo-fatigue in an easily digestible manner as a discipline interrelated with interdisciplinary sciences. Though its basic essence is the same with the textbook, just twenty pages disclose the methods of tribo-fatigue fully comprehensively. The text of the presentation in English will be specifically useful for students and instructors as it will add to their mastery of the English language.

The paper "Some Stages and Prospects of Progress of Tribo-fatigue" prepared for the 4th International Symposium (Ternopil, September 2002) is a brief chronicle of the most significant events in the progress of tribo-fatigue. It also outlines the main trends of further research in this domain formulated by a large group of scientists and specialists during the 2nd International Symposium on Tribo-Fatigue (Moscow, October 1996).

(4) A laboratory practical course in tribo-fatigue has been elaborated at the Belarusian State University of Transport (the first part of the coursebook on it has been published and the second is being prepared for print). PC-aided testing machines of SI-series are used for wear-fatigue studies (they are produced by the S&P Group TRIBOFATIGUE Ltd.; see Chapter 3). Also, a fatigue testing machine, □□□-6000-2 (produced by the factory of precise instruments in Ivanovo) so popular among the researchers in the former Soviet Union, has been modified into simpler workbenches to perform comprehensive tests (for mechano-sliding and mechano-rolling fatigue). Therefore, facilities for laboratory practical courses in tribo-fatigue can be provided relatively cheaply.

(5) Design-graphic (or designing) work of the course "Fundamentals of tribo-fatigue" can deal with the design of active system like *crankshaft /sliding bearing* (the textbook is in print), *wheel /rail, railway wagon axle / wheel pair,* toothed wheels (textbooks are being prepared) etc. The tasks for students should be selected taking into account major subjects they study. The set of training and systematic textbooks now in preparation will, in fact, serve as a basis of a special course in tribo-fatigue dealing with practical designing of typical general purpose active systems.

The text of the textbook does not contain any references to authors or studies with the exception of some experimental results that are specifically meaningful for the progress of tribo-fatigue. Almost all the information in the textbook can easily be retrieved from recommended publications.

I would like to express my profound appreciation of the help and encouragement of my colleagues, followers and students that I needed and enjoyed in research and lecturing in the domain of tribo-fatigue.

L Sosnovskiy

April 2002 *Gomel*

TO THE READER

Any author is glad that his book starts a new life in another (foreign) language.

By now several monographs on tribo-fatigue have already been published in the Russian language. The given book under the name "Fundamentals of Tribo-fatigue" was intended and written as a textbook for technical universities. The subject "Fundamentals of tribo-fatigue" was first introduced into the curriculum of the Belarusian State University of Transport at the suggestion of the dean of the mechanical faculty, Professor V. I. Senko (now Rector of the University).

As far as we know, there are no books at all in the English language dealing with this new field of knowledge, and only several papers have been published in English in contrast to those in Russian. Therefore the present book may serve as a monograph useful for any scientist and engineer who would like to have some information about the main ideas and achievements in tribo-fatigue.

A rudimentary knowledge of tribo-fatigue in science appeared long ago. Thus in the 1950s for specialists in mechanical fatigue it was common knowledge that wear was among numerous factors to reduce fatigue limit of constructional parts. At the end of the fifties and early in the sixties the first experimental works appeared in which it was reported that among many factors affecting the wear intensity in the friction pair there were cyclic stresses which were caused by noncontact loads. In the 1960s – 70s a great number of scientific papers were published that dealt with research of fretting as an important factor which decreased characteristics to fatigue resistance significantly (fretting fatigue). Yet it took more than 30 years to reach an understanding that friction and wear, on the one hand, and mechanical fatigue, on the other hand, are not the factors that affect each other but the phenomena which mutually interact (with each other) in a complicated way.

Essential difference of these two approaches to the analysis of damage and limiting state of materials is the following. The effect of the factors is always unambiguous. The boosting of this or that damaging factor leads to reducing strength or wear resistance. The interaction of the damaging phenomena turns out to be intricate and often unexpected. Thus the reliability and durability of the system can both be substantially reduced (as it was expected) or, on the opposite, increased (as it was in no way expected) or remain on the previous level (that could be "allowed") if contact load (wear and friction) is added to cyclic load (mechanical fatigue). Such results depend on the condition of interaction of the damaging phenomena. When it was realized, tribo-fatigue emerged at the interface between tribology and the mechanics of fatigue fracture. It happened in the middle of the 1980s. Thus, tribo-fatigue studies a complicated interaction of different damaging phenomena rather than mutual effect of separate damaging factors. As a result of this interaction a new type of degradation of materials – complex wear-fatigue damage is discovered. Tribo-fatigue studies conditions and regularities of such damages of specific objects – active systems of machines and equipment. All these problems are presented more comprehensively in this book.

I am grateful to the Springer Publishing House for their taking on an obligation to publish my book "Tribo-fatigue" in English.

Academician of the Russian Academy of Sciences and National Academy of Sciences of Belarus Konstantin V. Frolov and Corresponding Member of the Russian Academy of Sciences Nickolai A. Makhutov wrote the preface to the English edition of the book. I am deeply thankful to them for that. Having an opportunity I would like to emphasize their special role in the formation and development of tribo-fatigue. Professor K. V. Frolov and Professor N. A. Makhutov participated in the birth of tribo-fatigue. For 20 years they have been contributing to the organization solid research in this field of science and directing all International symposia on tribo-fatigue. Their personal contribution to the achievements in this science is significant.

I wish to express my deep appreciation to L. F. Burtsev, the translator of the book, and R. S. Sosnovskaya, the editor of the translation, who, in my opinion, were striving to overcome numerous difficulties in seeking the counterparts of some specialized terms and solve many other problems while translating the textbook in this new field of knowledge.

I would also like to thank A. M. Velikanova for her help in computer type – setting of the book in both (Russian and English) languages displaying much patience in doing this, it seemed there was no end to numerous corrections made by the author and editor of the translation. I wish to thank S. F. Goryachenko for his help in preparing all the illustrations for print.

L A Sosnovskiy

Gomel, June 2004

CONTENTS

1 VOLUME FRACTURE AND SURFACE DAMAGE

It is precisely... forces (of adhesion) that make everything strong that we erect on Earth using iron, stone and other durable materials.

Michael Faraday

1.1 General notions

1.1.1 Load

Load in the most general sense is any exposure of a body (or an object).

The *mechanical load* applied to a point, force (measured, e.g., in newtons – N) is its most essential characterizing parameter. *Force is a measure of mechanical interaction between bodies*. When interactions occur over the *contact surface* (or *area*), pressure is its characterizing parameter (measured by force per contact area unit – MN/m^2 = MPa). In a number of cases the notion of *linear loading* is characterized by force per unit of length (N/m); it is also termed the (distributed) *loading intensity*.

In case of *heat load*, *temperature* is its essential characterizing parameter (measured, e.g., in Kelvins). In case of *radiation load*, *irradiation intensity* is its essential characterizing parameter (measured, e.g., as the density of flux of neutrons per unit of time – neutron/ $(m^2 \cdot s)$). Various *chemical loads* are due to electrochemical interactions between the object and the environment.

The process of load action on the object is termed *loading*. In other words, *loading is the law of load variations in time.*

Any substance may be an object exposed to load: a solid (e.g., steel, polymer or a bone), a liquid (e.g., water, melt, blood), a gas (e.g., air, nitrogen, propane). Here a solid will be considered as the body under load. So, heat load affects gas turbine blades, radiation load affects the walls of a nuclear reactor; mechanical load affects the connecting rod of an engine (force), the walls of a hydraulic cylinder of a machine (pressure), the belt of a conveyer (load intensity).

1.1.2 Strength and stiffness

Any solid possesses two unique common properties: they are strength and stiffness.

Strength is the capability of the solid to take up and withstand load without fracture. Stiffness is its capability to retain dimensions and shape under the effect of mechanical load. Without these fundamental properties of solids nothing would exist on Earth: forests, or machines, houses, or man himself. Moreover, Earth itself would not exist, it is also a solid that has been undertaking and bearing various huge loads during an unimaginably long time, for billions of years, without failing or changing its size and shape in any substantial way... We will deal here primarily with some artificial solids, *metallic* or *polymeric* materials because they are used to produce structural components, machines, mechanisms, equipment, instruments.

1.1.3 Volume and surface strength

Every solid individually, such as any *machine part,* occupies in space some *solid volume* (*geometrical*). When a technical device operates, only some part of this volume undertakes load, so this part is termed the *working volume* of the loaded solid (a component). The geometrical and working volumes may be either similar or different. For instance, a beam on two pivoting bearings forms two cantilever arms from both ends, it is under a transverse load applied to the center of the span. In this case the working volume of the beam locates only between two bearings; its geometrical volume includes additionally two arms that are not under load. If the length of the beam coincides with the span between the bearings, its geometrical and working volumes coincide.

Since there is the working volume of a solid under load, the notion of *volume strength* should be introduced. So, the volume strength of a connecting rod (when a piston engine operates) can be implied because it bears an axial cyclic load and it will break into pieces unless its volume strength is sufficient. *This type of continuity disruption of a loaded solid leading to full loss of the load bearing capacity is termed volume fracture,* or just *fracture*.

Similarly, load in a *friction pair* acts on a specific working volume, not geometrical, this volume contacts some part of the surface of the solid, counteracts to the load and it is termed the *working surface*. Then the idea about the *surface strength* should be introduced. So, the surface strength of the (supporting) race of the rolling contact bearing takes up a radial load in the points where it contacts with balls (or rollers) along the so-called *rolling path*. If the surface strength of the material along the path is insufficient, *surface damage occurs,* or small flakes or metallic fragments spall and break off from the surface. As is common in rolling, and in relative sliding of two solids contacting under load, *friction surfaces demonstrate a specific type of gradual fracture termed as wear process*. It is characterized by continuous separation of particles and their removal from the friction zone, so the body's dimensions reduce in the direction perpendicular to the friction surface. Solids do not lose their bearing capacity in this case; only the performance of the surface layers of the material becomes impaired. In such cases the property of strength is termed *wear resistance.*

1.1.4 Crack growth resistance

Any fracture, surface or volume, may occur if it is preceded by nucleation and evolution of *local discontinuities or cracks* in a material. Hence, *fracture* should not be considered as some momentary event, it is *a kinetic process of loss by a solid of* (surface and /or volume) *continuity under the effect of loading*. In this case the property of strength of a material should be attributed to the resistance to nucleation and evolution of cracks, or briefly to its *crack growth resistance*. Since *cracks* can be both *surface* and *main*, so there can be both *surface* and *volume crack growth resistance*, apparently as a function of loading conditions.

Usually surface fracture precedes volume fracture. When a shaft is bent and rotated, small *scattered fatigue cracks* appear on its surface. Gradually some cracks grow quicker and quicker under the effect of an alternating load and turn into a *main* crack, its evolution ends, as a rule, in *fatigue fracture of the shaft*, i.e. it snaps into two pieces. In this case the process of fracture is determined by the *rate of growth of the* (main) *crack*.

1.1.5 Mechanical properties

There is a long-standing experience of assessing volume strength quantitatively by performing *mechanical tests* of special (including standard) *specimens* of materials. It was established that this property depended cardinally on the *pattern of load application*.

Figure 1.1 shows the classification of basic *laws of loading when an object is under mechanical load.*

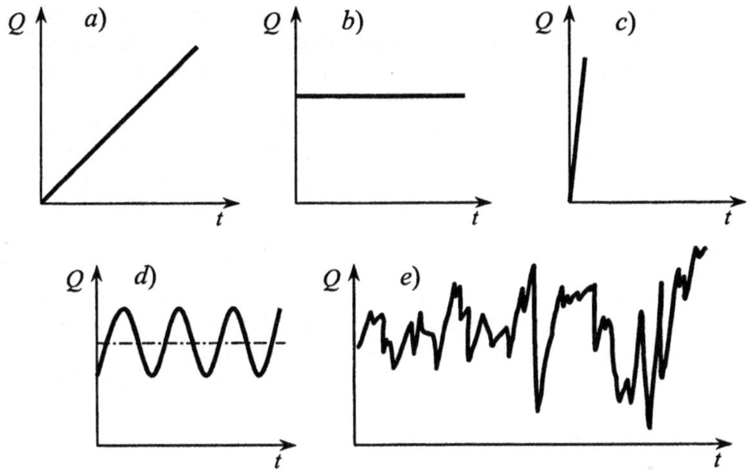

Fig. 1.1. Basic laws of loading of a solid

If load Q grows gradually and monotonously in time (Fig. 1.1, *a*), it is *static loading*. A typical example of such loading is when the load on the foundation increases as construction of a building progresses. Standard *tensile tests of specimens* are a typical example of static loading during mechanical tests of a material. In such cases a problem appears and is solved relating to the assessment of *static strength* of an object.

If load Q remains practically constant in time (Fig. 1.1, *b*), it is *sustained loading*. Pressure of overheated steam on the walls of a high-pressure boiler drum is a typical example because it remains practically constant in time. *Tests of specimens for sustained strength (and creep) at elevated temperatures* are a typical example of sustained loading during mechanical tests. In such cases a problem appears and is solved relating to the assessment of *sustained strength* of an object.

If load Q increases in time practically instantly (i.e. with a very fast rate), it is *dynamic*, or *impact loading* (Fig. 1.1, *c*). A contact between the wheel and the joint between two rails is a typical example of such loading. *Tests of a standard specimen for impact viscosity* during mechanical tests of a material are a typical example of dynamic loading. In such cases a problem appears and is solved relating to the assessment of *dynamic strength* of an object.

If load Q changes in time in some *cycle* (for example, sinusoidal, Fig. 1.1, *d*), it is *cyclic loading*. Pressure variations in the cylinder of the internal combustion engine during stationary operation are a typical example of such loading. *Fatigue tests* of cylindrical specimens by rotating bending during mechanical tests of a material are a typical example of cyclic loading.

Finally, if load Q changes in time then practically it cannot be predicted (Fig. 1.1, *e*), it is *random loading*. Variations of load on car wheels on a mud road are a typical example of such loading. Loading of a shaft performed as a narrow-strip *random process* is a typical example of random loading during mechanical tests. In such cases a problem appears and is solved relating to the assessment of *resistance* of an object to *fatigue*.

If relevant mechanical properties of a given material are analyzed in relation to the pattern of loading, it can be established that strength during cyclic loading is two times smaller than during static loading; when loading is impact-cyclic, the resistance to fatigue falls two times more.

Mechanical properties also significantly depend on the *type of the stressed state* of an object. The least resistance to fracture is observed during *torsion* (*shearing*), the most resistance during *bending*, and resistance is intermediate during axial *tension-compression*.

The surface strength property, or wear resistance, is assessed quantitatively by *friction (or wear) tests*. First, this property cardinally depends on the *type of* friction, whether it is *sliding, rolling or slip (fretting)*. Second, like volume strength, it is strongly governed by the pattern of *loading*. As a rule, wear processes intensify strongly under the effect of impacts. Third and last, it has been demonstrated that the process of surface damage depends on the *nature of interacting bodies* (substances). In this connection, for example, surface damage of a solid by impacts of solid particles or liquid drops is termed *erosion;* if a solid is exposed to radiation, it is termed *radiation erosion*. When a solid obstacle is affected by powerful laser irradiation or by high-temperature plasma, it is a special type of surface damage termed *ablation*. If a solid (a pipe, for example) is in continuous contact (under pressure) with a laminar liquid stream, *hydroerosion* of the pipe appears in its portion where the liquid stream has a non-stationary flow, i.e., *cavitation erosion* occurs.

Note that when the bearing capacity of surface layers of a solid is meant, it implies the *processes of damage* and not the property of strength. Of course, it is also possible to study resistance to relevant damage, e.g. *contact strength* (during rolling friction), *friction strength* (during sliding friction), *erosion, radiation, cavitation strength, etc.* It is exactly so: we have discussed above static, dynamic strength, fatigue resistance, etc. It is possible, however, to study static, dynamic, fatigue fracture, fracture in friction, etc. So, strength is a property to resist to surface and/or volume fracture, while damage and fracture are processes of corresponding loss of strength (bearing capacity). Hence, any damage is a partial loss of strength (bearing capacity).

In this connection, when volume fracture is discussed below, we assume that we deal with the property of strength of an object, irrespective of loading conditions. When surface damage of contacting bodies is discussed, it is assumed that we deal with the property of wear resistance, irrespective of the conditions of interactions.

The property of cracking resistance is rated quantitatively as a result of *tests for fracture toughness* during static, impact or cyclic loading of specimens. Methods of these tests have been developed and standardized relatively recently. So far there are no standard methods to rate cracking resistance in respect to surface damage during friction and wear.

1.1.6 Internal forces

If loading of a solid is characterized by external (effective) loads, its strength is governed by the *internal force*. The latter is resultant of the interatomic forces of attraction and (or) repulsion through any cross section of the solid under load.

It can be explained by a very simple example of the linear (uniaxial) state of stress.

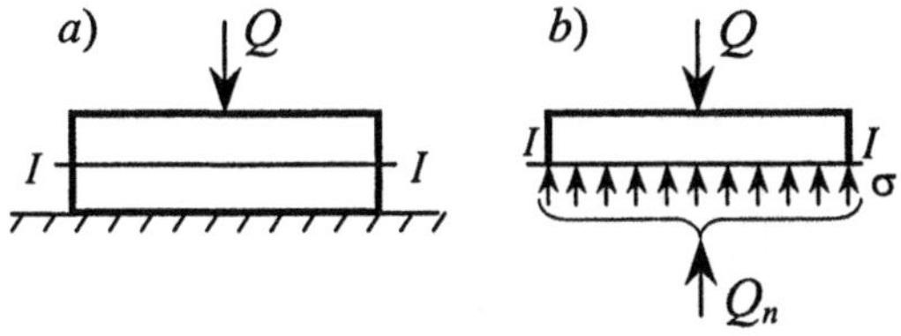

Fig. 1.2. Determination of internal force in brick

Assume a common red brick on the floor is subjected to a load Q = 750 N (Fig. 1.2, *a*). Let us see what is going on inside the brick by using the well-known *method of section*. Let us fancy that we cut the brick into two pieces through plane *I–I*. Let us ignore the lower part and study the upper part (Fig. 1.2, *b*). In order to put it into equilibrium a system of internal forces should be applied to plane *I–I* equivalent to the effect of the ignored part on the part in question. Assume that this system of internal forces σ is as Fig. 1.2, *b* shows it, i.e. forces distribute regularly through cross section *I–I*. Since these internal forces are perpendicular to cross section *I–I*, or directed normal to it, they are called *normal stresses*. So, stress *is an internal force per unit of area* (N/m^2, MPa, etc.). It is the resultant of the stresses through this cross section that is the *internal force Q_n in this cross section*. It follows from the apparent condition of equilibrium that

$$Q - Q_n = 0$$

in this case $Q_n = Q$. Designate the area of cross section *I–I* as A_0, then $Q_n = \sigma A_0$. Hence the internal force intensity is

$$\sigma = Q_n / A_0.$$

It is how normal stress is determined numerically.

What causes internal forces?

To answer the question it is worthwhile to recollect a "classic picture" (Fig. 1.3) from the course of physics. Assume any two atoms of an unloaded solid are in equilibrium with a spacing d_0 between them. Any external force to pull them apart (increase the spacing between them) or to compress them (to reduce this spacing) causes a corresponding (internal) elementary interatomic force of counteraction – attraction or repulsion (shown by arrows in Fig. 1.3). It is the resultant of these elementary interatomic forces through cross section *I–I* of the brick (cf. Fig. 1.2, *b*) that represents the internal force Q_n. Though displacements of individual atoms are extremely minute, the internal forces they produce may be very large since the number of atoms that displace across the section of a solid under load turns out to be huge. For instance, when a brick is compressed with a load $Q = 750$ N (cf. Fig. 1.2, *a*), the spacing between its atoms reduces by $\sim 2 \cdot 10^{-14}$ cm [1]. This displacement is hard to imagine since it is tremendously infinitesimal. Yet, when the same load displaces "all the atoms" of the brick, it turns out that its compression (i.e. the size reduction in the direction Q) amounts to ~1/20000 cm and that is already tangible. This degree of compression leads to a quite tangible resulting internal force $Q_n = 750$ N (cf. Fig. 1.2, *b*).

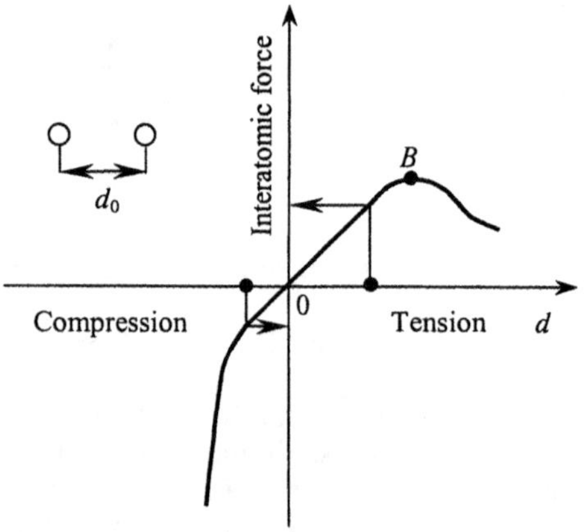

Fig. 1.3. Variations of interatomic force of interaction during relative displacement of two atoms

If in this simplest case the assertion that strength calculation should be based on the internal force Q_n rather than on the external loading Q seems to be quite sound (since $Q_n = Q$), it is not apparent in the other case (Fig. 1.4). When studying alternative systems of external transverse forces Q applied to one and the same cantilever beam (see Fig. 1.4, *a*, *b*, *c*), it can be assumed that the beam in Fig. 1.4, *b* is under heavier load than in Fig. 1.4, *a* (the sum of three forces $Q_1 + Q_2 + Q_3 = 2200$ N, that is over $Q = 1000$ N), while it is still larger in Fig. 1.4, *c* – (the sum of two forces $Q_1 + Q_2 = -3000$ N). Yet, when allowance is made for the external loading, the internal force is calculated in the dangerous section *I–I* – *the (internal) bending moment M*, then $M = 1000$ N · m = const *turns out to be for all three beams*, hence the maximum normal stress in the same section is also $\sigma_{max} = M/W$ = const (*W* – *the moment of resistance to bending*). It means that these three systems of external loading in Fig. 1.4, *a*, *b*, *c* carry the same danger for the beam, and its strength is determined by the maximum stress $\sigma_{max} = M/W$.

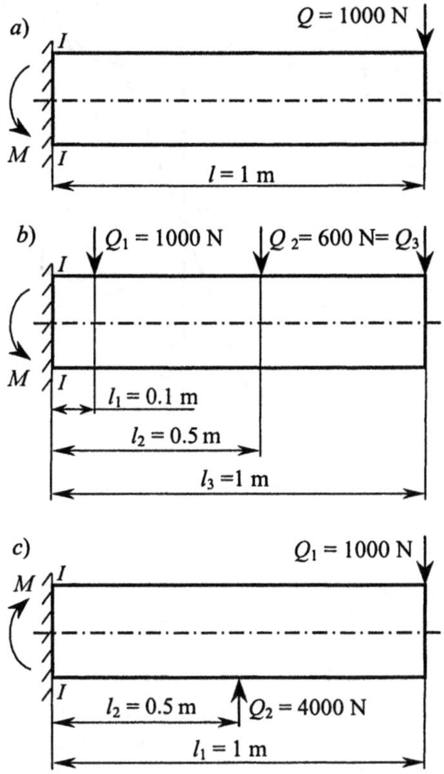

Fig. 1.4. Problem of calculation of beam bending strength

If the spacing d_0 between two atoms (cf. Fig. 1.3) grows gradually by increasing the loading, the interatomic force of their "bonding" can be exceeded, then an *elementary act of* fracture *or rupture of the atomic bonds occurs* (point *B* in Fig. 1.3). When standard tension tests of a steel specimen are performed, a similar phenomenon occurs, i.e. macroscopic (volume) fracture. One of the most essential characteristics of the material strength – *ultimate strength* corresponds to point *B* in Fig. 1.5 (curves *1* and 2) ($\sigma = Q/A_0$ on the scale of normal stresses)

$$\sigma_b = Q_{max}/A_0,$$

where Q_{max} – the maximum load that the specimen with the cross sectional area A_0 can withstand. Point B_1 on curve 2 corresponds to the compressive strength σ_b^{com}.

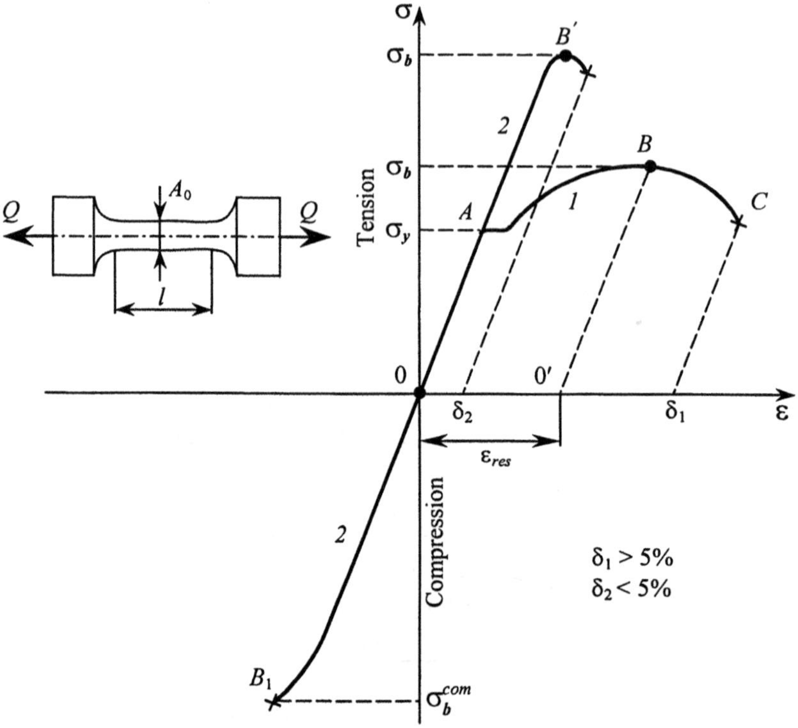

Fig. 1.5. Diagrams of tension of soft steel specimen (*1*) and tension-compression of high-strength steel specimen (2)

Fracture due to the internal forces of mechanical origin is discussed above. In reality there are also internal forces of thermodynamic origin; they are temperature flashes across atomic bonds or *thermofluctuations*. Since atoms are not fixed, they oscillate in a random manner in respect to some position, so thermofluctuations appear in an unloaded solid chaotically at any moment of time. If mechanical loading occurs, then the general temperature background increases (*a temperature field* appears) due to deformation, the thermofluctuations intensify both in respect to the time of their occurrence and their magnitude. The higher the mechanical load, the more frequent are the thermofluctuations and the stronger they are. Studies show that elementary ruptures of atomic bonds occur both due to the mechanical load and to its combined action with thermofluctuations. Mechanical load acts as an intensifier of thermofluctuations in this case. Under certain conditions (for example, at elevated temperatures) the latter may be governing the origination and evolution of primary damage of a material. Purely *thermodynamic* (heat) *fracture* or *melting of metals, softening of plastics* take place in extreme cases under the action of high temperature.

1.1.7 Basic types of fracture

In the general case *three basic types of volume fracture* of the deformable solid are identified, they are *brittle, viscous (plastic), fatigue*. As far as the surface fracture is concerned, it is also governed by the mechanisms of brittle, viscous (plastic) and fatigue fracture, yet it evolves within a limited contact area.

In case of static loading two basic types of fracture of materials are possible. They are illustrated in Fig. 1.6 and 1.7.

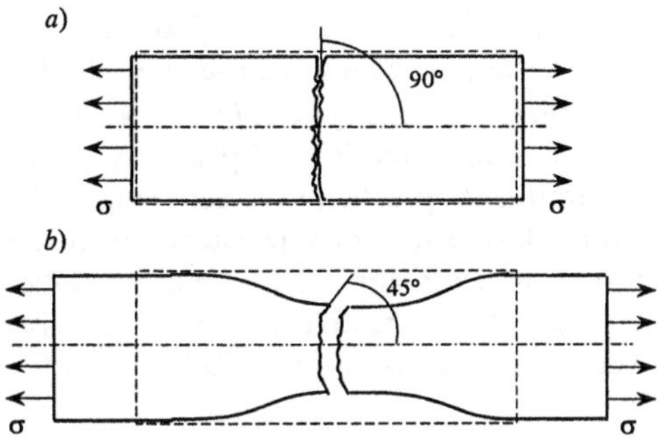

Fig. 1.6. Static fracture of metallic specimens in brittle (*a*) and plastic (*b*) states

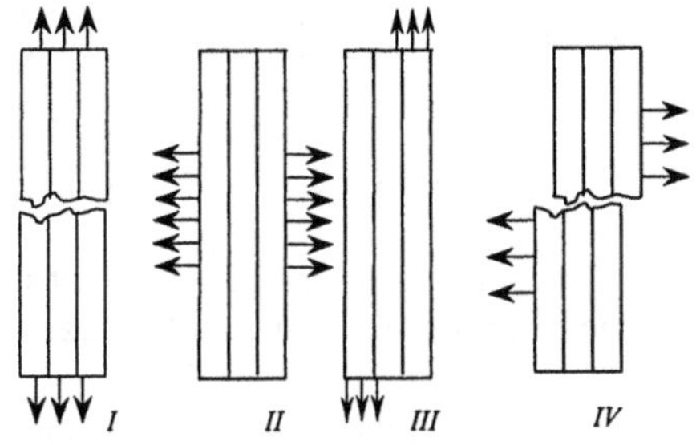

Fig. 1.7. Damage by separation (*I, II*) and shear (*III, IV*) of specimens of uniaxially oriented statically loaded polymer

As a rule, brittle fracture (Fig. 1.6, *a*; the dimensions of the original specimen are shown with a dotted line) is accompanied with very light plastic deformation (*elongation after rupture is* $\delta < 5\%$ – see also Fig. 1.5), it occurs suddenly and evolves practically instantly – usually by *rupture*, so that the *rupture plane* is perpendicular to the direction of tensile stresses σ. Viscous damage (Fig. 1.6, *b*) is preceded, as a rule, by strong plastic deformation ($\delta > 5\%$, see also Fig. 1.5); in case of soft steel a *neck* appears on the specimen, fracture results from shear over the sites affected by maximum *tangential stresses*; these sites are inclined in respect to the direction σ at an angle ~45°.

When a uniaxially oriented linear polymer is loaded statically (cf. Fig. 1.7), fracture evolves by separation (*I*, *II*) or shear (*III*, *IV*) across (*I*, *IV*) or along (*II*, *III*) the axis of fibers [2]. Two types of fracture, by way of separation or shear in any direction, remain in a non-oriented polymer because of the lack of macroscopic anisotropy.

Various factors may substantially change the type of fracture of one and the same material. For instance, when specimens become larger, the ambient temperature T reduces (towards the negative range specifically), stresses concentrate, deformation accelerates, etc., the material undergoes ***embrittlement*** and the so-called ***viscous-brittle transition*** (VBT) occurs (Fig. 1.8); some characteristics of fracture sharply reduce (cf. Fig. 1.8, *a*: KCV – impact strength), others, on the opposite, increase (cf. Fig. 1.8, *b*: σ_y – yield strength). Yet, viscous-brittle transition is always due to the loss of plasticity by a material. The plastic deformation work A_p diminishes practically to zero (Fig. 1.9) in case the thickness h of a compact specimen tested for crack resistance increases from 10 to 55 mm for steel 45 (*1*) and to 75 mm for steel 30 (2). When a critical thickness of steels 45 and 30 $h_k = 55$ or $h_k = 75$, respectively, is reached, viscous-brittle transition terminates; only brittle failure occurs at $h > h_k$ [3].

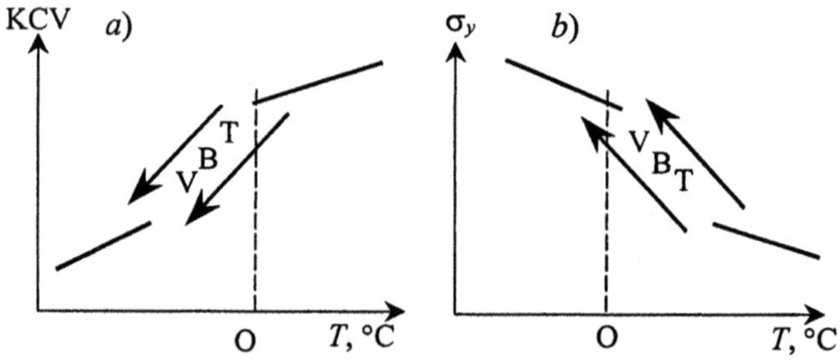

Fig. 1.8. Schemes of temperature dependence of impact strength KCV (*a*) and yield strength σ_y (*b*) of metallic materials

Thus, engineering materials are divided into brittle and plastic by convention because, as it is indicated above, any real mechanical state of a material is determined both by the conditions of loading and a whole number of other factors. The effect of some of these factors is such that a plastic material becomes brittle if put under definite conditions. Therefore, it seems better to imply *brittle* or *plastic state* rather than brittle or plastic materials.

Relation between the ultimate compressive and tensile strength can serve as a condition characterizing the state of a material

$$\chi = \sigma_b / \sigma_b^{com} .$$

If $\sigma_b = \sigma_b^{com}$, hence $\chi = 1$, the material is called *ideally plastic*. If $\chi = 0$, it is an *ideally brittle* material. The overwhelming majority of technical materials have $0 < \chi < 1$.

If an alternating load instead of a static load affects the specimen, and this load is *less than the yield strength*, volume fracture of the specimen may occur when the number of loading cycles becomes large enough, for example, 10^7 cycles.

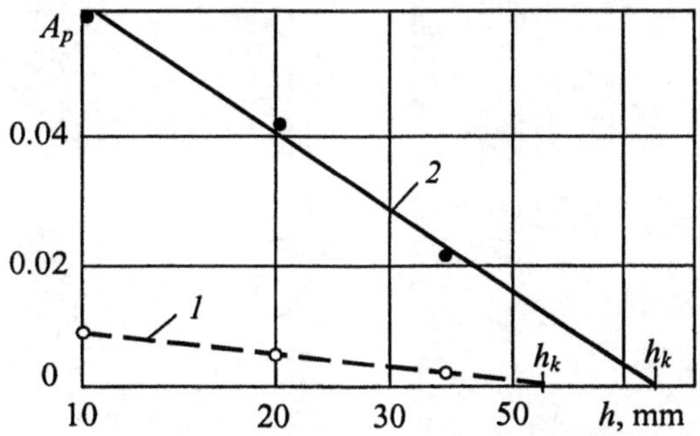

Fig. 1.9. Viscous-brittle transition during tests of compact specimens of steels 30 (*2*) and 45 (*1*) for cyclic crack resistance

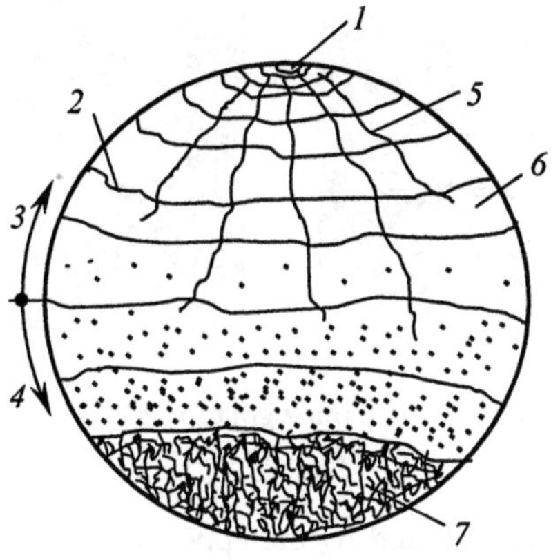

Fig. 1.10. Fatigue rupture

Figure 1.10 shows the pattern *of rupture of the metallic specimen* during *fatigue fracture*. A *primary fatigue crack* of macroscopic dimensions appears in *spot 1* where the material has the least resistance to fracture. When *front* 2 of fatigue crack propagates deeper, *fault lines 5* appear in the rupture that determine different *planes of separation 6*. *Zone 3 of steady crack evolution* transforms into *zone 4 of its unsteady growth* that terminates in *sudden complete fracturing* (*zone 7*).

Fatigue fracture in general has an intricate (complex) pattern. Yet, certain conditions allow to identify its dominating mechanism [4]. In case of viscous fracture a *cellular* (or *pitted*) structure is observed on the boundary surface (Fig. 1.11, *b*). There are spots of *spalling* in case of brittle fracture (Fig. 1.11, *a*).

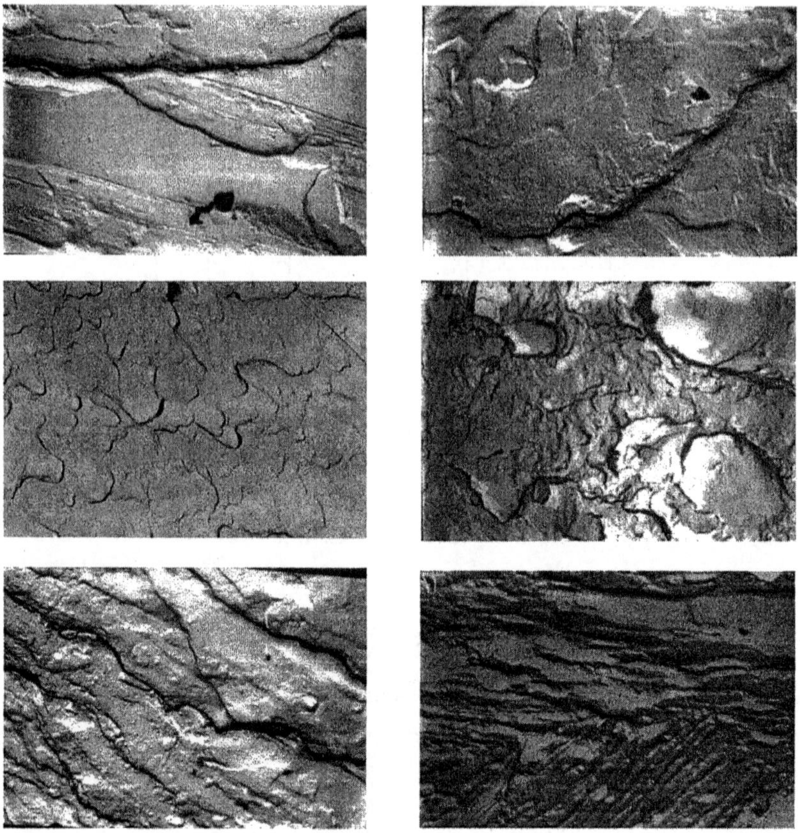

a)

b)

c)

Fig. 1.11. Typical electron microscopic fractograms of surfaces of fatigue rupturing of shaft from steel 45 (×2000)

Typically fracture is demonstrated by the appearance of periodic *fatigue grooves* over the boundary (Fig. 1.11, *c*). It is assumed that each groove appears during one loading cycle.

1.2 Static strength

1.2.1 Mechanical state

In accordance with the diagram of tension of soft steel represented by the curve *OABC* in Fig. 1.5, a deformable solid can be in one of three *mechanical states*.

The elastic state is characterized by a proportional relationship between stresses

σ and *relative longitudinal deformation* ε (the straight line *OA* in Fig. 1.5). The law of Hooke describes it as

$$\sigma = E\varepsilon, \tag{1.1}$$

where E – the *modulus of normal elasticity* or *Young's modulus.*

The elastic state is reversible: unloading from point A returns the body to the original point O in which $\varepsilon = 0$. The informative sense of the modulus of elasticity is established from law (1.1). If $\varepsilon = \Delta l/l = 1$, then $\sigma = E$, i.e. E is such a stress that elongates the tested specimen two times, providing, of course, that it does not fail before (the absolute elongation Δl should be equal to the original specimen's length l). It follows from geometrical considerations that the modulus of elasticity is the constant of a given material:

$$E = \sigma/\varepsilon = \text{Const} = \tan\alpha.$$

By comparison of two types of materials with different numerical values of the modulus of elasticity ($E_1 > E_2$) the following is established: at a similar relative elongation (ε = const) the first material has larger stresses ($\sigma_1 > \sigma_2$) (Fig. 1.12, *a*); on the opposite, when stresses coincide (σ = const), the first material deforms less than the second ($\varepsilon_1 < \varepsilon_2$) (Fig. 1.12, *b*). Hence, E is the *characteristic of rigidity of a material.*

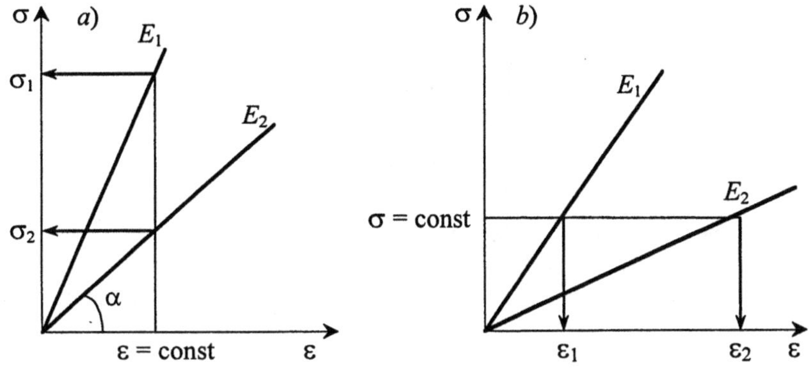

Fig. 1.12. Determination of sense of modulus of normal elasticity

An essential *elasticity parameter* is Poisson's ratio determined as a ratio between transverse ε_{tr} and axial ε deformations during tension:

$$\mu = |\varepsilon_{tr}|/\varepsilon = \text{Const}.$$

Plastic state is characterized by a non-linear relationship $\sigma(\varepsilon)$ between stresses and deformation (the curve *AB* in Fig. 1.5), with increasing σ so does ε:

$$\sigma = E_p\varepsilon\,. \tag{1.2}$$

Here the *modulus of plasticity*

$$E_p = \sigma/\varepsilon = \tan\alpha_p = \text{var}$$

changes within the interval $E > E_p \geq 0$ reaching the value $E_p = 0$ at point B (because in this case $\tan\alpha_p = 0$).

Plastic state is irreversible: unloading from any point along the line AB does not return the body to the original point O; it proceeds along the straight line (BO', for example) parallel to the elastic line OA in such a way that *residual (plastic) deformation* ε_{res} appears (cf. Fig. 1.5).

Characteristic stress $\sigma = \sigma_y$ (cf. Fig. 1.5) termed the *yield strength* corresponds to the transition from the state of elasticity to the state of plasticity. Characteristic stress $\sigma = \sigma_b$ (see Fig. 1.5) termed *ultimate strength* corresponds to the transition from the state of plasticity to the state of fracture.

The state of fracture might be described by an inverse non-linear relation $\sigma(\varepsilon)$ (curve BC in Fig. 1.5): rise of ε is accompanied by reduction of nominal stresses σ. Yet, in terms of true stresses and deformation in the neck, this relation turns out to be direct, on the opposite. Whence it follows that it is more proper to characterize the state of fracture by some other unambiguous parameter. In case of brittle fracture it is the *stress intensity factor*, its idea is disclosed in Sect. 1.3.4.

When a spatial system of forces characterized by *main stresses* $\sigma_1 \geq \sigma_2 \geq \sigma_3$ is active, the *generalized Hooke's law* describes the state of elasticity:

$$\left.\begin{aligned} \varepsilon_1 &= \frac{1}{E}\left[\sigma_1 - \mu(\sigma_2 + \sigma_3)\right]; \\ \varepsilon_2 &= \frac{1}{E}\left[\sigma_2 - \mu(\sigma_3 + \sigma_1)\right]; \\ \varepsilon_3 &= \frac{1}{E}\left[\sigma_3 - \mu(\sigma_1 + \sigma_2)\right], \end{aligned}\right\} \tag{1.3}$$

that establishes proportional relation between the *components of stresses* (σ_1, σ_2, σ_3) and the *components of deformation* (ε_1, ε_2, ε_3).

Law (1.3) can be resolved in relation to stresses and represented in the following form:

$$\begin{aligned} \sigma_1 &= 2G\left[\varepsilon_1 + \frac{3\mu}{1-2\mu}\varepsilon_m\right]; \\ \sigma_2 &= 2G\left[\varepsilon_2 + \frac{3\mu}{1-2\mu}\varepsilon_m\right]; \\ \sigma_3 &= 2G\left[\varepsilon_3 + \frac{3\mu}{1-2\mu}\varepsilon_m\right], \end{aligned} \tag{1.3a}$$

where the *mean deformation*

$$\varepsilon_m = \frac{1}{3}(\varepsilon_1 + \varepsilon_2 + \varepsilon_3)$$

and G is the *shear modulus (modulus of rigidity)*. This modulus is the elasticity parameter of a material in case of *simple shear* (Fig. 1.13) and it serves as the proportionality coefficient in the *Hooke's law during shear*

$$\tau = G\gamma, \tag{1.4}$$

where the *relative shear*

$$\gamma \approx \tan\gamma = \Delta a/a,\ \ \gamma << 1$$

and $\tau = Q/A$ – *tangential stress*, A – cross sectional area in the plane of which stresses τ distribute regularly, so that the shearing force $Q = \tau A$ is the resultant of stresses τ.

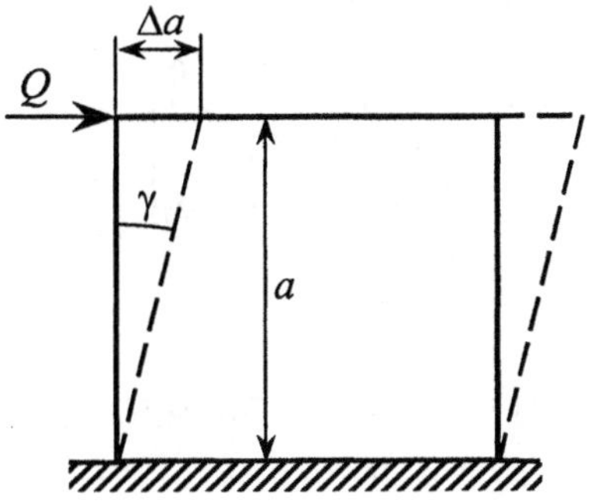

Fig. 1.13. Scheme of simple shear

Laws (1.1) and (1.4) have identical forms but different senses because resistance to shear, as a rule, is significantly less than resistance to separation (rupture during tension). It is because the shear modulus is $G \approx 0.4E$.

Note that three main parameters of elasticity G, E and μ are combined by the relation

$$G = \frac{E}{2(1+\mu)}. \tag{1.5}$$

Thus, it is practically enough to determine any two parameters and the third can be calculated from formula (1.5).

To describe the triaxial stressed state of a deformable solid the theory of elasticity uses the *stress intensity*

$$\sigma_{int} = \sqrt{\frac{1}{2}(\sigma_1 - \sigma_2)^2 + \frac{1}{2}(\sigma_2 - \sigma_3)^2 + \frac{1}{2}(\sigma_3 - \sigma_1)^2}. \tag{1.6}$$

It does not have any mechanical sense: there is no area where its effect can be detected. But the stress intensity (1.6) relates through the simplest dependence to the *octahedral tangential stress* τ_{oct}:

$$\sigma_{int} = \frac{3}{\sqrt{2}} \tau_{oct} ; \tag{1.6a}$$

$$\tau_{oct} = \frac{1}{3} \sqrt{(\sigma_1 - \sigma_2)^2 + (\sigma_2 - \sigma_3)^2 + (\sigma_3 - \sigma_1)^2} . \tag{1.7}$$

The latter appears on the site equally inclined to the *main sites* with stresses $\sigma_1 \geq \sigma_2 \geq \sigma_3$; it is called the *octahedral site.*

It is easy to observe that semidifferences of main stresses represent corresponding tangential stresses in (1.6).

If the main axes are assumed as the axes of coordinates, the stressed state is exhaustively described by the *stress tensor*

$$T_\sigma = \begin{vmatrix} \sigma_1 & 0 & 0 \\ 0 & \sigma_2 & 0 \\ 0 & 0 & \sigma_3 \end{vmatrix} \tag{1.8}$$

or its *invariants*

$$\left.\begin{aligned} I_1 &= \sigma_1 + \sigma_2 + \sigma_3; \\ I_2 &= -(\sigma_1\sigma_2 + \sigma_2\sigma_3 + \sigma_3\sigma_1) ; \\ I_3 &= \sigma_1\sigma_2\sigma_3, \end{aligned}\right\} \tag{1.8a}$$

in their turn, they can be combined by the octahedral stress (1.7)

$$\tau_{oct} = \frac{\sqrt{2}}{3} \sqrt{I_1^2 - 3I_2} . \tag{1.7a}$$

Taking into account (1.6a), (1.7) and (1.7a), the following chain of expressions for the stress intensity is recorded:

$$\begin{aligned} \sigma_{int} &= \frac{3}{\sqrt{2}} \tau_{oct} = \sqrt{\frac{1}{2}(\sigma_1 - \sigma_3)^2 + \frac{1}{2}(\sigma_2 - \sigma_3)^2 + \frac{1}{2}(\sigma_3 - \sigma_1)^2} = \\ &= \sqrt{I_1^2 - 3I_2} = \sqrt{\sigma_1^2 + \sigma_2^2 + \sigma_3^2 - \sigma_1\sigma_2 - \sigma_2\sigma_3 - \sigma_3\sigma_1} . \end{aligned} \tag{1.9}$$

According to (1.9), stress intensity is a generalized function of normal and (or) tangential stresses on any site, for example, octahedral, main, etc. It is due to the fact that it relates to tensor (1.8) (through its first and second invariants (1.8a)). Naturally this universal stress function may also have a variety of applications.

Thus, stress intensity (1.9) can be used to construct the simplest *theory of plasticity*.

In fact, let us assume that the plasticity during uniaxial tension of a specimen is *non-linear elasticity* (Fig. 1.14, *a*). Then the relation between stresses and deformation is defined by the formula

$$\sigma = E'\varepsilon, \tag{1.10}$$

where E' = tan α – the *secant modulus of deformation* depending on its degree: $E' = f(\varepsilon)$. Assume that in case of triaxial stressed state the relation between stress intensity σ_{int} and intensity of strain ε_{int} is similar (1.10) (Fig. 1.14, *b*). Then

$$\sigma_{int} = E'\varepsilon_{int},$$

where $E' = \Phi(\varepsilon_{int})$ and

$$\varepsilon_{int} = = \sqrt{(\varepsilon_1 - \varepsilon_2)^2 + (\varepsilon_2 - \varepsilon_3)^2 + (\varepsilon_3 - \varepsilon_1)^2}\,. \tag{1.11}$$

In this way, if, for example, a power relation is assumed between stresses σ and deformation ε in the state of plasticity during simple tension of the specimen from a given material

$$\sigma = a\varepsilon^m, \tag{1.12}$$

where a, m – some constants, then the law of plasticity in case of triaxial state of stress is represented in the similar form

$$\sigma_{int} = a\,\varepsilon_{int}^m, \tag{1.13}$$

with the constants a and m remaining.

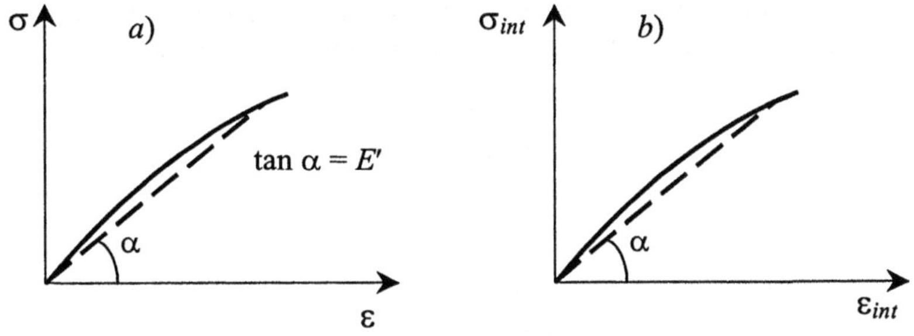

Fig. 1.14. Non-linear relations between stresses and deformation in uniaxial (*a*) and complex (*b*) states of stress

If properties of some other material differ from those of the first material, another relation between σ and ε may be applicable in the state of plasticity (during uniaxial tension), for example,

$$\sigma = E(1 - \omega)\varepsilon, \tag{1.14}$$

where E – Young's modulus and $\omega = f(\varepsilon)$ – some analytical function of relative elongation, while in case of triaxial state of stress of the same material the law of plasticity is

$$\sigma_{int} = E(1 - \omega)\varepsilon_{int}, \tag{1.15}$$

where $\omega = \Phi(\varepsilon_{int})$ – some function of deformation intensity.

Equations of state (1.12) and (1.13), (1.14) and (1.15) or any other with proper validation can serve as the basis of this or that theory of plasticity. Meanwhile, the values σ_{int} and ε_{int} can serve to analyze any mechanical state, whether elastic or plastic.

1.2.2 Condition of strength

The condition of strength for the linear state of stress, or *the condition of impossibility to reach the ultimate state*, is recorded in the following manner:

$$\sigma \leq [\sigma] = \frac{\sigma_{\lim}}{n}. \tag{1.16}$$

According to this condition, the maximum stress σ active in a part (a structure) should not exceed the *admissible* value $[\sigma]$, the latter being determined as the *ultimate stress* $\sigma_{\lim}$ reduced n times; the number $n > 1.0$ is termed the *strength safety factor*. If no transition into the plastic state is tolerated, then the ultimate stress is $\sigma_{\lim} = \sigma_y$. In case there is a risk of brittle fracture, it is assigned that $\sigma_{\lim} = \sigma_b$. In case of cyclic loading, condition (1.16) is assumed true at $\sigma_{\lim} = \sigma_{-1}$, where σ_{-1} is the *fatigue limit*.

Using condition (1.16) in this or other form, three *procedures of calculations of strength* are implemented (for example, a beam during bending with the moment M – see Fig. 1.4) based (*a*) on the tolerable or (*b*) ultimate stresses:

$$\left.\begin{aligned} &\sigma_{\max} \leq [\sigma]; \\ &W \geq M/[\sigma]; \\ &[\sigma] \geq M/W, \end{aligned}\right\} \tag{1.17a}$$

$$\left.\begin{aligned} &\sigma_{\max} \leq \sigma_{\lim}/n; \\ &W \geq Mn/\sigma_{\lim}; \\ &\sigma_{\lim} \geq (M/W)n, \end{aligned}\right\} \tag{1.17b}$$

that are respectively termed: *strength verification*; *determination of the cross sectional dimensions* of an element of a structure; *selection of a material* for its fabrication.

In case of the triaxial state of stress characterized by principal stresses $\sigma_1 \geq \sigma_2 \geq \sigma_3$, the routine of strength calculations is the following:

$$\begin{matrix} \sigma_1 \rightarrow \\ \sigma_2 \rightarrow \\ \sigma_3 \rightarrow \end{matrix} \boxed{\begin{matrix}\text{Theory}\\ \text{of}\\ \text{strength}\end{matrix}} \rightarrow f(\sigma_1, \sigma_2, \sigma_3, m_j, \ldots) = \sigma_{equiv} \leq [\sigma] = \frac{\sigma_{\lim}}{n}. \tag{1.18}$$

According to this routine based on the accepted *theory of strength* (*the theory of ultimate stressed states*), the following *function* is obtained:

$$f(\sigma_1, \sigma_2, \sigma_3, m_j, \ldots) = \sigma_{equiv} \tag{1.19}$$

for *reducing* the combined stressed state to the *equivalent* (*equally risky*) linear stressed state characterized by the *equivalent (or reduced) stress* σ_{equiv}. Then the condition of strength is

$$\sigma_{equiv} \le [\sigma] = \frac{\sigma_{\lim}}{n}, \tag{1.20}$$

that is practically similar to condition (1.16). In function (1.19) m_j are some parameters of the material.

Various functions (1.19) are known and practically used. *Classic theories of strength* are most popular. According to these theories, the equivalent stress is proportional to the first principal stress

$$\sigma_{equiv} = \sigma_1\, f\left(\frac{\sigma_2}{\sigma_1}, \frac{\sigma_3}{\sigma_1}\right). \tag{1.21}$$

Here the coefficient of proportionality is a certain function f of relations between principal stresses. For example, according to the first (*I*), third (*III*) and fourth (*IV*) theories of strength, it is obtained that

$$\left.\begin{aligned}
&\sigma^{I}_{equiv} = \sigma_1;\\
&\sigma^{III}_{equiv} = \sigma_1 - \sigma_3 = \sigma_1\left(1 - \frac{\sigma_3}{\sigma_1}\right);\\
&\sigma^{IV}_{equiv} = \sqrt{\sigma_1^2 + \sigma_2^2 + \sigma_3^2 - \sigma_1\sigma_2 - \sigma_2\sigma_3 - \sigma_3\sigma_1} =\\
&= \sigma_1\left[1 - \frac{\sigma_2}{\sigma_1}\left(1 - \frac{\sigma_2}{\sigma_1}\right) - \frac{\sigma_3}{\sigma_1}\left(1 - \frac{\sigma_3}{\sigma_1}\right) - \frac{\sigma_2}{\sigma_1}\frac{\sigma_3}{\sigma_1}\right]^{1/2}.
\end{aligned}\right\} \tag{1.21a}$$

The first (*I*) theory of strength is used successfully in a number of practical cases, especially in case of brittle static and fatigue fracture of parts *with stress concentrators* when $\sigma_1 >> \sigma_2 > \sigma_3$.

By comparing the last of formulas (1.121a) and (1.9), we establish:

$$\sigma^{IV}_{equiv} = \sigma_{int}.$$

Thus, it turns out that stress intensity possesses energy content because the fourth (*IV*) (classic) theory of strength is based on the analysis of energy needed to change the shape of a deformable solid.

1.2.3 Deformation energy

In the elastic state during simple tension of the specimen the *deformation energy is rated by the work of the internal force at a corresponding displacement*:

$$u = \frac{1}{2}\sigma\varepsilon = \frac{1}{2E}\sigma^2 . \qquad (1.23)$$

Geometrically energy (1.23) is numerically equal to the area under the corresponding segment of the straight line *OA* on the tension diagram *OABC* (cf. Fig. 1.5). Similarly, in the state of plasticity

$$u = k_{sh}\sigma\varepsilon \sim \sigma^2, \qquad (1.24)$$

where k_{sh} – the coefficient of the shape of the curve *OABC*.

In both cases, as it follows from (1.23) and (1.24), the *deformation energy is proportional to the square of the corresponding stress*.

In case of the triaxial stressed state and using (1.6) and (1.11), it can be recorded:

$$u = k_{sh}\sigma_{int}\varepsilon_{int} = k_{sh}\sigma_1\varepsilon_1\varphi\left(\frac{\sigma_2}{\sigma_1}, \frac{\sigma_3}{\sigma_1}, \frac{\varepsilon_2}{\varepsilon_1}, \frac{\varepsilon_3}{\varepsilon_1}\right) \sim \sigma_1^2 . \qquad (1.25)$$

Formula (1.25) confirms the conclusion made above about the proportionality of u and σ^2 for any general case of the stressed state of a solid.

1.3 Fatigue

1.3.1 Fatigue curve

The process of gradual damage accumulation under the effect of alternating stress leading to changes in the structure and properties of the material, nucleation and growth of cracks, and volume fracture is termed (*mechanical*) *fatigue*. Stresses may change in time *regularly* (cf. Fig. 1.1, *d*) or irregularly (cf. Fig. 1.1, *e*).

In case of cyclic loading the *fatigue curve* yields the most complete information about the *resistance to fatigue* of the components of a structure. This information is usually obtained experimentally as a result of time-consuming tests of a large number (or a series) of nominally identical specimens.

The full fatigue curve (Fig. 1.15) is a relation between the amplitude of (or maximum value) stresses σ and *cyclic durability* (the number of cycles until fracture) N_σ throughout the range of their possible changes: $\sigma \leq \sigma_b$ and $1 \leq N_\sigma < 10^9$ cycles [5]. If $\sigma = \sigma_b$, then $N_\sigma = 1$, i.e. static fracture may serve as a boundary case of fatigue fracture.

Volume fracture of the tested specimen (its separation into two pieces) or the moment when the fatigue crack reaches some (preset) length, for example, 0.5 or 1.0 mm (surface damage), may serve as the ultimate state criterion when assessing durability.

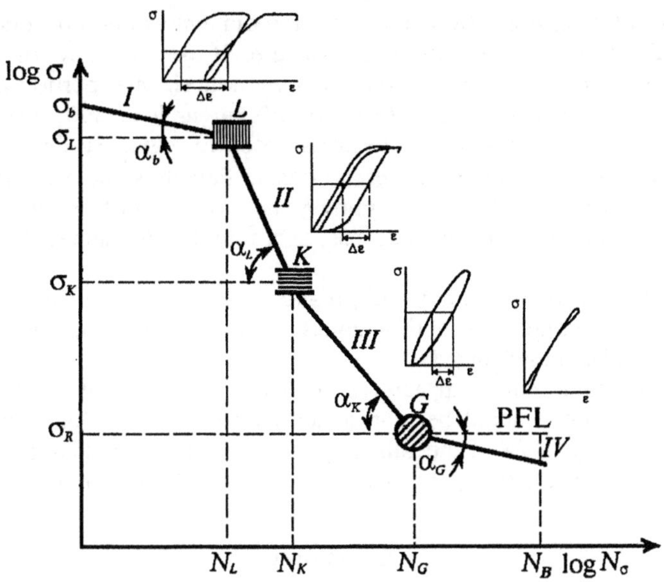

Fig. 1.15. Diagram of full curve of mechanical fatigue

If the fatigue curve is plotted in double logarithmic coordinates log σ – log N_σ, there appear four (*I, II, III, IV*) typical regions usually represented by segments of a straight line and having different angles of inclination α to the abscissa axis (cf. Fig. 1.15).

Relatively large changes in durability with little changes in stresses are typical for region *I* of *quasistatic fatigue*. Fracture during tests of soft steel is due to the evolution of strong plastic deformation: the relation σ–ε within one loading cycle represents an *open loop of plastic hysteresis*. Specimens usually withstand from several tens to several hundreds (sometimes up to a thousand) of cycles.

On the contrary, a relatively small change in the durability with a significant stress drop is typical for region *II* of *low-cycle fatigue*. Fracture in this case is due to the process of elastoplastic deformation: the relation σ–ε within one loading cycle is an *unclosed loop of elastoplastic hysteresis*. Low-cycle fatigue is observed within the range of durability of approximately $10^3...10^4$ cycles.

For curve *III* of *multicycle fatigue* the angle of slope α_K is less than the angle of slope α_L of the curve of low-cycle fatigue, but it is usually larger than the angle of slope α_b of the quasistatic fatigue curve. Fracture is due to the accumulation of *non-elastic deformation*: the relation σ–ε represents a *closed loop of mechanical hysteresis*. Since *microplasticity* developing in separate structural components of the material becomes responsible for fatigue fracture in this region, it has a quasibrittle pattern; it implies that the tested specimen does not manifest any measurable residual deformation. Yet, electron microscopic studies of fatigue ruptures during multicycle fatigue reveal the mechanisms of viscous and brittle fracture. The durability during multicycle fatigue is approximately within the interval of $5 \cdot 10^4$–$5 \cdot 10^6$ cycles.

Region *IV* of *gigafatigue* (tests in the air at room temperature) occurs only with the materials possessing an unstable structure. If the structure of the tested material is not subjected to *deformation ageing*, a horizontal portion appears on the fatigue curve corresponding to the (*physical*) *fatigue limit* (PFL, see the dotted line in Fig. 1.15). The mechanical hysteresis loop in region *IV* degenerates: the relation σ–ε becomes practically proportional, though some peculiarities are possible (secondary loops) in the beginning and at the end of one loading cycle. Fracture is due to *nanoplasticity* and has, as a rule, brittle nature; the durability exceeds 10^7 cycles.

Usually two *discontinuities of the fatigue curve* are present in the zones of transition from one region to another (zones *K*, *L*), they prove that the dominating fracture mechanism has changed.

If the full fatigue curve appears as shown in Fig. 1.15, it is clear that its analytical description cannot be represented as a single equation. On the other hand, since portions *I–IV* of the full fatigue curve plotted in double logarithmic coordinates are straight lines, they can be described by the simplest exponential equation

$$\sigma^{m_\sigma} N_\sigma = C_\sigma = \text{Const} \tag{1.26}$$

with its own (for each portion) *parameters* m_σ and C_σ. They are easy to find providing the coordinates of points *L*, *K* and *G* are known. No break of the fatigue curve at point *K* is found in some experiments, then portions *II* and *III* are approximated by a single smooth line.

Proceeding from the above-said, it can be considered that σ_L is the *quasistatic fatigue limit*, σ_K is the *low-cycle fatigue limit*, σ_G is the *multicycle fatigue limit* (for the case when gigafatigue appears). In case the latter does not appear, the multicycle fatigue limit is termed the *endurance limit* σ_R; here the index *R* designates the *coefficient of asymmetry of the stress cycle*. If the *cycle is pulsating*, then $R_\sigma = 0$; when the cycle is symmetric then $R_\sigma = -1$.

As a rule, the full fatigue curve is plotted using the *nominal stresses*, i.e. when calculating stresses any possible plastic deformation (in portions *I* and *II*) is disregarded (ignored).

In practice only the *multicycle fatigue curve* is obtained most frequently; it is simply called the *fatigue curve* (Fig. 1.16) or the *S–N-curve*, or the *curve of Whöler*. The *test base* numbers $N_B = 10^7$ (for ferrous metals) and $N_B = 2 \cdot 10^7$ (for non-ferrous metals) cycles. Tests are usually performed by bending round specimens with their rotation, i.e. with a symmetric cycle of changes of stresses in time. The endurance limit σ_{-1} corresponds to the horizontal line in Fig. 1.16 and divides the region of possible changes in the magnitude of cyclic stresses into two subregions: $\sigma > \sigma_{-1}$ (fatigue fracture takes place) and $\sigma < \sigma_{-1}$ (fatigue fracture does not take place until the test base N_B is reached). Hence, σ_{-1} is the *boundary* (based on stresses) *between the endurance and fatigue of the materials* (of the specimens). The *left branch of the fatigue curve* represented schematically by two intersecting lines in Fig. 1.16 may also include a part of the curve *LK* of low-cycle fatigue (cf. Fig. 1.15); Eq. (1.26) describes it in which the *parameter of slope*

$$m_\sigma = \cot\alpha = \frac{\log N_2 - \log N_1}{\log\sigma_1 - \log\sigma_2} \tag{1.27}$$

and the *constant of resistance to fatigue*

$$C_\sigma = \sigma_{-1}^{m_\sigma} N_{G\sigma}, \tag{1.28}$$

where $N_{G\sigma}$ – the *abscissa of the breakpoint of the fatigue curve*, and (σ_1, N_1), (σ_2, N_2) – coordinates of two points on its *left branch* (cf. Fig. 1.16). If $\log\sigma_1 - \log\sigma_2 = 1$, then according to formula (1.27), the parameter m_σ is the increment of the logarithm of durability of specimens when the logarithm of stresses is reduced by a unit.

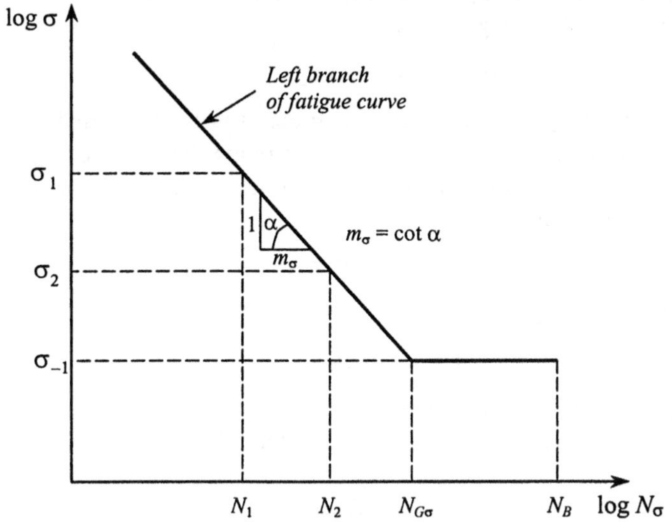

Fig. 1.16. Fatigue curve in multicycle region

Fatigue tests with a symmetric stress cycle are preferable due to several considerations. First, the symmetric cycle is more dangerous (Fig. 1.17) and the assessment of damage becomes more sensitive. Second, if the value σ_{-1} is known, it can be recalculated into the fatigue limit at any *cycle asymmetry* (cf. Fig. 1.17), if the dependence of the ultimate amplitude of stresses $\lim\sigma_a$ on the mean stress within the cycle σ_m is approximated by a straight line (dotted line) that is referred to the safety margin:

$$\sigma_{-1} = \frac{\lim\sigma_a}{1 - \dfrac{\sigma_m}{\sigma_y}} = \frac{\sigma_R}{1 - \dfrac{\sigma_m}{\sigma_y}}.$$

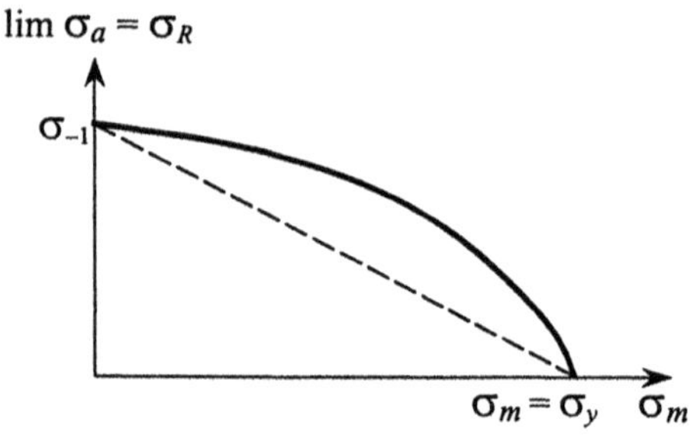

Fig. 1.17. Dependence of ultimate amplitude lim σ_a on mean value σ_m of stresses in cycle

Third, the fatigue limit during bending with torsion has a steady proportional relation to the ultimate strength of steel

$$\sigma_{-1} = (0.4...0.6)\ \sigma_b. \tag{1.29}$$

Hence, by knowing just the ultimate strength, it is always possible to make an approximate assessment of the fatigue limit assuming that $\sigma_{-1} \approx 0.5\sigma_b$. Fourth and last, there is a stable relation for steel between the limits of torsional strength τ_{-1} and during bending with rotation σ_{-1}:

$$\tau_{-1} = (0.5...0.6)\ \sigma_{-1}, \tag{1.30}$$

therefore, on the average

$$\tau_{-1} \approx 0.27\ \sigma_b. \tag{1.31}$$

The value σ_{-1} can be determined more precisely if basic characteristics of the mechanical properties of steel are known from the formula [3]

$$\sigma_{-1} = \frac{\sigma_b - \sigma_y}{0.75\left[\dfrac{\delta}{\psi}\left(7.5 - 4.167\dfrac{\delta}{\psi}\right) - 1\right]}, \tag{1.32}$$

that can also be recorded through Brinell hardness *HB* considering that

$$\sigma_b - \sigma_y = 0.35HB\left(1 - \frac{\sigma_y}{\sigma_b}\right). \tag{1.33}$$

Here δ and ψ – *relative elongation* and *contraction during rupture*. Formulas (1.29)–(1.33) have been validated experimentally.

Thus, the fatigue curve (cf. Fig. 1.16) serves to determine experimentally all basic characteristics of resistance of the material to alternating loads, including the fatigue limit as the most essential characteristic. At the same time Eq. (1.26) enables to calculate the cyclic durability N_σ as a function of normal stress σ:

$$N_\sigma = \left(\frac{\sigma_{-1}}{\sigma}\right)^{m_\sigma} N_{G\sigma} = \frac{C_\sigma}{\sigma^{m_\sigma}}. \tag{1.34}$$

According to this formula, durability is inversely proportional to cyclic stress (to the power of m_σ), meanwhile, if $\sigma = \sigma_{-1}$, then $N_\sigma = N_{G\sigma}$, and if $\sigma < \sigma_{-1}$ then $N \to \infty$ (i.e. it is assumed that fatigue fracture will not occur).

1.3.2 Mechanisms of fatigue of metals

The mechanisms of damage accumulation and fracture in different regions of the full fatigue curve differ substantially. The process of fracture in quasistatic region *I* (cf. Fig. 1.15) is determined by strong plastic deformation evolving under the effect of stresses approaching the ultimate strength. Significant displacements appear in a metal, in case whole grains displace, the route of displacement runs along their boundaries; when parts of grains displace, the route of displacement runs across grains themselves. Large displacements occur in one of the two ways: *sliding* and *twinning*. Displacement by sliding (Fig. 1.18, *a*) occurs along planes *1* because the spacing between adjacent planes with a large density of atoms (type 2) is the maximum and therefore the bonding between atomic planes is the least. Figure 1.18, *b* shows the result of displacement by sliding. The displacement by sliding occurs under the effect of tangential stresses τ.

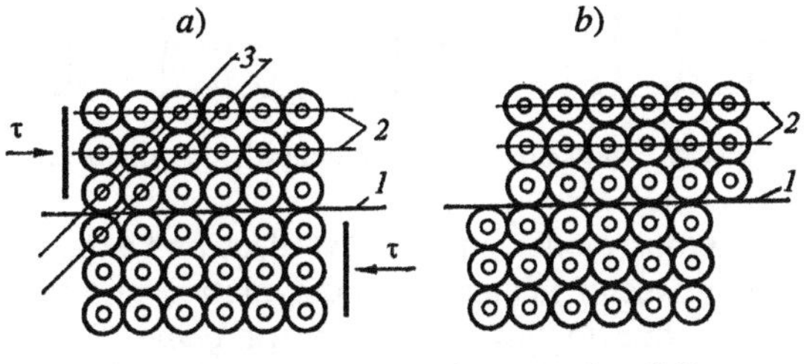

Fig. 1.18. Scheme of displacement by sliding

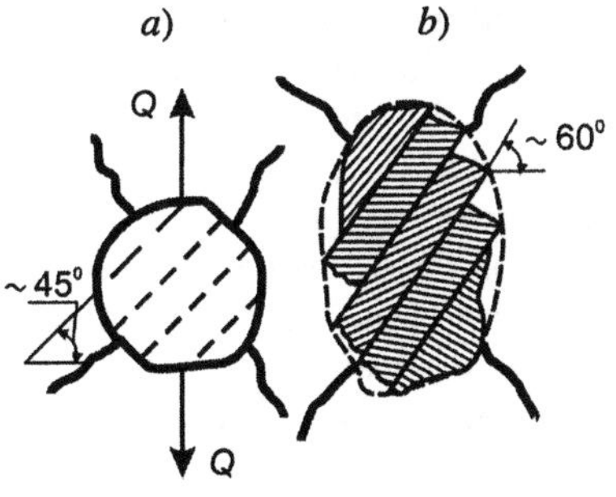

Fig. 1.19. Scheme of displacement by twinning

Figure 1.19 explains the displacement by twinning. If tensile force Q (Fig. 1.19, *a*) affects grains, then, concurrently with shear in the direction of action of maximum tangential stresses τ, parts of grains turn in the direction of tension (Fig. 1.19, *b*), i.e. the deformation is forced by the external force and the displacing parts cannot move freely in the direction τ.

Actual stresses in multicycle fatigue region *III* (cf. Fig. 1.15) are weak, hence they cannot produce any significant plastic deformation. Fatigue damage is determined by other mechanisms relating primarily to *local microplastic deformations*.

A real technical metal has *structural defects*, including *spot defects* (like *vacancies* and *interstitial atoms*), and *linear defects* (like *dislocations*). Such defects can travel under the effect of cyclic stresses. Fig. 1.20, *a* shows the crystalline lattice with one incomplete atomic plane with an *edge dislocation*. The dislocation has displaced by one parameter of the crystalline lattice under the effect of cyclic stresses (Fig. 1.20, *b*); after long deformation the dislocation emerges on the surface producing a shear step (Fig. 1.20, *c*). Displacement by shear takes place in this manner, yet this displacement is extremely localized, so that the body does not show any measurable plastic deformation.

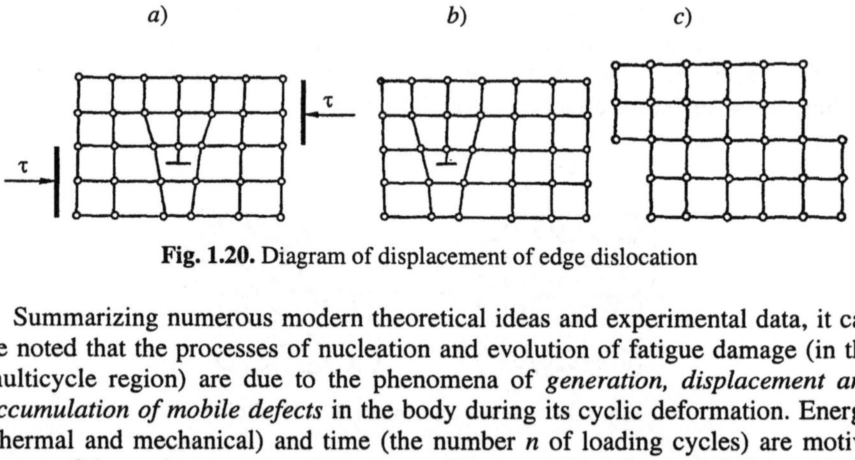

Fig. 1.20. Diagram of displacement of edge dislocation

Summarizing numerous modern theoretical ideas and experimental data, it can be noted that the processes of nucleation and evolution of fatigue damage (in the multicycle region) are due to the phenomena of *generation, displacement and accumulation of mobile defects* in the body during its cyclic deformation. Energy (thermal and mechanical) and time (the number n of loading cycles) are motive forces of these phenomena and processes. Since mobile defects of different kinds, congenital and generated by loading (deformation), exist in a metal, the phenomena of accumulation of fatigue damage can also relate to this or that type of a defect. Therefore, we can speak of *dislocation, vacation* and *thermofluctuation mechanisms of fatigue*.

According to the *dislocation ideas*, the period of fatigue incubation prepares the stage of nucleation of submicroscopic cracks and relates to the accumulation of a critical density of dislocations in local volumes of the metal. Cyclic shear stresses (deformations) $\tau(\gamma)$ give push to the dislocations, they begin to move and generate fresh dislocations in some regions of the metal. These regions are boundaries of grains and subgrains in the technical iron. When dislocations move, they cannot overcome all the obstacles. For example, if boundaries between adjacent grains form an insurmountable obstacle in a polycrystalline metal, the dislocations group into a *flat cluster* at a boundary. The larger the distance of free travel of dislocations usually corresponding to the radius of a grain, the more is their number in the flat cluster. The flat cluster of dislocations forms a *strip of sliding* that may become the site of nucleation of *primary submicrocracks*.

Figure 1.21 shows one and the same portion of nickel at different stages of fatigue tests [6]. Initially fine *strips of sliding* appear after 10^4 cycles. Their number grows and some initial strips widen noticeably as the number n of cycles increases. Steels demonstrate the same phenomenon. After a while the sliding strips transform into *bundles of sliding* (Fig. 1.21, *d*). Dark thin lines in the photograph are submicrocracks nucleating in the bundles of sliding [7].

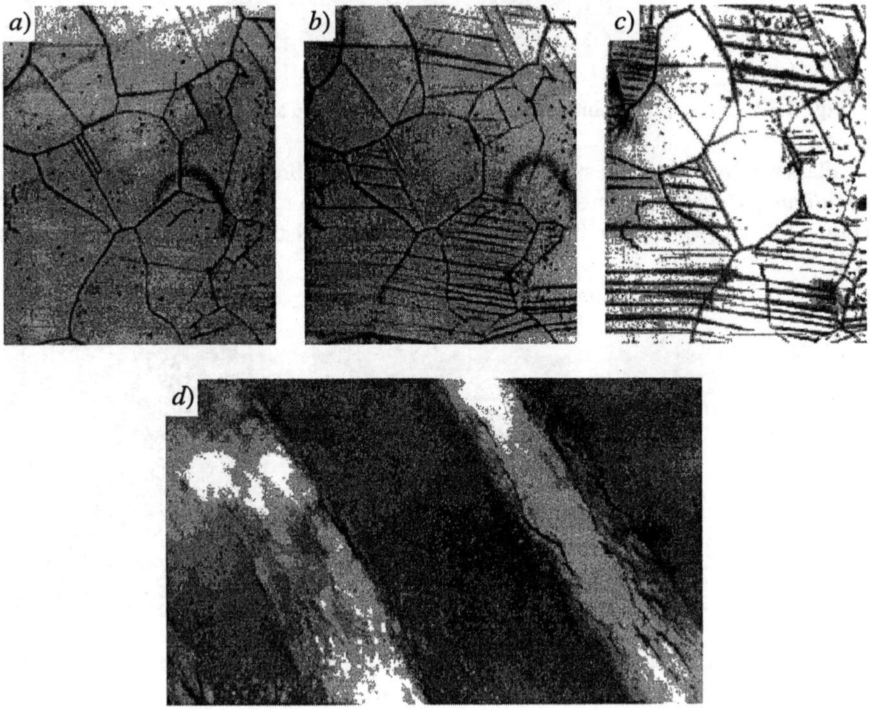

Fig. 1.21. Fine nickel at different stages of fatigue tests: ($a - n = 10^4$ cycles; $b - 5 \cdot 10^4$; $c - n = 2.7 \cdot 10^5$ cycles (×330) and *d* – electron diffraction pattern of sliding bundles in soft steel after $1.9 \cdot 10^6$ test cycles with stresses somewhat below fatigue limit (×5800)

If the sliding bundle meets with the specimen's surface, fine flakes of metal less than 1 μm thick are extruded from the deformed volumes of the body along the plane of cyclic sliding (Fig. 1.22) [8]. It is a phenomenon of *extrusion* that is in fact a process of local surface microfracture. Extrusions usually neighbor with *intrusions* or microgrooves stretching along the plane of sliding and alternating with the extruded flakes of the metal.

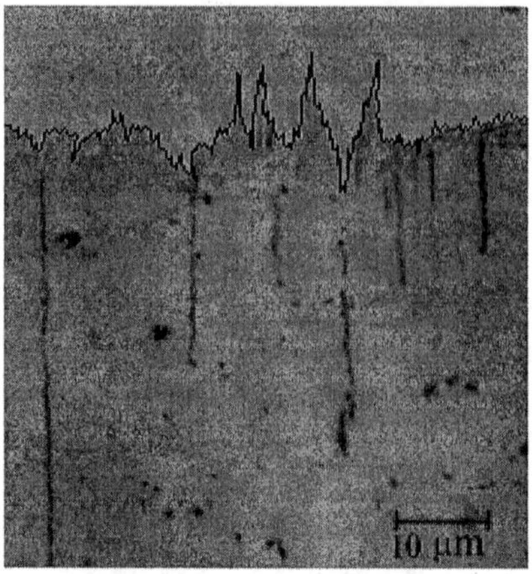

Fig. 1.22. Extrusion and intrusion on specimen's surface appearing during fatigue tests

Flakes extruded to the surface warp, crumble and break off. As a result *products of fatigue fracture* appear as loose differently shaped microparticles of the material (Fig. 1.23, *a*) [9]. Larger fracture particles appear also as cyclic stresses continue (Fig. 1.23, *b*, *c*).

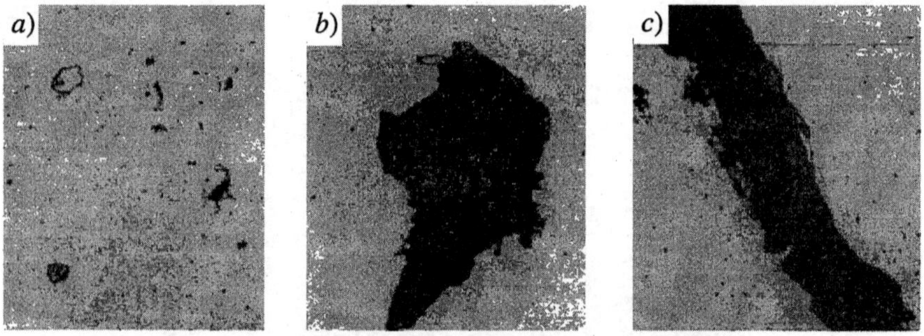

Fig. 1.23. Typical shapes of particles of fatigue fracture (×100): *a* – spot and flake-like globules, *b* – round plates, *c* – elongated plates

A multiple *statistical pattern of primary fatigue damage* should be emphasized specifically. A huge number of strips and bundles of sliding appear in the grains of a cyclically deformable body, with a significant portion of them being a source of submicrocracks. Kinetic interactions between them and their evolution under the effect of cyclic tangential stresses lead to the appearance of multiple fatigue microcracks, i.e. to *dissipated damage.*

The *vacation mechanism* of nucleation of fatigue submicrocracks is not universal, some metals demonstrate it under definite conditions. In order for fatigue cracks to nucleate by coagulation of vacancies, their concentration should be excessive and there should be favorable conditions for diffusion. In a number of cases no significant migration of vacancies is needed for nucleation of cracks. It can be observed when planes of sliding intersect. Then the vacancies appearing when moving dislocations intersect can quickly coagulate and produce *pores* in the nodes of intersection of planes of sliding. Figure 1.24 shows such pores observed in steel Г-13Л [10]. The pores locate in the intersections of the planes of sliding and have the shape of cylinders of an approximately equal diameter that was detected by consecutive etching of layers of the metal.

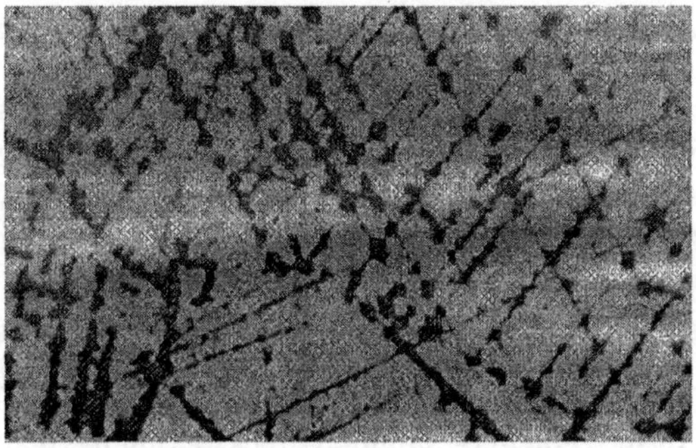

Fig. 1.24. Microstructure of steel Г-13Л tested for fatigue under repeated loads during $1.2 \cdot 10^5$ cycles (×500)

Further growth of a pore to microscopic dimensions may occur both by the gathering of vacancies around it and by the disappearance of dislocations on the surface of the pore. Diffusion of vacancies in the new pore is facilitated as the volumes of the metal near the strips of sliding become strongly loose.

The *kinetic theory of strength* is based on the theory of thermal motion in solids. According to the latter, the positions of atoms in the solid are not strictly fixed. Oscillating with a definite frequency near the position of equilibrium the atom has certain probability to break bonds with adjacent atoms and leave the place it occupies. The atoms acquire the energy needed for this at the expense of chaotic *thermal fluctuations*. These fluctuations being short-term concentrations of elevated kinetic energy of atoms in thermal motion appear from time to time of each atom. The probability of appearance of a thermal fluctuation of a given atom strongly depends on its magnitude: smaller fluctuations appear often, larger fluctuations rarely. The time of existence of the atom in bonds with adjacent atoms is the time of expectation of a fluctuation with the energy exceeding the energy of bond breaking. The stronger the atomic bonds and the lower the temperature, the less is the probability of fluctuation breaking bonds and the longer is the time that the atom remains in equilibrium. It allows to consider mechanical fracture of a solid as a time process of gradual accumulation of broken atomic bonds. A mechanical force applied to the specimen does not break bonds between atoms, it just deforms and excites atoms making them ready for bond breaking. The bond breaking is performed by thermal fluctuations due to the energy of thermal motion. Thus, according to the kinetic theory of strength, the process of appearance and accumulation of fatigue damage evolves in three stages:

1) excitation of atomic bonds by mechanical loading; 2) breaking of stressed bonds by thermal fluctuations generated by thermal motion; 3) accumulation of broken bonds resulting in multiple microcracks.

These multiple microcracks from the point of view of their evolution are divided into two types: the microcracks that each individually does not result in final fracture (the so-called *non-propagating cracks*) and the microcracks that grow into the *main crack* causing the body to fail.

In the process of fatigue the dissipated microcracks grow both in number and average length. In the course of time the *concentration* of these *cracks* reaches some critical value of the first order, and multiple microcracks of the next order appear. Then this process repeats until the density of multiple microcracks dissipated in the deformable body reaches the ultimate concentration.

Within the framework of the kinetic theory of strength the *concentration criterion of fracture* of solids had been advanced. According to this criterion, there is a relation at the moment of fracture between the mean statistical dimension d_T of multiple cracks and the mean spacing between them $C_*^{-1/3}$; this relation is the following: $C_*^{-1/3} = d_T e$, where *e* is the radix of natural logarithms [11].

If the direction of movement of submicrocracks along the planes of sliding with a high density of dislocations is governed by tangential stresses and coincides with the planes of the maximum cyclic microshearing, then the direction of migration of the microcrack diverges from the direction of the strip of sliding and coincides in general with the direction perpendicular to the direction of action of the maximum macroscopic tensile stress.

Regions along the strips of sliding, boundaries of grains, edges of non-metallic inclusions, interfaces between phases, boundaries of blocks and others may be sources of nucleation of a crack leading to fracture. Non-metallic inclusions in metallic materials become more often the zones of nucleation of the main fatigue crack when the level of the static strength is higher; the boundaries of the matrix structure become such zones more frequently when the material is softer. A single crack may cause final fracture of very hard materials (the brittle state). If a material has a sharp notch, the only crack nucleates and grows primarily at the bottom of the notch.

1.3.3 Cyclic hardening-softening

If changes in the *width* $\Delta\varepsilon$ of the *hysteresis loop per cycle* are studied in time (see curves σ–ε in Fig. 1.15) for various materials, it can be established that there are four types of relations $\Delta\varepsilon(n)$ (Fig. 1.25) under the effect of cyclic stresses $\sigma = \text{const}$ of different levels ($\sigma_1 > \sigma_2$) [12].

The *first type* (*I*): reduction of the width of the hysteresis loop per cycle $\Delta\varepsilon$ as the number of loading cycles *n* grows. The materials demonstrating this type of relationship of $\Delta\varepsilon(n)$ are called *cyclically hardenable*. Fine annealed metals (Cu, Ni and others) and solid solutions (alloys Д20, 30Х10Г10 and others) are among them. Their hardening is due to the appearance of effective barriers preventing sliding.

The *second type* (*II*): the hysteresis loop widens as the number of loading cycles grows. The materials demonstrating this type of relationship of $\Delta\varepsilon(n)$ are called *cyclically loosing strength*. They are the materials hardened by plastic deformation or by dispersed particles (copper in the deformed state, austenite steels of the type 1X18H10T, 0X14AГ11M, steels 40X, 12XH3, ЭИ612 and others).

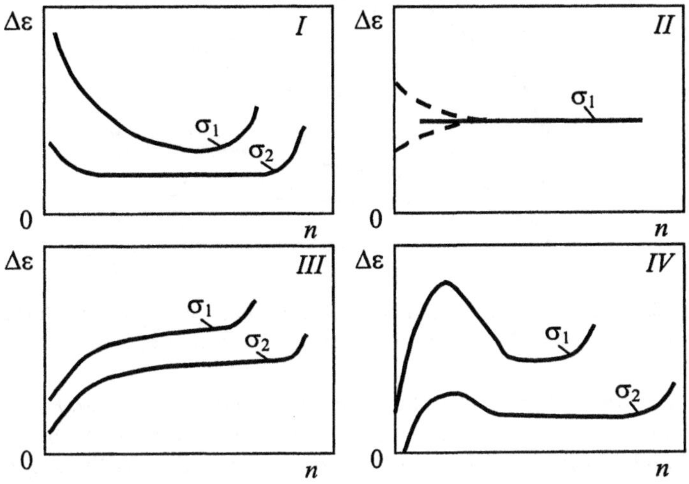

Fig. 1.25. Different types of relation between non-elastic deformation per cycle and number of loading cycles for metals ($\sigma_1 > \sigma_2$)

The *third type* (*III*): the width of the hysteresis loop remains practically unchanged during the entire time of loading. The materials demonstrating this type of relationship of $\Delta\varepsilon(n)$ are called *cyclically stable*. These materials include pig irons, some aluminium alloys, austenite steels at lower temperatures. This practical constancy of the value $\Delta\varepsilon$ in the process of cyclic loading is due to sufficiently large inclusions of the second phase favoring the evolution of non-elastic phenomena in the sites of concentration of stresses relating to inclusions.

The *fourth and the last type* (*IV*): *complex behavior in time* that is a combination of the curves of type *II* and then *I* (cf. Fig. 1.25). The materials demonstrating the complex behavior include carbon steels 30, 60, 45 and others, some alloyed steels (1X13, 15Г2АФДпс and others). Reduction of the value of $\Delta\varepsilon$ after the maximum is reached relates to the deformation ageing in the process of cyclic loading.

The *processes of hardening – softening* are also observed when studying accumulation of residual (non-elastic) deformation ε_{res} in time in a given material as a function of the level of cyclic stresses (Fig. 1.26) [13]. In the general case the *kinetic curve* $\varepsilon_{res}(t)$ *comprises three portions*: A, B, C (curve 2).

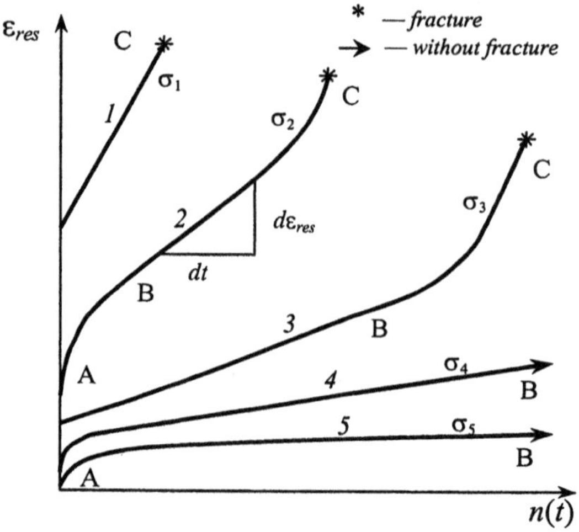

Fig. 1.26. Typical curves of accumulation of residual (non-elastic) deformation under effect of cyclic stresses $\sigma_1 > \sigma_2 > \sigma_3 > \sigma_4 > \sigma_5$

The first portion (A) relates to the *stage of unsteady deformation* when the rate of its accumulation decelerates as *n* grows due to the evolution of the processes of hardening and adaptation. The second portion (B) relates to the *stage of steady deformation*: a certain quantitative relationships between the processes of hardening-loosing strength sets in that remains unchanged for a long time and persists until deformation in the dangerous zone reaches some critical level of stresses. The third portion (C) is the *stage of fracture*: total deformation increases sharply as the specimen exhausts its bearing capacity, the main fatigue crack appears and grows.

In particular cases, as a function of the stress magnitude and test base, deformation accumulation is described by the curves of types *1*, *3*, *4*, *5* that can be derived from general curve *2* by truncating the latter from the right or from the left.

The kinetic straight line of type *1* is typical for the quasistatic fatigue region (cf. *I* in Fig. 1.15). The kinetic curves of types *2* or *3* are typical for the low-cycle fatigue region (cf. *II* in Fig. 1.15), curves *4* and *5* are typical for the multicycle fatigue region (cf. *III* in Fig.1.15) at $\sigma_1 > \sigma_{-1}$. If $\sigma_1 << \sigma_{-1}$, no measurable non-elastic deformation is observed.

According to Fig. 1.26, the higher the level of stresses, the shorter the time till fracture of specimens (or cyclic durability), and it is fully determined by the kinetics of residual deformation. The type of the kinetic curve is dictated by the *rate*

$$v_\varepsilon = \frac{d\varepsilon_{res}(\sigma)}{dt} \tag{1.35}$$

of accumulation of residual deformations. The rate (1.35) is maximum during quasistatic and minimum during multicycle fatigue; it is intermediate during low-cycle fatigue.

1.3.4 Cyclic resistance to cracking

In the general case the fatigue process has two stages: the *stage before crack nucleation* and the *stage of crack development.* The relation between duration of these stages varies within a broad range as a function of effective stresses, the scheme of loading, the dimensions and shape of an object, the state of the material, etc. In some cases development of the main crack can amount to 60...90 % of the total durability. It is specifically long in the objects with concentrators of stresses; this stage is termed *survivability.*

If N_I is the durability at the first stage (until a macroscopic crack appears) and N_{II} is the survivability at the stage when the main crack develops, then the total durability (from the start of loading until volume fracture) is

$$N = N_I + N_{II}.$$

Damage at the first stage is due to cyclic stresses σ. If $\sigma > \sigma_{-1}$, then durability is [14]

$$N_I = N_{G\sigma}\left(\frac{\sigma_{-1}}{\sigma}\right)^{m_\sigma}(1-\omega_1)^{m_c+1}, \tag{1.36}$$

where ω_1 – the *measure of structural damage of the material* due to the stressed state of the object during the first loading cycle; m_c – the parameter characterizing the intensity of damage acceleration as the level of cyclic stresses increases. It follows from formula (1.36) that durability N_I reduces as ω_1 and m_c increase. In order to preserve N_I = const when ω_1 increases, then m_c should be reduced correspondingly, i.e. a material should be selected with a stronger resistance to fatigue. If $\sigma = 0$, then $\omega_1 = 0$, it is predicted that $N_I \to \infty$ from formula (1.36). At $\omega_1 = 1$ we have $\sigma = \sigma_b$ and $N_I = 0$ according to (1.36).

Stresses do not govern the crack development at the stage of survivability, the parameter K does it and it is termed the *stress intensity factor*. Its sense is that in case the intensity factors for two different pieces are equal, the material has the same stress-strain state at the tip of the crack in both cases.

The parameter K depends on the magnitude and nature of external loads, the shape and dimensions of a body, location and length of the crack, loading conditions.

Depending upon the scheme of deformation of a body with a crack (Fig. 1.27) the following stress intensity factors are identified: K_I – during tension, K_{II} – during transverse shear, K_{III} – during longitudinal shear. In case of deformation according to scheme *I* (tension), the boundaries of the crack diverge; in case of deformation according to scheme *II* (transverse shear), the surfaces of the crack slide mutually in the transverse direction, and in case of deformation according to scheme *III* (longitudinal shear), the surfaces of the crack slide mutually in the longitudinal direction.

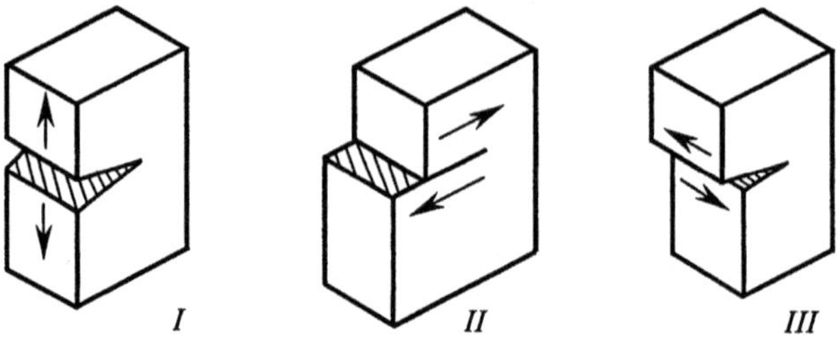

Fig. 1.27. Diagrams of deformation of body with crack

If the crack develops in the plates of unlimited dimensions, the stress intensity factors for the corresponding schemes of deformation (cf. Fig. 1.27) are

$$\left.\begin{aligned} K_I &= \sigma\sqrt{\pi l}; \\ K_{II} &= \tau_{II}\sqrt{\pi l}; \\ K_{III} &= \tau_{III}\sqrt{\pi l}, \end{aligned}\right\} \qquad (1.37)$$

where σ and τ – normal and tangent stresses, l – crack length (depth).

Formulas (1.37) for real (concrete) objects are recorded with the account of *correction functions* Y_I, Y_{II}, Y_{III}, that make allowance for the scheme of loading and the geometry of crack:

$$\left.\begin{aligned} K_I &= \sigma\sqrt{\pi l}; \\ K_{II} &= \tau_{II}\sqrt{\pi l}; \\ K_{III} &= \tau_{III}\sqrt{\pi l}, \end{aligned}\right\} \qquad (1.38)$$

For example, when *compact specimens* are tested for resistance to cyclic cracking following the scheme of off-center tension (Fig. 1.28, *a*), the correction function is

$$Y_1 = (l/B) = 29.6 - 185.5(l/B) + 655.7(l/B)^2 - 1017(l/B)^3 + 638.7(l/B)^4, \qquad (1.39)$$

and then

$$K_I = \frac{Q}{h_0\sqrt{B}}Y(l/B), \qquad (1.40)$$

where h_0, B – the dimensions of a specimen; l – the crack length counted from the line of load action Q (cf. Fig. 1.28, *a*, *b*).

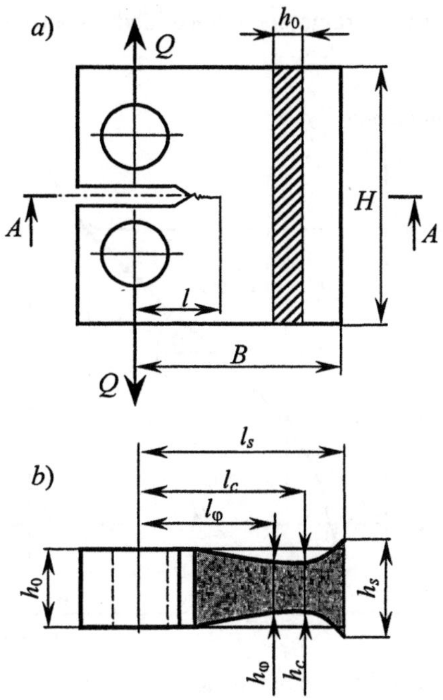

Fig. 1.28. Diagram of tests of compact cracked specimen (□) and measurement of its cross section contraction (*b*)

From the viewpoint of *linear mechanics of fracture*, the *kinetic diagram of fatigue fracture* is the integral characteristic of cracking resistance (Fig. 1.29). It is the relation between the rate of growth of a fatigue crack

$$v_K = dl/dn$$

and the maximum stress intensity factor K_{max} or its range $\Delta K = K_{max} - K_{min}$ per cycle. The diagram is plotted in double logarithmic coordinates log v_K – log K_{max} (log ΔK).

The areas can be identified in the diagram, each is characterized by its regularities of development of cracks: *I* – low ($0 < v_K < 5 \cdot 10^{-8}$ m/cycle), *II* – moderate ($5 \cdot 10^{-8} < < v_K < 10^{-6}$ m/cycle), *III* – high ($v_K > 10^{-6}$ m/cycle) rates of crack development.

The diagram (cf. Fig. 1.29) serves to determine two *characteristics of resistance to cyclic cracking*: *threshold* K_{th} and *critical* K_{fc} stress intensity factors; the latter is also referred to as *cyclic fracture toughness*. At $K = K_{th}$ the crack does not grow during 10^6 cycles; rupture of the object at $K = K_{fc}$ takes place. The most important area *II* of the kinetic diagram is described by a simple exponential equation [15]

$$v_K = C_K(\Delta K)^{n_K}, \qquad (1.41)$$

where C_K and n_K – the parameters determined experimentally; their sense is explained in Fig. 1.29.

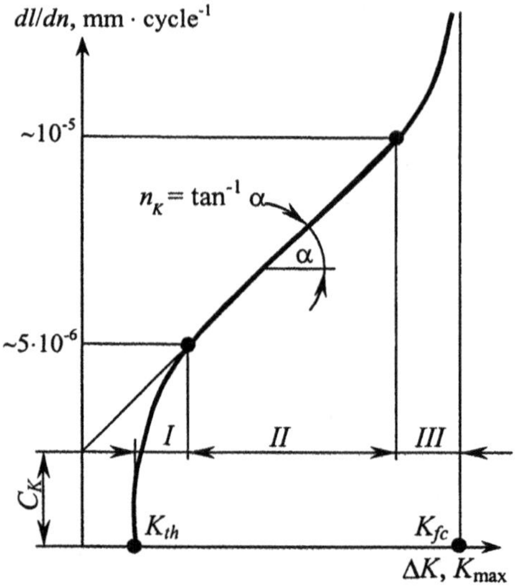

Fig. 1.29. Kinetic diagram of fatigue fracture of specimen with crack

Since volume fracture results from the development of the main fatigue crack over the entire dangerous cross section of the specimen, it is clear that actual stresses at the tip of the crack, hence the stress intensity factor, increase correspondingly as long as the effective cross section bearing the load reduces. It means that the rate of crack development accelerates too. Hence, the pattern of fatigue rupture (cf. Fig. 1.10) is significantly determined exactly by the rate of crack growth due to the growing stress intensity factor. The crack grows very slowly in center *1* of the fatigue rupture, it corresponds to area *I* of the kinetic diagram. Zone *3* of steady crack development (cf. Fig. 1.10) corresponds to area *II*, zone *4* of its unsteady development corresponds to area *III* of the kinetic diagram.

The rupture pattern (cf. Fig. 1.10) can also be assigned to the known fatigue fracture regions (cf. Fig. 1.15). Multicycle fatigue during stresses close to the endurance limit occurs in center *1* of the fatigue rupture. The region of transition from multicycle to low-cycle fatigue corresponds to the transition from zone *3* of steady crack development to zone *4* of unsteady crack development. Full rupture is in fact quasistatic fatigue. Thus, *rupture is a frozen picture of change of fracture mechanisms due to load growth* (the rates of crack development, the stress intensity factor). This picture is studied by special techniques of *fractography* in those cases when a structural component fails in operation in order to establish what loading conditions cause its failure. Fractographic studies reveal true causes of fracture.

If the critical crack length l_c is known that corresponds to the critical value K_{fc} of the stress intensity factor, the endurance limit can be determined from the formula

$$\sigma_{-1} = \frac{K_I}{Y_I \sqrt{\pi l_c}} . \tag{1.42}$$

If the specimen has a crack with the length l, the endurance limit with this crack is

$$\sigma_{-1K} = \sigma_{-1} \frac{\text{const}}{l^{1/m}} , \tag{1.43}$$

where σ_{-1} – the endurance limit of the specimen without a crack; m – the *parameter of mechanical homogeneity of the material.*

It follows from formulas (1.42) and (1.43) that the endurance limit of the cracked specimen is governed in the general case by the mechanophysical properties of the material (m, K_{th}) and the crack size (l, l_c), the relation between σ_{-1K} and l being inverse: the growing initial length of the crack implies a corresponding reduction of the endurance limit.

The survivability of the cracked specimen is determined from the formula [16] based on the equation of type (1.41):

$$N_{II} = \frac{1-\omega_0}{C_K \Delta K_\omega^{n_K} (n_K + 1)} , \tag{1.44}$$

where

$$\Delta K_\omega = \frac{K_{\max} - K_{th}}{1-\omega_0} ; \tag{1.45}$$

$$\omega_0 = l_0 / l ; \tag{1.46}$$

l_0 and $l > l_0$ – the initial and current crack length.

Equation (1.44) describes an inverse dependence of the survivability N_{II} on the value ΔK_ω, determined from formula (1.45). If initial damage is $\omega_0 = 0$ (the situation when $l_0 = 0$), then $\Delta K_\omega \to 0$, then $N_{II} \to \infty$. If $\omega_0 = 1$, then it means that the crack has become critically long ($l = l_c$), $K_{\max} = K_{fc}$, therefore, the specimen with such a crack fails during the first loading cycle ($N_{II} = 1$). Survivability in general is governed in many respects by the initial damage ω_0 (determined according to (1.46)): the smaller ω_0 the larger N_{II} (under other equal conditions).

Formulas (1.37)–(1.46) are true for elastic deformation providing the *conditions of flat deformation* are fulfilled. In case a compact specimen is tested, this condition requires that the relative residual *contraction* of the cross section (cf. Fig. 1.28, *b*) does not exceed 1.5%, i.e.

$$\Delta h_\varphi = \frac{h_0 - h_\varphi}{h_0} \cdot 100\,\% \le 1.5\,\%, \tag{1.47}$$

where $\varphi = h_0 - h_\varphi$ – absolute contraction; h_0 and h_φ – initial and current thickness of the compact specimen.

If condition (1.47) is not fulfilled (for example, for plastic steels), the characteristics of cracking resistance should be determined with the *correction for plasticity*.

From the standpoint of *non-linear fracture mechanics, the diagram of cyclic elastoplastic fracture of the cracked specimen* can serve as the integral characteristic of cracking resistance (Fig. 1.30, *a*) [3, 17]. The diagram is a relation

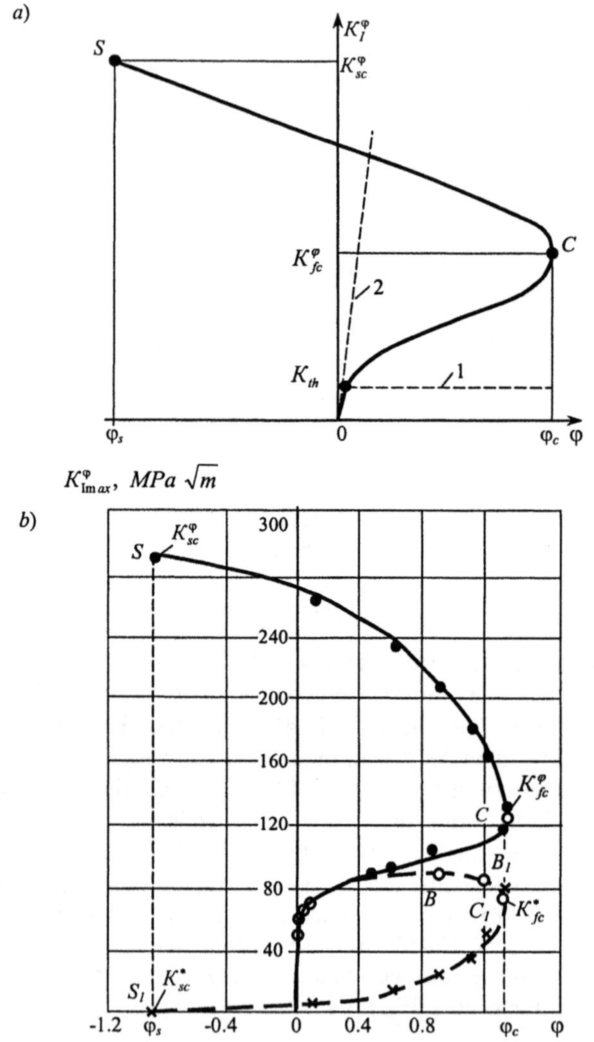

Fig. 1.30. Diagram of cyclic elastoplastic fracture of cracked specimen: *a* – scheme of *D*-diagram; *b* – experimental *D*-diagram and *Q*-diagram

between the stress intensity factor K_l^{φ} determined with the correction for plasticity and residual contraction of the cross section $\varphi = h_0 - h_{\varphi}$ (cf. Fig. 1.28, *b*).

The correction for plasticity is introduced into function (1.39) by substituting the relative crack length l/B in the elastically deformable specimen with the *true measure of damage*

$$\omega_l = A_l / A_0 \tag{1.48}$$

of the dangerous cross section with the allowance for its plastic deformation; here A_l – the true cross sectional area that the crack having length l occupies (with the account of contraction), A_0 – the nominal cross sectional (initial) area where the fatigue crack develops. Then formulas (1.40) and (1.39) with the account of (1.48) yield

$$K_l^{\varphi} = \frac{Q}{h_0\sqrt{B}}\left(29.6 - 185.5\omega_l + 655.7\omega_l^2 - 1017\omega_l^3 + 638.7\omega_l^4\right). \tag{1.49}$$

The diagram of cyclic elastoplastic fracture of the cracked specimen (Fig. 1.30, *a*) comprises two branches: 0C – the curve of steady fatigue crack development and *CS* – the curve of unsteady fracture (full rupture). The curve *0C* ends when the crack length reaches the value $l = l_c$ corresponding to the maximum contraction of the cross section φ_c (see also Fig. 1.28, *b*); the critical value of the stress intensity factor or cyclic elastoplastic fracture toughness K_{fc}^{φ} corresponds to them. The curve *CS* ends when the crack length reaches the value $l = B$ that corresponds to the maximum widening of the cross section $(-\varphi_s)$ (see also Fig. 1.28, *b*); the ultimate stress intensity factor or quasistatic fracture toughness K_{sc}^{φ} corresponds to them. Line *2* on the diagram (cf. Fig. 1.30, *a*) separates the region of quasibrittle from the region of elastoplastic fracture. Horizontal line *1* corresponds to the value K_{th} = const; the point of its intersection with line 2 has the abscissa corresponding to the relative contraction $\Delta h_{\varphi} = 1.5\%$ following condition (1.47).

Formulas (1.44) and (1.45) can be used for assessing the survivability of a cracked object in case of cyclic elastoplastic fracture providing a correction is made for plasticity when determining all the parameters:

$$N_{II} = \frac{1-\omega_l}{C_{\varphi}\Delta K_{\omega}^{n_{\varphi}}(n_{\varphi}+1)}; \tag{1.44a}$$

$$\Delta K_{\omega} = \frac{K_{\max}^{\varphi} - K_{th}}{1-\omega_l}. \tag{1.45a}$$

Here ω_l and $K_{\max}^{\varphi}$ are determined from formulas (1.48) and (1.49).

The diagram shown in Fig. 1.30, *a* as the *0CS* curve is plotted based on an assumption that load *Q* (cf. Fig. 1.28) remains nominal at the stage of full rupture of the specimen. It is called the *D-diagram* because its shape resembles *D* (with the account of the ordinate).

If it is re-plotted with the allowance for the real (true) value of load in the course of rupture, a *Q-diagram* is obtained ($0C_1S_1$ line in Fig. 1.30, *b*). An experimental *D*-diagram is also shown (line $0CS$ in Fig. 1.30, *b*) [3, 18]. Both diagrams are plotted using the results of tests of a compact steel 30 specimen (cf. Fig. 1.28) 10 mm thick. The maximum *Q*-diagram along the ordinate axis corresponds to the onset of *cyclic rupture of the specimen* (point B); *static rupture* begins at point B_1. Figure 1.30, *b* shows two values of cyclic elastoplastic fracture toughness, they are "nominal" (K_{fc}^{φ}) and "true" (K_{fc}^{*}), meanwhile $K_{fc}^{*} < K_{fc}^{\varphi}$. The true quasistatic fracture toughness is $K_{sc}^{*} = 0$ (see point S_1 in Fig. 1.30, *b*). It is essential to note that the relation $K_{fc}^{\varphi} / K_{fc}^{*} = \text{const}$ is for compact specimens of different dimensions made from one and the same material (providing their thickness is $t_0 \geq 20$ mm).

Publications describe other techniques of assessing resistance to cracking during static elastoplastic deformation (for example, using the *criterion of crack opening*, by calculating the *J-integral*), yet it has not been studied yet whether they are applicable to cyclic elastoplastic fracture.

1.3.5 Summing up damage

The relation between operating time (the number of loading cycles) n_j at a given level of effective stresses σ_j and *cyclic durability* $N_{\sigma j}$ under the effect of regular stresses of the same magnitude is the simplest and most popular *measure of fatigue damage*:

$$D_n = \frac{n_j}{N_{\sigma j}}. \tag{1.50}$$

For example, the durability $N_{\sigma j}$ is determined from the left branch of the fatigue curve (Fig. 1.31). So, if the stress σ_1 is effective during n_1 loading cycles, then according to (1.50), the accumulation of damage is $D_1 = n_1/N_{\sigma 1}$. Similarly, under the effect of σ_2 we obtain $D_2 = n_2/N_{\sigma 2}$. The measure in any of these cases (1.50) can change within the range

$$0 \leq D_n \leq 1. \tag{1.51}$$

If $D_n = 0$, then $n = 0$, hence there is no damage. If $D_n = 1$, fatigue fracture starts because $n = N_\sigma$. When a two-stage loading block (n_1 and n_2 – cf. Fig. 1.31) takes place, the *rule of linear summation of damage is true:*

$$\frac{n_1}{N_{\sigma 1}} + \frac{n_2}{N_{\sigma 2}} = 1,$$

according to this rule, *fatigue fracture occurs when the sum of damage under the effect of each stress* (σ_1 and σ_2) *reaches a unity.* This rule is easily expanded to any number *k* of stages within a *loading block* (Fig. 1.32)

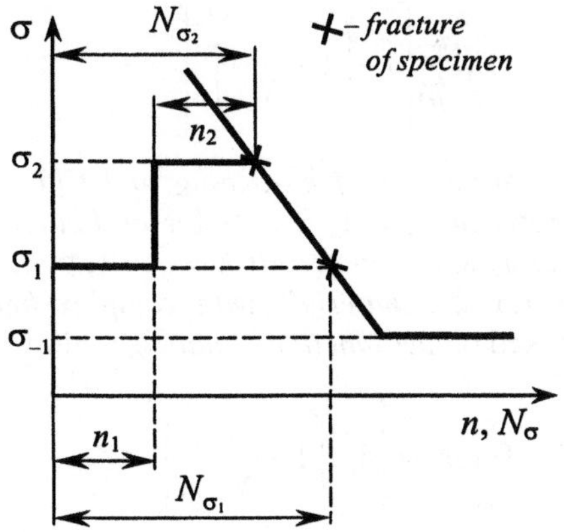

Fig. 1.31. Explanation of rule of linear summing up fatigue damage

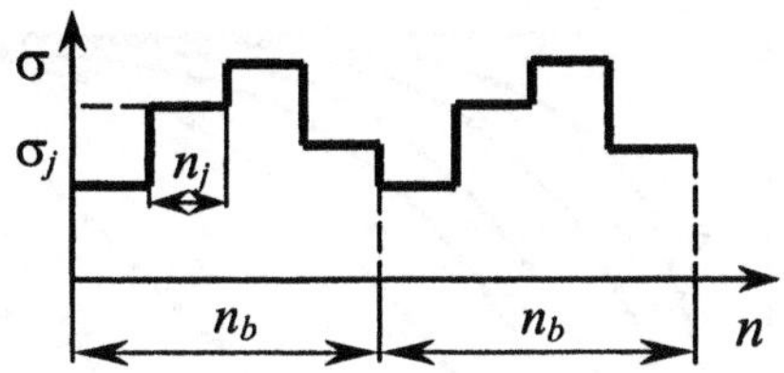

Fig. 1.32. Scheme of block loading

$$\lambda_b \sum_{j=1}^{k} \frac{n_j}{N_{\sigma j}} = 1, \tag{1.52}$$

where λ_b – the number of blocks until fracture of an object. It is easy to see that within each block a number of loading cycles takes place:

$$n_b = \sum_{j=1}^{k} n_j \,. \tag{1.53}$$

From (1.52) with the account of (1.53) and Eq. (1.26) of fatigue curve we obtain a formula to assess the *total number of cycles until fracture during block loading*

$$N_\Sigma = \lambda_b n_b = N_{G\sigma} \sum_{j=1}^{k} \left(\frac{\sigma_{-1}}{\sigma_j} \right)^{m_\sigma} . \tag{1.54}$$

The *non-linear hypothesis of damage accumulation* is one of the generalizations of rule (1.52):

$$\lambda_b \sum_{j=1}^{k} \left[1 - \left(1 - \frac{n_j}{N_{\sigma j}} \right)^h \right]^q = 1 , \qquad (1.55)$$

where q and h – the *parameters of hardening and softening*, respectively. For *cyclically softening materials* $q = 1$, $h > 1$; for *cyclically hardenable materials* $h = 1$, $q > 1$; for *cyclically stable materials* $h = q = 1$, therefore hypothesis (1.54) becomes rule (1.51); for the *materials with complex behavior* $h > 1$, $q > 1$. Hypothesis (1.55) is based on the *function of damageability* [14]

$$0 \le \omega_n = \left[1 - \left(1 - \frac{n}{N_\sigma} \right)^h \right]^q \le 1 . \qquad (1.56)$$

Graphs of function (1.56) for various classes of materials are presented in Fig. 1.33.

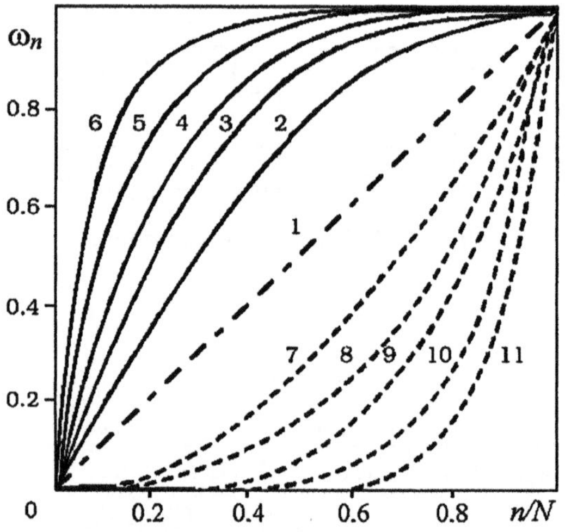

Fig. 1.33. Graphs of function (1.56) at different parameters h and q: straight line *1* – $h = q = 1$ (for cyclically stable materials); curves *2–6* – $q = 1$ = const and $h = 2, 3, 4, 6, 10$, respectively (for cyclically softening materials); curves *7–11* – $h = 1$ = const and $q = 2, 3, 4, 6, 10$, respectively (for cyclically hardenable materials)

In case of non-linear summation of fatigue damage, a formula follows from (1.55) for calculating total durability [14, 19]

$$N_\Sigma = \frac{n_b}{\sum_{j=1}^{k} \left[1 - \left(1 - \frac{n_j}{N_{G\sigma}} \left(\frac{\sigma_j}{\sigma_{-1}} \right)^{m_\sigma} \right)^h \right]^q} . \qquad (1.57)$$

Block loading is termed also *programmable* loading, it is a particular case of *irregular loading.* A *random process* is the most common law of irregular loading (cf. Fig. 1.1, *e*). Effective (actual) stresses $\sigma(t)$ change at a definite interval of time Δt in a random (i.e. unpredictable) manner within the range from the *lower* (σ_{min}) to the *upper* (σ_{max}) *dispersal limits* (Fig. 1.34). It is believed that *random stress* causes damage because its numerical value is $\sigma \geq \sim 0.5\sigma_{-1}$, where σ_{-1} – the fatigue limit that is established when plotting the fatigue curve (cf. Fig. 1.16).

Using the known *schematization procedures* the random process of loading $\sigma(t)$ can be replaced with the stage loading block (the *histogram* in Fig. 1.34, where $w(\sigma)$ – the *frequency ratio* for values of σ); the latter, in its turn, is approximated by a suitable *law (density) of distribution* $\varphi(\sigma)$ *of random value* σ. If dispersal of loading (σ) characteristic is subjected, for example, to the *normal law*, so it is easy to find its parameters, viz. the *mean value* $\overline{\sigma}$ and the *mean quadratic deviation* $S_{\overline{\sigma}}$.

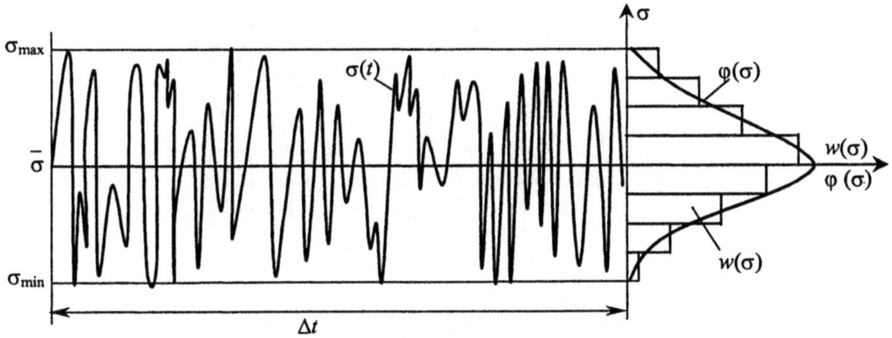

Fig. 1.34. Random process of loading $\sigma(t)$ and its schematization within limited interval of time Δt by law of distribution $\varphi(\sigma)$ of effective stresses σ or loading block $w(\sigma)$

Random fatigue occurs in case of irregular loading. It is described by the so-called *secondary fatigue curves,* while common fatigue curves during regular (cyclic) loading (cf. Fig. 1.16) are called *primary curves.*

Figure 1.35 shows primary (*1*) and secondary (*2* and *3*) fatigue curves for low-carbon steel [20]. The primary fatigue curves are plotted in the following coordinates: the amplitude of stresses σ_a – the number of cycles until fracture N_σ (*cyclic durability*); the secondary fatigue curves are plotted in the following coordinates: the maximum amplitude of stresses σ_{amax} – the total number of cycles until fracture N_Σ at random loading that is termed *fatigue durability.* It follows from experimental data that regular loading is the most dangerous; random loading becomes still more dangerous when the loading process approaches the *narrow strip* loading. While fatigue durability N_Σ during loading processes (*F*2) and (*F*3) differs approximately 4...6 times, cyclic durability N_σ during regular loading is approximately 200 times smaller than in case of the process (*F*3), providing the amplitude σ_a of stresses during regular loading coincides with the amplitude σ_{amax} of the maximum stress within the spectrum.

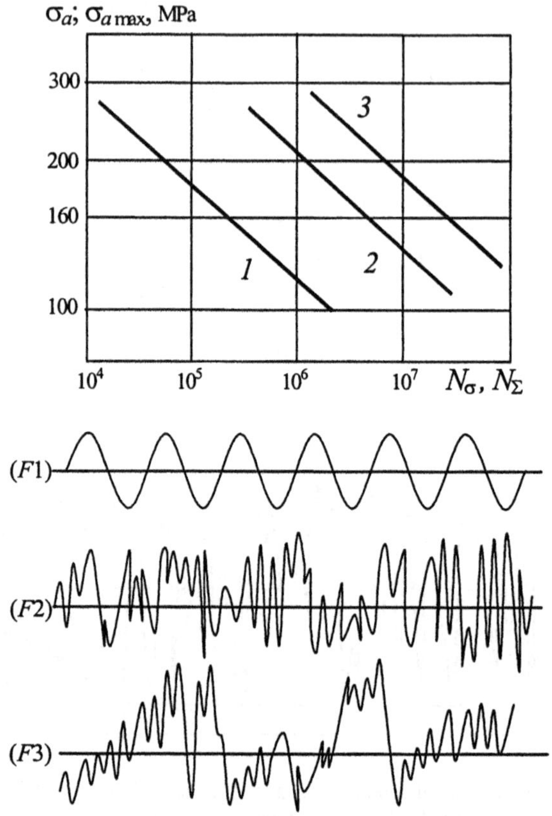

Fig. 1.35. Primary fatigue curve *1* during regular loading ($F1$) and secondary fatigue curves *2* and *3* respectively during random narrow-strip ($F2$) and wide-strip ($F3$) random loading processes

In case of block loading, a secondary fatigue curve is plotted similarly. Figure 1.36 shows the primary and secondary fatigue curves for medium-carbon steel. The loading block had 12 stages and represented a histogram of distribution of stresses that could be approximated by the normal law. Curve 2 in Fig. 1.36 characterizes in this way the resistance to fatigue during normal distribution of amplitudes of stresses. From the experimental data it follows that in case of the normal law of loading, the durability is approximately 20 times higher than during regular loading.

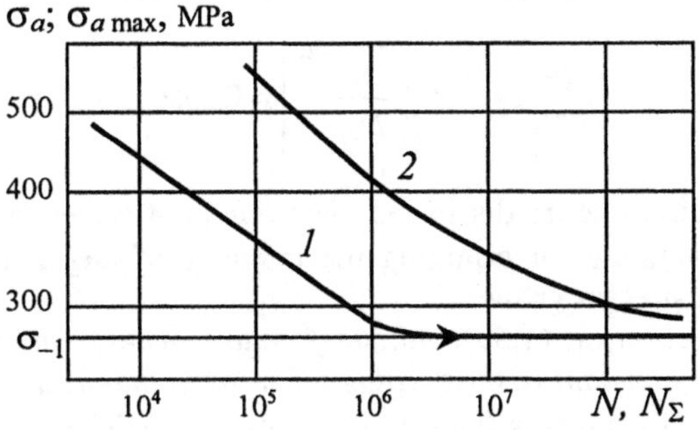

Fig. 1.36. Primary *1* (during regular loading) and secondary *2* (during block loading) fatigue curves for medium-carbon steel

It is noteworthy that according to the experimental data (see Figures 1.35 and 1.36), the primary and secondary fatigue curves are practically parallel; it means that their parameters of slope (the type m_σ – cf. Fig. 1.16) are similar. Exactly this circumstance justifies the summation of stresses σ_j at any j-th stage with the same exponent (m_σ) when calculating fatigue durability using formula (1.57) in case of block loading. Naturally this formula is justified during random loading too if summation is substituted with relevant integration:

$$N_\Sigma = \frac{n_b}{\int\limits_{0.5\sigma_{-1}}^{\sigma_{\max}} \left[1 - \left(1 - \frac{n_j}{N_{G\sigma}}\left(\frac{\sigma_j}{\sigma_{-1}}\right)^{m_\sigma}\right)^h\right]^q} . \tag{1.58}$$

Summation or integration in formulas (1.54), (1.57) and (1.58) is performed only in case stresses exceed the assigned bottom threshold $\sigma = 0.5\sigma_{-1}$ of damaging stresses.

1.3.6 Energy approach

The energy approach to describing fatigue damage accumulation is the most general since the *energy of deformation* is determined both by stresses and deformations.

Between *non-elastic deformation per cycle* $\Delta\varepsilon_n$ and irreversibly dissipated energy there is in the multicycle region (cf. Fig. 1.15) a relation of type (1.24):

$$D = k_s \sigma_a \Delta\varepsilon_n ,$$

where σ_a – the amplitude of stresses. There is the coefficient of shape $k_s = 1.33$ for the closed loop of mechanical hysteresis limited by the arcs of the parabola, for the loop in the form of an ellipse $k_s = 1.57$.

The criterion of reaching the ultimate state has the form [12]

$$\sum_{1}^{N_\sigma}\left[D - D_{-1}\left(\frac{D}{D_{-1}}\right)^{\alpha}\right] = \text{Const}\,, \tag{1.59}$$

where D_{-1} – specific energy dissipated per cycle at stress equal to the endurance limit, α – the parameter determining the intensity of augmentation of the safe energy portion as stresses grow.

According to criterion (1.59), *fatigue fracture occurs when the total energy reaches a certain constant* (Const) *value for a given material*; it does not depend on the level of cyclic stresses (hence, durability is N_σ). Theoretically the constant in (1.59) is not determined, therefore it is found from experimental results.

There is a power dependence between relations D/D_S and σ/σ_S, where σ_S – true resistance to rupture, D_S – ultimate work of deformation during static tension. Using the dependence the equation of the fatigue curve is obtained as following:

$$N_\sigma = \left(\frac{\sigma_s}{\sigma}\right)^{\frac{mk+1}{k}}, \tag{1.60}$$

where m, k – parameters determined experimentally.

Equations (1.34) and (1.60) of the fatigue curve are identical in the sense that both are exponential. Yet their content is different. While it is always $\sigma \geq \sigma_{-1}$, i.e. $\sigma/\sigma_{-1} \geq 1$ in Eq. (1.34), it is opposite in Eq. (1.60): $\sigma \leq \sigma_S$, i.e. $\sigma/\sigma_S \leq 1$. Moreover, while it is always $N_\sigma \leq N_{G\sigma}$ in Eq. (1.34), it is $N_\sigma \geq 1 = N_S$ in Eq. (1.60), where $N_S = 1$ – the "durability" under effect of the stress equal to the true stress of rupture (assumed equal to unity). Thus, curve (1.34) is limited from below with the point having the coordinates ($N_{G\sigma}$, σ_{-1}) and curve (1.60) is limited from above with the point having the coordinates ($N_S = 1$, σ_S).

1.3.7 The effect of various factors

The value of the endurance limit as the main characteristic of resistance to fatigue depends on many *factors*: concentration of stresses, dimensions of objects undergoing tests, surface roughness, ambient temperature and many others. The need to take these factors into consideration arises every time when proceeding from the analysis of resistance of a metal to fatigue (σ_{-1}) to evaluating the *endurance limit of a part* (σ_{-1D}). This evaluation is done from the relation

$$\sigma_{-1D} = \sigma_{-1}/K_0. \tag{1.61}$$

Here $K_0 \geq 1$ – the *factor of endurance limit reduction* calculated from the formula

$$K_0 = \left(\frac{K_\sigma}{K_d} + \frac{1}{K_F} - 1\right)\frac{1}{K_v K_A}. \tag{1.62}$$

In formulas (1.61) and (1.62): σ_{-1} – the endurance limit in bending with rotation of polished metallic specimens with the diameter $d = 7.5$ mm; σ_{-1D} – the endurance limit of a piece (a structural element) with the diameter $D > d$; $K_\sigma \leq 1$ – the *effective factor of concentration of stresses*; $K_D \leq 1$ – the *factor of the effect of absolute cross sectional dimensions*; $K_F \leq 1$ – the *factor of the effect of surface roughness*; K_A – the *factor of the effect of anisotropy of properties*; $K_V > 1$ – the *factor of the effect of surface hardening.*

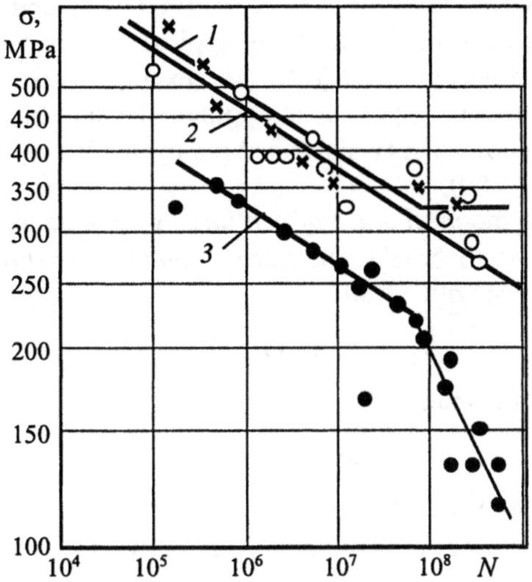

Fig. 1.37. Fatigue curves for alloy ЭИ437 at different temperatures of tests: *1* – 293 K; *2* – 973 K; *3* – 1073 K

The effect of some factors on the resistance of metals to fatigue can be quite intricate. Tests of steel at different ambient temperatures alter the pattern of the (multicycle) fatigue curve (Fig. 1.37). While the (physical) endurance limit (a horizontal line for curve *1*) is observed after 10^8 cycles at room temperature, fatigue curve *2* has no bend at all at the temperature 973 K. While at 1073 K curve *3* consists of two intersecting segments, the angle of slope in the gradual fatigue region being steeper than in the multicycle region (compare with the curve in Fig. 1.15 – region *IV* of gigafatigue at room temperature).

1.3.8 Calculations of fatigue

Calculations of strength under stresses variable in time are performed observing the condition of type (1.16), namely

$$\sigma \leq [\sigma] = \frac{\sigma_{-1}/K_0}{n_\sigma}. \qquad (1.63)$$

Here K_0 is determined from formula (1.62), n_σ – safety factor margin. If the loading cycle is asymmetric, one of the methods can be used (see, for example, Fig. 1.17) of reducing it to a symmetric cycle.

Calculations of durability are performed using some concept of its assessment.

If calculations are based on the theory of fatigue curve, durability is determined as some function of stresses: $N = \varphi_1(\sigma)$ (for example, see formula (1.34)). In those cases when the idea of linear (or non-linear) fracture mechanics is believed applicable, the durability is related to the range or the maximum stress intensity factor: $N = \varphi_2(\Delta K)$ (for example, see formulas (1.44) and (1.44a)). When the theory of fatigue damage accumulation is applied, the durability is governed by some measure of damage: $N = \varphi_4(D_n)$ (for example, see formulas (1.50) and (1.57)). When the energy approach is applied, the durability is determined by some function of specific energy dissipated per cycle: $N = \varphi_3(D)$ (for example, see formulas (1.59) and (1.60)).

Finally, assuming the *mechanics of continuous damage* as theoretical basis the durability is regarded to be *the structural damage measure*: $N_I = \varphi_5(\omega_1)$. For example, in accordance with (1.36)

$$N_I = N_\sigma (1-\omega_1)^{m_c+1}, \qquad (1.64)$$

where the function of damage is

$$(1-\omega_1)^{m_c+1} \le 1.$$

From (1.64) it follows that durability at the stage of nucleation of a fatigue crack is $N_I \le N_\sigma$ providing N_σ is the durability before the object disintegrates into pieces.

In the final account the formulas for calculating durability obtained from different considerations are treated as a function of effective stresses. It is natural because it is hard to imagine cyclic durability without dynamic conditions of loading.

A combined analysis of static and fatigue fracture can be performed using Fig. 1.38. We will consider *the safety factor based on the endurance limit*

$$n_{\sigma_{-1}} = \sigma_{-1} / \sigma \qquad (1.65)$$

as a criterion characterizing this or that mechanical state of the object.

If the inequality

$$n_{\sigma_{-1}} > 1,$$

takes place, it is equal to the *condition of unattainable ultimate state* recorded based on stresses as:

$$\sigma < \sigma_{-1}.$$

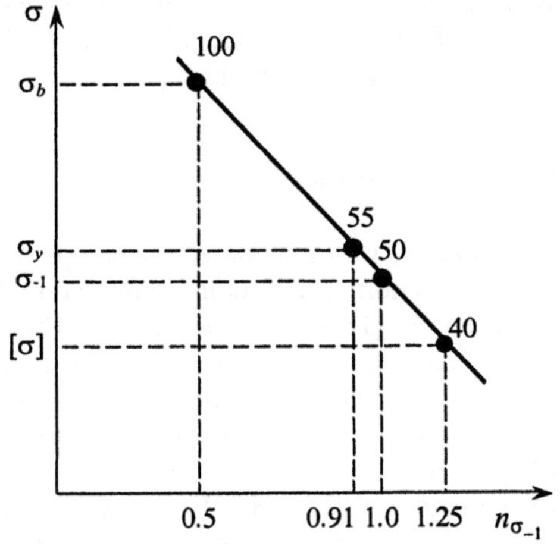

Fig. 1.38. Analysis of mechanical states of material

Equal *conditions of the onset of the ultimate state* based on the criterion of resistance to fatigue (1.65) are the equalities

$$\left.\begin{aligned} n_{\sigma_{-1}} &= 1.0; \\ \sigma &= \sigma_{-1}. \end{aligned}\right\} \tag{1.66}$$

Correspondingly the *condition of fatigue fracture* based on criterion (1.65) is

$$\left.\begin{aligned} n_{\sigma_{-1}} &< 1.0\,; \\ \sigma &> \sigma_{-1}. \end{aligned}\right\} \tag{1.67}$$

Conditions (1.67) allow to detect (in case the object is made from mild steel) two specific transitions of the material into the mechanical state. The first transition relates to the onset of yield:

$$\left.\begin{aligned} &\sigma = \sigma_y; \\ &n_{\sigma_{-1}} = \sigma_{-1}/\sigma_y = \text{const} \le 1.0. \end{aligned}\right\} \tag{1.68}$$

The second transition relates to full exhaustion of the bearing capacity – fracture during the first loading cycle:

$$\left.\begin{aligned} &\sigma = \sigma_b; \\ &n_{\sigma_{-1}} = \sigma_{-1}/\sigma_b = \text{const} \ll 1.0. \end{aligned}\right\} \tag{1.69}$$

The occurrence of conditions (1.66)–(1.69) is illustrated in Fig. 1.38 quantitatively. It is conventionally assumed that σ_b = 100 MPa, σ_y = 55 MPa, σ_{-1} = 50 MPa and admissible stress [σ] = 40 MPa.

So, the relations σ_{-1}/σ_y and σ_{-1}/σ_b are essential parameters of a material. The first governs the loss of rigidity (a transition from plastic to elastoplastic deformation), the second restricts the bearing capacity of an object.

1.3.9 Thermomechanical fatigue

The *thermodynamic state of solids* is characterized integrally by *temperature*. If temperature increases versus normal (or room) temperature, the thermal motion of atoms intensifies correspondingly and the *structure* of the material may undergo phase transitions. In extreme cases *thermodynamic fracture* (or *thermal degradation*) occurs, like *melting* of metals, *softening* of plastics. A different phenomenon is observed in the range of lower temperatures, such as *embrittlement* of metals, *ageing* of plastics.

Accumulation of damage and fracture of materials during cyclic temperature variations and mechanical stresses is termed *thermomechanical fatigue*. It is the most common case of fracture of metals under combined thermal and mechanical effects. Particular cases are *isothermal fatigue* (mechanical fatigue at a constant temperature), *mechanical fatigue* (at room temperature), *thermal fatigue* (during cyclic elevated temperature variations) and *long-term strength* (fracture under the effect of static stresses at a constant (elevated) temperature). Table 1.1 presents a comparison of durabilities of specimens from nimonic tested according to two schemes. It turns out that time until fracture reduces approximately 3 times (from 740 to 230 h) if static loading (mechanical or thermal) is substituted with cyclic loading of the same level. For equal values of cyclic plastic deformation the number of cycles until fracture during thermal fatigue is approximately 3–6 times less than during isothermal fatigue (if the maximum temperature is the same in both cases).

Table 1.1. Conditions and results of tests of nimonic

Scheme of tests	σ, MPa	T, °C	N, cycle	τ, h
Thermomechanical fatigue	$0 \leftrightarrows 270$	$550 \leftrightarrows 750$	31647	230
Long-term strength	270 = const	750 = const	–	740

Curves of thermal, thermomechanical, isothermal, mechanical fatigue look identically: they are slanting straight lines in corresponding double logarithmic coordinates. They are usually described by the parametric equations of the following type:

$$f(\sigma) = a \log N - bT, \tag{1.70}$$

where a, b – coefficients. Thus, taking into account the kinetic theory of strength, it can be obtained that

$$N = \frac{1}{C_U} \exp\left[\frac{U_0 - a_\gamma \sigma}{kT}\right], \qquad (1.71)$$

where C_U, U_0, a_γ – the equation parameters; k – the Boltzmann coefficient. Naturally the numerical values and the set of parameters differ when analyzing resistance to fatigue at different temperature and under different force conditions.

During isothermal fatigue the endurance limit σ_{-1} depends on the temperature T intricately, its dependence is governed by numerous processes and their interactions, including dispersion hardening, behavior of atoms of implantation and substitution, chemical interactions and other processes. The temperature dependence of endurance limits shown in double logarithmic coordinates (Fig. 1.39) in many metals reveals three rectilinear portions: *I* – the region of moderately elevated temperatures when a metal, as a rule, does not tend to *creep*, hence the regularities of resistance to fatigue in this region are similar to those at room temperature; *II* – the region of operating temperatures when the process of damage is determined by the effect of both cyclic stresses and the creep of the metal; *III* – the region of elevated temperatures when fracture is primarily due to creep. The level of temperatures characterizing transition from one region into another is different and it is determined both by homological temperatures and structural features of a metal. Fracture over boundaries of grains prevails at high temperatures, while fracture occurs predominantly within grains at operating temperatures.

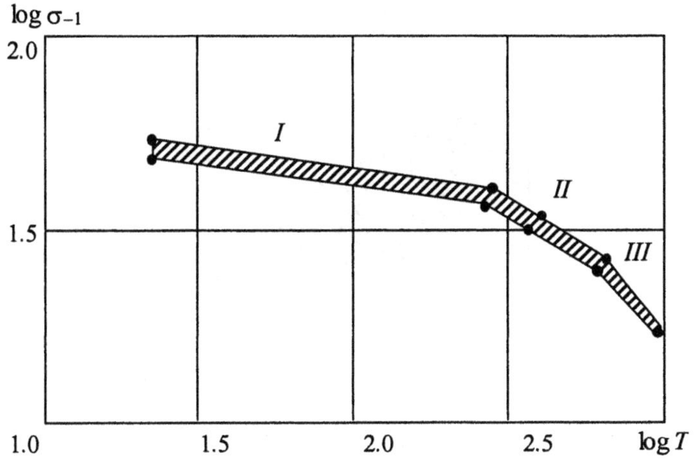

Fig. 1.39. Temperature dependence of endurance limits of Cr–Mo–steel

According to Fig. 1.39, the temperature dependence of endurance limits in each portion *I*, *II*, *III* can be approximated by the equation of the following type [14]:

$$\log\ \sigma_{-1} = B - m_T \log\,(T/T_n),$$

where T_n – nominal temperature, such as room temperature; B – coefficient; m_T – the parameter of heat activation of the processes of fatigue damage. The equation parameters differ in different portions (*I, II, III*) of the dependence σ_{-1} (T).

1.3.10 Impact mechanical fatigue

The peculiarity of *impact fatigue* when cyclic loading is dynamic is that *deformation waves* appear in the object that strongly reduce its resistance to fatigue. When 0.4 mm thick plates of steel 09X15H9Ю are loaded harmonically during symmetric cantilevered bending, no fatigue fracture occurs until 10^8 cycles of tests if the amplitude of vibrations of the free end of the plate does not exceed ±10 mm. Tests of the same plates by impact cyclic loading have revealed that the cyclic durability would amount to $4{\cdot}10^4$, 10^5 and 10^6 cycles, respectively, if the cantilever arm deviates by 7.5 and 3 mm.

Another feature of resistance of materials to impact fatigue is illustrated in Fig. 1.40: if the ultimate strength of steels (30X13 and 65Г, both in the improved condition) exceeds 1200 MPa, the resistance to fatigue rated by the maximum ultimate stress per cycle lim σ_{max} not just grows as σ_b augments, like it occurs during harmonic loading, on the contrary, it drops sharply. It is attributed to the fact that impact loading initiates the appearance of fine cracks and strongly intensifies their growth in the region of numerous microconcentrators of stresses.

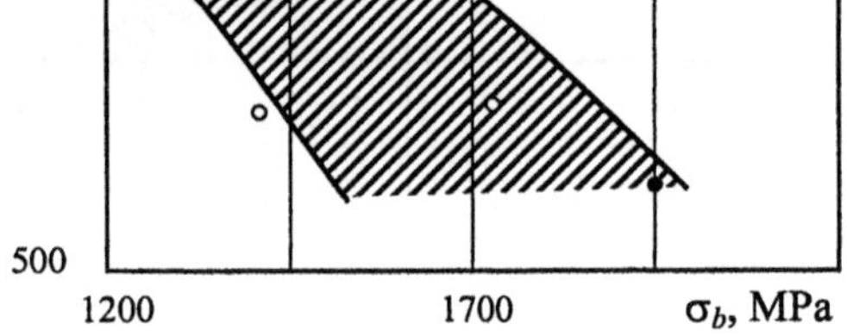

Fig. 1.40. Reduction of resistance of high-strength steels to impact mechanical fatigue at $\sigma_b > 1200$ MPa

If the cycle of harmonic stresses with the amplitude σ_a is supplemented with an impact impulse having the magnitude σ_{im} (Fig. 1.41, *a*), tests are called *combined*; the endurance limit is designated by σ^*_{-1} in this case. Fig. 1.41 [21] shows a change in σ_{-1} (line *1*) and σ^*_{-1} (line 2) in response to the temperature of testing the specimens of steel 15Г2АФДпс at σ_{im} = 350 MPa. The ratio between the number of cycles of application of impacts and the number of cycles of harmonic loading was 1:128. In both cases any temperature reduction results in growth of resistance to fatigue, yet the endurance limit σ^*_{-1} during impact fatigue is 1.5...2.0 times smaller than during usual fatigue. However, this difference becomes smaller as the temperature reduces (curves *1* and *2* mutually approach). Reduction of the endurance limit under combined loading is additionally due to the appearance of residual tensile stresses in the zone of compression at a level of total stresses ($\sigma_a + \sigma_{im}$) exceeding the yield limit.

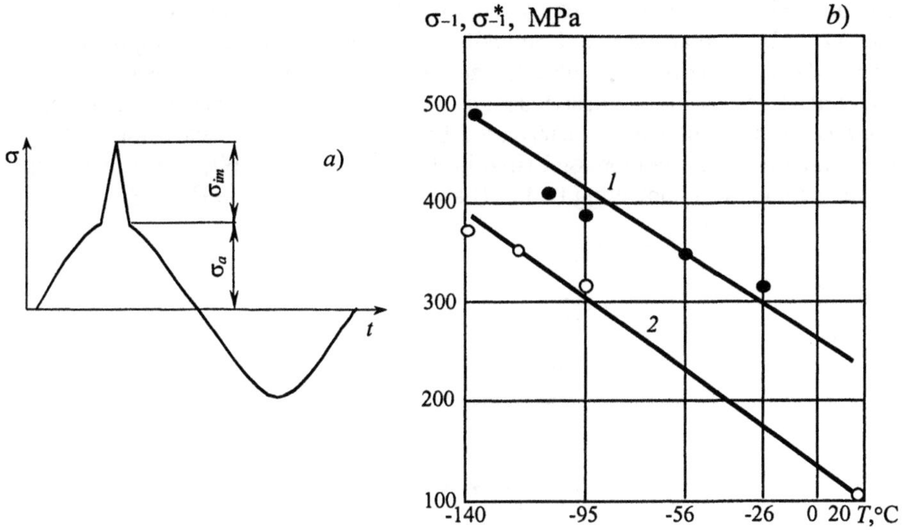

Fig. 1.41. Diagram of loading cycle (*a*) and results of low temperature tests of specimens of steel 12Г2АФДпс (*b*)

The curve of impact mechanical fatigue looks like a usual fatigue curve.

1.4 Friction and wear

1.4.1 Force and friction coefficient

When studying regularities of volume fracture we dealt with an individual *structural element* that we called a (deformable solid) body, a specimen or simply an object.

Surface damage occurs during relative motion of at least two interacting bodies *1* and 2, for example, during sliding (Fig. 1.42) or rolling (Fig. 1.43); in both cases they act as a *friction couple*. It is said that a friction couple (or *unit*) makes up a *specimen* and a *counterbody*, or alternatively a *body* and a *counterbody*. Their dynamic interaction is due to a specific *contact load* F_N. It is an external load that acts normal to the *contact site*. The contact site appears as a geometrical site of contact between two motionless solids due to their local (elastic or non-elastic) deformation when compressed by the load F_N. Depending on the geometrical shape and sizes of contacting bodies the contact site has *three shapes*: *round*, *strip*, *elliptic* (cf. Figs. 1.42 and 1.43). For example, a round contact site appears when two balls or two cylinders of the same diameters with mutually perpendicular axes are compressed; a strip site appears when two cylinders with parallel axes are compressed; an elliptic site appears when a ball and a cylinder or two cylinders having different diameters and mutually perpendicular axes are compressed. The *nominal contact area* A_a can be calculated if two determining dimensions (a, b) of the contact site are known designated, as a rule, as $a > b$. The size and shape of the nominal contact site can be determined both by deformation when the body and the counterbody are compressed (like in the case shown in Fig. 1.43) and simply by the geometry of contacting bodies (like in the case shown in Fig. 1.42).

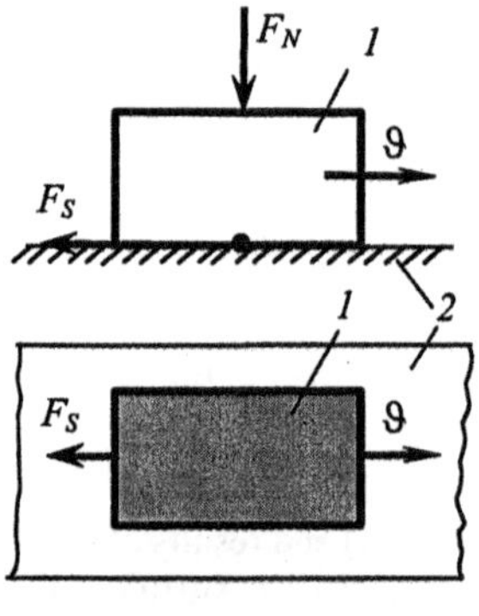

Fig. 1.42. Determination of friction force F_S in sliding

Division of the contact load F_N by the nominal contact area A_a that carries it (cf. Fig. 1.42) yields (mean) *contact pressure*

$$p_a = F_N / A_a, \tag{1.72}$$

that in fact is the *normal compressive stress* σ_z (the axis z is directed perpendicularly to the contact site, i.e. along the line of action of contact load F_N). In those cases when distribution of pressure over the contact site cannot be considered regular (cf. Fig. 1.43 – pressure distribution is usually described by an elliptic dependence); the maximum pressure $p_0 = \sigma_{z\max}$ is determined in the center of the contact site too.

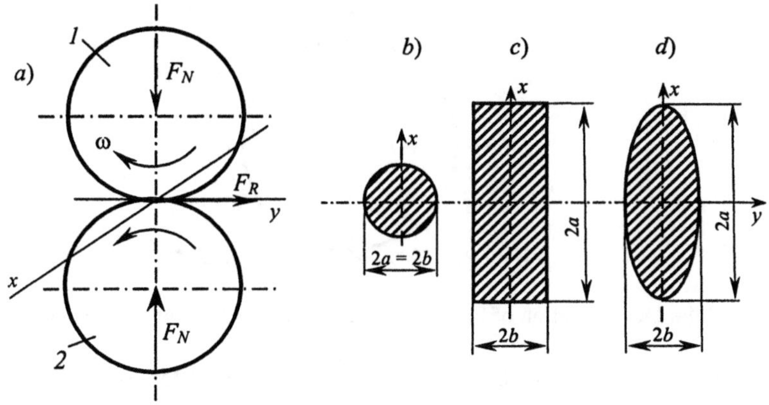

Fig. 1.43. Determination of friction force F_R in rolling (*a*) and typical configurations of contact site (*b*, *c*, *d*)

As soon as the body and the counterbody start moving mutually, an *internal force of resistance* F_f to this motion appears on the contact site; it is called *frictional force*. It is always directed opposite to motion during friction. We will designate the *frictional force in sliding* as F_S and in *rolling* as F_R (cf. Figs. 1.42 and 1.43). In both cases the frictional force is orthogonal to the direction of contact loading or, it is more proper, it is tangent to the contact site, i.e. it is in its plane. It means that the *frictional force is the force of displacement in the region of physical contact between the body and the counterbody*. Similarly to the specific (contact) pressure (1.72) the *specific frictional force* can be calculated

$$\tau_W = F_f / A_a, \tag{1.73}$$

that is called also *frictional stress*. In fact, it is tangential stress in the region of physical contact between the body and the counterbody appearing under specific conditions *exceptionally* in case their relative motion with friction.

The dimensionless relation between the frictional force in sliding and the contact load (cf. Fig. 1.42) is called *friction coefficient* (*in sliding*)

$$F_S / F_N = f_S = \text{Const.} \tag{1.74}$$

Its value does not depend (during elastic deformation) on the contact area A_a, or the magnitude of contact load F_N. If F_N grows, so correspondingly does the frictional force F_S, i.e. the value should be f_S = Const. for the friction coefficient to remain unchanged.

The *friction coefficient in rolling* is determined as a ratio between the *moment of resistance to rolling* called shortly the *friction torque* M_T, and the contact load F_N (Fig. 1.44):

$$f_R = M_T / F_N = FR/F_N. \tag{1.75}$$

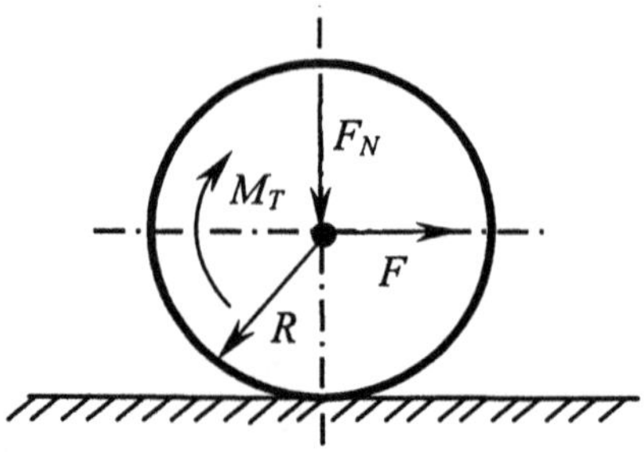

Fig. 1.44. Determination of friction coefficient and coefficient of resistance to rolling

Coefficient (1.75) has the dimensionality of a unit of length (determined by the dimensionality of the radius R). In order to avoid the dimensionality, the *coefficient of resistance to rolling is* introduced:

$$f_r = f_R / R = M_T / RF_N = F/F_N, \tag{1.76}$$

that is, in fact, similar to the friction coefficient in sliding (1.74). The numerical values of the coefficient of resistance to rolling ($f_r \approx 0.01...0.005$) are approximately one order of magnitude smaller than the sliding friction coefficient ($f_S \approx 0.1...1.0$). According to (1.74) and (1.76), it means that the resistance to sliding motion is approximately one order of magnitude stronger than the resistance to rolling motion.

In the general case the so-called *friction at rest* and *friction in motion* differ (Fig. 1.45). The force F_f appears during friction in motion, while a *non-full frictional force* F_n appears in the process of preliminary displacement; its maximum value

$$\max F_n = F_f^0 > F_f$$

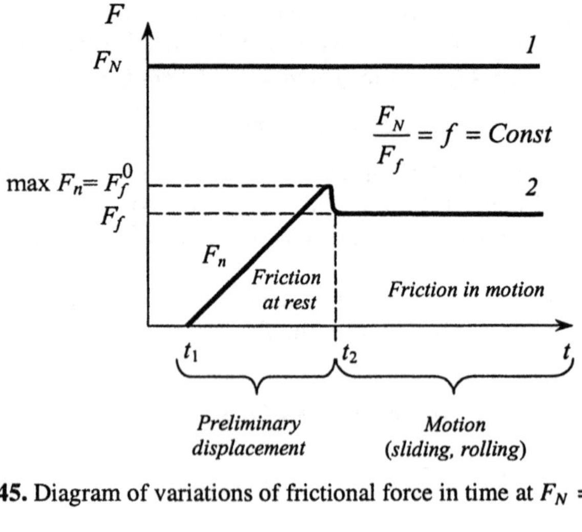

Fig. 1.45. Diagram of variations of frictional force in time at F_N = const

is called the *frictional force at rest* F_f^0. It is always larger than the frictional force in motion ($F_f^0 > F_f$). Transition from friction at rest to friction in motion is accompanied by a *jerk* (at the moment of time t_2 – cf. Fig. 1.45).

The *friction coefficient at rest* is determined as a relation between the maximum value of the non-full frictional force (i.e. the frictional force at rest) and the contact load:

$$f^0 = \max F_n / F_N = F_f^0 / F_N > F_f. \tag{1.77}$$

Depending on the contact load level and, therefore, the frictional force there is *friction during elastic deformation* (*elastic contact*) and *friction during elastoplastic deformation* (*elastoplastic contact*). In terms of tangential stresses (τ) and shear deformation (γ) the *mechanical state of the material in the contact region* in friction (at rest and in motion) can be described by the *shear curve* (Fig. 1.46) similar to the tensile curve (cf. Fig. 1.3). Elastic behavior is observed in the area OA, i.e. at $\tau_W < \tau_y$, where τ_y – *shear yield limit*; here the *law of Hooke in shear* is satisfied (1.7). If $\tau_W > \tau_y$, elastoplastic deformation in the contact region occurs. The relation between deformation and stresses becomes non-linear (Ш curve in Fig. 1.46)

$$\tau = G_p \gamma, \tag{1.78}$$

where G_p – *current modulus of plasticity in shear* similar to the modulus E_p in the law of plasticity (1.4).

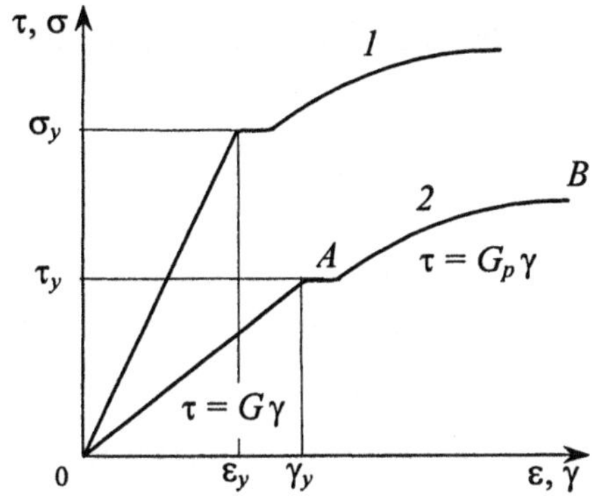

Fig. 1.46. Analysis of mechanical state of material in contact site region during friction: diagram of tension (*1*) and shear (*2*)

Record coefficient (1.74) through specific frictional force (1.73) and contact pressure (1.72):

$$f_S = \tau_W / p_a. \tag{1.79}$$

The law of Hooke in shear (1.7) is used in formula (1.79) for the case of elastic contact, then the *friction coefficient during elastic deformation*

$$f_{Se} = G\gamma_W / p_a = \text{Const}, \ \gamma_W < \gamma_y. \tag{1.80}$$

Law (1.78) is assumed for elastoplastic contact, then the *friction coefficient during elastoplastic deformation* is

$$f_{Sp} = G_p \gamma_W / p_a = \text{var}, \ \gamma_W > \gamma_y. \tag{1.81}$$

The *friction coefficient at yield limit* (when $\tau_W \approx \tau_y$) is a boundary between them

$$f_y = \tau_y / k\sigma_y,$$

where $k \approx 0.35...0.40$ for the materials with the same resistance to tension and compression and $k \approx 0.65...0.80$ for the materials with a significantly different resistance to tension and compression.

Based on (1.74) with the allowance for (1.80), we obtain a formula for determination of the *frictional force in the elastic contact*

$$F_{Se} = f_{Se} F_N = \frac{G\gamma_W}{p_a} F_N, \ \gamma_W < \gamma_y. \tag{1.82}$$

Based on (1.74) with the allowance for (1.81), we obtain a formula for determination of the *frictional force in the elastoplastic contact*

$$F_{Sp} = f_{Sp} F_N = \frac{G_p \gamma_W}{p_a} F_N, \ \gamma_W > \gamma_y. \tag{1.83}$$

The *boundary* value of the frictional force (during transition from the elastic to the elastoplastic contact) is

$$F_y = f_y F_N = \frac{\tau_y}{k\sigma_y} F_N .$$

It is quite obvious that the friction coefficients (1.80), (1.81) and the frictional forces (1.82), (1.83) are similar functions of the magnitude

$$\kappa = \gamma_W / p_a , \tag{1.84}$$

and can be interpreted as *rigidity in friction*: it is relative shear deformation on the contact site during friction in motion per unit of contact pressure.

Thus, the force and coefficient of friction are the most essential characteristics of resistance to friction in motion of a similar sense; their difference is that one of them is an absolute (i.e. dimensional) value, while the other is a relative (dimensionless) value.

Regarding the nature of friction then, according to the law of Amonton (cf. (1.74)),

$$F_S = f_S F_N , \tag{1.85}$$

the frictional force is a measure of mechanical (deformational) interaction (F_N) between two bodies during their relative motion. Law (1.85) can be supplemented with term A reflecting adhesive (molecular) interaction during compression of two bodies; then it is called the *law of Amonton-Coulomb:*

$$F_S = A + f_S F_N = F_a + F_d. \tag{1.85a}$$

Thus, the total frictional force is the sum of forces $F_a = A$ expended to break adhesive bonds and forces $F_d = f_S F_N$ expended to deform by shear a fine surface layer in the region of the contact site. When frictional force is determined experimentally, it is impossible to determine its adhesive (F_a) and deformational (F_d) components, that is why they deal practically with the friction law in form (1.85).

According to law (1.85a), it is possible to calculate the friction coefficient

$$f_S = f_a + f_d \tag{1.86}$$

as a sum of its *adhesive (molecular) component*

$$f_a = \tau_0/\sigma_{z\,\max} + \beta \tag{1.87}$$

and its *deformation (mechanical) component*

$$f_d = ka_g \sqrt{\delta_c / r}\,, \tag{1.88}$$

where τ_0 – resistance to shearing during simple shear, $\sigma_{z\,\max} = p_0$ – the maximum normal stress equal to the pressure in the center of the contact site under the effect of compressive load F_N; β – piezocoefficient; $k \approx 0.44$ – the coefficient depending on the surface roughness; $0 \le a_g \le 1$ – the coefficient of hysteresis losses; δ_c – mutual approach of the body and the counterbody under the effect of contact load F_N; r – a typical radius of irregularities (on the rough surface) that are modeled with semispheres. In plastic contact it is assumed that $p_0 = HB$, where HB – Brinell hardness, and $a_g = 1$. In a purely elastic contact $a_g = 0$. During transition from elastic to elastoplastic deformation the friction coefficient usually reduces. If the strength (and/or the elasticity modulus E) of the material grows, it leads to the reduction of the friction coefficient too.

1.4.2 Third body. Lubrication

As a rule, contact between two solids, or *frictional contact,* is not direct: there is an intermediate layer between two interacting surfaces (Fig. 1.47) consisting of a film of the *lubricating material*, oxides and other chemical compounds, adsorbed water vapors and a film of the degraded base material, i.e. finely dispersed *wear products*; this layer is called the *third body* [22]. It has little resistance to shear compared with the material of the body and counterbody, therefore a lubricant reduces the force and coefficient of friction.

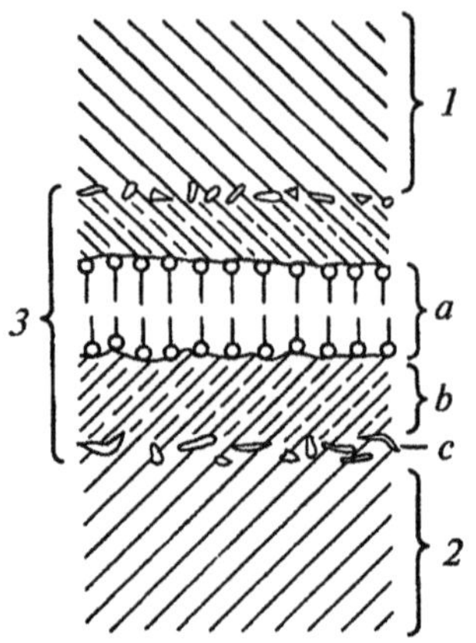

Fig. 1.47. Diagram of structure of zone of contact between two (*1* and *2*) solids; (*3* – third body); *a* – adsorbed layer; *b* – oxides and other chemical compounds; *c* – loose layer)

The curve (Fig. 1.48) [23] provides a general characteristic of operation of rubbing surfaces. The curve in its simplified version is the relation between the friction coefficient and the parameter $\eta v/F_N$, where η – the lubricant's viscosity, v – the rate of friction; *I* – the region of *hydrodynamic* or *elastohydrodynamic* (EHD) lubrication; *II* – the region of *partial EHD-lubrication* or *mixed lubrication*; *III* – the region of ***boundary lubrication***.

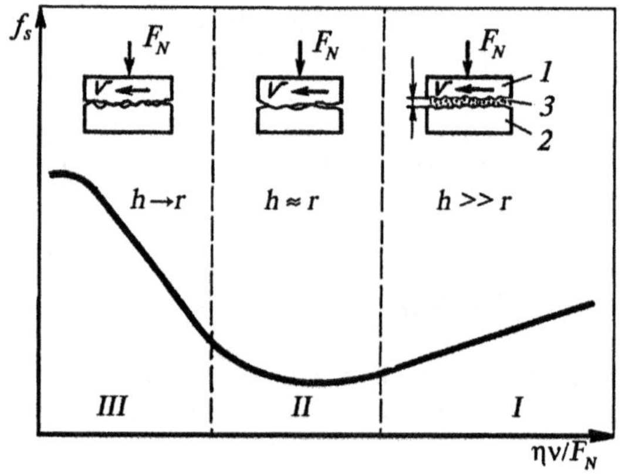

Fig. 1.48. Typical zones of lubrication modes: *I* – region of hydrodynamic or elastohydrodynamic (EHD) lubrication; *II* – region of partial EHD-lubrication or mixed lubrication; *III* – region of boundary lubrication

Solid surfaces in mode *I* are separated by the continuous layer of a lubricating material *3*, its thickness *h* is much larger than the measure *r* of roughness of the surfaces. The wear processes do not occur because there are no direct physical contact interactions between the surfaces of solids *1* and *2*.

If the viscosity of the lubricating material or speed reduces during hydrodynamic or EHD-lubrication while the contact load augments, the lubricating film becomes thinner and the clearance between the surfaces of bodies *1* and *2* diminishes. Mode *II* of partial EHD-lubrication, or mixed lubrication, begins when surfaces come into first contact interactions. Load in this mode is partially borne by the liquid film and partially by contacting surface irregularities, therefore, various wear mechanisms may come into action.

Volume rheological properties of the substance of the lubricating material become less essential in mode *III* of boundary lubrication because the film's thickness diminishes to several monomolecular layers and less, so the contact load is almost fully borne by deformations of irregularities. As a rule, wear processes intensify drastically in the extreme case of *friction without lubrication*.

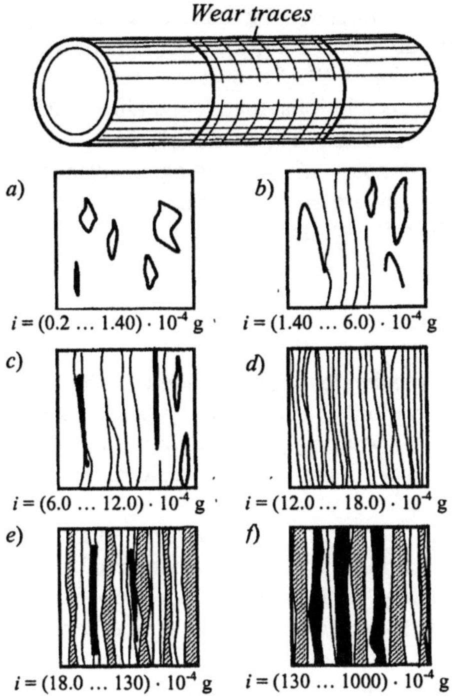

Fig. 1.49. Condition of surface of piston pin wear depending on type of lubrication

Lubrication influences the processes of surface damage and fracture of materials during friction in two ways. Oil produces negative effect in the contact zone under high pressure. Under the effect of the moving surface and due to the capillarity the oil penetrates into loose spots, among irregularities, into microcracks and expands them causing accelerated fracture of the surface layer of a metal (the *wedging effect*). This phenomenon is specifically pronounced in case one of the surfaces in the zone of higher pressure is subjected to tension that causes the opening of cracks. Water impurities in the lubricant cause saturation of the steel surface layer with hydrogen and lead to *hydrogen brittleness* and also to *corrosive damage*; both these processes additionally accelerate fracture of friction surfaces.

On the other hand, in case of moderate pressure in the contact zone the oil film favors a more regular distribution of contact pressure and increases the actual contact surface; it is an active heat-abstracting agent. Under definite conditions the liquid lubrication produces a *hydrodynamic effect*: the lubricant enters into the narrowing portion of the clearance between the body and counterbody and separates their surfaces, so that liquid friction appears in the tribocouple. It leads to stronger resistance of the material to fracture in friction.

Figure 1.49 [24] shows the condition of the surface of the body tested under similar conditions, excepting the type of lubrication. The lubricating oil was brand EP (Fig. 1.49, *a*); the oil intended for lubricating wheel gears (Fig. 1.49, *b*); the lubricating oil "Premium" (Fig. 1.49, *c*); basic oils of high (Fig. 1.49, *d*) and low (Fig. 1.49, *e*) quality. Fig. 1.49, *f* shows the pattern of fracture during friction without lubrication. It is obvious that, depending exclusively on the type of lubrication, wear i can increase approximately from $1 \cdot 10^{-4}$ to $1 \cdot 10^{-1}$ g, i.e. 1000 times, while the topography of the damaged surface transforms due to the presence of dissipated seizure spots (cf. Fig. 1.49, *a*) to rough grooves (cf. Fig. 1.49, *e*), typical for the microshearing process.

1.4.3 Wear processes

Wearing from the mechanical point of view is the process of fracture of fine surface layers in the region of frictional contact between two interacting solids under load. The result of wearing is called simply *wear*; it is characterized by the thickness (volume, mass) of the material separating and carried away from the friction zone.

The wear process in the determining manner is due to the nature of deformation and displacement of surface layers of the material. Fig. 1.50 [25] shows the oscillogram of undulatory motion of the surface of a polymeric body when a metallic indentor (a ball) slides over it. Three typical zones can be identified. Zone *I* (points *1–2* and *8–7*) corresponds to the period of rest: pulses from the perturbation source or loaded and moving indentor practically do not reach it. When the indentor moves, a *deformation wave* appears in zone *II* (points *6–5–4–3–2*), then follows zone *III* of restoration (points *6–7*). The deformation wave in the general case is not stationary: *oscillations of displacement* appear in all its zones or relatively small *vibrodisplacements* in the direction perpendicular to the friction surface. Under definite conditions such vibrodisplacements become the cause (the source) of *self-excited vibrations in friction*.

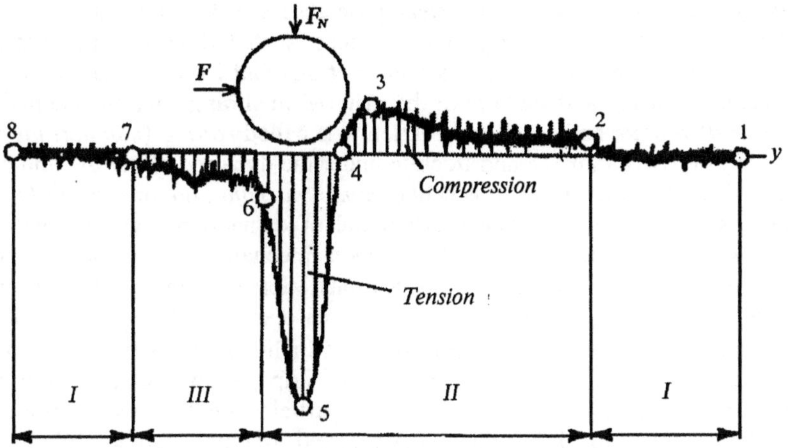

Fig. 1.50. Oscillogram of displacements of surface of polymeric material when metallic indentor slides over it

Scrutiny of Fig. 1.50 enables to make three important conclusions. First, though contact load is static (unchangeable in time: F_N = const), the surface layers of the material deform cyclically in the friction zone. The cycle of deformation ε_y (in the direction of motion) is most dangerous as it is sign-variable: before the indentor – the half-cycle of compression (points *4–2*), behind the indentor – the half-cycle of tension (points *4–5–6–7*). Second, the magnitude of the most dangerous tensile deformation exceeds many times compressive deformation that is less dangerous for initiating fracture. Third and last, the cycle of deformations and displacements expands over the surface zone of the body (from point *2* to point *7*), its stretch is considerably larger than the size of the ball's imprint (in the vicinity of point *5*) when it is indented by load F_N into the plane of the polymeric body. Therefore, the damage of the material is expected both over the contact site, along the friction track and beyond. Scrutinizing further Fig. 1.51, we see that the deformation wave is flat [22].

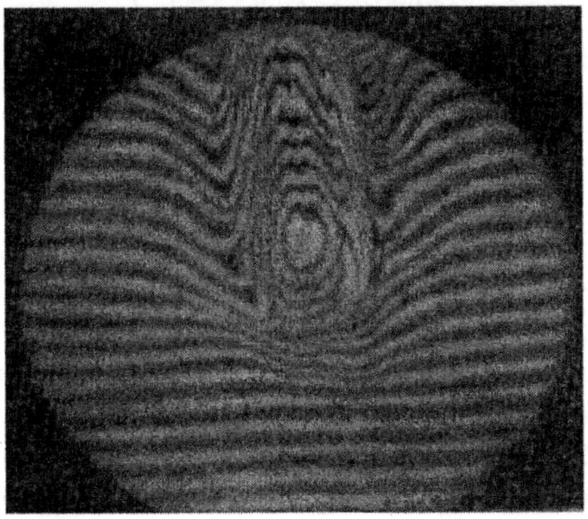

Fig. 1.51. Deformation waves excited by spherical indentor sliding over plane

A general conclusion is that wear processes are due to sign-variable cyclic deformations (stresses) during friction, hence, surface layers of the material fail under specific conditions of contact interactions that have fatigue nature. In this connection damage and fracture of the material in sliding friction is called *sliding* (*frictional*) *fatigue* and in rolling friction it is called *rolling* (*contact*) *fatigue*; the process of erosive damage and fracture under the dynamic effect of solid particles or liquid drops on an obstacle is called *impact microfatigue* or *erosion fatigue*. Of course, surface fatigue (such as frictional or contact) reveals its own specific features too compared with the common mechanical (volume) fatigue, but physical mechanisms of fracture (dislocations, vacancies, thermofluctuations, etc.) remain the same (see Sect. 1.3.2). As an example Fig. 1.52 [26] shows schematically the processes of appearance of inclined cracks (the angle φ) on the friction surface that originate in the sliding strips; growth of a system of these cracks leads to separation of a multitude of particles, i.e. to the wear of a material. Specific features of the surface damage and fracture during sliding, rolling, slippage (fretting) are discussed below; now some regularities of wear will be discussed and how a friction couple reaches the *ultimate* or *critical state*.

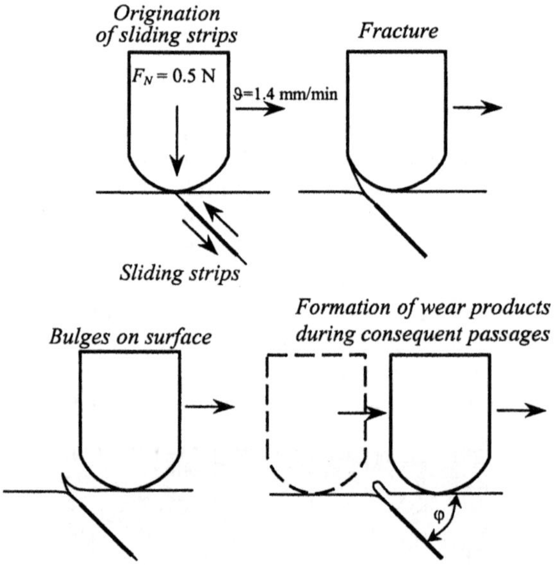

Fig. 1.52. Fracture origination and wear products formation

Such states are established based on various *indications*. For example, a *given density* or *pits depth of spalling* appearing along the *track of rolling* during rolling friction corresponds to the critical state. *Tolerable* [*i*] or *ultimate* $i_{\lim}$ *wear*, respectively, corresponds to the critical or ultimate state during sliding friction. These values are either calculated or determined experimentally using the kinetic curve of wear variations in time (Fig. 1.53, *a*). In the general case its pattern is similar to the curve of accumulation of residual deformation during cyclic loading (cf. Fig. 1.26, curve 2); wear rate ϑ_i changes in time identically too (cf. Fig. 1.53, *b*) and as well as the rate of accumulation of residual deformation v_ε (1.35). It confirms additionally (though indirectly) the common fatigue nature of fracture during friction and cyclic deformation. Hence, critical or ultimate states of a friction couple within a broad range of contact load variations can be described by the graph (Fig. 1.54) [27] similar to the full mechanical fatigue curve (cf. Fig. 1.15).

Specific frictional force, or frictional stress τ_W, is the most essential parameter of loading responsible for the process of wear in sliding. Therefore, the diagram of ultimate states or the *full sliding fatigue curve* (cf. Fig. 1.54) is plotted in double logarithmic coordinates log τ_W – log N_τ, where N_τ – the *number of cycles* (or *time*) before the *critical or ultimate state occurs.* In the general case the curve shows four characteristic regions: *quasistatic* (*I*), *low-cycle* (*II*), *multicycle* (*III*) and *high resource* (*gigacycle*) (*IV*) *fracture* (*wear*). The pattern of the "stress-deformation" cycle for different regions of mechanical fatigue (see hysteresis loops in Fig. 1.15) persists in similar regions of deformation in friction (cf. Fig. 1.54), yet it should be plotted using stresses τ and shear deformations γ (cf. Fig. 1.46).

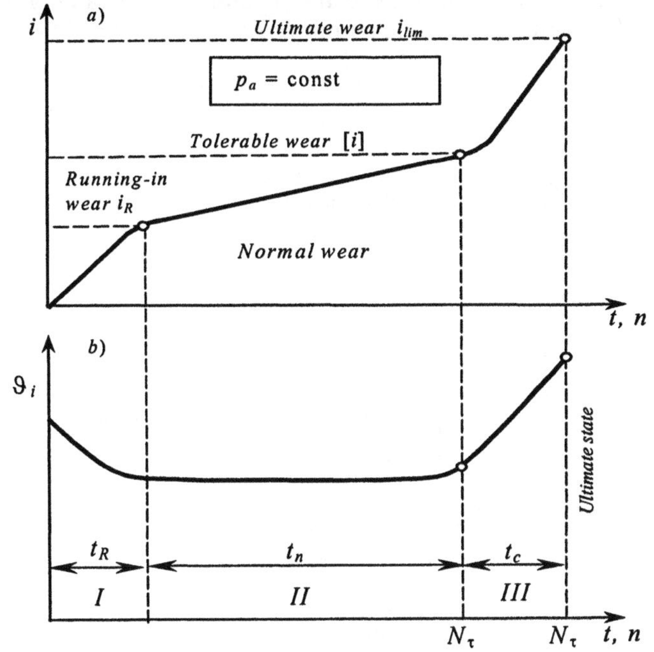

Fig. 1.53. Time wear characteristics: *a* – kinetic curve; *b* – wear rate graph (*I* – running-in period, *II* and *III* – periods of normal and catastrophic wear)

Main types of fracture in friction schematically (cf. Fig. 1.54) illustrate *motion* (with the speed ϑ) *of a single irregularity indented into the plane.* Figure 1.55 shows relevant microphotographs of the typical pattern of friction surface damage. *Microcutting* (*I* in Figs. 1.54 and 1.55) is quasistatic fracture by shear in friction;

its light form is called ab*rasive wearing*. It occurs under stresses $\tau_W >> \tau_L$. Usually in this case $h/r > 0.1$ during friction without lubrication and $h/r > 0.3$ during lubricated friction (h – the depth of penetration of a single irregularity with the radius r). The ultimate strength in shear τ_b corresponds to the number of cycles $N_\tau = 1$. *Plastic pushing* (*II* in Figs. 1.54 and 1.55) occurs under stresses $\tau_L > \tau_W > \tau_K$ (the value τ_K is close to the yield limit in shear). In this case *low-cycle* (*elastoplastic*) wear occurs, $h/r < 0.1$ during friction without lubrication and $h/r < 0.3$ during lubricated friction. *Elastic pushing* (*III* in Figs. 1.54 and 1.55) occurs under stresses $\tau_K > \tau_W >> \tau_G$. *Multicycle (quasibrittle) wear* is observed in this case, $h/r < 0.01$ during friction without lubrication and $h/r < 0.001$ during lubricated friction. *Fracture of films* (the third body) is typical for region *IV* (cf. Figs. 1.54 and 1.55) under stresses $\tau_W < \tau_G$ causing *oxidative wearing*. Cohesive separation of the material is observed in the intermediate zone (*III–IV* in Figs. 1.54 and 1.55) typical for *adhesive wearing*. It is possible only provided the shear resistance gradient (τ_0) in the surface layer is negative: $d\tau_0/dh < 0$. It can occur only when the shear resistance gradient in the surface layer is positive: $d\tau_0/dh > 0$. Ultimate stresses τ_L, τ_K, τ_G, with the corresponding wear durability $N_{L\tau}$, $N_{K\tau}$, $N_{G\tau}$ (cf. Fig. 1.54) govern the transition from one to another dominating wear type (fracture in friction).

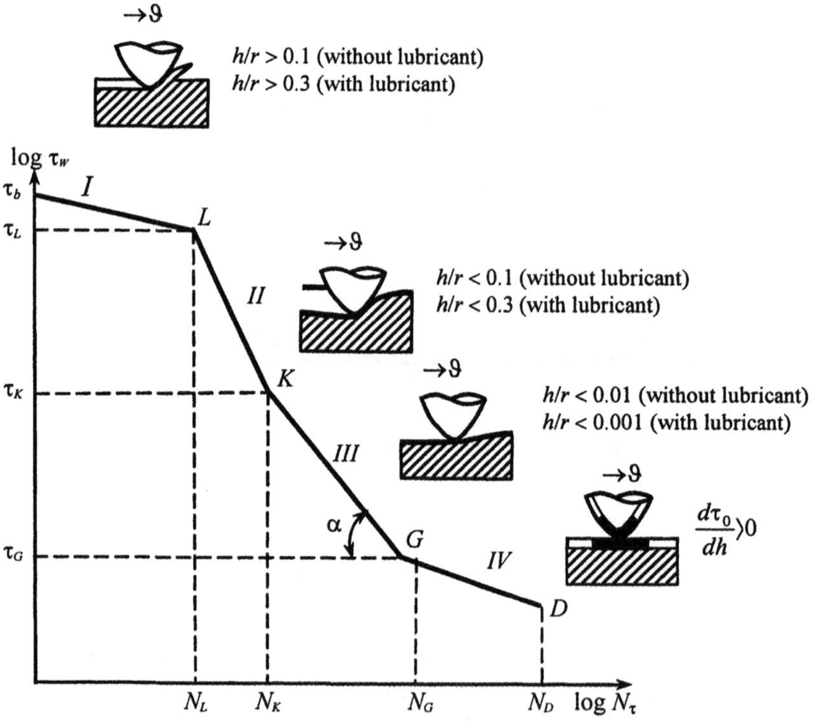

Fig. 1.54. Diagram of full sliding fatigue curve

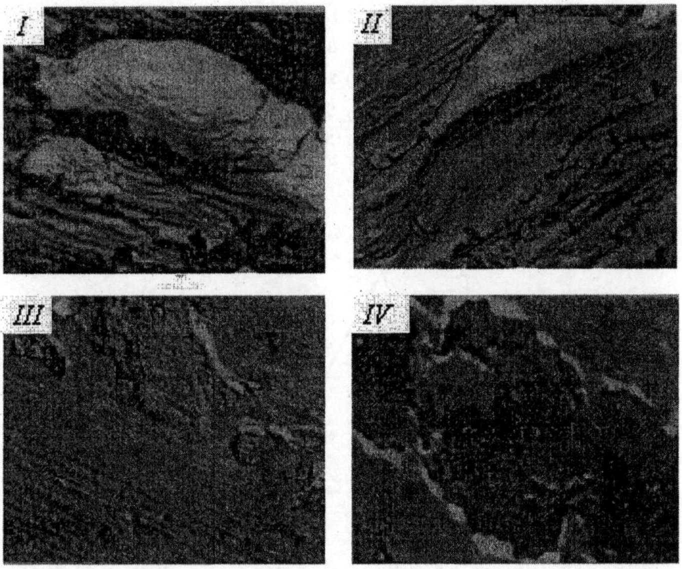

Fig. 1.55. Typical microphotographs (×5000) of damage and fracture of friction surfaces (*I*, *II*, *III* and *IV* correspond to regions *I*, *II*, *III* and *IV* in Fig. 1.54)

Two surface effects, the *effect of Rebinder* and the *effect of Roscoe*, strongly influence damage processes in region *IV* (cf. Fig. 1.54). Figure 1.56 shows the stress-deformation curve (the full line) for the material in its usual state and its changes (dotted lines) with the *film of a surfactant* on the surface (the effect of Rebinder) and with the *film of oxides* (the effect of Roscoe). These effects are principally different in respect to the normal state: either *surface hardening* or *softening* (*weakening*) occurs. Stresses needed to achieve plastic deformation of the same magnitude are higher in the first case (the effect of Roscoe) than under normal conditions and lower in the second case (the effect of Rebinder).

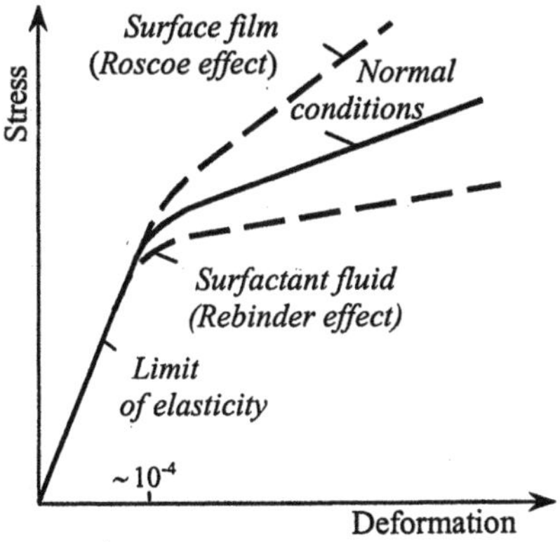

Fig. 1.56. Schematic representation of surface effects

If the breakpoint K of the full sliding fatigue curve is not found (cf. Fig. 1.54), region I makes no interest, then the sliding fatigue curve is plotted in the multicycle region and it is just called the *sliding fatigue curve* (Fig. 1.57). It serves to determine the *sliding fatigue limit* τ_f similarly like the endurance limit σ_{-1} is determined using the mechanical fatigue curve (cf. Fig. 1.16).

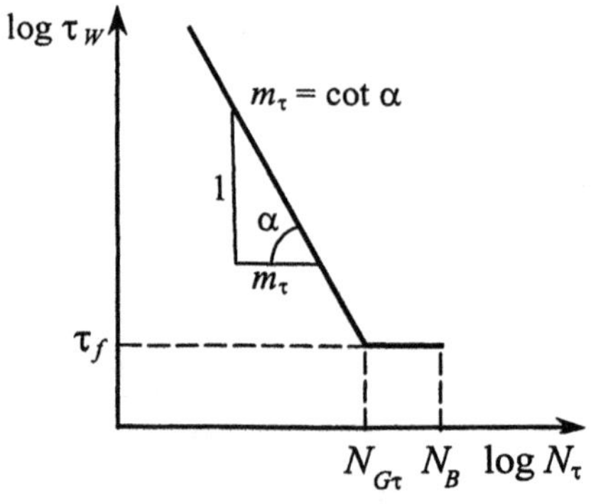

Fig. 1.57. Diagram of sliding fatigue curve in multicycle region

The term *sliding fatigue* is assigned the general sense of surface damage and fracture in sliding friction irrespective of its mechanisms (cf. Fig. 1.54), similarly like the term *mechanical fatigue* is assigned the general sense of volume damage and fracture during cyclic deformation irrespective of its mechanisms (cf. Fig. 1.15), and the term *rolling fatigue* is assigned the general sense of surface damage and fracture in rolling friction irrespective of its mechanisms (cf. Fig. 1.78).

The equation for any region (*I, II, III, IV*) of the sliding fatigue curve (cf. Figs. 1.54 and 1.57) has the form

$$\tau_W^{m_\tau} N_\tau = C_\tau \qquad (1.89)$$

where the indicator of slope (cf. Fig. 1.57) is

$$m_\tau = \cot\alpha; \qquad (1.90)$$

C_τ – a constant. The formula of wear durability follows from (1.89), for example, in the multicycle (*III*) region

$$N_\tau = C_\tau / \tau_W^{m_\tau}, \quad C_\tau = \tau_f^{m_\tau} N_{G\tau}, \qquad (1.91)$$

according to which the number of cycles until the friction couples reaches the critical or ultimate state is inversely proportional to the specific force of friction raised to the power m_τ. Equations (1.89)–(1.91) in their form are similar to (1.26) and (1.34) during mechanical fatigue.

Wearing intensity is an essential characteristic of the fracture process in friction. Though *linear* I_h, *volume* I_V, *mass* I_m and *energy* I_W *wearing intensities* are discriminated,

$$I_h = \Delta h / L_f,\ I_V = \Delta V / L_f,\quad I_m = \Delta m / L_f,\ I_h = \Delta V / F_s L_f, \tag{1.92}$$

it is practically enough to determine one (any) of them since they are all mutually proportional. Below we use the value I_h. The thickness, volume and mass of the layer L_f removed by wearing represent the *friction path* in formulas (1.92) Δh, ΔV, Δm, respectively.

The mean wearing intensity in any portion (*I, II, III, IV*) of the full sliding fatigue curve (cf. Fig. 1.54) can be estimated from the formula

$$I_h = K_0 \left(\frac{f p_a}{\tau_G} \right)^{1+\beta_0 m_\tau} \frac{K_{mo}}{N_\tau} K_{RR}, \tag{1.92a}$$

where K_0 – the empirical coefficient; K_{RR} – the *parameter of Roscoe-Rebinder* determined by the relation

$$\frac{\tau_f^{SUR}}{\tau_f^{OF}} = K_{RR} \tag{1.93}$$

of *sliding fatigue limits* (cf. Fig. 1.57) with the allowance for the *effect of surfactants* (τ_f^{SUR}) and *oxide films* (τ_f^{OF}).

In accordance with Eq. (1.92a), wearing intensifies together with the contact pressure p_a, the friction coefficient f, surface roughness (the parameter β_0), the coefficient of mutual overlap K_{mo}, the parameter m_τ and the parameter of Roscoe-Rebinder, but it reduces if resistance to fatigue fracture of material (τ_G) is increased. According to (1.92a), surfactants boost the wearing intensity, while an oxide film reduces it. In fact, formula (1.92a) takes into account a complex stressed state on the contact site because it is established [22] that an equivalent stress determined from the known (classical) theories of strength (cf. Sect. 1.2.1) is proportional to the specific friction force, i.e.

$$\sigma_{equiv} \sim f p_0 = \tau_W, \tag{1.94}$$

where p_0 – maximum pressure on the contact site.

In case of purely fatigue wearing (in multicycle region *III* – cf. Fig. 1.54), the wearing intensity is calculated from the formula [22]

$$I_h = \sqrt{\frac{h}{r}} \frac{A_r}{A_a} \frac{K_{mo}}{N_*} K_*, \tag{1.95}$$

where h/r – relative depth of the zone of deformation of single irregularities of the rough layer; A_r/A_a – relative contact area (A_r, A_a – *actual* and *nominal* contact areas); $K_* \approx 0.15...0.20$ – the coefficient determined by the arrangement of irregularities by their height; $N_* = N_\tau$ – wear durability.

Note that wearing intensity I_h in formulas (1.92a) and (1.95) is inversely proportional to durability N_τ, as it should be according to definitions (1.92).

Wearing intensity of polymeric materials is assessed using the equation

$$I_h = I_0 \exp\left[-\frac{U_0 - \gamma_\tau f_S p_a}{RT}\right], \qquad (1.96)$$

that follows from the kinetic (thermofluctuation) theory of strength of solids (cf. Sects. 1.32 and 1.39). Here U_0 – the energy of activation of breaking of chemical bonds; γ_τ – the coefficient depending on the structure of a polymer; RT – the energy of thermal motion of molecules, its fluctuations break chemical bonds weakened by the mechanical field; R – gas constant; I_0 – some constant.

Changes in the wearing intensity in time are usually described with a trough-shaped curve similar to the curve of changes in the wear rate in time (cf. Fig. 1.53, *b*) because these values are mutually proportional.

In practice a variety of types of wear and damage of materials in friction are observed. Figure 1.58 [28] provides their classification for metals and polymers; the intervals of possible variations of the wearing intensity are expressed approximately in technical units: the volume (mm^3) of the material removed from a unit of contact area (cm^2) per 1000 m of friction path. An extensive variety of the wearing processes and the intricacy of damage phenomena during friction highly complicate the problem of their calculation and assessment.

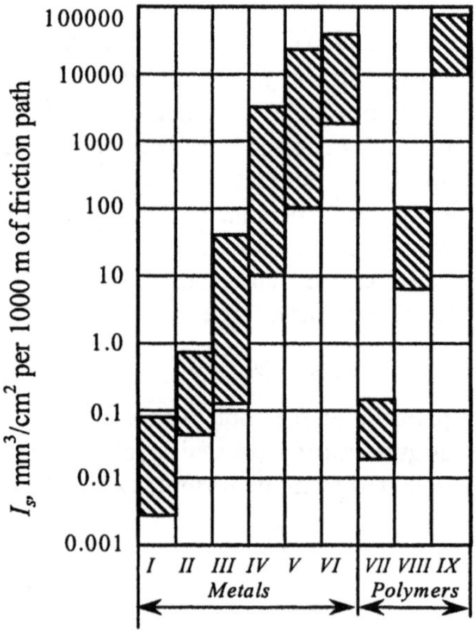

Fig. 1.58. Surface fracture intensity I_s during various types of wear and damage: *I* – normal mechanochemical wear of metals; *II* – mechanochemical type of abrasive wear of metals; *III* – fretting process; *IV* – seizure of kind *II*; *V* – seizure of kind *I*; *VI* – mechanical type of abrasive damage of metals; *VII* – normal mechanochemical wear of polymers; *VIII* – thermal damage of polymers; *IX* – abrasive damage of polymers

The *theoretical invariant approach* enables to assess the wearing intensity with the account of numerous mechanical, physical and chemical phenomena (Table 1.2) [29].

Table 1.2. Expressions for calculating wearing intensity

Wearing conditions	Formulas for calculating wearing intensity
Mechanical fracture of surfaces is dominant	$I_{mech} = K_M C_a^{X_1} C_{lubr}^{Y_1} C_Y^{Z_1} C_m^{\gamma_1}$
Allowance for the processes of sorption and chemical modification	$I_{phychem} = K_{phychem}(p_a v\tau_0 /(HBd_{fluc}))^m C_i^{n_1} L_{\tau_i}^{\Psi}$
Allowance for thermal processes	$I_{\mathrm{TF}} = F(t_k / t_{pla};\ q_0\delta_T /(\lambda_{1,2} t_{crit});$ $E\alpha\Delta t/((1-\nu)\sigma_{red})$
Nomenclature: K_M, $K_{phychem}$, X_1, Y_1, Z_1, γ_1, m, n_1 – coefficients and exponents having different physical sense and determined experimentally; $C_a = f_s p_a/\mathrm{HB}$ – the complex characterizing the stressed state of the contact and dimensionless area of actual contact of solids; $C_{lubr} = h_0/\chi$ – the complex determining the thickness of the lubricating layer; h_0 – the absolute thickness of the lubricating layer; χ – the characteristic dimension (the diameter of a cutting abrasive particle, the reduced dimension of roughness); $C_Y = \xi_1 p_a/\sigma_0$ – the complex characterizing resistance of rubbing surfaces to fatigue; ξ_1 – the coefficient depending on the value f_s and the stressed state in the contact; σ_0 – the endurance limit of the material under given conditions of friction; $C_m = R_{\max} / R_{red} b_1^{1/\nu_1}$ – the complex making allowance for the effect of surface roughness; $R_{\max}$ – the maximum height of profile irregularities; R_{red} – the reduced radius of irregularities; b_1 and ν_1 – the reference curve parameters; $C_i = C_{adh}$ the complex characterizing the properties of boundary lubrication of adsorption nature, or $C_i = C_{add}$ – the complex making allowance for chemical modification and appearance of protective films due to the action of additives; L_{τ_i} – the time simplex (or several simplexes); $p_a v\tau_0/HBd_{flu} = C_{kin}$ – the kinetic factor, dimensionless time of chemophysical transformation in the contact zone; d_{flu} – the average diameter of the actual contact spot; v – the speed of relative motion (rolling v_k or sliding v_c); $\tau_0 \approx 10^{-12}$ sec – the period of thermal oscillations of atoms; $t_c/t_{melt} = C_{ct}$ – the factor of contact temperature effect; t_c and t_{melt} – the temperature in the zone of contact between bodies and the temperature of melting of materials; $q_0\delta_T/\lambda_{1,2}t_{crit} = C_{grad}$ – the factor determining the effect of the temperature gradient and thermal boundary layer ; q_0 – the specific heat flux affecting a given body (the heat flux density); $\lambda_{1,2}$ – the coefficient of heat conductivity of the material; δ_T – the thickness of the thermal boundary layer; t_{crit} – the critical temperature (for example, the homological temperature, the temperature of chemophysical, structural transformations of the material of rubbing bodies); $E\alpha\Delta t/(1-\mu)\sigma_{red} = C_{th\,s}$ – the factor characterizing thermal stress in the surface layer; E – the elasticity modulus; α – the coefficient of linear thermal expansion; Δt – the temperature increment; σ_{red} – the ultimate stress; μ – the Poisson coefficient.	

Thus, the main process appearing in friction and leading to wearing is mechanical interaction between surfaces of solids during alternating deformation by shear. This main process is accompanied by many derivative phenomena that have a mechanical, physical and chemical nature. These phenomena include:

a) multiform *chemical processes* (for example, appearance of oxide films; dissolution of the surface of one rubbing body; embrittlement of the metal by atomic hydrogen released by the lubricant, etc.);

b) *thermomechanical processes* (for example, alteration of the properties of the lubricating material due to temperature rise in the friction zone; occurrence of momentary temperature flashes on the actual contact spots that may cause local phase transformations of the metal, etc.);

c) *hydrodynamic effects* of interaction of a material with rough surfaces in relative motion, including the wedging effect of a liquid when it penetrates into cracks;

d) the processes of *physical transfer of the substance* from one surface to the other (for example, *selective atomic transfer, smearing* or transfer of the film of the softer material to the harder material as a result of molecular seizure; transfer of steel or iron as a result of hydrogenation of their surfaces to the softer counterbody, bronze, plastic, etc.).

All *wearing types* can be divided into three basic groups:

1) *mechanical wearing* that results exceptionally from mechanical interactions between rubbing surfaces;

2) *molecular mechanical wearing* that is additionally accompanied by the action of molecular and/or atomic forces;

3) *corrosive mechanical wearing* that occurs during friction of the material that entered into chemical reactions with the environment.

Wear resistance of rubbing bodies is determined by the value inverse to the wearing intensity, i.e.

$$\varepsilon_h = 1/I_h \, . \tag{1.97}$$

Table 1.3. Classes of wear resistance of friction couples

Class	ε_h	I_h	Class	ε_h	I_h
3	$10^3...10^4$	$10^{-3}...10^{-4}$	8	$10^8...10^9$	$10^{-8}...10^{-9}$
4	$10^4...10^5$	$10^{-4}...10^{-5}$	9	$10^9...10^{10}$	$10^{-9}...10^{-10}$
5	$10^5...10^6$	$10^{-5}...10^{-6}$	10	$10^{10}...10^{11}$	$10^{-10}...10^{-11}$
6	$10^6...10^7$	$10^{-6}...10^{-7}$	11	$10^{11}...10^{12}$	$10^{-11}...10^{-12}$
7	$10^7...10^8$	$10^{-7}...10^{-8}$	12	$10^{12}...10^{13}$	$10^{-12}...10^{-13}$

Ten *classes of wear resistance* (from the 3rd to the 12th) are identified, each having a different value ε_h from the next one (or the preceding one) and the difference being of an order of magnitude (Table 1.3) [30]. The larger the class, the higher the wear resistance of the material, hence the less is the wearing intensity. The classes of wear resistance can be arranged in accordance with the characteristic regions of fracture on the full fatigue curve in friction (cf. Fig. 1.53). While region *I* of wearing intensity has an order of magnitude $10^{-3}...10^{-4}$ (class 3 of wear resistance), it reduces to $10^{-11}...10^{-13}$ in region *IV* (classes 11 and 12 of wear resistance).

1.4.4 Energy analysis

All the processes in friction appear and evolve as a result of struggle between two basic phenomena – *activation* (growth) *of free energy in materials of a tribosystem* and *passivation* (reduction) of this energy. *Damage* appears when the energy of activation is excessive and may be due to various causes, such as deformation, heating, etc. That is why dynamic equilibrium is the requisite condition to normalize the processes of friction and surface fracture

$$G_A = G_{SS}$$

of the energy of activation G_A and the energy G_{SS} needed for the appearance of *secondary* protective *structures* [31]. Such structures possess extreme properties and protect the base material of rubbing surfaces from direct contact and thus from fracture. In this connection all the variety of processes occurring in friction can be divided into two groups: (1) normal (theoretically unavoidable and practically tolerable) and (2) practically intolerable phenomena of damage. As an example Figs. 1.59 and 1.60 [28] show patterns of normal wear and damage of the bearing insert of the internal combustion engine (ICE).

Figure 1.61 [31] presents the general regularity of the processes of self-organization in friction. The energy of activation G_A here is the function of many parameters: contact pressure, friction rate, temperature, properties of bodies involved in the processes, etc. While the energy of passivation G_P is the function of wearing intensity and the friction coefficient. The main point of the general regularity shown in this Figure is the following: there is a range of loads and speeds of motion for all the materials and operating media within which the indicators of friction (f) and wear (I_h) are steady (region *II* on Fig. 1.61) and an order of magnitude less than outside this range. Its limits are determined by the critical values of the activation energy G_{A1} and G_{A2}. Region *I* is typical for unsteady processes at relatively small parameters of loading. Region *II* is due to the dynamic equilibrium of mechanochemical processes of formation and fracture of secondary structures. Evolution of external effects causes the transition of a friction couple from the stationary state into state *III* of unsteady damage.

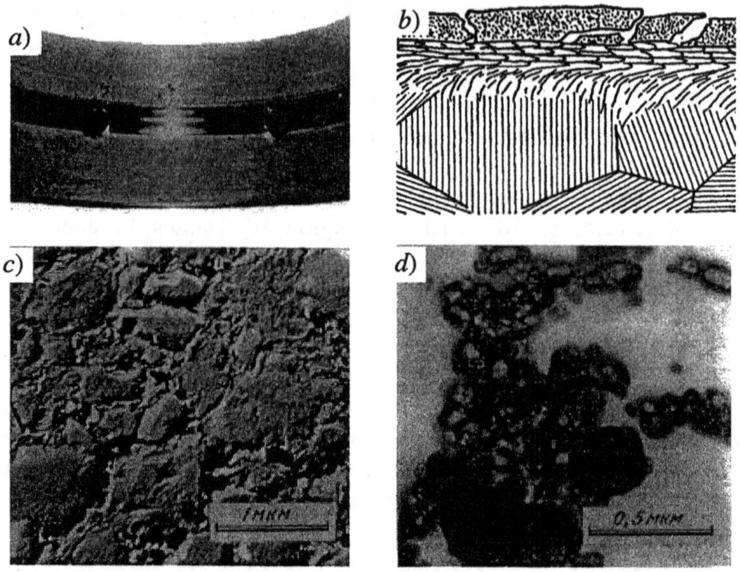

Fig. 1.59. Normal mechanochemical wear of the bearing insert of the internal combustion engine: *a* – general view; *b* – diagram of structure of surface layers; *c*, *d* – electron photographs of secondary structures on friction surface (*c*), debris (*d*)

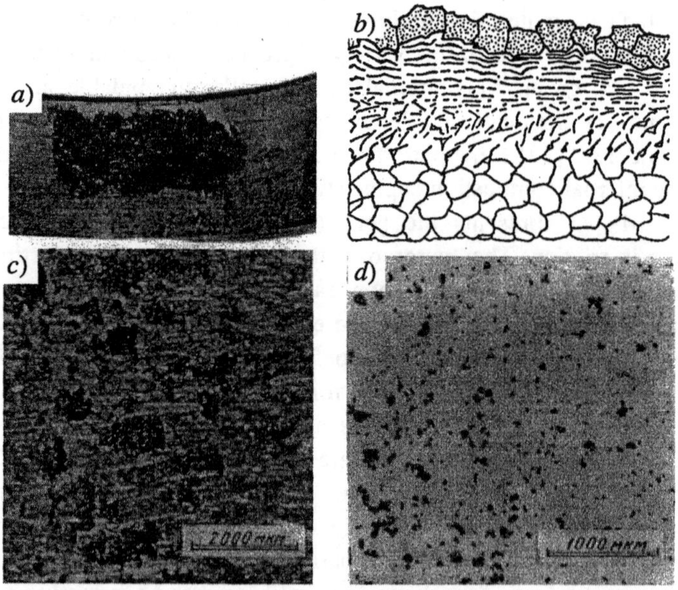

Fig. 1.60. Damage of bearing insert of ICE during fretting process: *a* – general view; *b* – diagram of structure of surface layers; *c*, *d* – photographs of secondary structures on friction surface (*c*), debris (*d*)

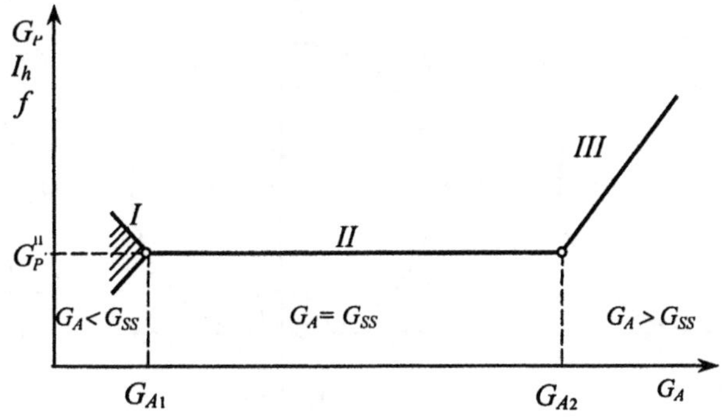

Fig. 1.61. General regularity of processes in friction: *I* – region of unsteady processes at $G_A < G_{SS}$, $\Delta E/A_{fr} \rightarrow$ min; *II* – range of normalization at $G_A = G_{SS}$, $\Delta E/A_{fr} \rightarrow$ min; *III* – region of damaging at $G_A > G_{SS}$, $\Delta E/A_{fr} \rightarrow$ max ($A_{fr} = Q + \Delta E$ – external mechanical energy in friction and energy of internal processes: thermal (Q) and structural (ΔE))

In accordance with the energy approach [32] to describing wear in friction, the critical energy density is calculated

$$e_R^* = \frac{W_R}{\Delta V} = \frac{N_K e_{\mathrm{Re}}}{\nu_V}, \tag{1.98}$$

that causes surface fracture of bodies. Here W_R – the work of friction; ΔV – the worn material volume; e_{Re} – the elementary energy density (the ratio between the work of frictional forces and the deformable volume); $N_K \approx N_\tau$ and

$$\nu_V = \Delta V/V_D, \tag{1.99}$$

where V_D – the deformable volume in which friction energy accumulates.

The wearing intensity I_h and the critical energy density e_R^* are interrelated through the specific frictional force

$$e_R^* = \tau_W/I_h, \tag{1.100}$$

because this energy analysis practically reduces to the assessment of the linear wearing intensity ($I_h = \tau_W / e_R^*$).

1.4.5 Sliding

Let us consider the problem of contact between a cylinder and a plane. If there are no external forces, these two bodies contact along the line forming a cylinder (Fig. 1.62, *a*). A compressive force F_N produces a contact site in the form of a strip having the dimensions $2b \times 2a$ (Fig. 1.62, *b*, *c*). In case the cylinder slides over the plane under the effect of force F, the force of resistance to motion appears on the contact site, i.e. the frictional force F_S.

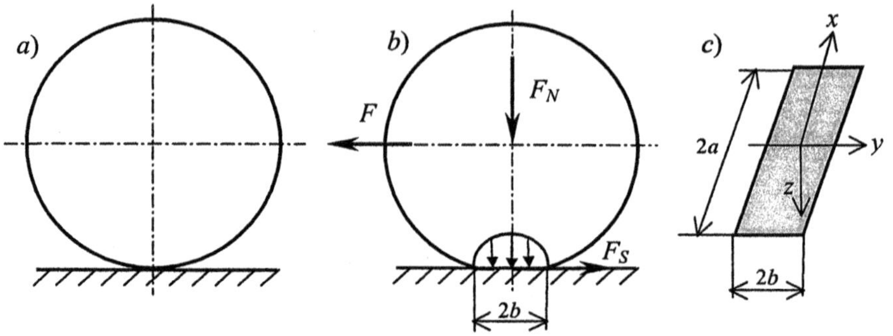

Fig. 1.62. Diagram of contact between cylinder and plane before (a) and after (*b*) loading

Set out the solution of the problem of the stress-strain state in the zone of contact between the cylinder and the plane under the effect of both contact loading and tangential force directed perpendicularly to the line of initial contact between the bodies in this friction couple [33]. The stress components are sought for as a function of the combination of relative rectangular (ξ, ψ) and elliptic (α, β) coordinates (Fig. 1.63, *a*) combined by the relationships

$$y = b \operatorname{ch} \alpha \cos \beta; \quad z = b \operatorname{sh} \alpha \sin \beta; \quad \psi = y/b; \quad \xi = z/b,$$

where y, z – rectangular coordinates; b – half-width of the contact strip. Formulas for stresses have the form

$$\left.\begin{aligned}
\sigma_y &= p_0\left\{f\left[2\left(\sqrt{\xi^2+\sin^2\beta}-\xi\right)\cot\beta-\frac{\xi\sin^3\beta\cos\beta}{\xi^2+\sin^4\beta}\right]+2\xi-\right.\\
&\left.-\left(1+\frac{\xi^2}{\xi^2+\sin^4\beta}\right)\sqrt{\xi^2+\sin^2\beta}\right\};\\
\sigma_z &= \frac{p_0\sin^3\beta}{\xi^2+\sin^4\beta}\left(f\xi\cos\beta-\sqrt{\xi^2+\sin^2\beta}\sin\beta\right);\\
\tau_{zy} &= p_0\left[f\left(\sqrt{\xi^2+\sin^2\beta}-2\xi+\frac{\xi^2\sqrt{\xi^2+\sin^2\beta}}{\xi^2+\sin^4\beta}\right)-\xi\frac{\sin^2\beta\cos\beta}{\xi^2+\sin^4\beta}\right],
\end{aligned}\right\} \quad (1.101)$$

where $p_0 = \sqrt{P_l E_{red}/(\pi D)}$ – maximum pressure on the contact site; P_l – normal linear loading; E_{red} – reduced elasticity modulus; D – the diameter of the cylinder.

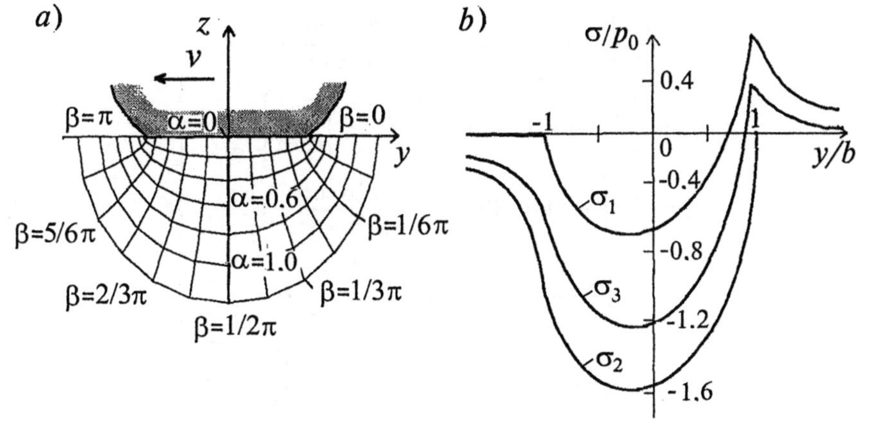

Fig. 1.63. Elliptic coordinates (*a*) and distribution of principal stresses over breadth of contact between cylinder and plane (*b*)

Principal stresses are determined from formulas

$$\left.\begin{aligned} \sigma_1 &= p_0 \exp(-\alpha)\,[f(1+\cos\beta) - \sin\beta]; \\ \sigma_2 &= -p_0 \exp(-\alpha)\,[f(1-\cos\beta) + \sin\beta]; \\ \sigma_3 &= \mu(\sigma_1 + \sigma_2). \end{aligned}\right\} \qquad (1.102)$$

Principal stresses in the contact zone ($\alpha = 0$) are

$$\left.\begin{aligned} \sigma_1 &= p_0\,[f(1+\cos\beta) - \sin\beta]; \\ \sigma_2 &= -p_0\,[f(1-\cos\beta) + \sin\beta]. \end{aligned}\right\} \qquad (1.102a)$$

Ahead of the contact site ($\beta = \pi$) it is

$$\sigma_1 = 0, \;\; \sigma_2 = -2p_0 f \exp(-\alpha),$$

and behind it

$$\sigma_1 = 2p_0 f \exp(-\alpha), \;\; \sigma_2 = 0.$$

Figure 1.63, *b* shows the distribution of principal stresses along the friction path (y-axis). It is apparent that the material is exposed to uniform compression ahead of the contact site and biaxial tension behind the contact site.

Based on (1.101) and using the law of Hooke (1.5), we obtain the formulas for deformation components

$$\left.\begin{aligned}
\varepsilon_y &= \frac{p_0}{2G}\left\{\left[-(\operatorname{cth}\alpha-1)(1-2\mu)-\frac{\operatorname{sh}^3\alpha\operatorname{ch}\alpha}{\operatorname{sh}^4\alpha+\xi^2}+1\right]\xi+\right.\\
&\quad\left.+f\left[2(1-\mu)(\operatorname{cth}\alpha-1)-\frac{\xi^2}{\operatorname{sh}^4\alpha+\xi^2}\right]\frac{\psi}{\operatorname{cth}\alpha}\right\};\\
\varepsilon_z &= \frac{p_0}{2G}\left\{\left[-(\operatorname{cth}\alpha-1)(1-2\mu)-\frac{\operatorname{sh}^3\alpha\operatorname{ch}\alpha}{\operatorname{sh}^4\alpha+\xi^2}-1\right]\xi-\right.\\
&\quad\left.-f\left[2\nu(\operatorname{cth}\alpha-1)-\frac{\xi^2}{\operatorname{sh}^4\alpha+\xi^2}\right]\frac{\psi}{\operatorname{cth}\alpha}\right\};\\
\gamma_{zy} &= \frac{p_0}{2G}\left\{f\xi\left[(\operatorname{cth}\alpha-2)+\frac{\operatorname{sh}^3\alpha\operatorname{ch}\alpha}{\operatorname{sh}^4\alpha+\xi^2}\right]-\xi^2\frac{\sqrt{\operatorname{sh}^2\alpha+\xi^2}}{\operatorname{sh}^4\alpha+\xi^2}\right\}.
\end{aligned}\right\}\quad(1.103)$$

Figures 1.64 and 1.65 present the analysis of formulas (1.103) and experimental results. It is apparent that the material is exposed to sign-variable deformations in two mutually perpendicular directions and shear deformation too. The material is compressed in the direction of motion (along the y-axis) ahead of the contact site (negative values ε_y in Fig. 1.64, *a*), it is lengthened in the contact zone (positive values ε_y). Tensile deformation is again observed behind the contact site. Thus, the material is subjected to two cycles of sign-variable deformations ε_y during one passage of the cylinder over the plane. On the contrary, the material is lengthened in the direction perpendicular to the friction surface (z-axis) ahead of contact site, while it is compressed in the contact zone (cf. Fig. 1.64, *b*). Hence, the material is subjected to sign-variable deformations $\pm\varepsilon_z$ in the direction of z-axis. Shear deformations γ_{zy} (cf. Fig. 1.64, *c*) have opposite signs ahead and behind the contact site.

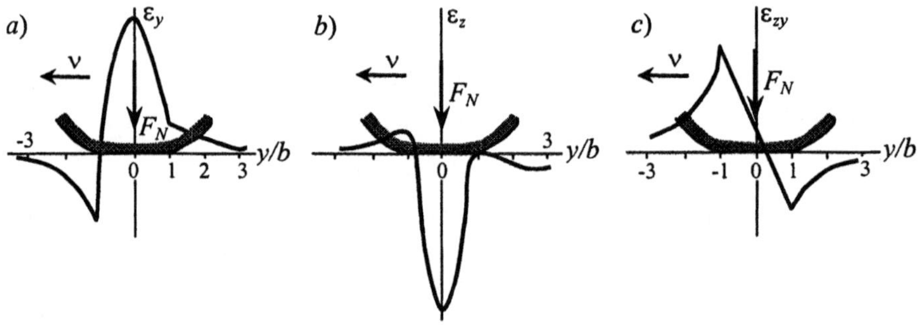

Fig. 1.64. Distribution of deformations along friction path (y-axis)

Figure 1.65 provides a pictorial idea about redistribution of deformations of the material under the contact site's surface (along the y-axis). The element A does not suffer from deformation, the element B undergoes shear and compression in the direction of the y-axis and somewhat lengthens in the direction z. The element C in the center of the contact site is compressed along the axis z and lengthens along the axis y. The element D does not undergo practically any normal deformation, yet it is subjected to shear in the direction opposite to the shear deformation of the element B.

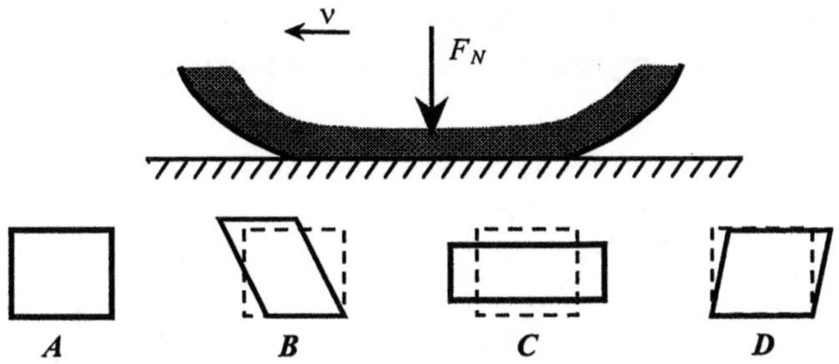

Fig. 1.65. Diagram of deformation of material under surface when cylinder slides over it

According to formulas (1.101)–(1.103), as the friction coefficient f grows, the deformation of both signs augments. Yet the friction coefficient affects more strongly the value of tensile deformation. Generally, when contact loading and the friction coefficient rise, it leads to the corresponding rise of the amplitude values of deformations.

The analysis of the stress-strain state of the material in sliding friction in the elastoplastic region is highly intricate and is omitted here (approaches to constructing the theory of plasticity see in Sect. 1.2.1). Hence, we will show the analysis of processes in sliding friction by constructing the sliding fatigue curve in the coordinates "contact loading F_N – ultimate wear $i_{\lim}$".

The advantage of the loading parameter F_N is that it is not any calculated value, it is a physical value assigned and measured during tests and it remains as it is under any contact conditions (elastic, elastoplastic deformation, microshearing, seizure, appearance and fracture of films, etc.). If the specific frictional force is assumed as the loading parameter, as it is shown in Fig. 1.54, problems appear of assessing (measuring) the friction coefficient as its numerical values are different in regions *I*, *II*, *III*, *IV* of the sliding fatigue curve; they can also change within each region.

A friction couple *steel 45 / polymer Ф4-BM* was tested in sliding [34]. The steel shaft was 10 mm in diameter and rotated with the speed 3000 min^{-1}. The polymeric specimen was a cube with the dimensions $10\times10\times10$ mm^3 which was pressed in the process of tests against the steel shaft by contact loading F_N, variable within a broad range. The value of wear $i_{\lim} = 1$ mm was assumed as the ultimate state of the polymer. All tests lasted until the polymeric specimen reached the ultimate state. The base of tests was $8\cdot10^7$ cycles.

Totally 12 friction units were tested under 12 various contact loads within the range from 10 to 440 N. The test results were used to plot the sliding fatigue curve (Fig. 1.66) in semilogarithmic coordinates: contact loading F_N expressed in Newtons – wear durability N determined as the logarithm of the number of loading cycles until the value of wear i_{lim} was reached. It has turned out that this curve has four typical regions: *I* – the region of quasistatic fracture (approximately up to $N = 4 \cdot 10^5$ cycles), *II–III* – the region of low- and multicycle fracture ($N = 4 \cdot 10^5 ... 5 \cdot 10^6$ cycles), *IV* – the region of gigacycle fracture in operation ($N > 5 \cdot 10^6$ cycles). Transition from region *I* to region *II* occurs under the contact load $\sim F_L = 330$ H, transition from region *III* to region *IV* occurs at $F_G \approx 80$ H. The boundary between the multicycle and low-cycle is weakly pronounced (at $F_K \approx 200$ H), therefore regions *II* and *III* are approximated with single dotted straight line *II–III*.

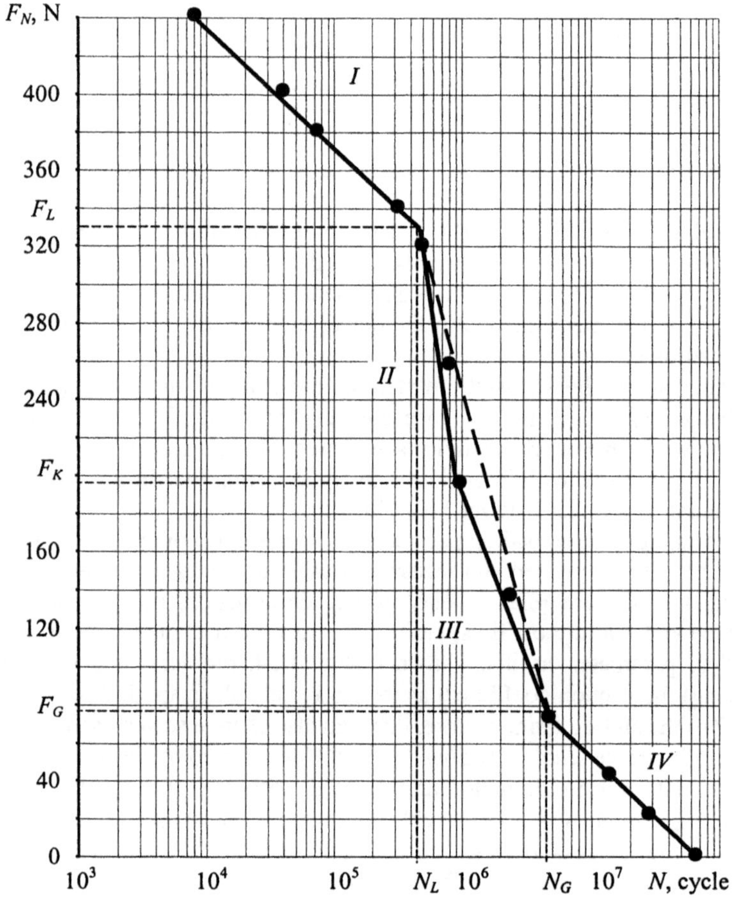

Fig. 1.66. Sliding fatigue curve for *steel 45 / polymer Φ4-BM* tribocouple

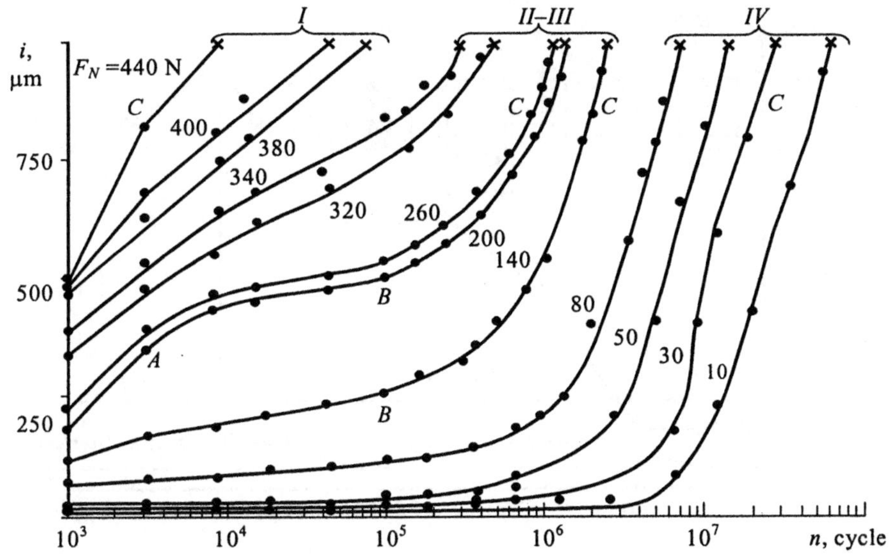

Fig. 1.67. Kinetic wear curves *i* of polymer Φ4-BM under various contact loads F_N

Figure 1.67 shows the kinetic curves of wear changes in time at assigned levels of contact loads. It is quite apparent that three types of kinetic curves are identified. Type *I* occurs under heavy contact loads (F_N = 380...440 N) and it is practically a straight line. Comparison with the sliding fatigue curve (cf. Fig. 1.66) shows that this process of wear accumulation in time is typical for the region of quasistatic fracture. Type *II–III* is a complex curve consisting of three (*A, B, C*) regions (curves at F_N = 200 and 260 H). Comparison with the sliding fatigue curve in this case (cf. Fig. 1.66) evidences that this process of wear accumulation in time is typical for the region of low- and multicycle fracture (F_N = 140...340 N). Finally, type *IV* is a curve consisting of two regions (*B* and *C*); such curves are typical for the region of gigacycle fracture in operation (F_N = 10...80 N). Thus, the regularities of wear accumulation in time during sliding friction happen to be similar to the regularities of accumulation of residual deformation during mechanical fatigue (cf. Fig. 1.26). Quasielastic pushing of the material dominates in the region of gigacycle fracture *IV*; it is characterized by fine wear (approximately 5 μm) that is not accompanied by any plastic pushing. This conclusion is confirmed by region *IV* in Fig. 1.68: the value $c/c_0 = 1$ at F_N up to 80 N; (c and c_0 – the size of the working zone of the rubbing specimen in the direction of motion before and after the tests, respectively). Plastic pushing in region of low- and multicycle fracture *II–III* ($1.0 < c/c_0 < 1.15$ at $80 \le F_N \le 340$ N) plays the main role in the processes of wearing, the heavier the contact loading, the stronger is plastic pushing. Finally, microcutting plays a decisive role in causing wear in quasistatic fracture region *I*, it is accompanied by thermal damage; plastic pushing in this case (c/c_0 at $F_N > 340$ N) becomes lesser as the value F_N becomes larger.

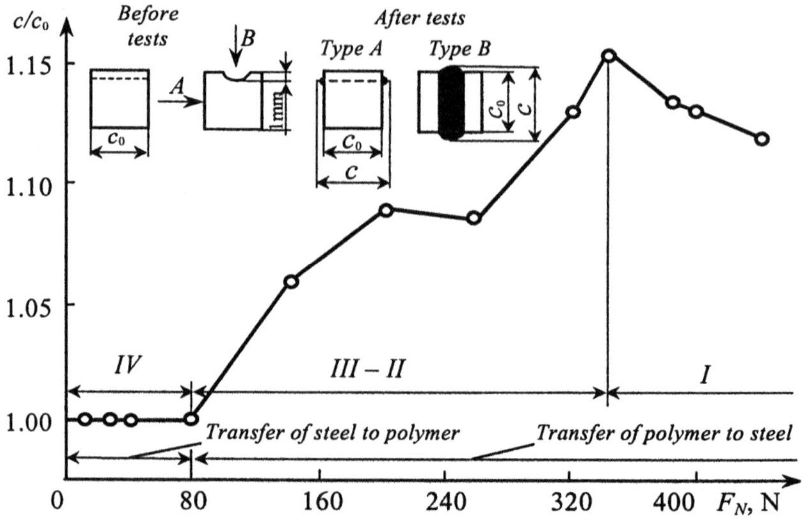

Fig. 1.68. Development of plastic pushing of polymer in direction of motion during sliding friction

Visual examination of the friction surfaces has revealed that the polymer actively smears over steel in low- and multicycle regions. This phenomenon does not practically occur in the high resource region; on the contrary, impregnations of finely dispersed steel particles are clearly seen even at slight increases, i.e. back transfer (from the hard steel to the relatively softer polymer) occur. Both friction surfaces become significantly rougher. Apparently fresh finely dispersed loose particles of polymer appear in the friction zone in the process of protracted contact interactions and these particles act as surfactants. They facilitate and accelerate the formation, migration and multiplication of dislocations on the steel surface, with the steps (extrusions) easily breaking off (cf. Figs. 1.22 and 1.23). Then they charge into the relatively soft polymer and thus not all are carried away from the contact zone. Friction of the polymeric surface charged with metallic particles over steel impairs the roughness of both contacting surfaces.

Table 1.4 presents the equations for the typical regions of the sliding fatigue curve with their indicators of slope m_N obtained by the method of least squares. It is seen that this indicator may vary more than 10 times. The equations themselves are similar to formula (1.89).

Table 1.5 and Fig. 1.69 present the results of analysis of the mean wearing intensity of the polymer in operation of this tribocouple. The calculation is performed using two formulas (1.92).

Table 1.4. Equations of sliding fatigue curves

Regions	Equations	Indicators of slope m_N
I	$\log F_N = -0.0763 \log N + 2.950$	13.11
II–III	$\log F_N = -0.6125 \log N + 6.037$	1.63
IV	$\log F_N = -0.9528 \log N + 8.551$	1.05

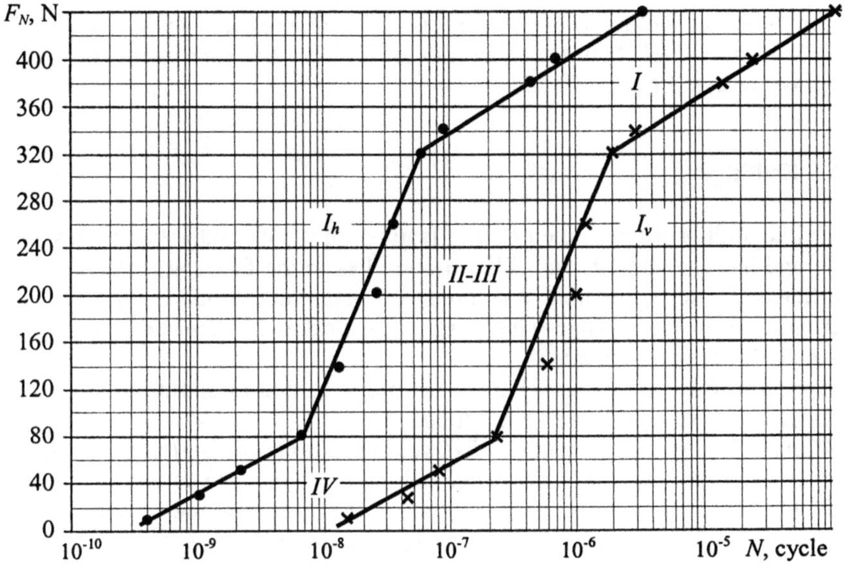

Fig. 1.69. Wearing intensity curves: I_h [1/cycle]: I_v [mm^2/cycle]

Table 1.5 presents the equations of relation between contact loading and wearing intensity and the indicators of slope m_I derived by the method of least squares. It is apparent in Tables 1.4 and 1.5 that indicators m_N and m_I for the similar portions of the sliding fatigue curves and wearing intensity curves practically coincide. Moreover, from Fig. 1.69 it follows that the wearing intensity curves plotted as a function of the level of contact loading have the same three regions (*I*, *II–III* and *IV*) that the sliding fatigue curve has in Fig. 1.66. According to the data of Table 1.5, the indicators of slope of similar portions of both curves are similar too. In this case the loading coordinates F_L and F_G of the inflection points of the curves of both types are stable and they also coincide.

Table 1.5. Equations of wearing intensity curves

Regions	I_h	m_I	I_v	m_I
I	$\log F_N = 0.076 \log I_h + 3.06$	13.16	$\log F_N = 0.077 \log I_v + 3.645$	12.99
II–III	$\log F_N = 0.627 \log I_h + +7.07$	1.59	$\log F_N = 0.615 \log I_v + 11.553$	1.63
IV	$\log F_N = 0.745 \log I_h + 8.07$	1.34	$\log F_N = 0.794 \log I_v + 14.4$	1.26

Addressing further the analysis of wearing intensity it can be established that in region *I* it varies within the range $I_h \geq 5.8 \cdot 10^{-8}$ and in region *IV* within $I_h \leq 7 \cdot 10^{-9}$; there is region *II–III* between these values. Volume wearing intensity is usually much larger than linear.

1.4.6 Rolling

Unlike sliding friction, the peculiar features of contact interaction between the body and the counterbody in rolling friction lies in the fact that, first, the friction coefficient is approximately one order of magnitude smaller than in sliding friction; second, the contact site is strongly localized, therefore it should bear relatively high specific loads.

Appearance of the frictional force in rolling is due to sliding of the coupled surfaces and hysteresis losses in the solid.

Mutual slip of surfaces can be observed when a ball rolls along a trough (Fig. 1.70, *a*). The circumference *AB* of the ball moves along the center of the trough, while the circumference *CD* touches its sides. It is clear that the circumference *AB* passes longer distance per rotation than the circumference *CD*. It is this difference that causes slip of the friction surfaces. *Hysteresis losses in rolling friction* will be considered using the example of rolling of a solid ball on a flat rubber surface (Fig. 1.70, *b*). When the ball makes one rotation and passes a shorter path than the length of the circumference of its diametrical cross section, causing slip with corresponding energy dissipation. The conditions of rubber deformation are different in point *C* from those in points *B* and *D*. A depression appears ahead in point *E* and behind in point *A* the forces of elasticity restore the deformable material. As a result the ball performs the work of deformation that may be different in portions *DE* and *AB*.

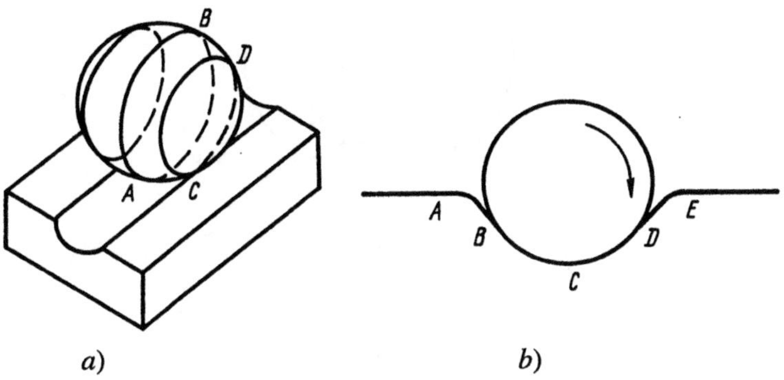

Fig. 1.70. Diagram of ball rolling along groove (*a*) and over plane (*b*)

Though when two solids roll (Fig. 1.71, *a*), some slip does occur, it is usually called *free rolling friction* or simply rolling friction, while the notion *rolling friction with slip* refers to the cases when slip is caused by loading conditions. So, when two cylinders roll (Fig. 1.71, *b*), the slip is due to the braking torque M_T.

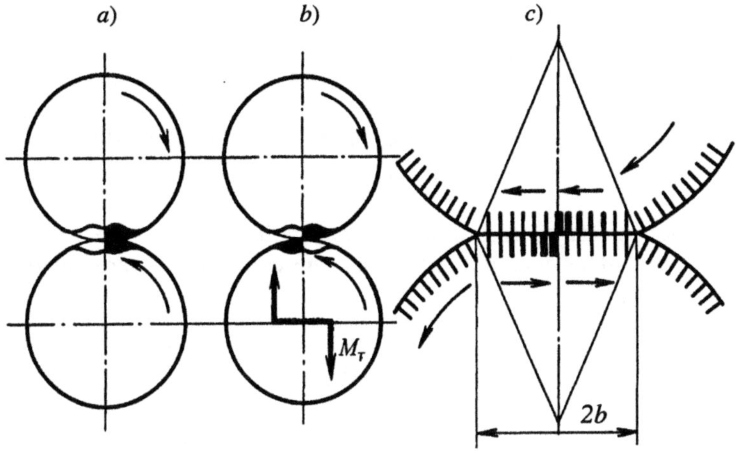

Fig. 1.71. Diagram of deformation of surface metallic layers when two cylinders roll

When two cylinders roll (cf. Fig. 1.71, *a*) and roll with slip (Fig. 1.71, *b*), the metal in the contact site zone is subjected to tensioning (light areas) and compression (dark areas). Metallic fibers approach and displace in the direction shown by arrows in the zone of compression of the leading surface (Fig. 1.71, *c*). Fibers stretch elastically in the zone of tension and displace in the same direction, while those of the lagging surface displace in the opposite direction.

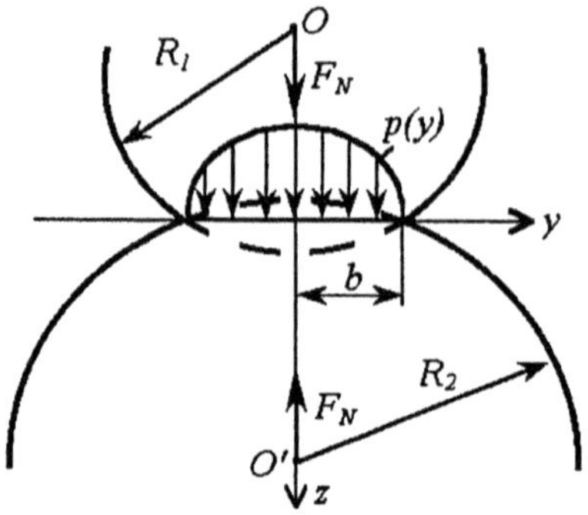

Fig. 1.72. Pressure distribution over site of contact between two cylinders with parallel axes

Let us study the stressed state in the region of contact between two cylinders (rollers) having radii R_1 and R_2 and compressed by load F_N (Fig. 1.72). Normal pressure distribution p_a over the width $2b$ of the contact site (the axis y) is described by the elliptic law

$$p(y) = \frac{2F_N}{\pi bl}\left(1 - \frac{y^2}{b^2}\right)^{1/2}, \tag{1.104}$$

so that it reaches maximum in the center of the contact site ($y = 0$):

$$p_0 = 2F_N / \pi bl. \tag{1.104a}$$

The components of stresses in an arbitrary point with coordinates y, z (in this case the stressed state does not depend on the coordinate x) are calculated with equations

$$\left.\begin{aligned}
\sigma_x &= -2\mu p_0 \frac{z}{b}\left(\sqrt{\frac{b^2+\lambda}{\lambda}} - 1\right); \\
\sigma_y &= -p_0 \frac{z}{b}\left[\sqrt{\frac{b^2+\lambda}{\lambda}}\left(2 - \frac{b^2 z^2}{\lambda^2 + b^2 z^2}\right) - 2\right]; \\
\sigma_z &= -p_0 \frac{bz^3}{\lambda^2 + b^2\lambda^2}\sqrt{\frac{b^2+\lambda}{\lambda}};
\end{aligned}\right\} \tag{1.105}$$

$$\left.\begin{aligned}
\tau_{yz} &= \tau_{zy} = -p_0 \frac{byz^2}{\lambda^2 + b^2\lambda^2}\sqrt{\frac{\lambda}{b^2+\lambda}}; \\
\tau_{zx} &= \tau_{xz} = 0; \ \tau_{xy} = \tau_{yx} = 0,
\end{aligned}\right\} \tag{1.106}$$

where λ – the maximum root of the equation

$$\frac{y^2}{b^2+\lambda}+\frac{z^2}{\lambda}=0\,. \tag{1.107}$$

The tangential stresses $\tau_{yz}=\tau_{zy}$ in (1.105) at $y = 0$, i.e. for the points on the plane perpendicular to the contact strip plane and passing through its centerline (the plane xy), become zero, Eqs. (1.105) pass respectively into formulas

$$\left.\begin{aligned}
\sigma_x &= -2\mu p_0\left[\sqrt{\left(\frac{z}{b^2}\right)^2+1}-\frac{z}{b}\right];\\
\sigma_y &= -p_0\left[\frac{1+2\left(\frac{z}{b}\right)^0}{\sqrt{\left(\frac{z}{b}\right)^0+1}}-2\frac{z}{b}\right];\\
\sigma_z &= -p_0\frac{1}{\sqrt{\left(\frac{z}{b}\right)^2+1}}\,.
\end{aligned}\right\} \tag{1.105a}$$

The maximum tangential stresses in accordance with (1.106) are

$$\tau_{zy}=p_0\frac{z}{b}\left(1-\frac{1}{\sqrt{\left(\frac{z}{b}\right)^2+1}}\right). \tag{1.106a}$$

The maximum normal stresses occur at $z = 0$, i.e. for the points of the centerline of the plane we have

$$\sigma_{x\,\max}=-2\mu p_0,\ \sigma_{y\,\max}=-p_0,\ \sigma_{z\,\max}=-p_0=2F_N/\pi bl\,. \tag{1.108}$$

Normal stresses (1.105a) reach their maximum (1.108) on the surface of the contact. When moving away the value σ_y recedes considerably faster than σ_z; the values $\sigma_x << \sigma_z$ (Fig. 1.73, *a*). The maximum tangential stresses according to (1.106a) are detected at a depth 0.786*b*, depending on the contact conditions their numerical values may reach (0.4...0.6)p_0 and more (Fig. 1.73, *b*). When a tangential load *F* is applied, considerable tangential stresses change approximately from 0.4p_0 to 0.8p_{0c} providing the friction coefficient increases from $f = 0.2$ to $f = 0.4$; they occur over the boundaries of the contact site and have opposite signs in these points (Fig. 1.74).

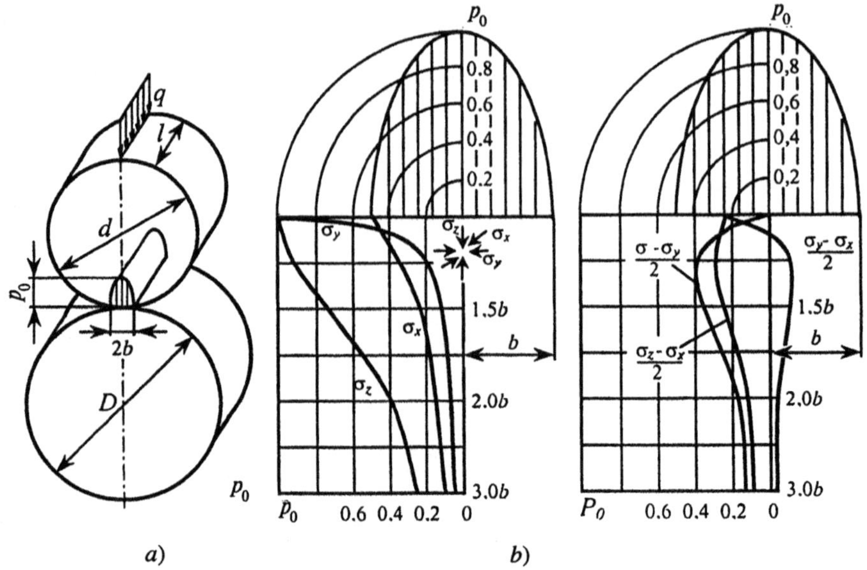

Fig. 1.73. Distribution of contact stresses during initial linear contact in points lying along axis of pressures ($\mu = 0.3$): *a* – normal stresses; *b* – tangential stresses

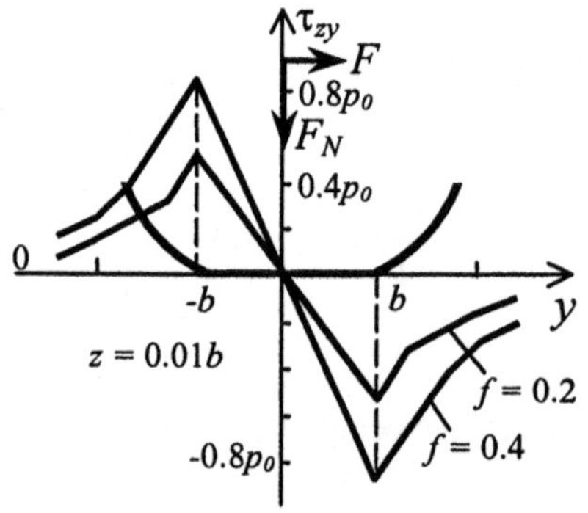

Fig. 1.74. Distribution of tangential stresses $\tau_{xy} = \tau_{yz}$ along line of contact between cylinder and semiplane (axis *y*)

Figure 1.75 shows the results of calculation of fields of equal stresses with (1.106) and (1.107) when two cylinders with radii $R_1 = 6$ mm and $R_2 = 50$ mm contact and when they are compressed with load $F_N = 600$ N (it is assumed that $E = 2 \cdot 10^6$ MPa, $\mu = 0.3$; the length of the cylinders is $l = 3$ mm). Figure 1.76 shows the surfaces of stresses for the components σ_x, σ_y, σ_z and $\tau_{xy} << \tau_{yx}$. In this case the half-width of the contact strip is equal to $b \approx 0.11$ mm, the maximum pressure in its center is $p_0 = -1142$ MPa; the maximum tangential stress is $\tau_{max} = \pm 280$ MPa.

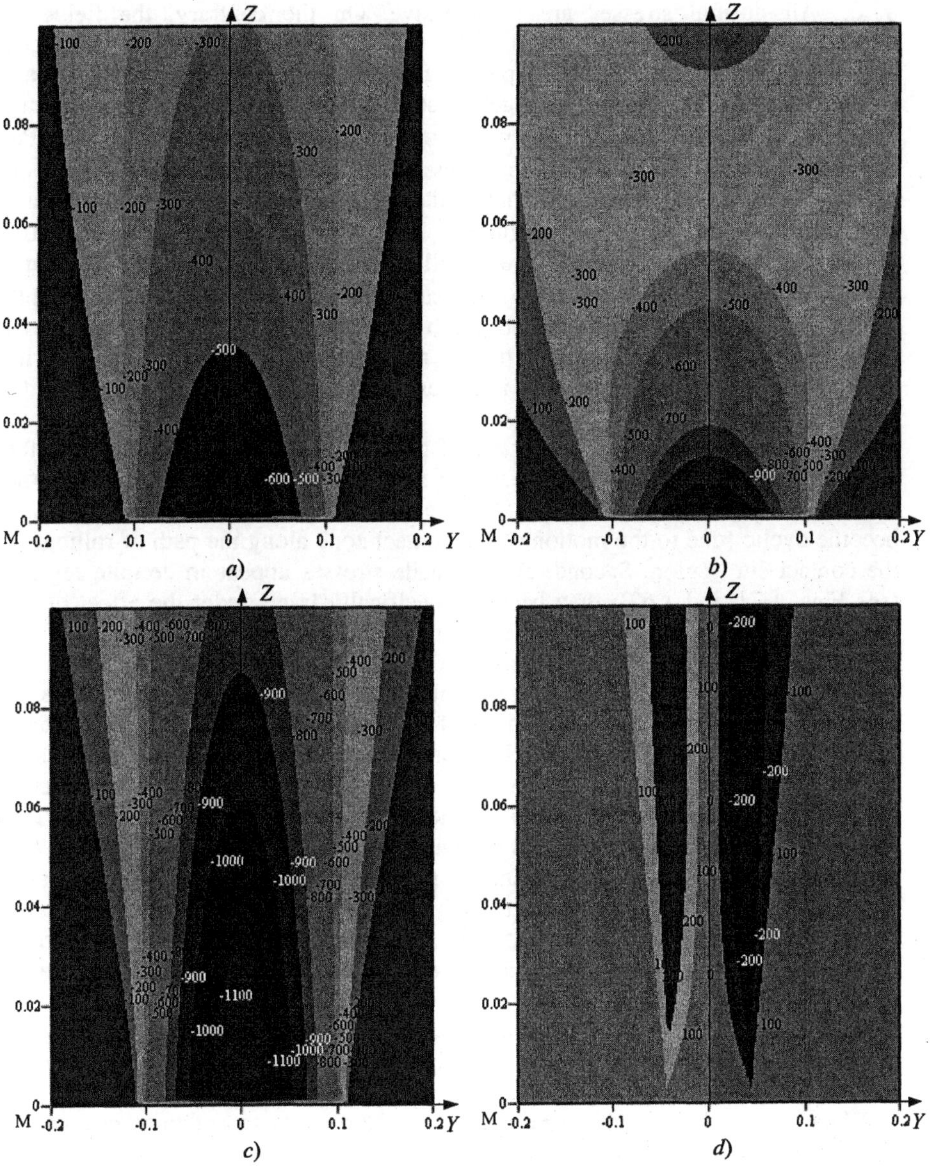

Fig. 1.75. Fields of stresses of equal level based on components σ_x (*a*), σ_y (*b*), σ_z (*c*) and $\tau_{xy} = \tau_{yx}$ (*d*) at contact load $F_N = 600$ N

From Figs. 1.75 and 1.76 it follows that the field of maximum stresses based on the components σ_x, σ_y, σ_z occurs always on the surface in the vicinity of the contact site center, $\sigma_{x\,max}$ being approximately two times smaller than $\sigma_{z\,max} = \sigma_{y\,max}$. All normal stresses are compressive. On the contrary, the fields of maximum tangential stresses occur under the contact surface, they are arranged symmetrically on both sides in respect to the contact site center; when they pass through the plane z, x the sign of tangential stresses becomes opposite. The gradients of normal stresses is very high. The values σ_y and σ_z reduce from 1100 to 100 MPa, i.e. approximately 10 times, and the values σ_x reduce from 600 to 100 MPa, i.e. 6 times, within the half-width of the contact site (i.e. at a distance $b = 0.1$ mm). Values σ_z reduce by ~20%, σ_y – almost 6 times and σ_x reduces 2 times at a depth $z = 0.1$ mm. Thus, the processes of damage in rolling friction should localize either in a very fine surface layer (σ_y and σ_z) or in subsurface zones ($\tau_{xy} = \tau_{yx}$), their depth can exceed the size of the contact strip half-width ($z > b$). In general, the material at some depth under the surface during rolling friction of two cylinders with parallel axes deforms similarly like it was established for the sliding friction of the cylinder on the plane (cf. Fig. 1.65).

The pattern of the stress-strain state of the material changes principally in the contact region as soon as the processes of rolling friction starts. First, though the contact loading remains constant (F_N = const), all the components of loading become cyclic (due to the motion of the contact zone along the path of rolling) in the contact site region. Second, cyclic tensile stresses appear in definite regions (see Figs. 1.71 and 1.63); they become specifically large under the effect of the tangential force in the contact. In this connection the process of fracture of the surface layer in rolling friction is described as *rolling (contact) fatigue*.

The process of rolling fatigue resembles in many respects the process of common fatigue (appearance and gradual propagation of cracks, the dependence of the durability and the endurance limit on a number of factors, etc.), yet it has its own specific features. They are due to the fact that the volume stressed state takes place in the contact zone, sharp gradients of stress components occur and maximum stresses localize in small volumes of the metal (cf. Fig. 1.75 and 1.76). It causes a sharp change in the extent and pattern of deformation of the metal as it penetrates deeper into its surface. While significant plastic deformation is observed in the surface layer (specifically on the tips of microprojections), normal stresses amount to just tenth or hundredth fractions of the elasticity limit at a depth exceeding just a few times the size of the contact site. Moreover, presence of two dangerous zones is typical for rolling fatigue; one zone is a fine surface layer on the contact site, the other zone is the subsurface region of maximum tangential stresses lying at a depth frequently less than the size of the contact site.

When two cylinders with parallel axes roll, the contact conditions govern the process of cracking.

Cracks appear in the subsurface zone under the conditions of pure rolling (the friction coefficient is $0.005 < f < 0.025$), they develop and emerge on the surface. Appearance of such cracks is attributed to the action of maximum tangential stresses τ_{max} at a depth $z \approx 0.7...0.8b$.

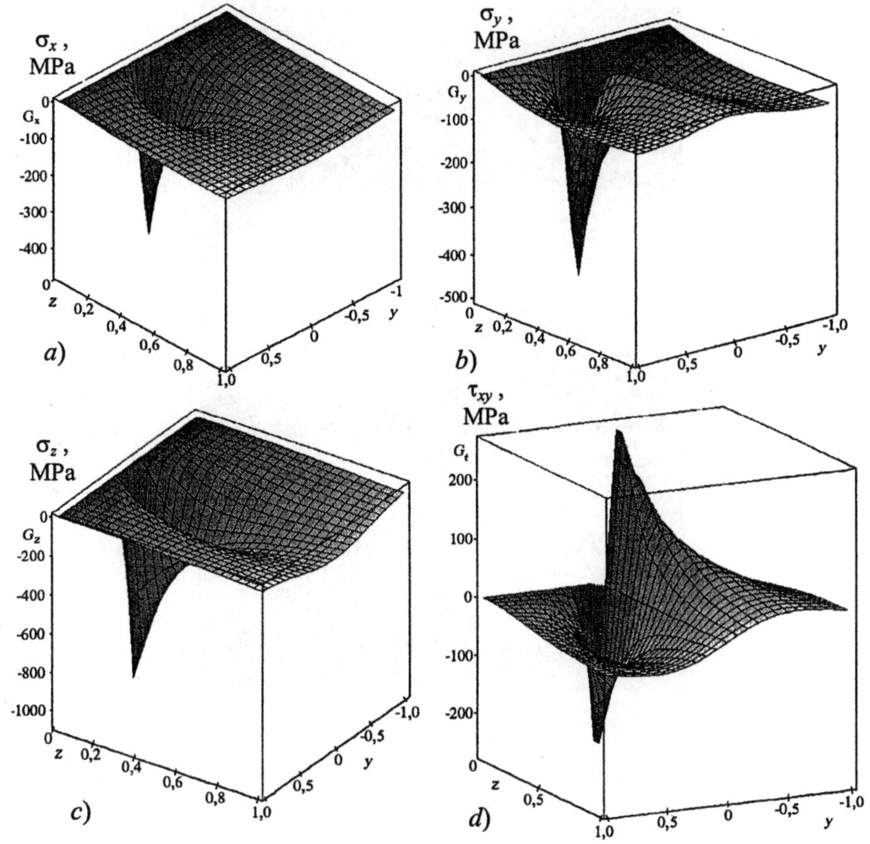

Fig. 1.76. Surfaces of stresses based on components σ_x (*a*), σ_y (*b*), σ_z (*c*) and $\tau_{xy} = \tau_{yx}$ (*d*) at $F_N = 600$ N (coordinate *y* varies within range (–1; +1) mm, coordinate *z* varies within range (0; 1) mm

During rolling with slip ($0.025 < f < 0.05$) the metal is damaged both on the surface and in depth. Though the value τ_{max} in the surface layer may exceed the stresses on the surface, the surface fracture is still intensified by oxidative and thermal processes.

As a rule, the crack grows from the surface during free sliding ($f > 0.05$).

Two typical types of surface fracture are observed during tests for rolling fatigue: pitting and wear by spalling. The latter manifests separation of fine flakes or plates of the embrittled metal. This fracture is enabled by the appearance of a subsurface crack parallel to the plane of rolling, its development completes with the emergence on the surface. Pitting represents the spalling of separate spots on the surface; sometimes it is accompanied by the breaking off of quite large metallic fragments. The sizes of the pits of spalling (and their number) grow together with the loading cycles (Fig. 1.77) [35]. Pitting is possible if a system of inclined cracks develops.

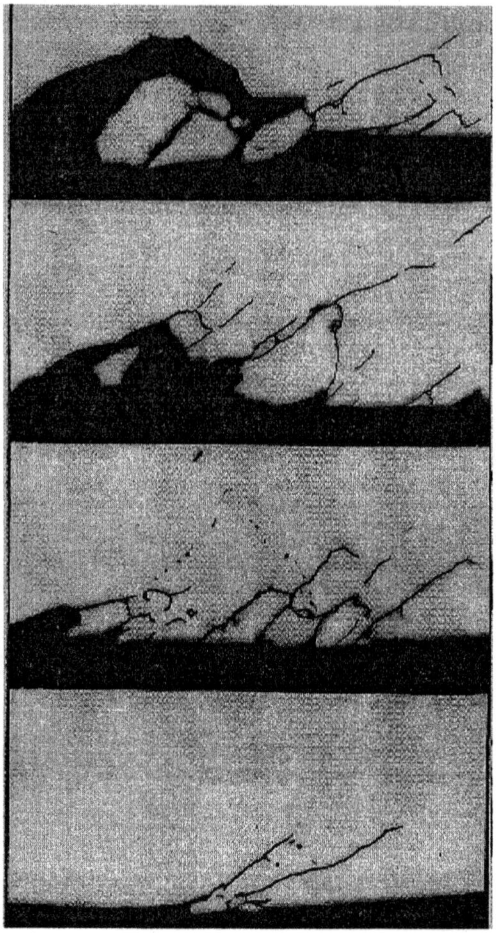

Fig. 1.77. Spalling on surface of roller as number of loading cycles grows

The rolling fatigue curve, like the mechanical fatigue curve (cf. Fig. 1.15) is plotted in double (or semi-) logarithmic coordinates "the maximum pressure in the contact site center p_0 – the number of cycles N_p until the friction couple reaches the ultimate state" (Fig. 1.78). The latter is established based on two criteria: 1) appearance of the pits of spalling of critical density or critical depth along the rolling path; 2) approach between the body and the counterbody to a specified extent (due to residual deformation and/or wear of the rubbing surfaces). In case the rolling fatigue curve has a horizontal portion corresponding to the rolling (contact) fatigue limit p_f, the following base of tests is assigned: $N_B = 10^7$ cycles for metals with the hardness HB ≤ 200; $5 \cdot 10^7$ cycles for the metals with the hardness HRC ≤ 40; 10^8 cycles for the metals with the hardness HRC > 40. $N_B = (2...5) \cdot 10^8$ cycles are specified if the rolling fatigue curve has no horizontal portion.

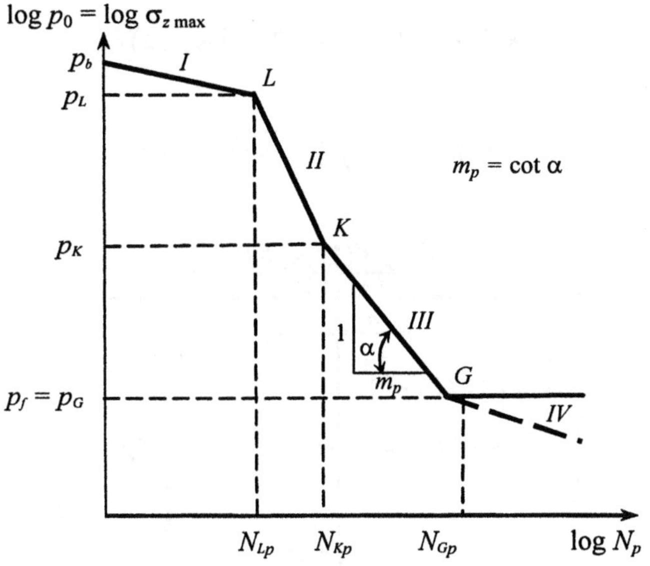

Fig. 1.78. Diagram of full rolling fatigue curve

Any region *I, II, III, IV* of the rolling fatigue curve is described satisfactorily with a power equation of type (1.26)

$$p_0^{m_p} N_p = C_p = \text{const}\,, \tag{1.109}$$

that serves to determine durability, for example, during rolling fatigue in region *III*:

$$N_p = \left(\frac{p_f}{p_0}\right)^{m_p} N_{Gp} = \frac{C_p}{p_0^{m_p}}, \tag{1.110}$$

and it is similar to Eq. (1.34).

We will analyze some regularities of development of cracks in the surface layer during contact fatigue using the linear mechanics of fracture (cf. Sect. 1.3.4).

First we will consider the process of wearing by spalling [36]. From the theoretical point of view, as indicated above, wearing by spalling is possible only when a horizontal subsurface crack appears, develops and comes across a vertical crack growing from the surface (pitting) or emerges on the surface in case the crack front contorts (Fig. 1.79, *a*). Based on the linear mechanics of fracture the CIN for a horizontal crack is determined using the formulas

$$K_{\text{I}}(-b-l_1) = \lim_{x \to (-b-l_1)} \sqrt{2[x-(-b-l_1)]}\;\sigma_z(x,h);$$

$$K_{\text{II}}(-b-l_1) = \lim_{x \to (-b-l_1)} \sqrt{2[x-(-b-l_1)]}\;\tau_{xz}(x,h);$$

$$K_{\mathrm{I}}(-b) = \lim_{\substack{x \to (-b) \\ +}} \sqrt{2[x-(-b)]}\,\sigma_z(x,h);$$

$$K_{\mathrm{II}}(-b) = \lim_{\substack{x \to (-b) \\ +}} \sqrt{2[x-(-b)]}\,\tau_{xz}(x,h),$$

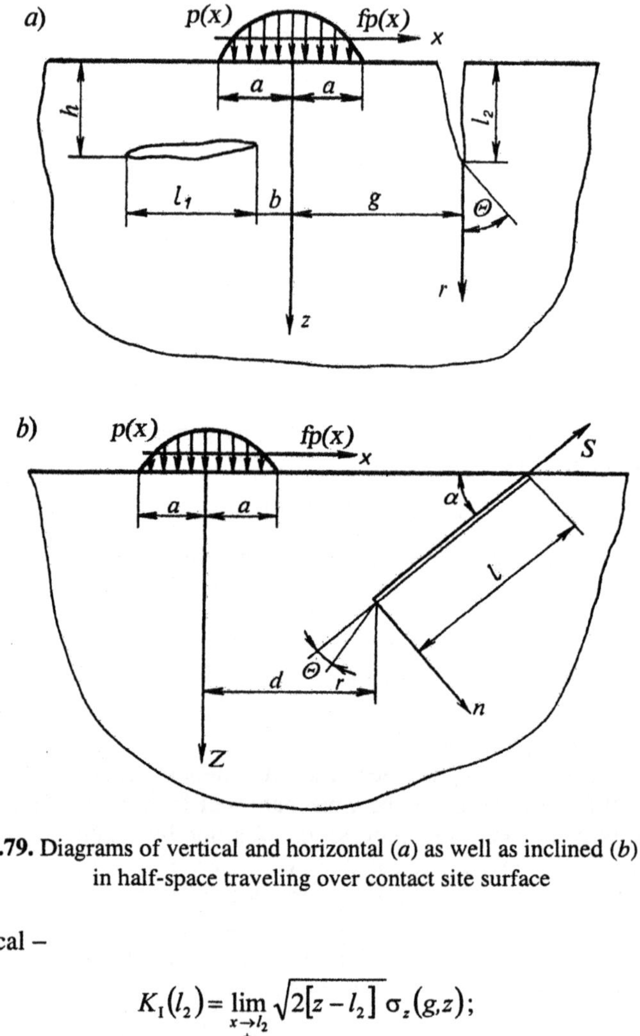

Fig. 1.79. Diagrams of vertical and horizontal (*a*) as well as inclined (*b*) cracks in half-space traveling over contact site surface

for vertical –

$$K_{\mathrm{I}}(l_2) = \lim_{\substack{x \to l_2 \\ +}} \sqrt{2[z-l_2]}\,\sigma_z(g,z);$$

$$K_{\mathrm{II}}(l_2) = \lim_{\substack{x \to l_2 \\ +}} \sqrt{2[z-l_2]}\,\tau_{xz}(g,z),$$

here

$$\left.\begin{aligned}\sigma_z &= -p(x) = -\frac{p_0}{a}\sqrt{a^2 - x^2};\\ \tau_{xz} &= -fp(x);\end{aligned}\right\}\quad \begin{aligned}&z = 0;\\ &-a \le x \le a,\end{aligned}$$

where f – the friction coefficient, the sign "minus" under the arrow of the limit corresponds to the negative displacement along the axis x, the sign "plus" corresponds to the positive displacement (similarly for the axis z).

The angle of deviation Θ of the crack from the initial location can be assessed with an approximated formula

$$\left(\frac{3}{8\sqrt{2r}}\right)^{1/2} \cdot K_{\mathrm{I}} \sin\Theta - K_{\mathrm{II}}(1 - 3\cos\Theta) = 2A\cos\Theta \sin\frac{\Theta}{2},$$

where

$$A = \lim_{z \to l_2}\left[\sigma_z(g, z) - \frac{K_{\mathrm{I}}}{\sqrt{2r}}\right].$$

The analysis of these solutions leads to the following conclusions regarding the horizontal crack growth (Fig. 1.80, *a*):

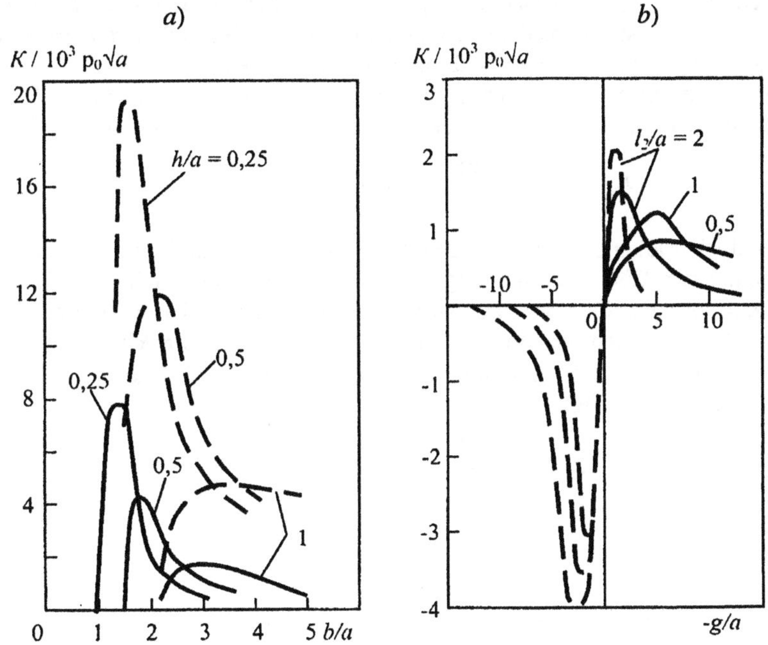

Fig. 1.80. Dependence of coefficients K_{I} (full curves) and K_{II} (dotted curves) on dimensionless distance between crack end and contact site axis: *a* – horizontal crack; *b* – vertical crack

(1) if the crack is under the contact site, it shuts under the effect of compressive stresses;

(2) when the contact site moves from left to right, tensile stresses appear that stimulates the crack's growth up;

(3) when the contact site moves away from the crack's edge, the stress intensity coefficient acquires larger values;

(4) the closer the crack to the surface (h/a diminishes), the higher the stress intensity coefficient values;

(5) if $K > K_c$ (K_c – the critical stress intensity coefficient value), the crack grows.

The pattern of growth of the vertical crack is highly intricate and has the following features (Fig. 1.80, *b*):

(1) the maxima K_{II} can appear if the contact site is located at some distance to the right or to the left of the crack. Consequently the front of development of the crack can change its direction;

(2) the vertical crack may be closed from the surface (its edges are joined) but open at some distance under the surface;

(3) the angle of deviation of the crack from the vertical line is determined in a significant manner by the value of lagging of the contact site from the axis. The maximum angle of deviation Θ is $\approx -69°$, i. e. $\sim -21°$ in respect to the surface of the half-space.

Now about the pitting process [37]. From the theoretical point of view the pitting is growth of an inclined surface crack (cf. Fig. 1.79, *b*) that can either twist or emerge on the surface or meet with other similar cracks.

The stress intensity coefficient in the Cartesian coordinate system (n, S) with the center at the tip of an inclined crack is assigned in the following form:

$$K_I = \lim_{S \to 0} \sigma_n \sqrt{2S}\,; \quad K_{II} = \lim_{S \to 0} \tau_{nS} \sqrt{2S}\,.$$

The analysis indicates that K_{II} depends on the crack's length and remoteness d of the contact site. When the right edge of the contact site approaches the tip of the crack ($d/a > 1$), the SIC (stress intensity coefficient) reaches its maximum, then it diminishes to zero when the center of the site overhangs the crack's tip, it reaches its minimum when the left edge approaches the tip ($d/a \to 1$). The larger d/a, the higher the value of the SIC.

The number of cycles N_c, after which the crack reaches its critical length l_c and spalling occurs, can be determined from the formula

$$N_c = N_0 + \frac{1}{B} \int_{l_0}^{l_c} \frac{dl}{(\Delta K_{II})^m}\,,$$

where N_0 – the number of cycles needed for nucleation of a crack embryo that has the length l_0; $\Delta K = K_{max} - K_{min}$; B and m are the parameters. Calculations are performed by numerical integration.

1.4.7 Fretting

An intricate combination of mechanical, physical, chemical, thermal and electrical processes is observed in the process of *fretting*. These processes evolve *in the zones of contact of coupled bodies at small vibratory displacements of one surface in respect to the other.* Figure 1.81 shows one typical scheme of tests in fretting. Bridge 2 is pressed against the surface of specimen *1* with contact load *Q*, so that it can displace tangentially with a relatively small amplitude *a* under the effect of alternating force *F* with a frequency ±ν.

Figure 1.82 [38] shows two typical curves the *tangential force – displacement* in the case of fretting of nickel-silver wire in a couple with phenol resin. When the amplitude is 1.88 μm, the curve is a non-distorted sinusoid implying that motion is purely elastic, hence, there is no slip. When the amplitude is 12.5 μm and more, this motion becomes complex with the occurrence of slip and microvibrations, so that centers of *seizure* appear and disappear on the actual contact sites during each cycle.

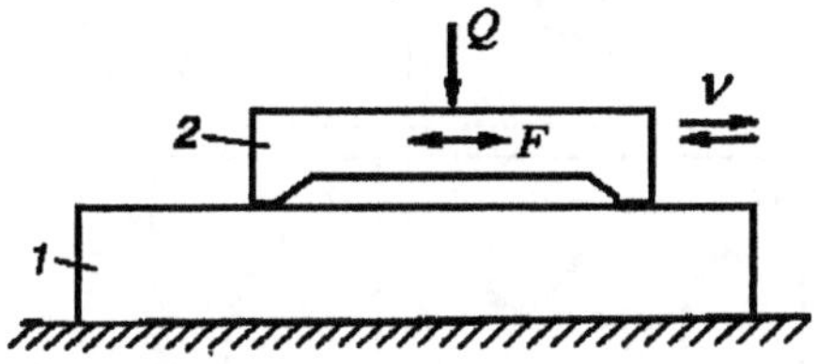

Fig. 1.81. Principal diagram of tests for fretting: *1* – specimen; *2* – fretting bridge

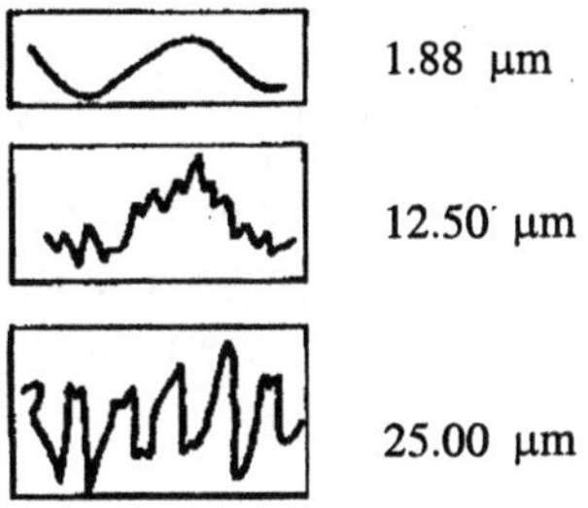

Fig. 1.82. Oscillograms of tangential forces at different displacement amplitudes

Figure 1.83 [39] shows the oscillograms the *tangential forces – displacement* (curves *I*), *force – time* (*II*) and *displacement – time* (*III*). The Arabic numerals on the curves designate identical points in the motion cycle. The amplitude of displacement is about 20 μm.

When there is no slip, the oscillogram is an ellipse; it transforms into a parallelogram in case of slip. Indents appear on both sides of the parallelogram in places of slip in case of motion with slip and seizure. The force reaches its maximum when the coefficient of friction at rest is reached and slip begins, while the speed of motion happens to be maximum when the slip occurs.

Fretting process can lead to produce the following results:

– wear, if mechanical surface fracture dominates (*fretting wear*);

– corrosive damage, if chemical and electrochemical processes dominate (*fretting corrosion*);

– combination of fretting corrosion and fretting wear.

Wear in fretting is strongly localized on the actual contact sites because of a small amplitude of relative slip of contacting surfaces, the products of fracture of surface layers are unable to leave freely the two-dimensional space between rubbing bodies. Hence, they are ground and accumulate near the actual contact sites intensifying their abrasive effect.

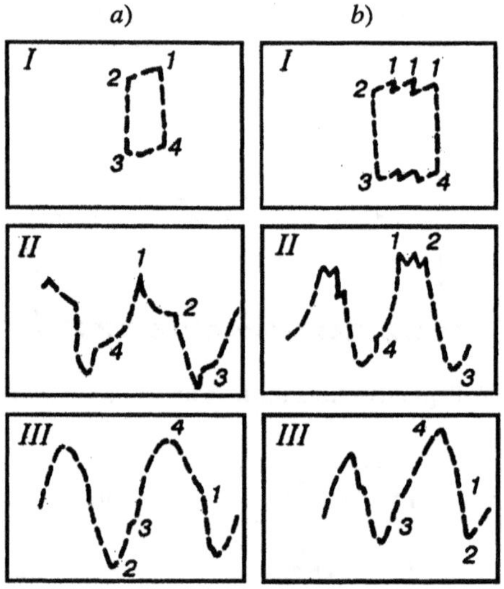

Fig. 1.83. Curves of tangential forces and displacements during slip without seizure (*a*) and during slip with seizure (*b*): *I* – slip loop; *II* – curve *force – time; III* – curve *displacement – time*

Chemical or electrochemical interactions with the environment are accompanied by the appearance of particles harder than the base metal, such as oxides, intensifying the wearing processes additionally and adding the products of corrosion to the volume of the worn material.

If it is assumed that wear in fretting is a simple sum of losses of the mass Δm due to the mechanical fracture and corrosive damage of the surface under the effect of normal load Q, then it is possible to obtain the equation [40]

$$\Delta m = \left(k_0 Q^{1/2} - k_1 Q\right)\frac{n}{\nu} + k_2 Q l\, n, \qquad (1.111)$$

where k_0, k_1, k_2 – coefficients determined experimentally; n – number of fretting cycles; ν – frequency; l – the distance an irregularity on the friction surface travels within one cycle semiperiod.

Equation (1.111) is linear in respect to the number of loading cycles, parabolic in respect to the contact load and hyperbolic in respect to the frequency of testing.

Figure 1.84 [41] shows the *fatigue curve in fretting* plotted using the experimental results (in double logarithmic coordinates). The criterion of reaching the ultimate state at any point of the curve was assumed the appearance of scar of a specified width in the process of fretting wearing. It is apparent that the pattern of this curve is similar to the mechanical fatigue curve (cf. Fig. 1.15). The *fatigue limit* in fretting based on the data in 1.84 was q_f = 80 MPa, the top point corresponds to the bearing capacity of the couple during static loading.

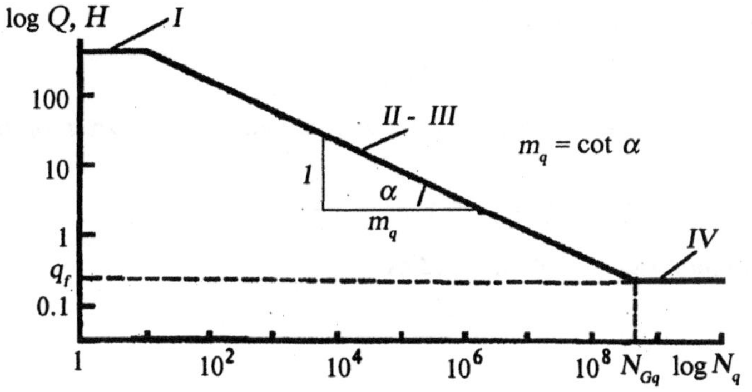

Fig. 1.84. Experimental fatigue curve in fretting

Durability *in fretting* in region *II–III* (cf. Fig. 1.84) is determined from the equation

$$N_q = \left(\frac{q_f}{q}\right)^{m_q} N_{Gq} = \frac{C_q}{q^{m_q}}, \tag{1.112}$$

where q – contact pressure in fretting , m_q – the parameter of the slope of the curve of fatigue in fretting, N_{Gq} – the abscissa of the inflexion point of the curve.

1.4.8 Calculations of friction and wear

Two types of wear are identified: *zero* and *measurable.* If wear does not exceed the height of surface roughness, it is zero wear corresponding to practically *wearless friction.* In the opposite case it is *friction with wear.* So, according to Fig. 1.67, zero wear occurs in portion B of the kinetic curves at F_N = 10...50 H, i.e. in region *IV* of longer-term fatigue, according to Fig. 1.66. Hence, according to the diagram of the full fatigue curve in friction (cf. Fig. 1.54), the *zero wear condition* is [42]

$$\tau_W = fp_a \leq \frac{\tau_G}{n_\tau} = [\tau], \tag{1.113}$$

where $[\tau]$ – *allowable specific frictional force (allowable frictional stress)* determined with the *load factor* $n_\tau > 1$; p_a – nominal (design) contact pressure.

The *measurable wear condition* is opposite to inequality (1.113):

$$\tau_W = fp_a > [\tau] = \frac{\tau_G}{n_\tau}. \tag{1.114}$$

Similarly to the use of condition (1.16), three procedures (1.7a) or (1.7b) of calculations of strength are recorded, let us establish similar *procedures of calculations of friction and wear*:
verification of wear resistance –

$$\tau_W \leq [\tau], \tag{1.113a}$$

determination of the dimensions of the nominal contact area of the friction couple –

$$A_a \geq F_S/[\tau] = F_S n_\tau / \tau_G, \tag{1.113b}$$

selection of materials of rubbing bodies –

$$[\tau] \geq \tau_W; \tag{1.113c}$$

$$f \leq [\tau]/p_a. \tag{1.113d}$$

Unlike the calculations of strength, conditions (1.13) and (1.113a)–(1.113c) should be used twice – for each body in the friction couple if they are made from unlike materials and/or have unlike shapes and dimensions. Moreover, a proper selection of the materials of rubbing bodies can be validated by satisfying condition (1.113d) according to which the friction coefficient cannot exceed a specified value. Finally, it is noteworthy that all the conditions of calculating friction and wear can also be recorded based on the contact pressure taking into account that $\tau_W = fp_a = fF_N/A_a$ and $F_S = fF_N$.

Then, the durability in friction with wear is assessed using the formula of type (1.91) depending on the loading level, properties of materials, type of friction, etc. (for example, cf. formulas (1.110), (1.112) and others). Knowing N_τ, the wearing intensity is calculated using formula (1.93) or other formulas reflecting main conditions of operation of a given friction couple (for example, cf. formulas (1.92a), (1.95), (1.96), formulas in Table 1.2 and others). The obtained value I_h is compared with the (normative) wear resistance $[\varepsilon_h]$ to satisfy the *condition of wear resistance*:

$$\frac{1}{I_h} \leq [\varepsilon_h]. \tag{1.115}$$

A given friction couple can thus be referred to the established class of wear resistance k_i (according to the data in Table 1.3):

$$10^{k_b} < 10^{k_i} < 10^{k_t},$$

where k_b, k_t – the top and bottom values for the established class of wear resistance; they are determined on the basis of feasible calculations or experience of operating typical friction couples.

Calculations of wear (including microcutting) and durability in regions *I, II, III, IV* of the full wearing curve are performed similarly using the corresponding parameters of the curve $N(\tau_W)$ (cf. Fig. 1.54). Thus, the *condition of friction in plastic contact* is

$$\tau_L > \tau_W > \tau_K,$$

if it is discovered that $\tau_K \approx \tau_y$, and the *condition of microcutting* is

$$\tau_W > \tau_L.$$

If it is necessary, relations (1.92), (1.111) and others are used to assess the absolute wear value.

1.5 Reliability

1.5.1 Model of failures

Failure is when an object (a structural element, a friction couple (pair), etc.) reaches the ultimate state according to the corresponding criterion of resistance to fatigue, wear resistance, etc.

Construction of *failure models* will be considered using a specific example of operation (or tests) of a friction couple. Let there be a sufficiently large number (k_0) of nominally identical friction couples observed when they operate under a constant contact load F_N = const. Then each separate friction couple has its own way how the process of wear accumulation evolves in time (Fig. 1.85 shows only three of them). The ultimate state is when wear i reaches the critical value $i_{\lim}$. If none of the friction couples reaches the ultimate state within the interval of time $(0, t_0)$, it means a *probability of failure-free operation* $Q(t_0) = 1.0$ (for the whole integrity of the studied friction couples), hence, the *probability* of their *failure* is $P(t) = 1 - Q(t) = 0$. Friction couples fail within the interval of time (t_1, t_2); the number k of failed couples grows with time. Hence, the probability of failure-free operation

$$Q(t) = \frac{k_0 - k}{k_0} \tag{1.116}$$

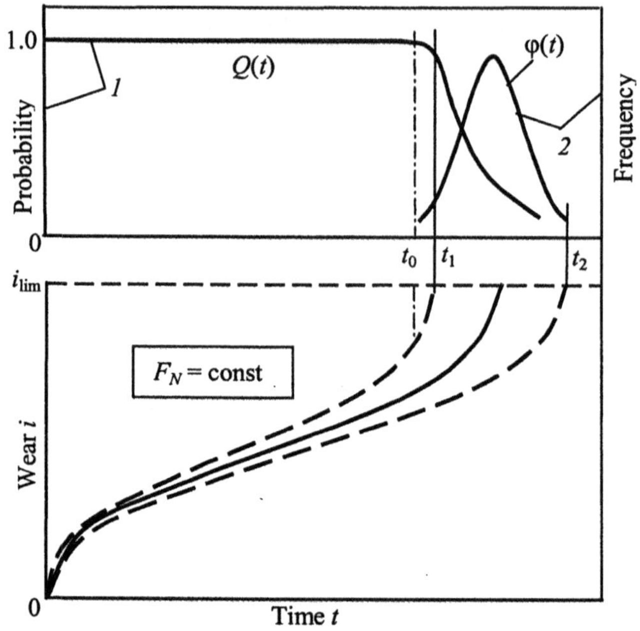

Fig. 1.85. Curves of wear and their relation to probability of failure-free operation (*1*) and function of distribution of operating time before failure (*2*)

will reduce correspondingly within this time interval (cf. curve in Fig. 1.85). If the frequency of failures is known within short intervals Δt_i of time

$$a(\Delta t_i) = \frac{k(\Delta t_i)}{\Delta t_i}\frac{1}{k_0},$$

it is easy to determine the *density of distribution* $\varphi(t)$ *of operating time before failure* (cf. curve 2 in Fig. 1.85). The relation between the density of failures and the probability of failure-free operation is called *intensity of failures*:

$$h(t) = \varphi(t)/Q(t). \tag{1.117}$$

Borrow curve 2 from Fig. 1.85 and place into separate Fig. 1.86. According to this Figure, the probability of failure during time $t < t_n$ is

$$P(t \le t_n) = P(t) = \int_0^{t_n} \varphi(t)dt\,, \tag{1.118}$$

then the probability of failure-free operation is

$$Q(t > t_n) = Q(t) = \int_{t_n}^{\infty} \varphi(t)dt = 1 - P(t) \tag{1.119}$$

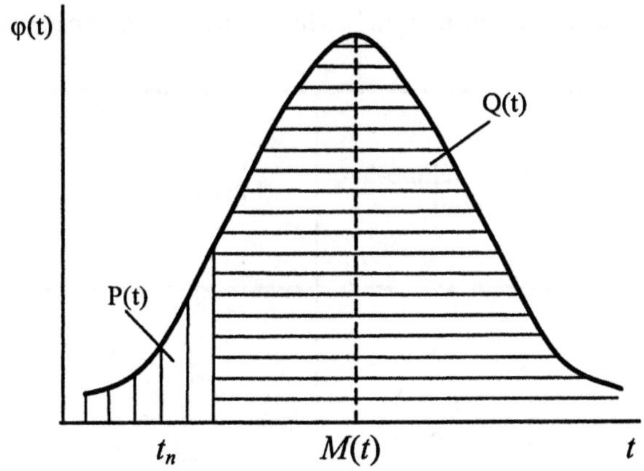

Fig. 1.86. Determination of probability of failure $P(t)$ and probability of failure-free operation $Q(t)$

Mathematical expectation (or mean time) of operating time before failure is

$$M(t) = \int_{-\infty}^{\infty} t\varphi(t)dt. \tag{1.120}$$

There are functional relationships between the *main indicators of reliability* ($\varphi(t)$, $Q(t)$, $P(t)$, $h(t)$) (Table 1.6), so that in practice it is enough to determine one of them and the others are obtained by simple recalculation using formulas in Table 1.6.

Table 1.6. Relationship between reliability indicators

Indicators	$P(t)$	$Q(t)$	$\varphi(t)$	$h(t)$
$P(t)$	$P(t)$	$1 - Q(t)$	$\int_0^t \varphi(t)dt$	$1-\exp\left[-\int_0^t h(t)dt\right]$
$Q(t)$	$1 - P(t)$	$Q(t)$	$\int_t^\infty \varphi(t)dt$	$\exp\left[-\int_t^\infty h(t)dt\right]$
$\varphi(t)$	$\frac{d}{dt}P(t)$	$-\frac{d}{dt}Q(t)$	$\varphi(t)$	$h(t)\exp\left[-\int_0^t h(t)dt\right]$
$h(t)$	$\frac{\frac{d}{dt}P(t)}{1-P(t)}$	$-\frac{d}{dt}[\ln Q(t)]$	$\frac{\varphi(t)}{\int_t^\infty \varphi(t)dt}$	$h(t)$

Table 1.7. Formulas for calculating probability of failure-free operation and intensity of failures

Density of distribution of operating time before failure	Probability of failure-free operation	Intensity of failures	Distribution parameters
Normal distribution	$Q(t) = P\left(z_p > \frac{t-\mu}{S_t}\right)$	$h(t) = \frac{\varphi\left(\frac{t-\mu}{S_t}\right)}{S_t\Phi\left(\frac{\mu-t}{S_t}\right)}$	$\mu = \bar{t}$ – mathematical expectation (mean value); S_t – mean quadratic deviation
Logarithmically normal distribution	$Q(\ln t) = P\left(z_p > \frac{\ln t-\mu}{S_{\ln t}}\right)$	$h(t) = \frac{\varphi\left(\frac{\ln t-\mu}{S_{\ln t}}\right)}{\ln S_{\ln t}\Phi\left(\frac{\mu-\ln t}{S_{\ln t}}\right)}$	$\mu = \overline{\ln t}$ – mean value; $S_{\ln t}$ – mean square deviation
Exponential distribution	$Q(t) = \exp(-\lambda t)$	$h(t) = \lambda$	λ – intensity of failures
Distribution of Waybull (three-parameter)	$Q(t) = \exp\left[-\left(\frac{t-t_{\min}}{t_W}\right)^{\eta_t}\right]$	$h(t) = \begin{cases} \frac{\eta_t}{t_W}\left(\frac{t-t_{\min}}{t_W}\right)^{\eta_t-1} \\ 0, \text{ in the remaining cases} \end{cases}$	t_W – scale parameter ; η_t – shape parameter ; $t_{\min}$ – shear parameter

The abscissa scale in Fig. 1.85 can be substituted for the number of loading cycles n at stress σ = const. Then, for example, residual deformation (ε_{res}) or the length of the fatigue crack (l) can be plotted along the ordinate axis; ε_{lim} or the critical crack length l_c will correspond to the ultimate state – fatigue fracture. A change in the size of the rolling path (its width or depth), for example, is plotted along the ordinate axis in relation to contact fatigue, while the critical density of the pits of spalling can correspond to the ultimate state. In any of these cases (and other similar cases) the procedure of estimating main reliability indicators remains the same. Yet the functions of operating time distribution before failure $\varphi(t)$ can be different for different objects operating under specific conditions. For example, the distribution of fatigue durability of shafts follows the logarithmically normal law, while the durability of rolling bearings is well described by the law of

Waybull. Naturally, specific (engineering) formulas for calculating main reliability indicators are different and they depend on what law of distribution is used. Table 1.7 lists formulas for calculating the probability of failure-free operation and intensity of failures of objects when using four (commonly used in engineering) laws of distribution $\varphi(t)$, viz. normal, logarithmically normal, exponential and Waybull.

1.5.2 The load-strength model

The *load-strength model* will be exemplified with the assessment of the probability of fatigue fracture (the probability of failure) in case of the linear stressed state.

Assume the *characteristic of loading – cyclic stress* σ is a random value (cf. Fig. 1.34) and follows *normal distribution* with parameters $\overline{\sigma}$ (the *mean value*) and S_σ (its *mean square deviation).* Assume also the *characteristic of strength* – the endurance limit is a random value and also follows normal distribution with parameters $\overline{\sigma}_{-1}$ (the mean value) and $S_{\sigma_{-1}}$ (the mean square deviation). Depict graphically the density of distribution of both values $\varphi(\sigma)$ and $\varphi(\sigma_{-1})$ in a single coordinate system; the result is the model *load – strength* (Fig. 1.87). If the functions $\varphi(\sigma)$ and $\varphi(\sigma_{-1})$ intersect as it is shown in Fig. 1.87, then the *condition of fatigue fracture* exists

$$\sigma > \sigma_{-1}\,, \tag{1.121}$$

since cyclic stresses may exceed the endurance limit; the *probability of fracture* is rated by the area under both functions of distribution (cross-hatched in Fig. 1.87).

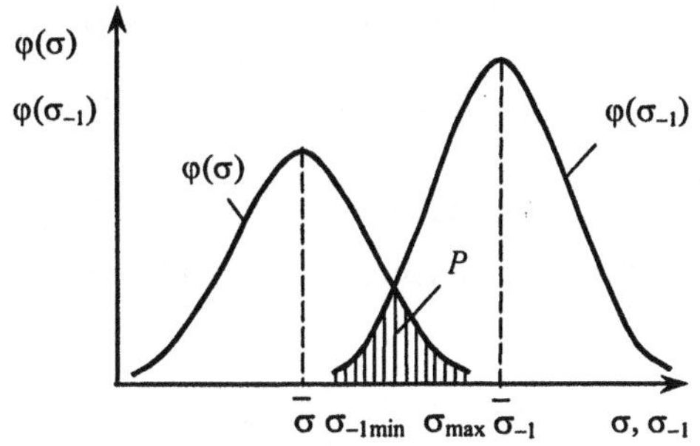

Fig. 1.87. *Load –strength* model

The procedure of calculation of this area is simple if a new random value is taken into consideration, viz.

$$\eta_\sigma = \sigma_{-1} - \sigma, \tag{1.122}$$

that also follows normal distribution with the mean value

$$\bar{\eta}_\sigma = \bar{\sigma}_{-1} - \bar{\sigma} \tag{1.123}$$

and the mean square deviation

$$S_{\eta_\sigma} = \sqrt{S_{\sigma_{-1}}^2 + S_\sigma^2}\,. \tag{1.124}$$

Figure 1.88 shows the integral function of distribution $P(\eta_\sigma)$ of the random value η_σ. Since the inequality equivalent to (1.121) is an apparent condition of fatigue fracture,

$$\eta_\sigma = \sigma_{-1} - \sigma \le 0, \tag{1.122a}$$

then the probability of fatigue fracture in accordance with (1.118) is

$$P = P(\eta_\sigma \le 0) = \int_{-\infty}^{0} \varphi(\eta_\sigma) d\eta_\sigma = \frac{1}{\sqrt{2\pi} S_{\eta_\sigma}} \int_{-\infty}^{0} \exp\left[-\frac{1}{2}\left(\frac{\eta_\sigma - \bar{\eta}_\sigma}{S_{\eta_\sigma}} \right)^2 \right] d\eta_\sigma \,. \tag{1.125}$$

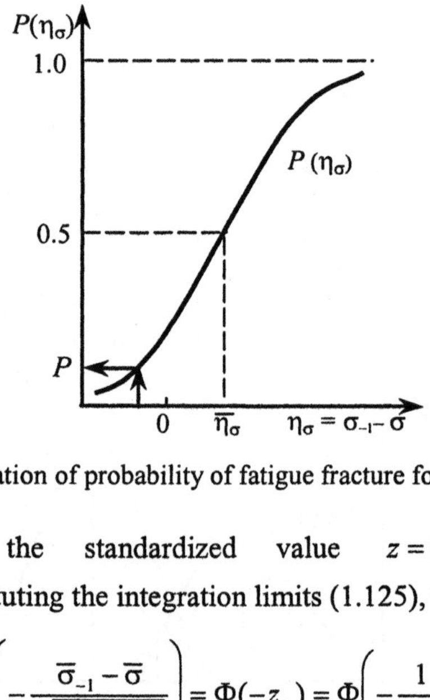

Fig. 1.88. Determination of probability of fatigue fracture for linear stressed state

By introducing the standardized value $z = (\eta_\sigma - \bar{\eta}_\sigma)/S_{\eta_\sigma}$ and by correspondingly substituting the integration limits (1.125), we obtain

$$P = \Phi\left(-\frac{\bar{\sigma}_{-1} - \bar{\sigma}}{\sqrt{S_{\sigma_{-1}}^2 + S_\sigma^2}} \right) = \Phi(-z_p) = \Phi\left(-\frac{1}{v_{\eta_\sigma}} \right), \tag{1.126}$$

where Φ – standardized function of normal distribution; z_p – the quantile of normal distribution corresponding to probability P,

$$z_p = -\frac{\bar{\sigma}_{-1} - \bar{\sigma}}{\sqrt{S_{\sigma_{-1}}^2 + S_\sigma^2}}, \tag{1.127}$$

and v_{η_σ} – the *coefficient of variation* of the random value η_σ; taking into account (1.123) and (1.124), we obtain

$$v_{\eta_\sigma} = \frac{S_{\eta_\sigma}}{\bar{\eta}_\sigma} = \frac{\sqrt{S_{\sigma_{-1}}^2 + S_\sigma^2}}{\bar{\sigma}_{-1} - \bar{\sigma}}. \tag{1.128}$$

Formula (1.126) is also recorded with the *mean coefficient of the strength margin*

$$\bar{n}_p = \frac{\bar{\sigma}_{-1}}{\bar{\sigma}} \tag{1.129}$$

in the following manner:

$$P = \Phi\left(-\frac{1-\bar{n}_p}{\sqrt{v_{\sigma_{-1}}\bar{n}_p^2 + v_\sigma}}\right), \tag{1.130}$$

where $v_{\sigma_{-1}}$ and v_σ – the coefficients of variation of the characteristics of strength and stress loading, respectively,

$$v_{\sigma_{-1}} = S_{\sigma_{-1}} / \bar{\sigma}_{-1}, \; v_\sigma = S_\sigma / \bar{\sigma}. \tag{1.131}$$

The mean coefficient of safety margin (1.129) relates to the probability of fatigue fracture P through quantile (1.127):

$$\bar{n}_p = \frac{1 + \sqrt{1-(1-z_p^2 v_{\sigma_{-1}}^2)(1-z_p^2 v_\sigma^2)}}{1 - z_p^2 v_{\sigma_{-1}}^2}. \tag{1.132}$$

Thus, *probabilistic calculation of strength* (under time-variable stresses) can be performed using the traditional idea about safety factor (cf. Sect. 1.2.2).

The above model is also applicable to the assessment of the probability of reaching the ultimate state (the probability of failure) based on any other criteria, for example, static strength, wear resistance, etc. Other (main) reliability indicators are easy to calculate using the formulas in Table 1.6.

While the characteristics of loading and strength obey different laws of distribution, similarly corresponding calculation formulas are obtained for the probability of fracture (failure).

Application of the *load – strength* model to a combined stressed state yields the following *scheme of assessing the probability* P_c of fatigue fracture [43]:

$$\left.\begin{array}{l} \left.\begin{array}{l}\varphi(\sigma_1)\\ \varphi(\sigma_{-1})\end{array}\right\rangle P_1 \\ \left.\begin{array}{l}\varphi(\sigma_2)\\ \varphi(\sigma_{-1})\end{array}\right\rangle P_2 \\ \left.\begin{array}{l}\varphi(\sigma_3)\\ \varphi(\sigma_{-1})\end{array}\right\rangle P_3 \end{array}\right] \rightarrow F(P_1, P_2, P_3) = P_c. \tag{1.133}$$

According to this scheme, each independent principal random stress σ_i, $i = 1, 2, 3$ causes damage of an object with fatigue cracks with the probabilities P_i that are calculated using the formulas of type (1.126). The assessment of the P_c of the probability fracture (failure) of the object operating in the conditions of the combined stressed state is obtained assuming a suitable function $F(P_1, P_2, P_3)$ of combination of particular probabilities P_i. A satisfactory compliance with experience is provided by the function

$$F(P_1, P_2, P_3) = (P_1 + P_2 + P_3 - P_1P_2 - P_2P_3 - P_3P_1 + P_1P_2P_3)a_c, \tag{1.134}$$

that we represent as

$$P_c = \left[1 - \prod_{i=1}^{3}(1 - P_i)\right]a_c\,. \tag{1.134a}$$

These formulas correspond to the known in the theory of probability scheme of concurrent and independent events, their particular probabilities are reduced to a total assessment. The parameter a_c, determined experimentally, is introduced in order to allow for the statistical interdependence of concurrent events of fatigue damage on the effect of each main stress.

Note that formula (1.134) contains *the invariants of the probability of fatigue damage* [14]

$$\left.\begin{aligned} J_{P1} &= P_1 + P_2 + P_3\,; \\ J_{P2} &= -(P_1P_2 + P_2P_3 + P_3P_1)\,; \\ J_{P3} &= P_1P_2P_3\,, \end{aligned}\right\} \tag{1.135}$$

that correspond to the first, second and third *invariants of the stressed state* (1.8a).

It is exactly this circumstance that dictates practical applicability of formula (1.134a) since the invariants I_1, I_2, I_3 are known to describe exhaustively the triaxial stressed state of any object.

Approach (1.133) and formula (1.134a) are also used in case of any biaxial stressed state assuming one of particular probabilities of damage in them is equal to zero (for example, $P_2 = 0$ or $P_3 = 0$).

A comparative analysis of the danger of various stressed states from the point of view of changes in the probability of fracture of an object is disclosed in accordance with (1.134) in Fig. 1.89. It has been performed for the limiting case when three main stresses are assumed equal ($\sigma_1 = \sigma_2 = \sigma_3$), so that the damage probabilities they dictate are also equal ($P_1 = P_2 = P_3$). The increment ΔP of the probability of fracture of the object is plotted along the ordinate axis for the case of biaxial (P_{12}) or triaxial (P_{123}) stressed state versus the linear stressed state (P_1), or the increment of the damage probability in case of triaxial stressed state versus the biaxial stressed state (ΔP_{123-12}). Relations between ΔP and P_1 or $P_1 = P_2$ have one maximum: it is always possible to indicate the only combined stressed state that is most dangerous compared with the other stressed state type. The biaxial stressed state for which $P_1 = P_2 = 0.5$ is the most dangerous; $\Delta P_{12-1} = 0.25$ in this case. Compared with the linear combined stressed state the triaxial stressed state for which $P_1 = P_2 = P_3 = 0.40$ is the most dangerous; $\Delta P_{123\text{-}1} = 0.384$ in this case. Any other combinations of equal principal stresses $\sigma_1 > \sigma_2 > \sigma_3$ lead to a smaller value ΔP. Compared with the biaxial stressed state the triaxial stressed state for which $P_1 = P_2 = P_3 = 0.35$ is the most dangerous; $\Delta P_{123\text{-}12} = 0.15$ in this case. In any other case the danger of the triaxial stressed state reduces compared with the biaxial stressed state.

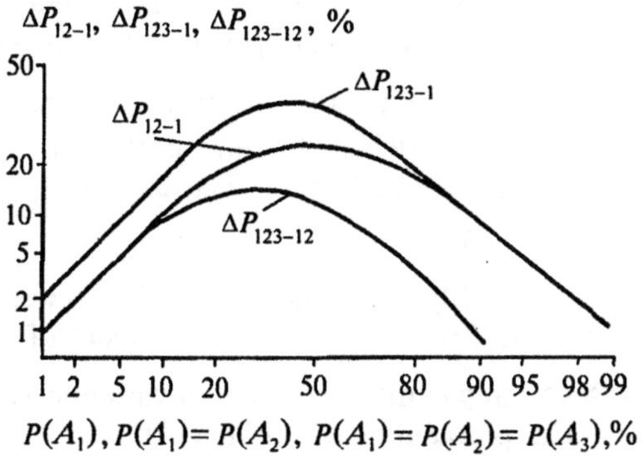

Fig. 1.89. Comparison of danger of fatigue fracture in different stressed states

So, *introduction of the second principal stress significantly increases the probability of fatigue fracture of an object than subsequent introduction of the third principal stress.* Also, it is exactly the first principal stress that presents the maximum danger. Introduction of any second principal stress is less dangerous: fracture probability can increase not more than 25%. The maximum contribution into the probability of fracture of an object is still less, 15%, when the third principal stress is introduced.

1.5.3 Calculations of reliability

Calculations of reliability using the criteria of resistance to fatigue and/or wear resistance are performed by comparing the calculated failure probability or the probability of failure-free operation with a corresponding standard value determined as a result of a technical and economic study.

Let us show an example of the procedure of calculation of reliability of a structural element using the criterion of fatigue fracture. The condition of reliability is

$$\begin{array}{ccccc} & & \bar{\sigma}_{-1},\ S_{\sigma_{-1}} & & \\ & & \searrow\ \swarrow & & \\ [P] \ge P & \longleftarrow & v_{\eta_\sigma} & \longrightarrow & \bar{n}_p \ge [n]. \\ & & \swarrow\ \searrow & & \\ & & \bar{\sigma},\ S_{\bar{\sigma}} & & \end{array} \qquad (1.136)$$

Calculation is performed with a simultaneous assessment of both failure probability P (for example, using formulas (1.126) or (1.130)) and the corresponding safety factor $\overline{n}_p$ (for example, using formula (1.132).

Condition (1.136) enables to solve the following problems:

(1) to establish what function of distribution of endurance limits (parameters σ_{-1}, $S_{\sigma_{-1}}$) a structural element should possess under given loading (the parameters of the distribution function $\overline{\sigma}$, S_σ) to achieve the design fracture probability $P \leq [P]$ (the safety factor is $\overline{n}_p \geq [n]$, where $[n]$ and $[P]$ – normative values);

(2) to determine what load (the parameters of the distribution function $\overline{\sigma}$, S_σ) can be tolerated to achieve the design fracture probability $P \leq [P]$ (the safety factor is $\overline{n}_p \geq [n]$) at a given function of distribution of endurance limits of structural element ($\overline{\sigma}_{-1}$, $S_{\sigma_{-1}}$);

(3) to calculate the probability of fatigue fracture P of the structural element for which resistance to fatigue (for example, the coefficient of variation $V_{\sigma_{-1}}$) and loading parameters (for example, the coefficient of variation V_σ) are known;

(4) to establish what probability of fatigue fracture P corresponds to the assigned safety factor $\overline{n}_p$.

The technical and economic assessment of the normative fracture probability $[P]$ can be performed in the following manner.

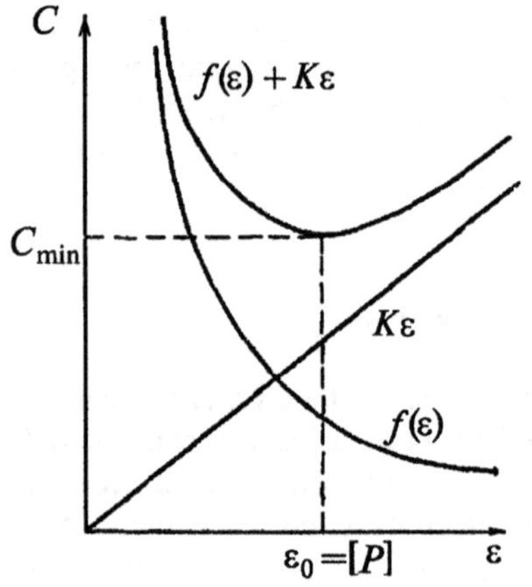

Fig. 1.90. Validity of normative fracture probability for costly pieces

The fracture probability P is some small numeral ε. If it is required to reduce the fracture probability, the cost of production C_1 necessary to satisfy this requirement grows following the dependence $C_1 = f(\varepsilon)$ (Fig. 1.90). On the other hand, a higher fracture probability leads to proportional rise of operation cost $C_2 = K\varepsilon$ (K – the coefficient of proportionality) needed to replace or recondition damaged parts. The total cost amounts then to $C = C_2 + C_1 = K\varepsilon + f(\varepsilon)$. The value

$[P] = \varepsilon_0$ should be selected from the condition of the minimum of this sum (cf. Fig. 1.90).

When operation damaged parts may lead to severe accidents, not just to economic losses, the value $[P]$ should be reduced accordingly.

The above procedure of calculation of reliability is applicable to friction couples also if parameters $\overline{\sigma}_{-1}$, $S_{\sigma_{-1}}$ and $\overline{\sigma}$, S_σ in formulas (1.126)–(1.132) are substituted for relevant parameters $\overline{\tau}_f$, S_{τ_f} and $\overline{\tau}_W$, S_{τ_W} in friction ($\overline{\tau}_f, \overline{\tau}_W$ – mean values, S_{τ_f}, S_{τ_W} – their mean square deviations). The latter parameters can be recorded in terms of contact pressure too ($\overline{p}_f, S_{p_f}$ and $\overline{p}_a$, S_{p_a}). Thus, similarly with formula (1.126), the failure probability of a friction couple is

$$P = \Phi\left(-\frac{\overline{\tau}_f - \overline{\tau}_W}{\sqrt{S_{\tau_f}^2 + S_{\tau_W}^2}} \right), \tag{1.137}$$

where $\overline{\tau}_f$ is determined in accordance with Fig. 1.57 (however using the results of statistical tests).

For calculation of friction couples to wear a formula is derived to estimate the probability of failure-free operation:

$$Q = \Phi\left(-\frac{\Delta_{max} - \overline{\Delta}_0 - v_i t}{\sqrt{S_{\Delta_0}^2 + S_{v_i}^2 t^2}} \right), \tag{1.138}$$

where Δ_{max} – the maximum tolerable clearance between the shaft and the sliding bearing insert; $\overline{\Delta}_0$ – the mean clearance before operating; v_i – the wear rate of the couple; t – the time of operation.

1.5.4 Reliability and safety; risk

The terms *reliability* and *safety*, on the one hand, and *danger* and *risk*, on the other hand, are generally speaking hard to discriminate, their connotations frequently overlapping. Hence, for example, the notions *safety analysis* and *reliability analysis* are used as equivalents.

Safety implies reliability in respect to human life and health, the condition of the environment [44]. Whence it follows that main quantitative indicators of safety in the mathematical respect are similar to the relevant indicators in the theory of reliability. Thus, the *safety function* $S(t)$ is the probability of a random event implying that emergency situation never occurs within the interval of time $[0, t]$, i.e. no "failure" happens. A complement of the safety function to unity is called *the risk function*

$$H(t) = 1 - S(t), \tag{1.139}$$

that is equivalent to the idea of probability of a failure. Similarly to the intensity of break-downs the risk intensity (specific risk) is introduced:

$$h(t) = H'(t)/[1 - H(t)] = -S'(t)/S(t) \approx H'(t) = S(t), \tag{1.140}$$

since

$$1 - H(t) = S(t) \to 1.$$

Another idea about risk is that safety analysis presupposes the possibility of a system's break-down, therefore consequences of accidents are determined in the sense of damage of equipment and in the sense of injuries of people servicing it. Risk r_s in this case is calculated as a product of frequency v_r by the mean $\overline{S}$ of break-downs of a given system [45]:

$$r_s = v_r \overline{S}. \tag{1.141}$$

It follows from formula (1.141) that risk due to a large number of light emergencies can equal to risk due to a single emergency that entails grave consequences. A strong disadvantage of formula (1.141) is the following: similar values r_s do not mean identical consequences.

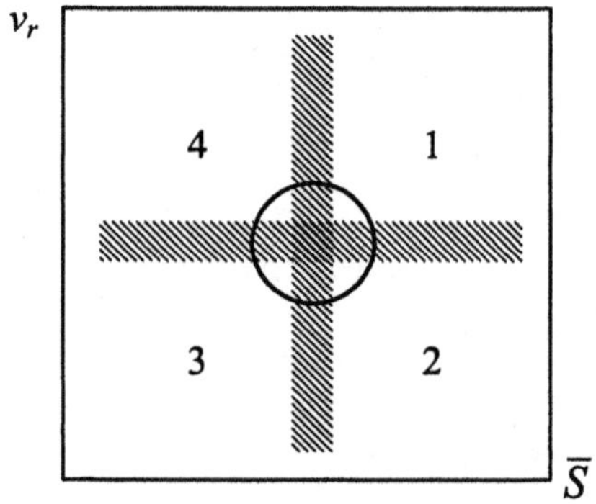

Fig. 1.91. Characteristic regions of risk field

The field $v_r - \overline{S}$ is arbitrarily divided into four regions 1, 2, 3, 4 (Fig. 1.91) [46]. The risk is high in region 1 and the aim is to reduce v_r and $\overline{S}$. Region 3 in the left bottom corner is uninteresting due to insignificant risk. Regions 2 and 4 require a precise assessment of risk in order to identify the necessity and type of a decision to be made to reduce risk. Moreover, risk analysis within the conventional circle is uncertain. So, field $v_r - \overline{S}$ contains several uncertainties that naturally complicate both risk analysis and identification of means of combating it. On the other hand, risk function (1.141) is of practical significance and it is illustrated in Fig. 1.92 [45] as a dependence of v_r on $\overline{S}$ for car accidents per year

(1 – all transport vehicles, 2 – defective transport vehicles). If r_s is measured by the number of accidents per year and $\overline{S}$ is the mean number of fatal outcomes per accident, then risk r_s has the dimensionality of mortality per year. According to the data in Fig. 1.92, the following conclusions can be made:

– most intensive accidents happen with lesser frequency than less intensive ones;

– quantitatively the distribution of the intensity of car accidents does not depend whether the accident is a consequence of a car fault or a driver's error;

– the main share of risk is due to the operation of defective transport vehicles.

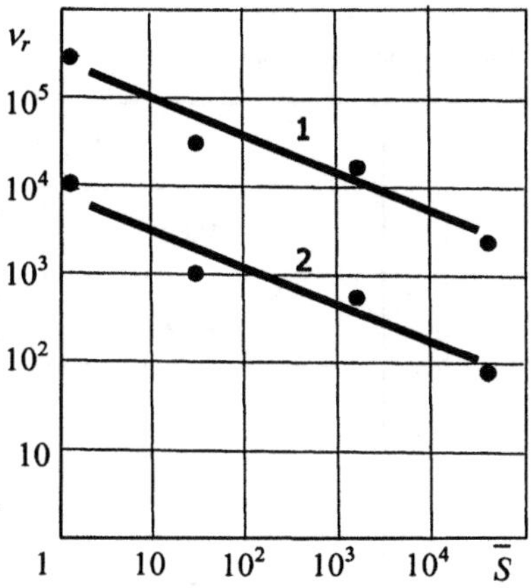

Fig. 1.92. Relation between frequency and intensity of car accidents (per year)

It is highly essential to be able to obtain detailed information, yet it should be noted that more or less accurate assessment of the consequences of accidents (i.e. risk in the sense of (1.141)) is possible only after such accidents have happened.

Thus, risk r_s, or danger due to operation of this object depends on the frequency v_r and intensity $\overline{S}$ with which safety violation (accidents) upset safety and which occur as a result of use of the object. Safety violations that lead to negligibly small material losses and do not cause injuries of personnel are called ($\overline{S} \to 0$) incidents. An emergency causing severe (human and/or material) consequences ($\overline{S} >> 0$) is called an accident.

The value r_s is not a probability, therefore risk analysis based on (1.141) is also possible for rare events.

1.6 Strength of materials in structures

The characteristics of volume and/or surface strength determined experimentally with laboratory specimens, as a rule, turn out strongly *overrated*: in fact, resistance of relevant structural elements or mechanical systems to fracture may lower by

30...100% and more. A general explanation of this fact is that *the strength of materials in structures* changes strongly under the effect of a complex of design, technological, metallurgical and operation factors.

Let us analyze specific examples of this effect.

The constructive embodiment of a mechanical system can change its resistance to fatigue within a broad range. A shaft 30 mm in diameter made from carbon steel ($\sigma_b = 600$ MPa) has the fatigue limit $\sigma_{-1} = 260$ MPa. A mechanical coupling of the shaft with a bush may lead to more than a double loss of fatigue resistance (Table 1.8), with a spline joint being least strong and a molded coupling most efficient, while a key joint being in between.

Table 1.8. Effect of design on fatigue resistance

Article	Fatigue limit	
	MPa	%
Shaft	260	100
shaft / bush (steel):		
molded joint	156	~60
splined joint	130	~50
keyed joint	117	~45

Table 1.9. The effect of material on fatigue resistance

Material and its ultimate strength	Article and its fatigue strength		$K_f = \sigma_{-1}/\sigma_{-1P}$
	Shaft, σ_{-1}, MPa	Shaft / bush, σ_{-1P}, MPa	
High-strength steel , $\sigma_b = 1100...750$ MPa	460	120	3.8
Alloyed steel, $\sigma_b = 700...1200$ MPa	300	110	2.7
Carbon steel, $\sigma_b = 550...570$ MPa	200	100	2.0

The metallurgical factor is as essential. When switching over from carbon to high-strength steel, the shaft's fatigue limit more than doubles, yet the fatigue resistance of its coupling with the bush yields a just 20% gain in the fatigue limit (Table 1.9). Moreover, unlike with the separate shaft the fatigue strength of the shaft/bush system reduces 2 times if the shaft is made from carbon steel ($\sigma_b \approx 560$ MPa) and 3.8 times if it is made from high-strength steel ($\sigma_b \approx 925$ MPa).

One of the *technological factors*, such as surface roughness, enables to change the fatigue limit of shafts from alloyed steel 40X approximately 15%.

Conditions of operation frequently influence the resistance of mechanical systems to fatigue as strongly as other factors. For example, when *a shaft/bush joint* operates under different contact pressures, it may lead to the lowering of fatigue strength by 15–20% in case the shaft and the bush are made from carbon or alloyed steel (Table 1.10).

Table 1.10. Contact pressure effect on resistance to fatigue

Material of shaft and bush, ultimate strength	Contact pressure, MPa	σ_{-1P}	
		MPa	%
Carbon steel, $\sigma_b = 520$ MPa	0	224	100
	0.6	155	69
	113	112	50
Alloyed steel, $\sigma_b = 520$ MPa	0	550	100
	10	310	56
	100	260	47

Wear resistance of a friction couple may change by an order of magnitude and more depending on the type of friction and design features.

The analysis of these and a variety of other similar examples leads to a general conclusion: one general principal feature appears in all such cases that in addition to cyclic loading *concurrently* the friction process occurs in all its manifestations (fretting, sliding, rolling, etc.). It is namely the combination of two damaging processes of fatigue and wear that determines in many respects the nature and conditions of *additional* loss of strength by the material in devices. It is truly so because special experimental studies have demonstrated that, for example, the processes of sliding friction lead to the reduction of resistance of steel shafts to fatigue by 20...40% and more (depending on the conditions of operation). It is explained by the negative effect of three factors: 1) reduction of the shaft's cross section as a consequence of wear; 2) appearance of numerous microconcentrators of stresses, such as marks, pitting, etc., on the friction surface; 3) irregular wear of bearings of multiple-bearing shafts leading to excitation of vibration in the system.

Further progress of studies in this direction has revealed that a concurrent combination of the damaging processes of friction, wear and mechanical fatigue in joints is capable to lead to an unambiguous result: their resistance to fatigue and /or wear resistance can both drop and grow substantially. New ways were thus found how to improve reliable operation of most essential and highly loaded mechanical systems of machinery and equipment for which *complex wear-fatigue damage* is typical; they acquired a special name: *active systems* (GOST 30638–99). The latter became the object of study of tribo-fatigue, the science of wear-fatigue damage and fracture.

Bibliography

1. Gordon G, Why do not we fall through the floor?, Mir, Moscow, 1971, 272 p. (in Russian).
2. Bartenev G M, Strength and mechanism of fracture of polymers, Moscow, 1984, 280 p. (in Russian).
3. Sosnovskiy L A, Mechanics of fatigue fracture. Handbook of terms, two volumes, Gomel, 1994, 668 p. (in Russian).
4. Sosnovskiy L A, Marchenko V G, Physics of metals and metal science, 1976, V. 42, No. 2, 377–383. (in Russian).
5. Sosnovskiy L A, Makhutov N A, Zavodskaya Laboratoriya, Moscow, 1995, No. 5, 33–35. (in Russian).
6. Smith G C, Proc. Roy. Soc., 242A, 1957, 189–197.
7. Hempel M, Int. Conf. on Fatigue of Metals, Inst. Mech. Engineers, London, 1956.
8. Forsyth P J E, Proc. Roy. Soc., 242A, 1957, 198–202.
9. Sosnovskiy L A, Electric stations, 1968, No. 7, 14–18. (in Russian).
10. Oding I A, Ivanova V S, Strength of metals under alternating loads, Moscow, 1963, 3–13. (in Russian).
11. Zhurkov S N, Kuksenko V S, Petrov V A, Future of science. Int. Year-book, Moscow, 1983, Issue 16, 100–111. (in Russian).
12. Troshchenko V T, Fatigue and non-elasticity of metals. Kiev, 1971, 268 p. (in Russian).
13. Kudryavtsev I V, Sosnovskiy L A, Strojirenstvi. 1967, No. 11, 826–829. (in Hungarian).
14. Sosnovskiy L A, Statistic mechanics of fatigue fracture. Minsk, 1987, 286 p. (in Russian).
15. Paris P, Erdogan F, J. of Basic Eng., Trans. ASME, Dec. 1963, 528–534.
16. Sosnovskiy L A, Problems of Strength. Kiev, 1986, No. 11, 16–20. (in Russian).
17. Sosnovskiy L A, Proc. of Academy of Sciences of Belarus, Series of physico-technical sciences, 1990, No. 2, 3–7. (in Russian).

18. Sosnovskiy L A, Bogdanovich A V, Problems of Strength, Kiev, 1995, No. 5–6, 39–61. (in Russian).
19. Sosnovskiy L A, Bogdanovich A V, Theory of wear-fatigue damage accumulation, Gomel, 2000, 60 p. (in Russian).
20. Pfeiffer F, Verwendung von Werk, 1976, Nos. 1–2, 19–30. (in German).
21. Troshchenko V T, Strength of materials and parts of constructions in extreme conditions, Kiev, 1980, 219–326. (in Russian).
22. Kragelsky I V, Dobychin M N, Kombalov V S, Basic analysis of friction and wear, Moscow, 1977, 526 p. (in Russian).
23. Striblek R, VDI-Zeitschrift, 1902, Bd. 46. (in German).
24. Bartel A, VDI-Berichte, 1957, Bd. 20, 79–90. (in German).
25. Sysoev P V, Bogdanovich P N, Lizarev A D, Wear and deformation of polymers in friction, Minsk, 1985, 240 p. (in Russian).
26. Buckley D N, Surface effects in adhesion, friction, wear and lubrication, Elsevier Sc. Pub. Co., 1981.
27. Sosnovskiy L A, Proc. of IV Int. Symp. on Tribo-fatigue, Ternopil, 2002, V. 1, 9–23. (in Russian).
28. Surface strength of materials in friction, Ed. by B Kostetsky, Kiev, 1976, 295 p. (in Russian).
29. Drozdov Yu N, Pavlov V G, Puchkov V N, Friction and wear in extreme conditions, Moscow, 1986, 223 p. (in Russian).
30. Fundamentals of tribology, Ed. by Chichinadze, Moscow, 2001, 663 p. (in Russian).
31. Kostetsky B I, Reliability and durability of machines and structures. Kiev, 1982, No. 1, 72–82. (in Russian).
32. Fleischer G, Schmierungstechnik, 1973, Bd. 4, 9. (in German).
33. Saverin M M, Contact strength of materials in the conditions of concurrent action of normal and tangential loads, Moscow, 1946, 148 p. (in Russian).
34. Sosnovskiy L A, Marchenko A V, Problems of Strength. Kiev, 2001, No. 1, 48–61. (in Russian).
35. Mazin A, Strojirenstvi. 1985, V. 35, No. 8, 447–451. (in Hungarian).
36. Keer L M, Briant M D, Horitos G K, J. Lubr. Tech., Trans. ASME, 1982, V. 104, No. 3, 347–351.
37. Keer L M, Briant M D, J. Lubr. Tech., Trans. ASME, 1983, V. 105, No. 2, 198–205.
38. Mason W P, White S D, Bell Syst. Tech. J., 1952, V. 31, 469 p. (in Russian).
39. Sackmann V W, Rightmire B G, Nasa Tech. Note, 1948, 1492.
40. Uhlig H H, Appl. Mech., 1954, V. 21, 401.
41. Pittroff H, Trans. ASME, 1965, V. 87, 713.
42. Sosnovskiy L A, Vysotsky M S, Makhutov N A, Troshchenko V T, Contemporary methods of machines design. Minsk, 2002, V. 1, Issue 1, 12–18. (in Russian).

43. Troshchenko V T, Sosnovskiy L A, Problems of Strength, Kiev, 1979, No. 7, 3–11. (in Russian).
44. Bolotin V V, Prediction of life of machines and constructions. Moscow, 1984, 312 p. (in Russian).
45. Henly E D, Kumamoto Ch, Reliability of technical systems and analysis of risk, Moscow, 1984, 528 p.
46. Tetelman A, Besuner P, Mechanics. New facts in science abroad, Moscow, 1980, 7–30. (in Russian).

2 ACTIVE SYSTEMS. WEAR-FATIGUE DAMAGE

Problems are more variable than solutions.
Solutions may become obsolete, while problems remain.

Niels Bohr

2.1 Active systems and their damage

The active system *is any mechanical system that reacts to and transmits the alternating workload with simultaneous occurrence of the friction process with its any manifestation: in sliding, rolling, slippage, impact, etc.* Some typical examples of the system are the following: the *shaft / hub* (of the gear wheel, the flywheel, etc.) with the stationary or running fit (molded, keyed, splined and other joints); the *crankpin / connecting rod with the sliding bearing* (bolted, riveted or threaded connections that transmit cyclic loading); the *wheel / rail* and many others. As a rule, they are the most essential and widely used units of modern machinery and equipment. Complex *wear-fatigue damage* (WFD) is typical for them with the main types being *mechano-sliding, mechano-rolling, mechano-erosion and fretting fatigue*. Table 2.1 gives the definitions of these types of WFD in respect to the typical examples of the active systems.

Figure 2.1 shows the calculation diagrams of three typical active systems *solid / solid* with three WFD types accompanying their operation.

Two different types of loads are observed in the active system of the *shaft / sliding bearing* type, viz. F_N – static *contact load* and M – the cyclic bending moment which represents the *off-contact* load (Fig. 2.1, *a*).

Contact interactions between the components of an active system can be integrally characterized, like in a corresponding friction pair, either by the maximum pressure in the contact site center $p_0 = \sigma_{z\,max}$ (see, for example, formulas (1.108)) or nominal mean pressure (1.72)

$$p_a = F_N / A_a, \tag{2.1}$$

or by *specific frictional force* in sliding (1.73)

$$\tau_{Ws} = f_S p_a = F_S / A_a = f_S F_N / A_a, \tag{2.2}$$

that is also called *frictional* stress.

Cyclic bending of the shaft can be characterized by normal stresses that reach the maximum on the surface

$$\sigma = M / W, \tag{2.3}$$

where W – the moment of resistance.

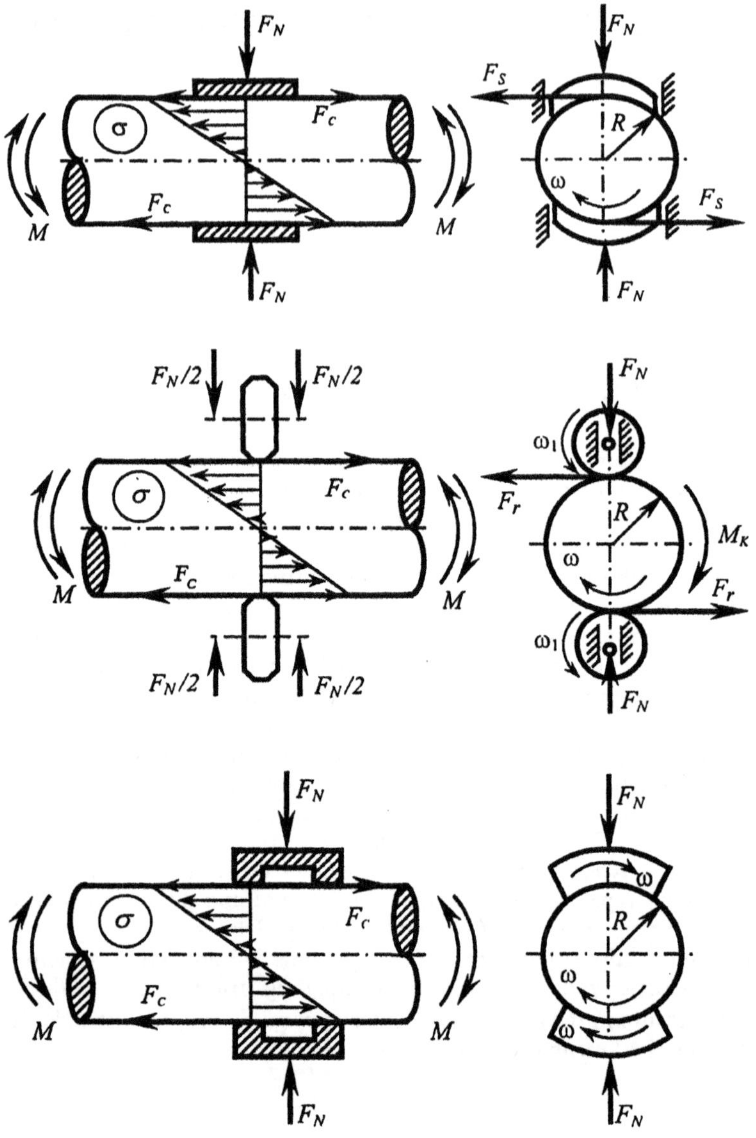

Fig. 2.1. Design schemes of typical active systems *solid / solid*

Thus, orthogonal stresses, axial (2.3) and circumferential (2.2), appear in the zone of contact between the bearing and the shaft. They cause diverse modifications of chemophysical and mechanical properties of materials (see Sects. 1.3.2, 1.3.3, 1.4.3). Frictional stresses (2.2) initiate surface damage of both components of the active system when they slide one in respect to the other. The damage appears as fatigue microcracks distributed over contacting surfaces. Evolution of multiple cracks leads to the separation of a large number of particles, i.e. to wear. Cyclic stresses (2.3) initiate both surface damage with scattered microcracks and volume fracture (disintegration into pieces of the shaft) due to the development of the main fatigue crack. Exactly due to this reason the process of WFD of the components of the active system of the *shaft / sliding bearing* type is called *mechano-sliding fatigue*.

Table 2.1. Main types of wear-fatigue damage

Typical active system	Complex damage and fracture	Definition
Crank-pin / connecting rod end with sliding bearing	Mechano-sliding fatigue	Wear-fatigue damage due to the effect of kinetic interaction between the phenomena of mechanical fatigue and sliding friction
Wheel / rail	Mechano-rolling fatigue	Wear-fatigue damage due to the effect of kinetic interaction between the phenomena of mechanical fatigue and rolling friction (rolling friction with slippage)
Spline shaft / bushing	Fretting fatigue	Wear-fatigue damage due to the effect of kinetic interaction between the phenomena of mechanical fatigue and fretting
(Screw) propeller shaft / sea water	Mechano-corrosion fatigue	Fatigue of the material under the simultaneous effect of alternating stresses and corrosive environment
Turbine blades / fluid or gas stream carrying solid particles	Mechano-erosion fatigue	Wear-fatigue damage due to the effect of kinetic interaction between the phenomena of mechanical fatigue and erosion

Two different types of loads are also observed in the active system of the *shaft / wheel* (or *roller*) type (Fig. 2.1, *b*), viz. contact F_N and off-contact M loads. Yet, in this case it is rolling friction, therefore the frictional stress (see Sects. 1.4.1 and 1.4.6) is

$$\tau_{Wr} = f_r p_0 = F_r / A_a = f_r F_N / A_a \,. \tag{2.4}$$

Hence, an orthogonal system of axial (2.3) and circumferential (2.4) stresses appears in the zone of contact between the shaft and the roller. It causes in the active system of the *shaft / wheel* (or *roller*) type the WFD process called *mechano-rolling fatigue*.

Again, two different types of loads are observed in the active system of the *shaft / bushing* type (Fig. 2.1, c), viz. contact F_N and off-contact loads ▯. Yet, slip friction occurs in this case, therefore the frictional stresses (see Sects. 1.4.1 and 1.4.7) are

$$\tau_{Wq} = f_q p_a = F_q / A_a = f_q F_N / A_a, \tag{2.5}$$

where F_q and f_q – the force and coefficient of friction in fretting.

Normal (2.3) and friction (2.5) stresses are in one and the same (axial) direction z in the zone of contact between the shaft and the bushing (if there is no torque). They initiate the wear-fatigue damage process usually called *fretting fatigue*.

Since damages due to friction (2.2) or (2.4) or (2.5) and normal (2.3) stresses appear and develop in one and the same zone of contact between the components of an active system, naturally these damages interact. Hence, on the one hand, the effect of friction and wear processes on changes in the characteristics of fatigue resistance of the shaft (the *direct effect*) is studied. On the other hand, the effect of cyclic stresses on changes in the characteristics of wear resistance of both components (the *back effect*) is studied.

Occurrence of the direct effect and / or back effect in active systems brings about the *limiting state* with a single or several concurrent *manifestations*:

- wear of the ultimate value;
- critical dimensions or critical concentration of fine surface fatigue cracks and / or pitting spots;
- residual deformation of inadmissible value;
- appearance of surface undulatory damage;
- a main crack of inadmissible length or area;
- fracture (disintegration into pieces) of components, etc.

In case of the direct effect the limiting state is reached by predominant development of the main fatigue crack (volume fracture), while the wear processes are concomitant. On the contrary, in case of the back effect the limiting state is based on the wear criteria (surface fracture), while the appearance of microcracks of mechanical fatigue is concomitant damage.

Shafts usually operate in a combined state of stress, for example, with bending and torsion. In these cases it is possible to use instead of cyclic stress (2.3) the equivalent stress σ_{eqviv} that is calculated from one of the theories of strength (see Sect. 1.2.2). The general procedure of analyzing WFD remains unchanged in this case. Note that if $\sigma_{eqviv} = \sigma_1$, where σ_1 is the first principal stress, the first theory of strength (maximum normal stresses) is assumed automatically. It is easy to demonstrate that the equivalent stress assigned by any theory of strength is directly proportional to the value of the first principal stress:

$$\sigma_{eqviv} = \sigma_1 f\,(\sigma_2/\sigma_1,\ \sigma_3/\sigma_1,\ \sigma_b / \sigma_b^{com}, \ldots). \tag{2.6}$$

The function $f(\cdot)$ may depend both on the relevant principal stresses (see (1.21a)) and on the properties of the material, for example, on the ratio between the strength limits in tension (σ_b) and compression (σ_b^{com}). For example, if the theory of strength of More is used, then

$$\sigma_{eqviv} = \sigma_1\left(1 - \frac{\sigma_3}{\sigma_1}\frac{\sigma_b}{\sigma_b^{com}}\right).$$

On the other hand, it can be demonstrated that frictional stress τ_W determined from formulas (2.2), (2.4) or (2.5) is also proportional to the equivalent stress, for example:

$$\sigma_{eqviv} \sim \tau_W = f p_0. \tag{2.7}$$

Hence, characteristics (2.2)–(2.5) of the state of stress of shafts in cases of cyclic loading and contact interaction in friction are sufficient if (2.6) and (2.7) are taken into account.

Thus, active systems (their main types are shown in Fig. 2.1) share the specific feature of demonstrating *interacting damages* that are caused by different types of loads (contact and off-contact), but they appear and evolve in one and the same region of deformable components, namely, around the contact site.

There is another type of active systems in which damage evolves in different regions under the effect of one and the same load. Gearings can serve as a typical example (Fig. 2.2, *a*), their model is shown in Fig. 2.2, ***b***. Basic conditions of operation of cylinder gear wheels are simulated: rolling friction appears in the pitch of engagement (a contact between two rollers) under the effect of contact load F_N; in the zone of passage of the tooth to the ring (the fillet passage with the radius R) cyclic bending appears (during cantilever bending) under the effect of the same load F_N. Though wear and fatigue damage is observed in different regions of the deformable component (tooth of the gear wheel or its model) considerably spaced one from the other, they turn out to be *interrelated damages* because, if, for example, load F_N increases, both damaging phenomena intensify due to concurrent and proportional growth of cyclic and contact stresses

a) *b)*

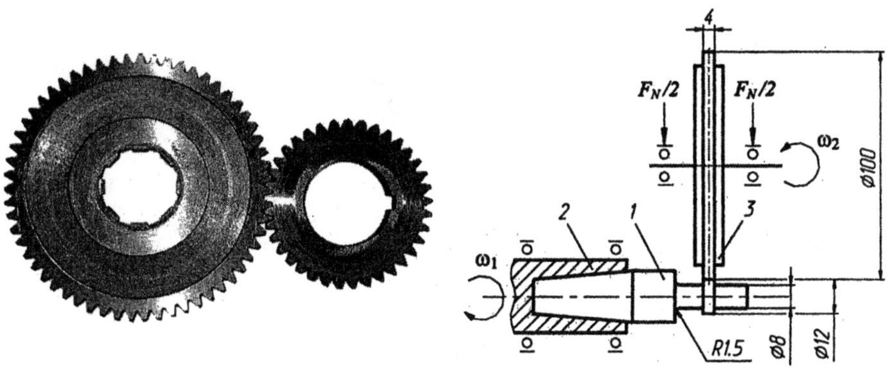

Fig. 2.2. Cylindrical gear wheels (*a*) and model of their engagement (*b*)

$$\sigma = F_N \frac{l}{W}, \quad \tau_W = F_N \frac{f_R}{A_a}.$$

It has been noted that Figs. 2.1 and 2.2 show typical, yet simplest diagrams of active systems. More intricate or *combined active systems* are typical in real machines. Figures 2.3 and 2.4 show two examples of such systems.

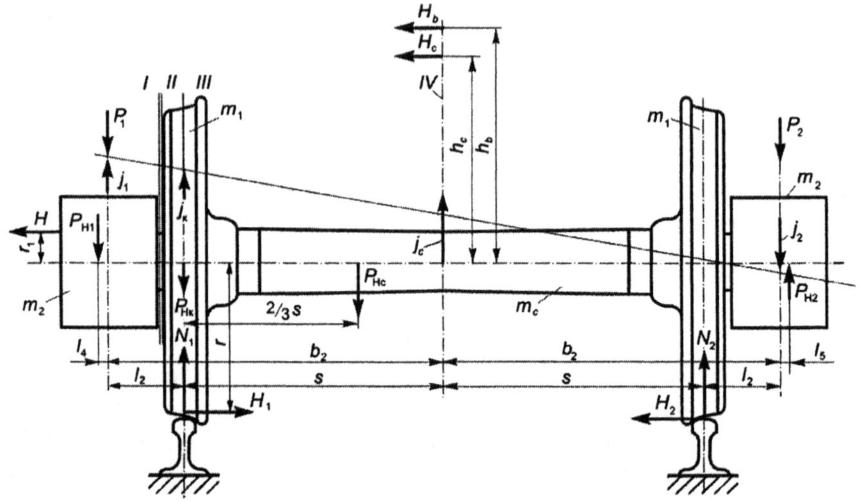

Fig. 2.3. Calculation scheme of car axle

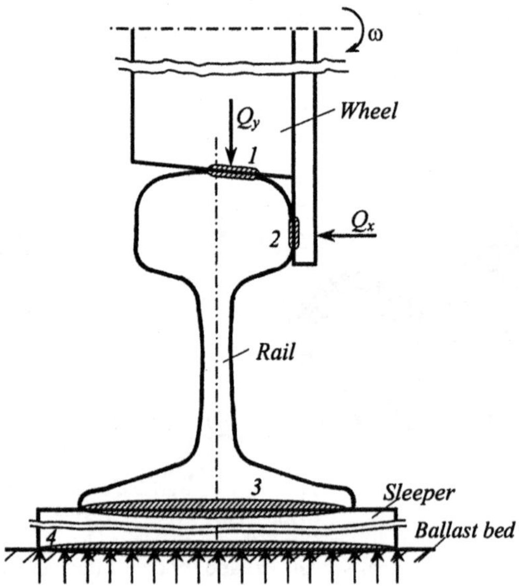

Fig. 2.4. Combined active system *wheel / rail / sleeper / railway bed*

Usually four dangerous cross sections (zones) are identified when designing the railway car axle (cf. Fig. 2.3) that are determined by one system of loads:

(*a*) cross section *I* near the internal rear roller bearing; there is an active system of the *shaft / bushing* type characterized by fretting fatigue;

(*b*) cross section *II* through the internal crankpin fillet (at a distance of one third of the length of passage to where the fillet begins); there is a peculiar active system of the *shaft / environment* type characterized by mechano-corrosion fatigue;

(*c*) cross section *III* in the plane of the circle of rolling with the system of the *wheel / rail* that is analyzed in more detail below;

(*d*) cross section *IV* in the axle center with the system of the *shaft / environment* type characterized by mechano-corrosion fatigue.

Now let us consider a combined active system made up by the upper structure of the railway and the wheel (cf. Fig. 2.4).

Assume in the first approximation that in motion there is an effective vertical load Q_y applied off-center to the axis of symmetry of the rail cross section and there is also a lateral force Q_x. Four risky zones can be indicated conventionally shown by hatching:

(*a*) zone *1* of contact between the wheel and the rail head (in the vicinity of the rolling path) where mechano-rolling fatigue occurs (bending and torsion of the rail + rolling friction);

(*b*) zone *2* of contact between the wheel flange and the lateral surface of the rail head where mechano-sliding fatigue occurs (bending and torsion of the rail + sliding friction);

(*c*) zone *3* of contact of the rail foot with the sleeper where fretting fatigue occurs (fretting due to alternating deformation of both the rail and the sleeper);

(*d*) zone *4* of contact between the sleeper and the railway bed where fretting fatigue occurs (fretting due to alternating bending of the sleeper on the elastic bed).

Moreover, the components of the active system are in contact with the environment in all these zones. The air environment (i.e. the *atmosphere*) can produce a damaging effect in zones *1*, *2* and *3*; the railway bed (the geosphere, in the general case) can produce a damaging effect in zone *4*. When it rains too long, water (the *hydrosphere,* in the general case) can produce an additional damaging effect. Hence, a gas, a fluid or soil can act as the environment in respect of the *solid body / environment* active system as extreme simplification. The environment in the general case is multicomponental (for example, *biphase*, like a mixture of soil and water). The phenomena of contact between the environment and the deformable solid body are various, with those of the electrochemical nature being most dangerous. Corrosion is the basic electrochemical processes damaging metallic materials. Combination of cyclic stresses with the corrosive environment causes WFD of metals called above the mechano-corrosion fatigue. In this case the active system can reach the limiting state both with a complex of mechanical signs listed above and due to the appearance of:

– an inadmissible thick layer of corrosive damage (in case of *general* or *even* corrosion);

– corrosive pittings and pits of an inadmissible depth (in case of *crevice* or *local* corrosion).

Active systems of the *solid / stream of particles* represent a separate class (Fig. 2.5). Again two types of loads exist: off-contact (M) and contact impact loads determined by the speed ϑ_i of individual liquid drops and solid particles (Fig. 2.5, *a*) or by the total speed ϑ of the stream of particles (Fig. 2.5, *b*) that can produce a continuous jet (a medium). A peculiar contact load is due to the effect of electrical charges or charged particles, for example, radiation flux (electrons, neutrons, etc.), powerful laser emission, etc.; the flux intensity usually characterizes such loads. Exposure of the surface of the solid to impacts by solid particles, liquid drops or to a multiphase environment is accompanied by *erosion*, i.e. by irreversible alterations of the mechano-physical properties of a material when microcracks appear, the mass is carried away from the surface in the form of separate particles, fragments or even whole layers. *Radiation erosion* occurs in case of exposure to a flux of charged particles. The *swelling of the material* occurs as a typical mechanical phenomenon, i.e. the volume grows and atomic bonding forces weaken at the same time. A peculiar type of erosive damage called *ablation* occurs when the surface of the solid body is exposed to an impacting jet of high-temperature gas or to a powerful laser emission: the material is ablated due to evaporation.

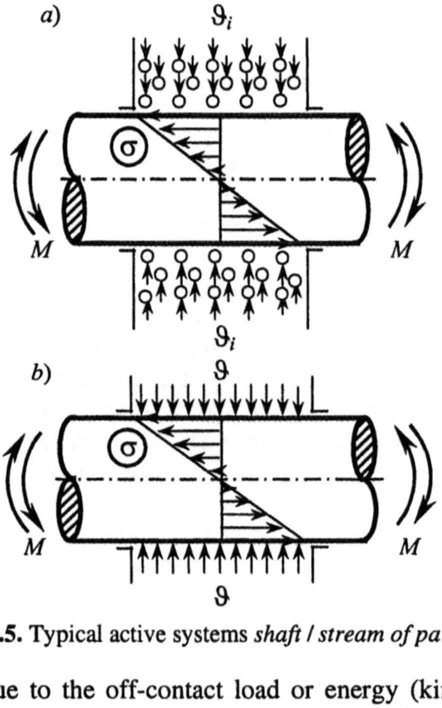

Fig. 2.5. Typical active systems *shaft / stream of particles*

Cyclic stresses due to the off-contact load or energy (kinetic) contact of the body with the stream of solid particles, the jet of liquid particles, gas or a fluid or with the multiphase environment or exposure to emission, initiate the WFD process called *mechano-erosion fatigue*.

Let us consider a simplified example of impact contact of the *solid body / stream of particles* active system. If the particle with mass m_ϑ at the moment of impingement has the speed $\vartheta(t)$, a dynamic force $F_\vartheta = ma(t)$ appears, where $a(t)$ – acceleration. Then dynamic pressure p_ϑ appears in the "point of impact" (Fig. 2.6), also, because the shaft rotates with the frequency ω, the *dynamic specific frictional force* τ_ϑ, appears called *dynamic frictional stress*. The contact pressure appearing when liquid drops modeled as spheres impinge the plane is approximately rated with the formula

$$p_v = \gamma_{liq}\vartheta_{liq}c_{liq}\left(1+\frac{\gamma_{liq}c_{liq}}{\gamma_s c_s}\right)^{-1}, \tag{2.8}$$

where γ_{liq} – the liquid density; ϑ_{liq} – the speed of a drop normal to the surface; c_{liq} – the sound speed in the given liquid medium; γ_s – the density of the solid; c_s – the speed of sound in the material from which the solid is made. The *dynamic frictional stress* then is

$$\tau_\vartheta = f_d\, p_\vartheta, \tag{2.9}$$

where f_d – the *dynamic friction coefficient.*

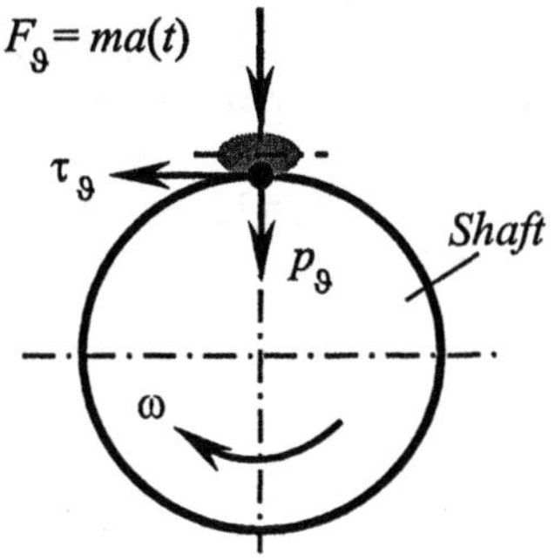

Fig. 2.6. Scheme of impact interaction in *shaft / particle* active system

It is exactly the kinetic interaction of damages due to the combined and simultaneous effect of cyclic stresses (2.3) and dynamic stresses (2.9) or (2.8) that governs the process of mechano-erosion fatigue in this case.

A technical example of the combination *solid / stream of particles* active system is an oil pipeline line section (Fig. 2.7). When oil is pumped through, the loading process of the pipeline with the internal pressure p turns out to be alternating. The maximum (circumferential) cyclic stress is determined from the formula

$$\sigma_\theta = \frac{p(d-h)}{2h}, \tag{2.10}$$

where d and h – the diameter and thickness of the pipe walls.

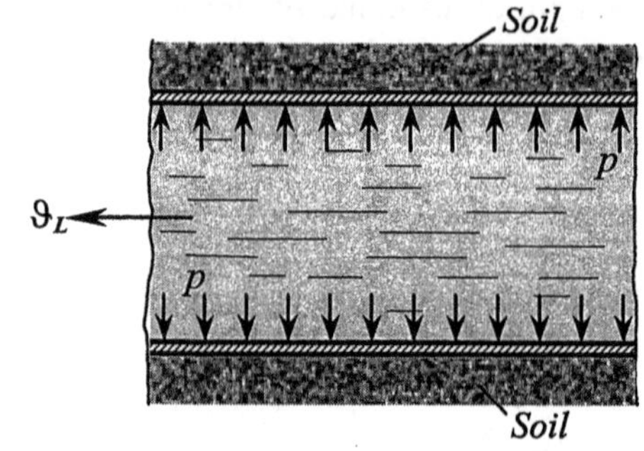

Fig. 2.7. Combined active system – linear section of oil pipeline

The stream of oil moving with the linear speed ϑ_L contacts with the internal surface of the pipe causing its hydroerosion. This contact in the first approximation can be characterized by friction stresses

$$\tau_L = a_L f_L \vartheta_L^2 / 2\,, \tag{2.11}$$

where a_L – the dimensional parameter; f_L – the coefficient of friction of oil against steel.

It is the kinetic interaction of damages due to the combined and simultaneous effect of cyclic (2.10) and frictional (2.11) stresses that determines the process of mechano-erosion fatigue. When oil is pumped through, it corrodes the internal surface of the pipe, hence the pipe operates in the conditions of mechano-corrosion fatigue. Summarizing it should be mentioned that three damaging effects are typical for the internal surface of the pipe:

erosion ⇔ mechanical fatigue ⇔ corrosion.

So, WFD in the general case is the most intricate process of complex damage and fracture of the components of an active system that can be treated as a result of dialectic interaction between separate (simpler) damaging phenomena, such as mechanical fatigue, friction and wear, erosion, corrosion, etc.

2.2 Practical analysis

We will illustrate the analytical procedure using the moving mechanisms of the fodder harvester produced by the Gomel Agricultural Machinery Group. The task is to identify all typical active systems and indicate main types of their possible damage in operation. Introduce the following designations: Δ – fretting fatigue, O – mechano-rolling fatigue, □ – mechano-sliding fatigue.

Some specific examples analyzed are given below.

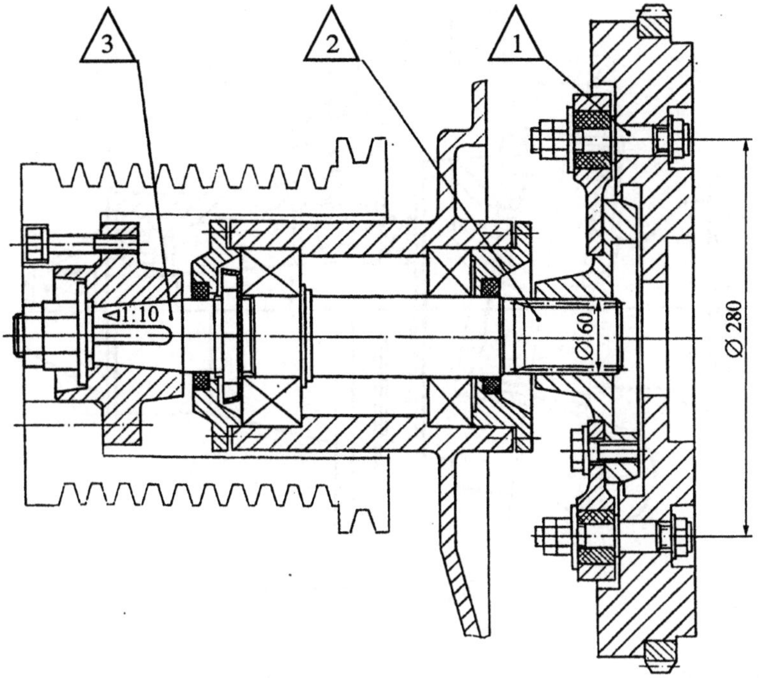

Fig. 2.8. Drive shaft

Table 2.2. Characteristic of active systems in drive shaft unit

Position and WFD type	Active system	Design features	Typical conditions of operation
△1	Shaft / bushing	Gapless connection $\varnothing 20 \frac{H8}{u7}$ Shaft – steel 35, (27...32) HRC_e Bushing (flywheel) – cast iron	Bending + fretting
△2	Shaft / bushing	Splined joint $60 \times 3{,}5 \times \frac{11H}{10d}$ GOST 6033–80 Shaft – steel 20X ГHP, (57...64) HRC_e Bushing – steel 25XГT, (56...63) HRC	Bending + fretting
△3	Shaft / bushing	Cone keyed joint ØH 52 ◁ 1:10; Shaft – steel 20X ГHP, (57...64) HRC_e Bushing – steel 40 (forging)	Torsion + fretting

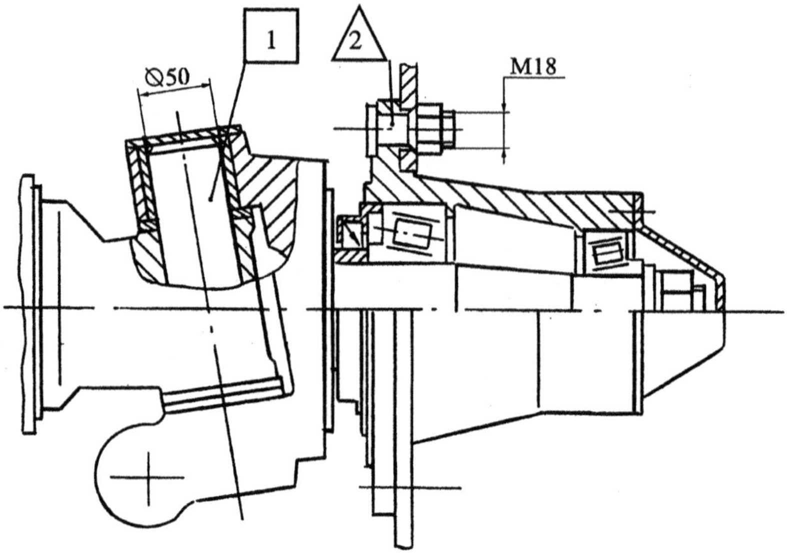

Fig. 2.9. Steering knuckle

Table 2.3. Characteristic of active systems in steering knuckle unit

Position and WFD type	Active system	Design features	Typical conditions of operation
1 (square)	*Shaft / bushing*	Sliding bearing $\varnothing 50\frac{H8}{h9}$ Shaft – steel 40X, (51...63) HRC_e Bushing – brass ЛО 90-1	Bending + sliding friction
2 (triangle)	*Shaft / bushing*	Gapless joint $\varnothing 18.5\frac{H9}{X8}$ Shaft – steel 40X, (32...41) HRC_e Bushing – cast iron КЧ 38-3-Ф	Bending + fretting

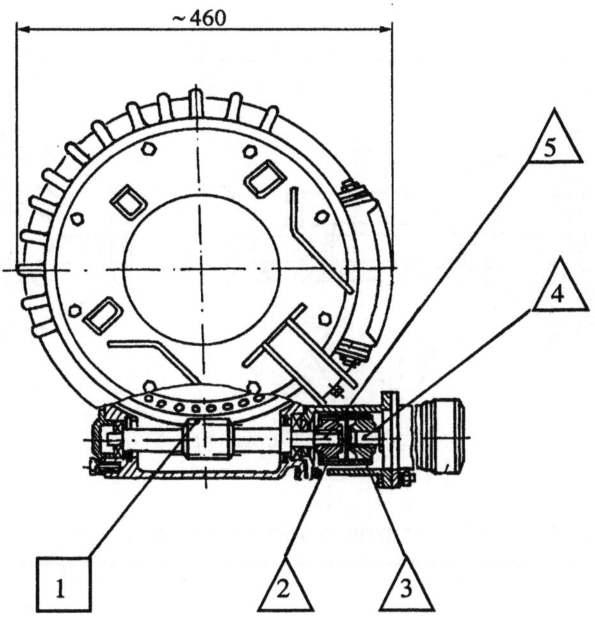

Fig. 2.10. Turning drive of silo feeder

Table 2.4. Characteristic of active systems in turning drive of silo feeder

Position and WFD type	Active system	Design features	Typical conditions of operation
1	*Shaft / bushing*	Worm gear $m = 4$; $z_1 = 2$; $z_2 = 103$ Shaft – steel 40X, 46 HRC_e h 1.0 ... 4.5 mm Bushing – cast iron B445	Bending + sliding friction
2 3	*Shaft / bushing*	Splined joint $55\times2.5\times\frac{11H}{10d}$ GOST 33–80 Shaft – steel 25ХГТ, (56...64) HRC_e h 0.4 ... 0.8 mm Bushing – steel 40X, (107...229) HB	Bending + fretting
4	*Shaft / bushing*	Keyed joint $\varnothing\,25\frac{H9}{h9}$ Shaft – steel 40X, (167...207) HB Bushing – steel 25ХГТ (56...64) HRC_e h 0.4 ... 0.8 mm	Torsion + fretting
5	*Shaft / bushing*	Keyed joint $\varnothing\,32\frac{H9}{h6}$ Bushing – steel 25ХГТ (56...64) HRC_e h 0.4 ... 0.8 mm	Torsion + fretting

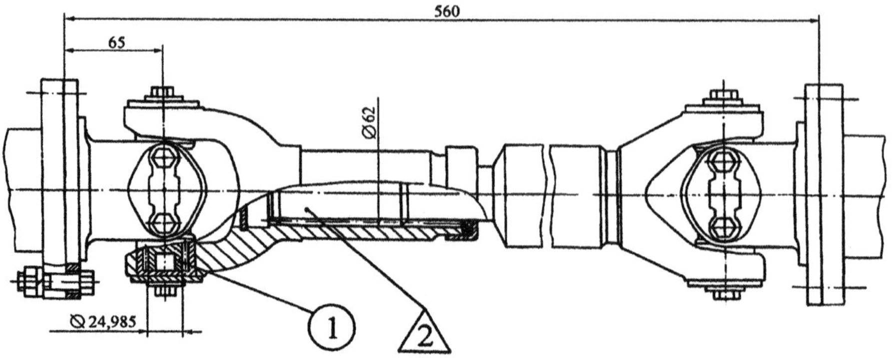

Fig. 2.11. Cardan shaft

Table 2.5. Characteristic of active systems in cardan shaft unit

Position and WFD type	Active system	Design features	Typical conditions of operation
(1)	*Shaft / roller*	Crossbar pin in needle bearing Shaft – steel 20X, (50...65) HRC_e Roller – bearing steel	Bending + rolling friction
△2	*Shaft / bushing*	Splined joint $m = 2.5$; $z = 22$; $\alpha = 15°$ ØH62 Shaft – steel 45, (45...57) HRC_e Bushing – steel 45, (255...302) HB	Bending + fretting

Figures 2.8–2.11 show the drawings of the simplest units, Tables 2.2–2.5 list the characteristics of the typical active systems that are registered in these units. More complex cases are analyzed in Figs. 2.12 and 2.13 and corresponding Tables 2.6 and 2.7. A more exhaustive analysis of the moving mechanisms of the "Polessye" fodder harvester leads to a conclusion that it comprises 125 standard dimensions of active systems that can be referred to three types of WFD (Table 2.8): fretting fatigue determines operation of 85 active systems, mechano-corrosion fatigue determines operation of 36 and mechano-sliding fatigue determines operation of 4.

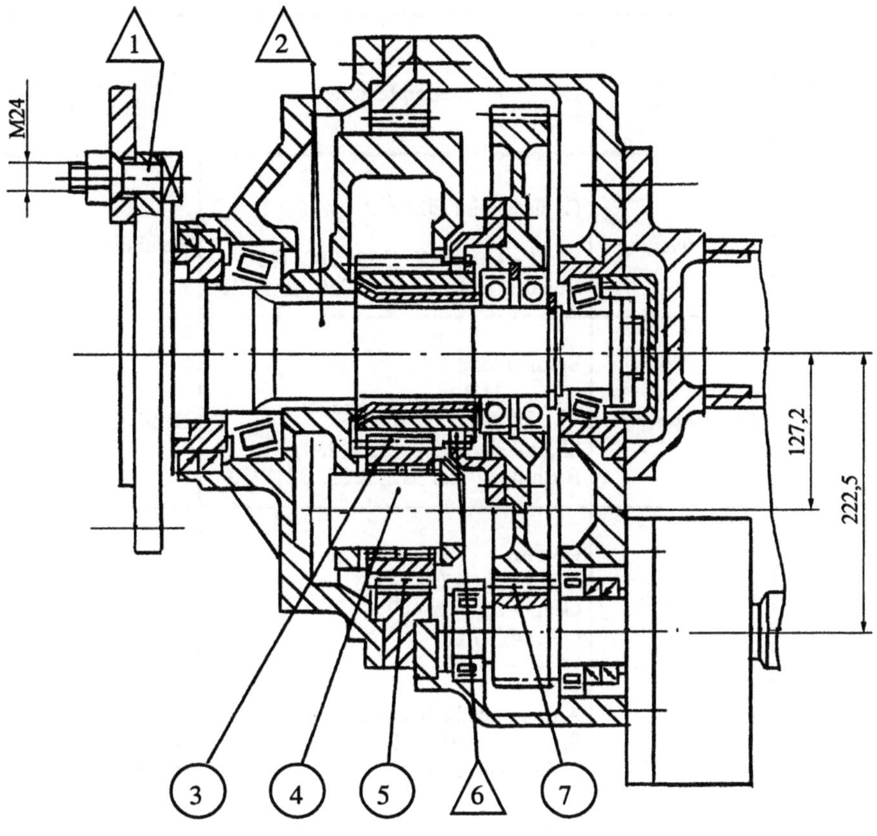

Fig. 2.12. Reduction gear box

As a rule, couplings operate with fretting fatigue. 5 types of joints are employed in the moving mechanisms of the fodder harvester: splined, gear, keyed, gapless, cone keyed joints. Gearings and crosspiece pins in needle bearings operate with mechano-rolling fatigue. Worm gears and units with sliding bearings operate with mechano-sliding fatigue

Eleven steel grades, one pig iron grade and one brass grade are used to fabricate all the active systems. According to the data in Table 2.8 all gears without exception are made from the steel of one grade (25ХГТ).

Regarding the geometrical dimensions of active systems the governing geometrical parameter varies basically within the range from 10 to 1500 mm, i.e. over two orders of magnitude.

Table 2.6. Characteristic of active systems of reduction gear-box

Position and WFD type	Active system	Design features	Typical conditions of operation
△ 1	*Shaft / bushing*	Gapless joint $\varnothing 25 \frac{H9}{z8}$ Shaft – steel 40Х, (255...302) HB Bushing – steel 40Х , (229...269) HB	Bending + fretting
△ 2	*Shaft / bushing*	Splined joint $95\times 5\times \frac{11H}{11C}$ GOST 6033–80 Shaft – steel 40Х, (229...269) HB Bushing – steel 35Л, (225...321) HB	Bending + fretting
○ 3	*Roller/ roller*	Gearing $m = 5$; $z_1 = 27$; $z_2 = 22$; Gears – steel 25ХГТ, (57...64) HRC_e	Bending + rolling friction
○ 4	*Shaft / roller*	Axle in needle bearing Shaft – steel18ХГТ, (47...64) HRC_e Roller – bearing steel	Bending + rolling friction
○ 5	*Roller/ roller*	Gearing $m = 5$; $z_1 = 22$; $z_2 = 72$ Gears – steel 25ХГТ, (59...64) HRC_e	Bending + rolling friction
△ 6	*Shaft / bushing*	Gear coupling $m = 5$; $z_1 = 22$; $z_1 = z_2 = 22$ Shaft – steel 25ХГТ, (57...64) HRC_e Bushing – steel 40Х, (228...268) HB	Bending + fretting
○ 7	*Roller / roller*	Gearing $m = 5$; $z_1 = 14$; $z_2 = 75$ Gears – steel 25ХГТ, 51 HRC_e Wheel – steel 25ХГТ, (57...64) HRC_e	Bending + rolling friction

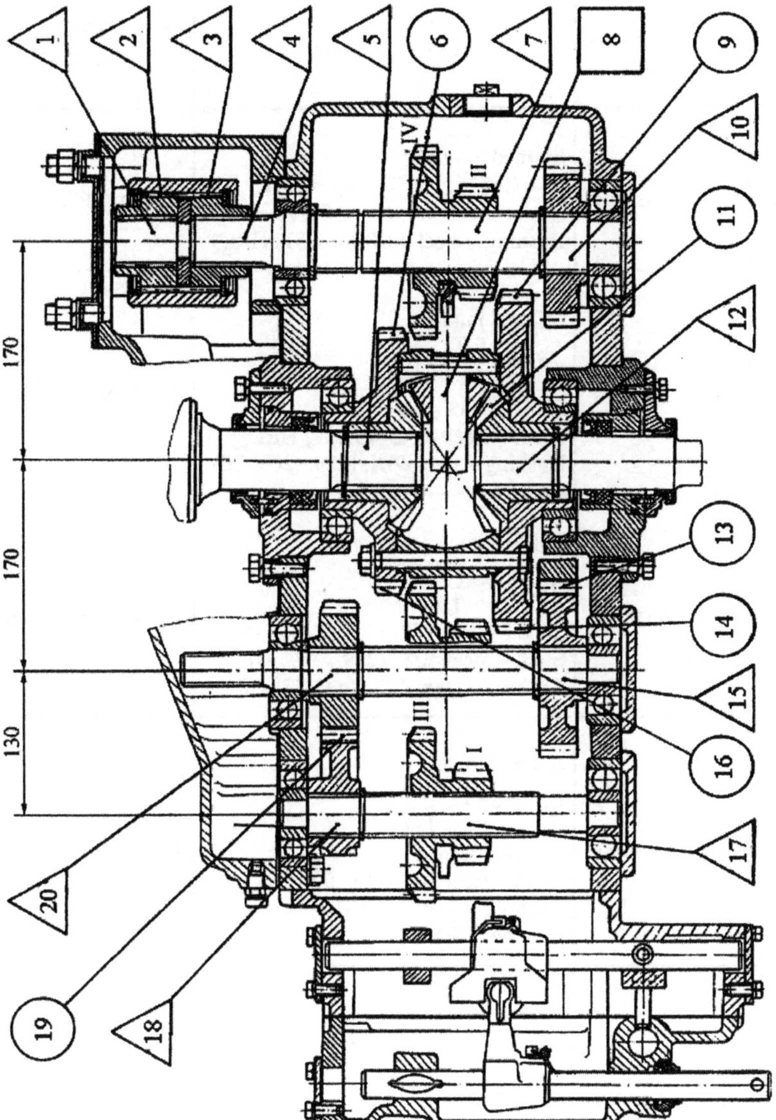

Fig. 2.13. Speed gear-box

Table 2.7. Characteristic of active systems of speed gear-box

Position and WFD type	Active system	Design features	Typical conditions of operation
△1	*Shaft / bushing*	Splined joint $m = 1.5875$; $z = 23$; $D_H = 38.1$; Hydraulic motor shaft МП-90 Bushing – steel 40X, (43...51) HRC_e	Bending + fretting
△2 △3	*Shaft / bushing*	Splined joint $75 \times 3 \times \frac{9H}{9g}$ GOST 6033 – 80 Shaft – steel 40X, (43...51) HRC_e Bushing – steel 40X, (229...269) HB	Bending + fretting
△4	*Shaft / bushing*	Splined joint $40 \times 2{,}5 \times \frac{11H}{11a}$ GOST 6033 – 80 Shaft – steel 40X, (229...269) HB Bushing – steel 40X, (43...57) HRC_e	Bending + fretting
△5 △12	*Shaft / bushing*	Splined joint $\text{Ev.}\,40 \times 2{,}5 \times \frac{11H}{11a}$ GOST 6033 – 80 Shaft – steel 40X, (40...50) HRC_e Bushing – steel 25XГТ, (57...64) HRC_e	Bending + fretting
⑥ ⑯	*Roller / roller*	Gearing $m = 5$; $z_1 = 27$; $z_2 = 39$ Gears – steel 25XГТ (57...64) HRC_e	Bending + rolling friction

Table 2.7 continued

Position and WFD type	Active system	Design features	Typical conditions of operation
△7 △10 △15 △17 △20	*Shaft / bushing*	Splined joint $d-8\times 46\frac{H8}{e8}\times 50\frac{H12}{a11}\times 9\frac{C11}{d9}$ Shaft – 40Х, (31...41) HRC_e Bushing – steel 25ХГТ, (57...64) HRC_e	Bending + fretting
□8	*Shaft / bushing*	Sliding bearing $\varnothing 28\frac{H8}{h7}$ Axle – steel 20Х, (57...64) HRC_e Satellite – steel 25ХГТ, (57...64) HRC_e	Bending + sliding friction
○9 ○14	*Roller / roller*	Gearing $m=5$; $z_1=15$; $z_2=51$ Gears – steel 25ХГТ, (57...64) HRC_e	Bending + Rolling friction
○11	*Roller / roller*	Gearing $m=6$; $z_1=18$; $z_2=12$ Gears – steel 25ХГТ, (57...64) HRC_e	Bending + rolling friction
○13	*Roller / roller*	Gearing $m=5$; $z_1=23$; $z_2=27$ Gears – steel 25ХГТ, (57...64) HRC_e	Bending + rolling friction
○19	*Roller / roller*	Gearing $m=5$; $z_1=29$; $z_2=21$ Gears – steel 25ХГТ, (57...64) HRC_e	Bending + rolling friction

Table 2.8. Data of active systems of fodder harvester

Wear-fatigue damage	Active systems (real)	Material		Model
		Shaft	Bushing	
Fretting fatigue	Joints: – splined; – gearing; – keyed; – cone keyed; – gapless	40X, 45X, 18XГТ, 25XГТ, 20X23H18	20, 40, 40X, 18XГТ, 25XГТ, 12X18H10T	*Shaft / bushing*
	Total 85 standard sizes	Total 8 steel grades		
Mechano-rolling fatigue	Gearing	25XГТ		*Roller/ Roller*
	Crosspiece pins in needle bearings	20X (shaft)		*Shaft / roller*
	Total 36 standard sizes	Total 2 steel grades		
Mechano-sliding fatigue	Worm gears	40X	ВЧ 55 (cast iron)	*Shaft / bushing*
	Units with sliding bearings	20X 40X	25XГТ ЛО90-1 (brass)	
	Total 4 standard sizes	35	35Л	
		Total 6 grades		

Since the active systems are, as a rule, most essential and heavily loaded units, provision of their reliable performance is the top-priority task.

Reliability is known to be incorporated into designs and ensured by the technology. Still, tests play a specific role in its quantitative evaluation at the stages of designing and manufacture.

The *concept* of priority of *full-scale tests* is extensively used in machine building, from tests of individual pieces to tests of units and machines in their integrity.

It is sometimes clear at the initial stages of designing what units are to be tested with special stands. Usually they are gearboxes, components of hydraulic systems and power units. It is ambiguous whether other units should be tested; a clear list of objects for bench tests appears only after tests of a pilot design. Full set tests performed with unique stands in lab conditions enable to reveal such units much quicker and undertake their improvement before operational tests. However these tests and labour and time consuming are highly costly. Besides, they posses a serious engineering drawback: in case any unit, assembly or separate part of the machine fail, the tests are to be suspended until causes are identified and decision is made how to restore operability. It strongly reduces the efficiency of the test bench for full-set tests. Hence, full-set tests are obligatorily supplemented with the tests of units on stand-alone benches. It adds to the cost of tests and requires more time for their execution. As a result, still more time is needed to perfect a machine to satisfy the reliability criteria.

Bench tests of units and assemblies, in their turn, manifest another substantial disadvantage: the results have limited significance and relate just to those standard sizes of the object for which the stand is adapted. Otherwise a new bench is to be designed and constructed in case different standard sizes appear with corresponding consequences.

Model tests are therefore useful for relatively small objects. The effectiveness of this method is demonstrated above: just three standard size models can be subjected to quick laboratory tests instead of 125 active systems of the fodder harvester. Yet, there appears another rather intricate problem of how to implement *modelling* and *proceed back to the full-scale machine* in order to obtain valid prediction of resistance of real active systems to WFD? It is apparent that the tests of models of active systems contain more significance (and they are more popular) if they enable to assess the reliability indicators of full-scale units more accurately in the conditions close to those in operation.

2.3 Methodology of tribo-fatigue

Let us see how in practice the performance of a specific mechanical system is rated that is typically exposed to WFD.

Assume we deal with the mechanical system such as a *connecting rod* (with a polymeric sliding bearing) / *a steel crankpin.* In *tribology* this system is considered as a common *friction pair.* Since the steel shaft does not wear in contact with a polymer, the wear resistance of the polymeric bearing is assessed (by calculation or experimentally). For example, this assessment can be performed as shown schematically in Fig. 2.14 below. If p is contact pressure, the failure probability $F(p)$ in the general case is based on the wear resistance criteria assumed usually as the wearing intensity $I(p)$ and / or wear durability $N(p)$.

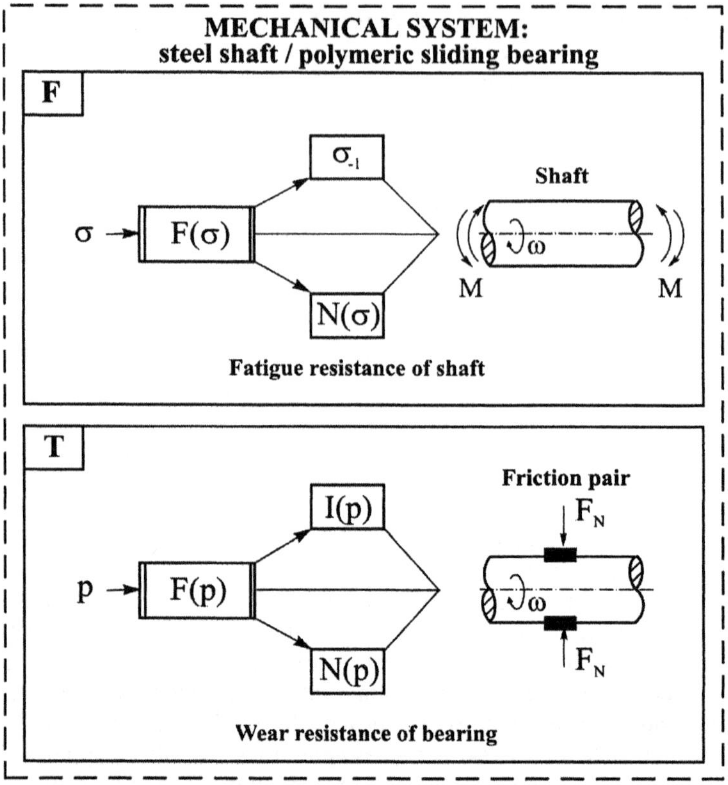

Fig. 2.14. Traditional approach to calculating mechanical system

From the viewpoint of the *resistance of materials*, we have a crankpin as a *structural element* (cf. Fig. 2.14, top). If σ is the effective cyclic stress, then the failure probability $F(\sigma)$ is determined in the general case (by calculation or experimentally) using the criteria of fatigue resistance usually assumed as the endurance limit σ_{-1} and / or fatigue durability $N(\sigma)$.

So, it is assumed that, in case of the traditional assessment of reliability of a given mechanical system based on individual criteria (either fatigue resistance or wear resistance), the relation between its components (the crankpin ↔ the sliding bearing) is either weak (tribology studies only the friction pair) or is entirely lacking (the mechanics of fatigue fracture studies only an individual structural element). In reality, *from the viewpoint of tribo-fatigue*, in this case we deal with a *metal-to-polymer active system* that operates in the conditions of mechano-sliding fatigue. It means that there is dynamic interaction between the components of the system due to concurrent and combined effect of both contact pressure ▯ and cyclic stresses σ excited by off-contact loads. Then the system's operability should be assessed as it is schematically shown in Fig. 2.15.

The reliability of an active system can be described with the following *complex indicators*: $F(\sigma, p)$ – the probability of failure of the system due to the probability of failure of either the shaft or the bearing or of both these components simultaneously; $N(\sigma, p)$ – the durability of the system due to the durability of either the shaft or the bearing or of both these components simultaneously; σ_{-1p} – the endurance limit of the shaft with the allowance for the effect of the processes of friction and wear under the effect of contact pressure p; $I_\sigma(p)$ – the intensity of wearing of the bearing under the effect of cyclic stresses σ.

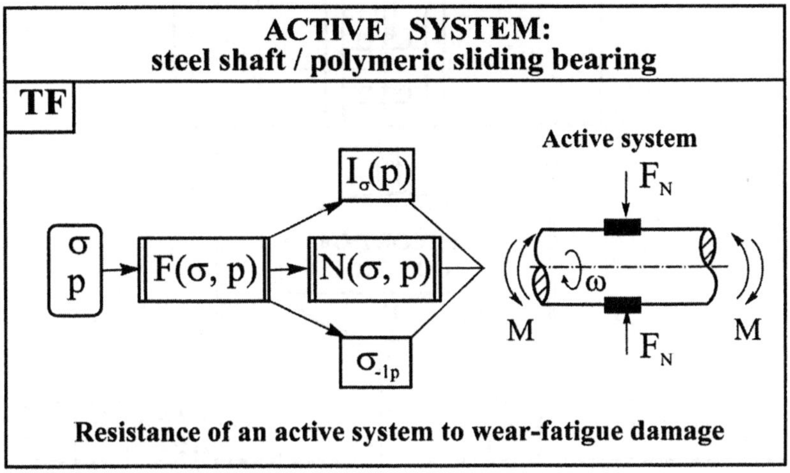

Fig. 2.15. To designing active system

Since the processes of friction and wear affect the characteristics of resistance of the components of the system to fatigue and their effect is direct, it is clear that the characteristic σ_{-1p} describes this effect quantitatively. Since cyclic stresses produce a back effect on the characteristics of friction and wear, it is clear that the characteristic $I_\sigma(p)$ describes this effect quantitatively.

It is quite apparent in Figs. 2.14 and 2.15 *that only the active system is truly adequate to the mechanical system in question* while a friction pair or a structural component are just its particular schematizations. The indicators of the reliability of the studied object determined by the methods of tribo-fatigue reflect sufficiently fully the real conditions of its operation, while similar indicators determined by the methods of tribology or fatigue fracture mechanics describe the behavior of the object in the idealized conditions.

The above ideas are summarized in Fig. 2.16. On the one hand, the theories of friction, wear, and lubrication are integrated into tribology as a single scientific discipline (T). It is natural because the processes of friction and wear, including lubrication, really combine and interact in friction pairs.

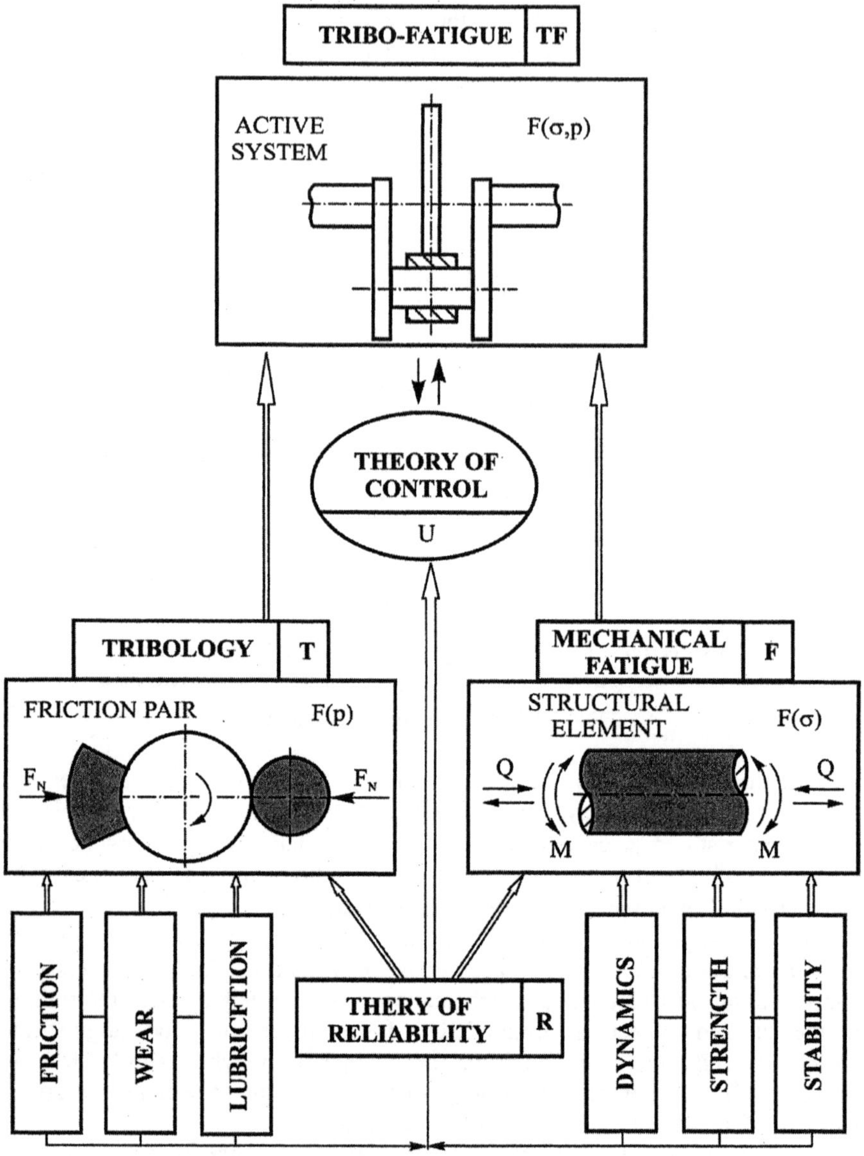

Fig. 2.16. Tribo-fatigue as a complex scientific discipline

On the other hand, fatigue fracture mechanics (F) has come apart among other general problems of dynamics, strength and stability as a discipline having specific practical significance for modern machine building. Creation of tribo-fatigue (TF) became indispensable and urgent in order to solve effectively the complex problems of reliability (R) of the most essential active systems of machines and equipment based on the most important criteria of serviceability.

So, *tribo-fatigue is the science of wear-fatigue damage and fracture of active systems* (GOST 30 638–99).

Since it deals with the WFD mechanics, the term *tribo-fatigue* reflects fully, though very briefly, its contents: tribos – *friction from* Greek symbolizing modern tribology, including tribological reliability; fatigue – *fatigue* from French symbolizing modern fatigue fracture mechanics, including strength reliability. It is apparent that the term tribo-fatigue consists of two words (notions) that are completely equivalent. In the mechanophysical processes determined by these notions the dialectic interactions and interference of the processes should be born in mind: *friction* (and *wear*) $\Leftrightarrow$ *fatigue* = *tribo-fatigue.*

Table 2.9 shows a comparative analysis of methods of research and designing of objects that tribology, fatigue fracture mechanics and tribo-fatigue deal with.

First let us consider experimental methods.

Table 2.9. Methods of studying objects

Discipline	Object of study	Basic methods of studies		Scale of damage
		experimental	theoretical	
T (tribology)	Friction pair	Friction Tests	Mechanics of contact interactions	Surface damage (wear, pitting, etc.)
F (mechanical fatigue)	Structural element	Fatigue Tests	Mechanics of deformation and fracture	Volume (fatigue) fracture
TF (tribo-fatigue)	Active system	Wear-fatigue Tests	Mechanics of wear-fatigue damage	Complex surface damage and volume fracture

Specialists in the mechanical fatigue develop and employ methods and machines for testing structural components under a variety of conditions of cyclic loading. Tribologists develop and employ methods and machines for testing friction pairs under a variety of conditions of contact interactions. Specialists in tribo-fatigue develop and employ methods and machines for *complex wear-fatigue tests* of models of active systems. The difference is the following. Friction machines preclude studies of resistance of structural components to fatigue. Machines for fatigue tests preclude studies of the processes of friction and wear. Meanwhile *machines for wear-fatigue tests* (SI series) enable to study both and what is important that they enable to perform complex tests with any combinations of cyclic and contact loads acting simultaneously. It naturally provides an opportunity to obtain fundamentally new experimental results.

Now let us consider the *methods of theoretical studies* (see Table 2.9). Theory is known to be based on practical experience. That is why tribologists use their experience to develop primarily the mechanics of contact interactions. Specialists in strength use their experience to develop the mechanics of deformation and fracture. Of course, specialists in tribo-fatigue use both inseparably. But they have a more complex object of study and have to seek for new approaches to investigating complex phenomena.

A non-traditional approach to analyzing contact problems and problems of the mechanics of deformation and fracture is developed in tribo-fatigue. The approach is based on the use of a *statistical model of the deformable solid with a damaged (dangerous) volume* (the SDV model). According to this model the strength of a body (including surface strength) is determined by a region of finite dimensions with a critical level of stresses. This region is called *dangerous (damaged) volume.*

Conceptually the essence of the approach is the following.

Assume that the steel *shaft is cyclically bent* by the moment M so that the field of normal stresses σ causes damage in some its region. It means that a *damaged volume* $V_{P\gamma} > 0$ (cases F in Fig. 2.17) appears on the surface of the shaft, the region is limited by the condition $\sigma \geq \sigma_{-1\min}$, where $\sigma_{-1\min}$ – the lower boundary of scatter of the endurance limit.

Assume also that the process of *rolling friction* or sliding friction occurs in this dangerous zone. Assume also that the contact pressure field p in this active system (the shaft / *the counterbody*) is such that it produces the appearance of the *damaged volume* $S_{P\gamma}$ in the contact site region (cases T in Fig. 2.17). If the sliding friction process occurs, the damaged volume $S_{P\gamma} = S_{WP}$ appears within a fine surface layer on the shaft (the case *T-1*). If the rolling friction process occurs, the damaged volume $S_{P\gamma} = S_{WP}$ (the case *T-1*) and / or $S_{P\gamma} = S_{VP}$ (the case *T-2)* appears both on the surface and under it. In all the cases it is limited by the damaging level of tangent stresses $\tau \geq \tau_{f\min}$, where $\tau_{f\min}$ – the lower boundary of scatter of endurance limits determined in shear (or during friction fatigue).

Now let us treat the wear-fatigue tests as *a superposition of cyclic bending and friction* (sliding or rolling). Two situations are possible then when damaged volumes appear (the right column of the diagrams in Fig. 2.17). First, volumes $V_{P\gamma}$ and $S_{P\gamma}$ coincide on the surface (the case *TF-1*). Second, they coincide under the surface (the case *TF-2*)

In both these cases a *complex dangerous (damaged) volume* $W_{P\gamma} > 0$ appears with its magnitude determined by the volumes $V_{P\gamma}$ and $S_{P\gamma}$, i. e.

$$W_{P\gamma} = \varphi_{SV}(V_{P\gamma}, S_{P\gamma}), \tag{2.12}$$

where φ_{SV} – the *function of interaction*; it may be assumed that it is enough to consider it the *parameter of interaction* in some cases.

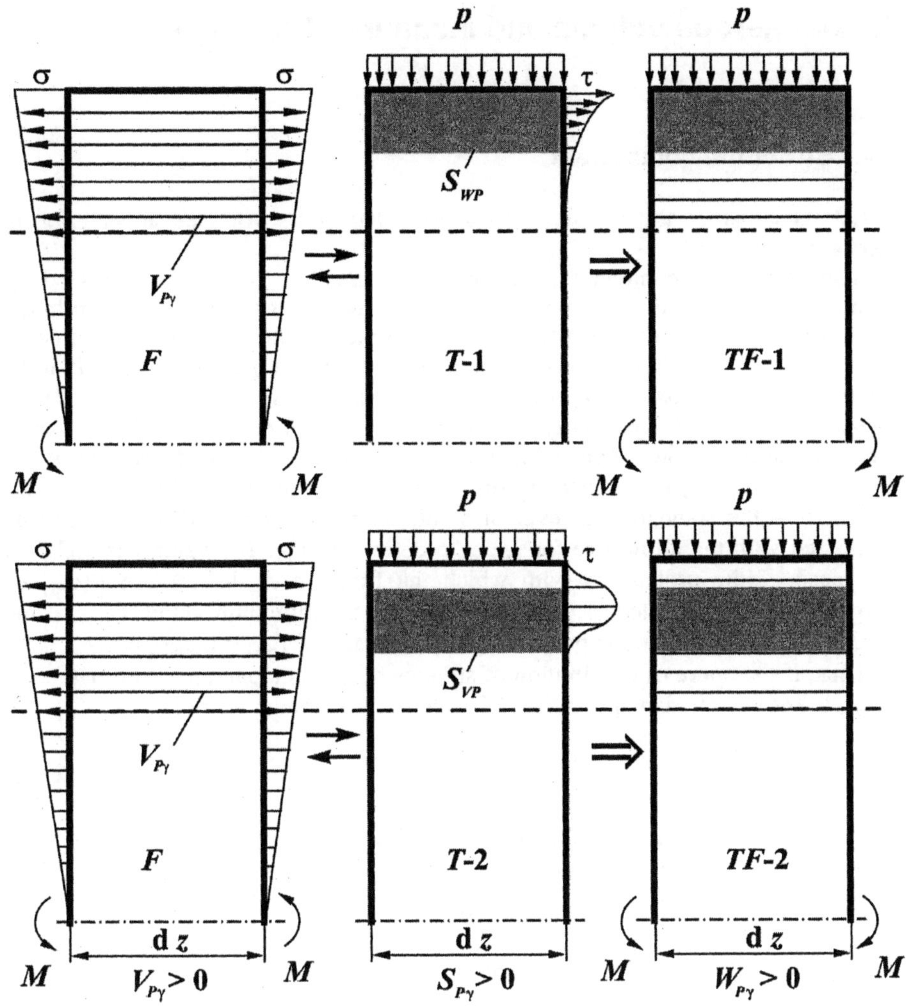

Fig. 2.17. Scheme of appearance of damaged volumes during tests for friction ($S_{P\gamma} > 0$), mechanical fatigue ($V_{P\gamma} > 0$) and wear-fatigue damage ($W_{P\gamma} > 0$)

After that the problems of reliability, strength, wear resistance and durability are solved with the consideration of the fundamental dependence of the relevant characteristics on the scale of damageable volumes of the object, i.e. on the magnitude of the damaged volume.

2.4 Dangerous volume and measure of damage

2.4.1 Structural component

Let us consider the simplest cases how dangerous volumes appear during mechanical fatigue.

If the structural component, such as a deformable solid, is characterized by such a state of stress that it is possible to determine its fatigue fracture, then in the general case it consists of two regions: a volume of safe state and a volume of unsafe state. The volume is named *dangerous* (*damaged*) *volume* V_P *with some probability* P *of possible appearance of cyclic stresses* σ, *above the lower boundary* $\sigma_{-1\min}$ *of dissipation of endurance limits* σ_{-1} *in the body*. It follows from this definition that the value P has a sense of the probability of fracture of the body. By assigning a confidence estimate to it with the probability $\gamma = 1 - \alpha_0$, where α_0 – the standardized level of significance, the damaged volume can be calculated with the allowance for γ, i.e. to determine $V_{P\gamma}$. The value γ is assigned the sense of the probability with which the function of distribution $P(\sigma_{-1})$ of fatigue limits is restricted in order to determine the minimum value $\sigma_{-1\min}$ of the random value σ_{-1} so that the probability is practically $P\,(\sigma_{-1} < \sigma_{-1\min}) = 0$.

Thus, if the epure of distribution of stresses $\sigma(x, y, z)$ along the coordinate axes x, y, z is known, the damaged volume is calculated from the formula

$$V_{P\gamma} = \iiint\limits_{\sigma(x,y,z)>\sigma_{-1\min}} dxdydz\,. \tag{2.13}$$

From (2.13) the *generalized condition of fatigue fracture* follows (the condition of failure, the condition of attainment of the limiting state)

$$V_{P\gamma} > 0 \tag{2.14}$$

with some probability P at the confidence probability γ.

If

$$V_{P\gamma} = 0, \tag{2.15}$$

the fatigue fracture is impossible physically (because $\sigma < \sigma_{-1\min}$ in this case), therefore, (2.15) is the *generalized condition of non-fracture* (the condition of failure-free operation, the condition of non-attainment of the limiting state).

Figure 2.18 explains it. Assume that an integral function of distribution $P(\sigma_{-1})$ of the endurance limits σ_{-1} is obtained for a given structural component and it is used to establish the lower boundary $\sigma_{-1\min}$. If condition (2.14) of fatigue fracture occurs, it means that according to (2.13) the effective stress should be $\sigma > \sigma_{-1\min}$; in such a case it is equal to the endurance limit σ_{-1P} with the probability of occurrence P. A definite value of the probability of fatigue fracture $P(\sigma)$ corresponds to the stresses $\sigma = \sigma_{-1P}$. If $\sigma < \sigma_{-1\min}$, then $P(\sigma) = 0$ because the condition of non-fracture (2.15) occurs.

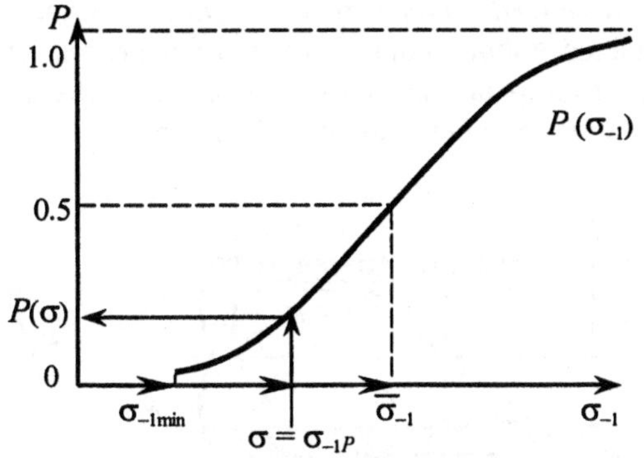

Fig. 2.18. Scheme of assessment of probability $P(\sigma)$ of fatigue fracture of structural component with damaged volume $V_{P\gamma} > 0$ under effect of stress $\sigma > \sigma_{-1\min}$

Calculate the damaged volumes in the simplest structural components, like cylindrical and prismatic specimens during bending. The state of stress in this case is uniaxial and distribution of normal stresses across the cross section is described with a linear relation. Integral (2.13) can be calculated proceeding from simple geometrical constructions.

If the shaft with the radius r_0 undergoes pure bending (with the moment M) with the frequency of rotation ω, it is easy to plot the epure of normal stresses in the cross section. Assume the maximum stress on the shaft's surface is $\sigma > \sigma_{-1\min}$ (Fig. 2.19, *a*). The depth of the damaged layer $c_{P\gamma}$ is then determined from the similarity of triangles with the sides r_0, σ and $(r_0 - c_{P\gamma})$, $\sigma_{-1\min}$, namely:

$$c_{P\gamma} = r_0\left(1 - \frac{\sigma_{-1\min}}{\sigma}\right) = \frac{\sigma - \sigma_{-1\min}}{G_r}, \tag{2.16}$$

where $G_r = \sigma/r_0$ – the gradient of normal stresses through the cross section of the shaft.

A *dynamically damaged volume forms within one revolution of the shaft* shaped as a ring with the radius r_0, wall thickness $c_{P\gamma}$ and length l_0 equal to the length of the portion of the shaft subjected to pure bending (shaded in Fig. 2.19).

Now assume that a plate with the cross dimension $2r_0$ is subjected to cantilever bending in one plane (Fig. 2.19, *b*). Then the maximum depth $c_{P\gamma}$ of the damaged layer in the dangerous cross section is calculated with formula (2.16), the length $l_{P\gamma}$ of the damaged layer is limited by the condition $\sigma = \sigma_{-1\min}$ (on the surface of the beam), hence, it can be calculated with the formula

$$l_{P\gamma} = l_0\left(1 - \frac{\sigma_{-1\min}}{\sigma}\right) = \frac{\sigma - \sigma_{-1\min}}{G_l}, \tag{2.17}$$

where G_l – the gradient of stresses along the beam.

A *dynamically damaged volume appears within a single cycle of cantilever bending;* this damaged volume consists of two flat prisms arranged in height on the opposite sides of the plate (shaded in Fig. 2.19, *b*); the dimensions $c_{P\gamma}$, $l_{P\gamma}$, b (the plate's width) determine the volume of the prisms.

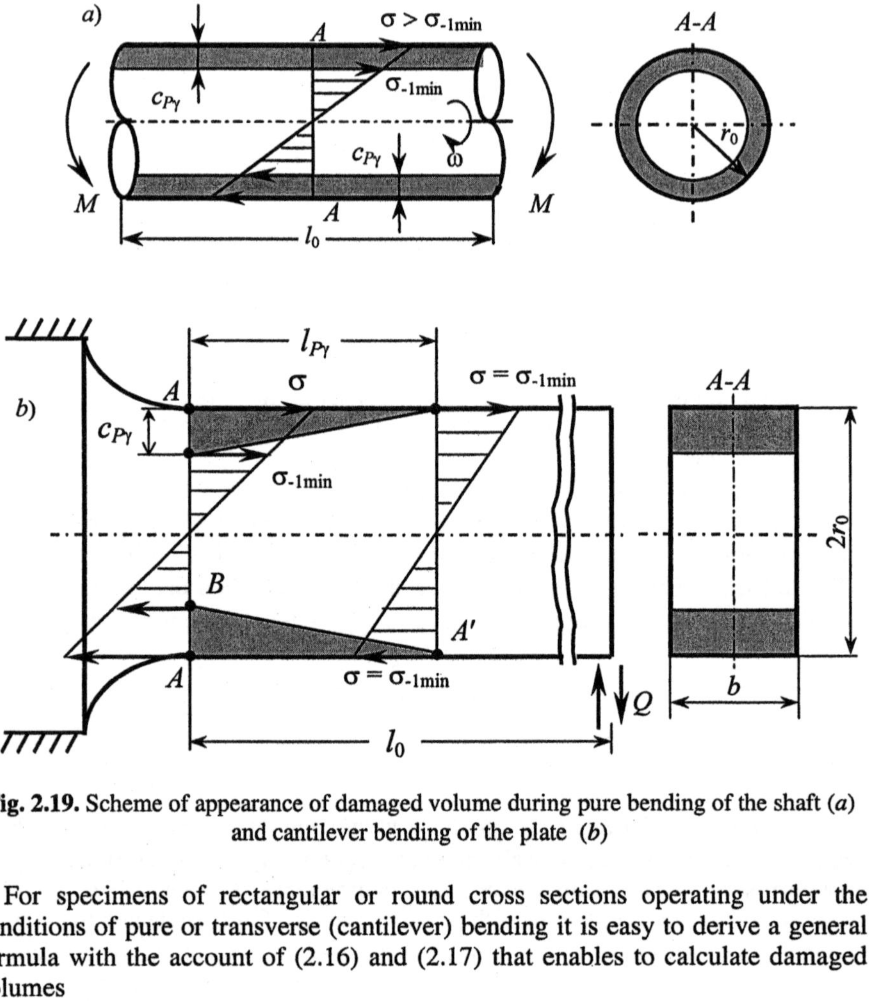

Fig. 2.19. Scheme of appearance of damaged volume during pure bending of the shaft (*a*) and cantilever bending of the plate (*b*)

For specimens of rectangular or round cross sections operating under the conditions of pure or transverse (cantilever) bending it is easy to derive a general formula with the account of (2.16) and (2.17) that enables to calculate damaged volumes

$$\frac{V_{P\gamma}}{V_0} = \lambda\left(1 - \frac{\sigma_{-1\min}}{\sigma}\right)^{\beta}\left(1 + \frac{\sigma_{-1\min}}{\sigma}\right)^{\alpha_1}\left(2 + \frac{\sigma_{-1\min}}{\sigma}\right)^{\alpha_2}, \qquad (2.18)$$

where numerical values of coefficients β, α_1, α_2 determine a priori the shape of the specimen and its loading diagram; they are listed in Table 2.10.

Table 2.10. Values of the coefficients λ, β, α_1, α_2 for popular schemes of fatigue tests of variously shaped pieces

Scheme of fatigue tests	Cross-sectional shape	λ	β	α_1	α_2
Circular pure bending	Round beam	1	1	1	0
Pure bending in one plane	Prismatic beam	1	1	0	0
	Round beam	$8/3\pi$	$\frac{3}{2}$	$\frac{1}{2}$	0
Cantilever or transverse circular bending	Round beam	$\frac{1}{3}$	2	0	1
Cantilever or transverse bending in one plane	Prismatic beam	$\frac{1}{2}$	2	0	0
	Round beam	$8/9\pi$	$\frac{5}{2}$	$\frac{1}{2}$	0

So, a damaged volume is a region of a cyclically deformable solid within which fatigue damage originates and initial fatigue cracks develop. The region is non-equivalent in the sense that the probability of appearance of fatigue cracks is different in each "point" of the damaged volume; it is higher there where actual stresses are stronger. If Fig. 2.19, *b*, is analyzed, it is easy to establish that initial cracks are more probable in the vicinity of *A*. When moving away from □ towards the plane of equal stresses *BA′* (where $\sigma = \sigma_{-1\min} = \text{const}$), the probability of their appearance reduces to the minimum, the cracks will sooner appear on the surface □□′ than in the subsurface layer *AB* due to the fact that the state of the surface is known to be energy specific (atoms on the surface have bonds only from the side of the basic portion of the metal, therefore they are unbalanced, possess excessive energy and therefore react more actively with the environment). Nevertheless, subsurface fracture is possible too because the material has somewhat inferior strength properties in the depth than near the surface. Special observations have revealed that this prediction is in agreement with the real pattern of small fatigue cracks propagated by scattered damage.

Formula (2.18) is derived with the provision that cyclic stresses σ are determined values, while the endurance limit is a random value, its function of distribution $P(\sigma_{-1})$ has the lower boundary $\sigma_{-1\min}$. If σ is a random value too, it is necessary to determine the upper boundary σ_{max} of its distribution $P(\sigma)$. Then the formula for calculating damaged volumes similar to formula (2.18) is

$$\frac{V_{P\gamma}}{V_0} = \lambda\left(1-\frac{\sigma_{-1\min}}{\sigma_{\max}}\right)^{\beta}\left(1+\frac{\sigma_{-1\min}}{\sigma_{\max}}\right)^{\alpha_1}\left(2+\frac{\sigma_{-1\min}}{\sigma_{\max}}\right)^{\alpha_2}, \qquad (2.18a)$$

while the coefficients β, α_1, α_2 remain the same.

If the values $\sigma_{-1\min}$ and σ_{max} with the lower $t_{\sigma_{-1}}(\alpha_{\sigma_{-1}}, \gamma_{\sigma_{-1}})$ and upper $t_\sigma(\alpha_\sigma, \gamma_\sigma)$ tolerance coefficients determined with the confidence probability γ at a level of significance α, formula (2.18a) can be recorded with the mean values $\overline{\sigma}$ and $\overline{\sigma}_{-1}$ of actual σ and ultimate σ_{-1} stresses:

$$\frac{V_{P\gamma}}{V_0} = \lambda\left(1-\frac{\overline{\sigma}_{-1}}{\overline{\sigma}}\kappa\right)^{\beta}\left(1+\frac{\overline{\sigma}_{-1}}{\overline{\sigma}}\kappa\right)^{\alpha_1}\left(2+\frac{\overline{\sigma}_{-1}}{\overline{\sigma}}\kappa\right)^{\alpha_2}, \qquad (2.18b)$$

where

$$\kappa = \frac{1-t_{\sigma_{-1}}(\alpha_{\sigma_{-1}}, \gamma_{\sigma_{-1}})v_{\sigma_{-1}}}{1+t_\sigma(\alpha_\sigma, \gamma_\sigma)v_\sigma};$$

$v_{\sigma_{-1}} = S_{\overline{\sigma}_{-1}}/\overline{\sigma}_{-1}$; $v_\sigma = S_{\overline{\sigma}}/\overline{\sigma}$ – the coefficients of variation of ultimate and actual stresses; $S_{\overline{\sigma}_{-1}}$ and $S_{\overline{\sigma}}$ – mean quadratic deviations of corresponding stresses.

We proceeded from an assumption that actual and ultimate stresses change in a symmetric cycle. If the cycle of changes of stresses is asymmetric, then, for example, the known formula of reduction should be used:

$$\lim \sigma_a = \sigma_{-1} - \psi_\sigma\sigma_m,$$

where σ_a, σ_m – the amplitude of stresses and mean stress in the cycle; ψ_σ – the coefficient of reduction of the asymmetric cycle into the symmetric cycle (see also Fig. 1.17).

The damaged volume value is presented in formulas (2.18), (2.18a) and (2.18b) in the relative form (V_0 – the working volume). The *relative dangerous volume is a probabilistic measure of damage*

$$\omega_{VP} = \frac{V_{P\gamma}}{V_0}, \qquad (2.19)$$

that according to (2.14) and (2.15) can change within the interval of values

$$0 \le \omega_{VP} \le 1. \qquad (2.20)$$

When $\omega_{VP} = 0$, the material of the structural component is not damaged in operation or during tests, i.e. a certain event takes place: the probability of fatigue fracture of the object is $P = 0$.

When $\omega_{VP} = 1$, the structural component fails during the first loading cycle, i.e. a certain event takes place: $P = 1$. While, if $0 < \omega_{VP} < 1$, a random event of fatigue fracture (failure, the limiting state is reached) takes place with some probability P. The larger the damaged volume the higher the probability of fracture. The probability of fatigue fracture changes within the same interval of values

$$0 \leq P \leq 1, \tag{2.21}$$

like the probabilistic measure of damage (2.20). Formally each value ω_{VP} can be put in agreement with some value of P:

$$0 \leq \omega_{VP} \rightarrow P \leq 1. \tag{2.21a}$$

Since the values of damaged volumes are governed according to (2.13) both by the level of cyclic stresses and by the geometrical dimensions of structural components, their shape and loading diagram, it is apparent that $V_{P\gamma}$ can serve as a parameter for describing the *scale effect* or the relation between the endurance limits and the dimensions (volume) of a body. Moreover, it is established experimentally that the relation

$$\frac{V_{P\gamma}}{V_{st}} = \Pi_{VP} \tag{2.22}$$

is an *integral criterion of similarity of fatigue fracture.* The content of criterion (2.22) is the following: if the numerical value Π_{VP} for different standard dimensions of objects is similar, it means that the ultimate stresses in them (or their durabilities) coincide. Naturally, this assertion is realizable with the probability P. V_{st} in formula (2.22) is the dangerous (or working) volume of the object assumed as standard; usually $V_{st} << V_{P\gamma}$.

Similarity criterion (2.22) gives an opportunity to adequately proceed from the full-scale design to the low-sized model and vice versa. It yields huge savings for tests both at the stage of designing and in the process of engineering development of new machinery.

We will show the sense and significance of the relative measure of fatigue damage (2.19) and similarity criterion (2.22) of fatigue fracture by analyzing experimental data (at $P = 0.5$). Multidiameter shafts 10, 20, 30 and 40 mm in diameter from steel 45 were tested for fatigue by cantilever bending with torsion, the radius of the fillet portion was variable within a broad range from 0.05 to 60 mm. The scale dependence of the endurance limits was plotted in double logarithmic coordinates (Fig. 2.20, *a*) $\log \sigma_{max} - \log (V_{0.5\gamma} / V_{st})$, where $\sigma_{max} = \alpha_\sigma \sigma_{-1}$, α_σ – the theoretical coefficient of concentration of stresses. Two straight lines limited the field of scatter of experimental results. The sense of this dependence is the following: if similarity criterion (2.22), $\Pi_V = V_{0.5\gamma} / V_{st} = \text{const}$ in this case, then the endurance limit is similar for shafts of different sizes with different levels of concentration of stresses. If the experimental results are re-plotted in the coordinates $\log \sigma_{max} - \log (V_{0.5\gamma} / V_0)$ (Fig. 2.20, *b*), then the dependencies of damage $\omega_{-1} = V_{0.5\gamma} / V_0$ on $\alpha\sigma_{-1}$ split into separate straight lines, each corresponding to a definite diameter (i. e. the volume V_0) of the shafts. It means that the measure of damage $\omega_{-1} = V_{0.5\gamma}/V_0$ is sensitive both to the diameter of the shafts (lines *1–4* in Fig. 2.20, *b* correspond to different diameters of the shafts) and to the level of concentration of stresses (movement from right to left along any line *1–4* corresponds to the augmentation of the fillet portion radius of the multidiameter shafts). The extent of *damage* ω_{-1} determines the *endurance limit of the shaft*. This conclusion from the data in Fig. 2.20, *b* can be described by the equation of the bundle of straight lines in double logarithmic coordinates

$$\log(\alpha_\sigma \sigma_{-1}) - \log \sigma_{-1\min} = m_{-1} \log \frac{V_{0.5\gamma}}{V_0},$$

where m_{-1} characterizes the inclination of the relation between $\alpha_\sigma \sigma_{-1}$ and ω_{-1} in relation to the abscissa axis. The graphs show that the equation correlates quite satisfactorily with the experimental results.

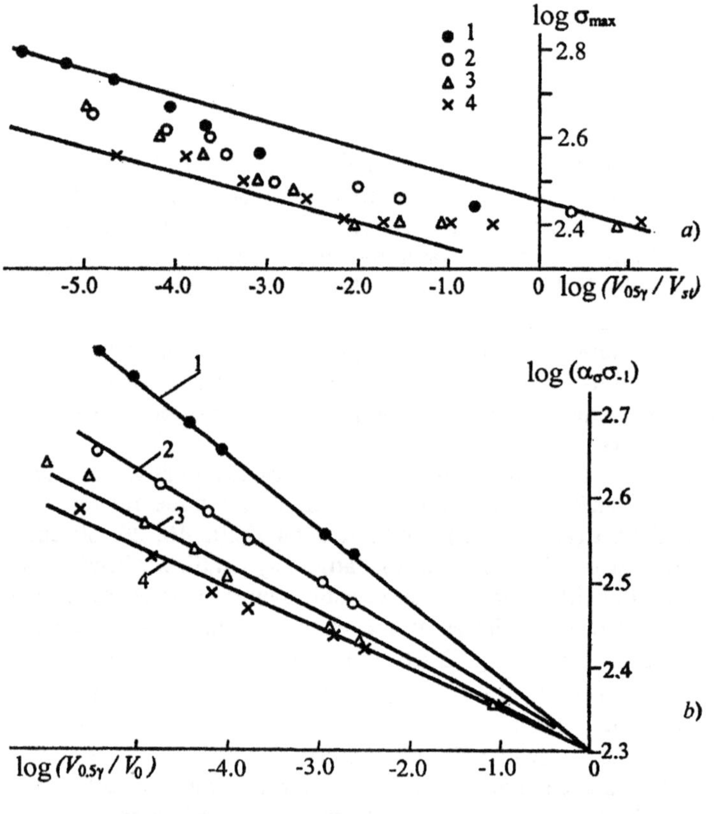

Fig. 2.20. Graphic analysis of criterion of similarity of fatigue fracture (*a*) and relation between the endurance limit and extent of damage of shafts with diameters 10 (straight line *1*), 20 (*2*), 30 (*3*) and 40 mm (*4*)

If the state of stress of the object is complex, then the corresponding components of the damaged volume can be determined providing the distribution of each stress component is known. The procedure of these calculations will be demonstrated using the analysis of a friction pair.

2.4.2 Friction pair

First we will analyze damage in sliding. Reasoning similarly like it was done when the conditions of fatigue damage and fracture were studied (see Sect. 2.4.1), we record the *general formula for calculating dangerous volumes using friction (tangent) stresses*

$$S_{P\gamma} = \iiint\limits_{\tau_W(x,y,z)>\tau_{f\min}} dxdydz\,, \tag{2.23}$$

the probabilistic condition of occurrence of the limiting state (failure)in friction with any manifestation

$$S_{P\gamma} > 0 \tag{2.24}$$

and the condition of *failure-free operation of the friction pair*

$$S_{P\gamma} = 0. \tag{2.25}$$

Formula (2.23) and conditions (2.24), (2.25) are similar to formula (2.13) and conditions (2.14), (2.15), respectively. If S_k is the working volume of the body in the friction pair, the *probabilistic measure of damage in friction is*

$$0 \le \omega_{SP} = \frac{S_{P\gamma}}{S_k} \le 1 \tag{2.26}$$

and the *integral similarity criterion in friction* is

$$\frac{S_{P\gamma}}{S_k} = \Pi_{SP}\,. \tag{2.27}$$

Criteria (2.26) and (2.27) in friction have the content similar to criteria (2.19), (2.20) and (2.22) during cyclic loading.

It is noted in Sect. 2.4.1 that damaged volumes can be calculated using all the stress components, i.e. to obtain the *components of damaged volumes.*

Let us consider the example of appearance of damaged (dangerous) volumes when two cylindrical rollers with different radii r_1 and r_2 of the same width (thickness) l roll (Fig. 2.21). The damaged volume components are detected both in the body and in the counterbody in this case.

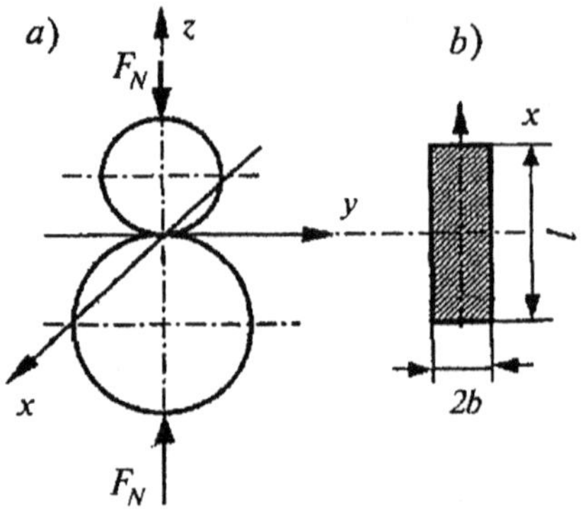

Fig. 2.21. Scheme of static contact between two rollers with parallel axes (*a*) and dimensions of the contact site (*b*)

Now we show how the problem is solved of calculating damaged volumes V_x, V_y, V_z due to normal stresses σ_x, σ_y, σ_z, respectively; their distribution is set by formulas (1.105). In order to calculate the volumes we introduce the criterion of their restriction by the critical stress $\sigma_* = p_{f\min}$, equal to the lower boundary of scatter of contact fatigue limits p_f expressed by pressure in the center of the contact site. Volumes V_x, V_y, V_z are called *normal damaged volumes*. Based on the tangent stresses $\tau_{yz} = \tau_{zy}$ with the distribution according to formula (1.106), the *tangent damaged volumes* S_τ can be calculated that are limited by the critical stress $\tau_* = \tau_{f\min}$.

Based on the stress components σ_x, σ_y, σ_z, $\tau_{yz} = \tau_{zy}$, the equations of surfaces of damaged volumes are

$$\sigma_x = \sigma_*, \quad \sigma_y = \sigma_*, \quad \sigma_z = \sigma_*, \quad \tau_{zy} = \tau_*, \tag{2.28}$$

where σ_x, σ_y, σ_z, τ_{zy} are determined by Eqs. (1.105) and (1.106), respectively.

It is quite obvious that the configurations of the damaged volumes determined with the system of equations (2.28) are very intricate and it is impossible to obtain simple (engineering) formulas for their calculation following procedures (2.13) or (2.23). Hence, the statistical Monte Carlo method can be useful for determining the values of the damaged volumes. The essence of the method is that a cube is selected which includes the entire surface of the damaged volume (Fig. 2.22). A generator of random numbers serves to obtain a set of points regularly distributed within the cube. This operation can be arranged in the following manner. Assume that the length of an edge of the cube is equal to L and all three coordinates of the points it includes change from zero to L. By addressing the generator of random numbers triply we obtain three numbers y_1, y_2, y_3 within the interval (0, 1). They are used to plot the coordinates of the first point inside the cube with the help of formulas $X_1 = Ly_1$, $Y_1 = Ly_2$, $Z_1 = Ly_3$. After repeating the procedure Q times we obtain Q points that regularly fill up the cube on the average. Assume Q_1 is the number of points that happen to be inside the surface. Since the points distribute regularly, the number Q_1 characterizes the volume limited by the surface. Namely, if the number Q is sufficiently large, the required volume is equal to $L^3 Q_1/Q$.

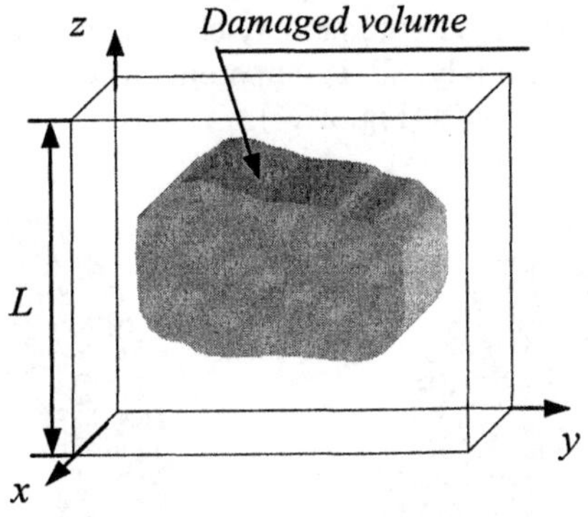

Fig. 2.22. To determination of the damaged volume with the Monte Carlo method

Table 2.11 provides the parameters for numerical modelling; they correspond to the rolling friction pair shown in Fig. 2.21.

Table 2.11. Input parameters for numerical modelling

Radius of the 1st roller r_1, mm	Radius of the 2nd roller r_2, mm	Thickness of rollers l, mm	Poisson coefficient, μ	Elasticity modulus E, MPa	Loading F_N, H
6	50	3	0.3	$2 \cdot 10^5$	1200

In the process of modelling the following critical levels of stresses are assumed: $\sigma_* = -750$ MPa and $\tau_{*yz} = \tau_{*zy} = \pm 300$ MPa.

Table 2.12 lists the results of calculation.

Table 2.12. Calculation values of the components of the damaged volume

p_0, MPa	b, mm	V_x	V_y	V_z	S_τ
		mm^3			
1615	0.16	$3.84 \cdot 10^{-4}$	$6.34 \cdot 10^{-4}$	5.19	$5.21 \cdot 10^{-2}$

Figure 2.23 shows graphically the damaged volumes. They have a typical configuration and according to the numerical values (in this case) they arrange in the following manner: $V_z > S_\tau > V_y > V_x$ (see also Table 2.12). The ratio between the components of the damaged volume may change in other cases when the input parameters change accordingly. It is apparent when all the components are combined in one coordinate grid (Fig. 2.24).

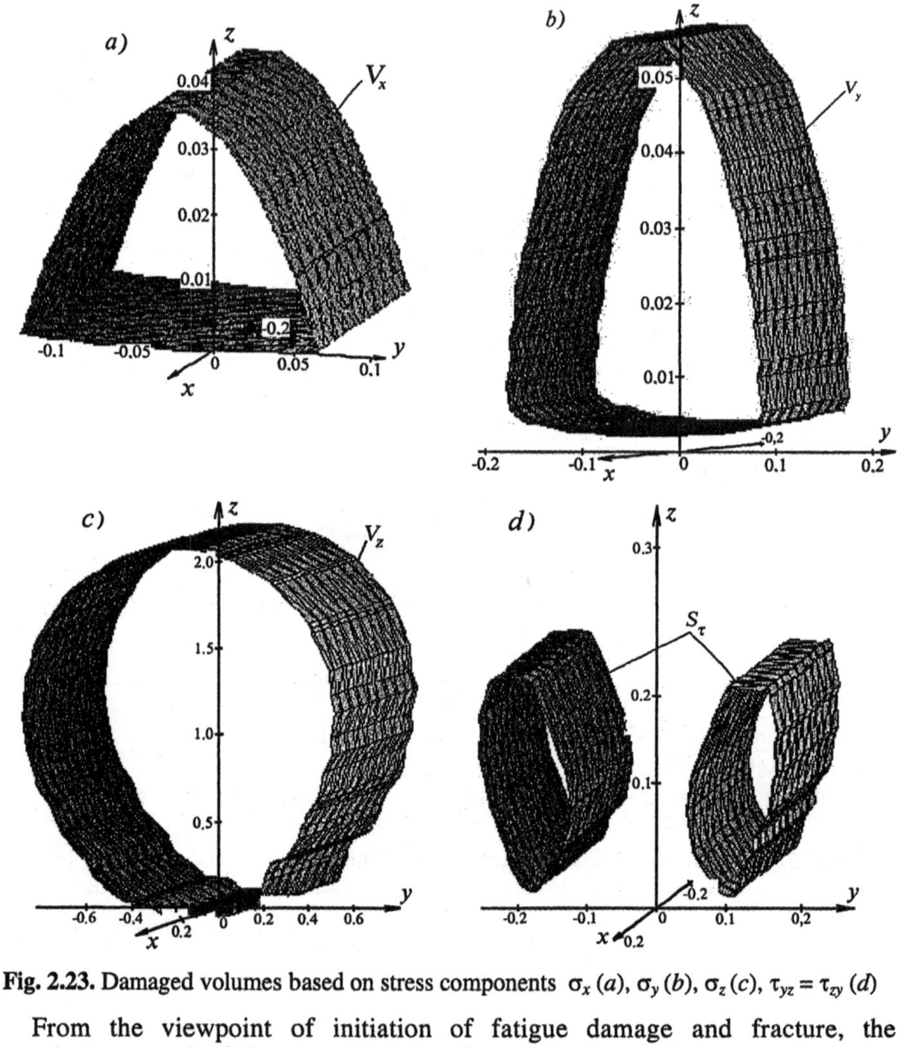

Fig. 2.23. Damaged volumes based on stress components σ_x (*a*), σ_y (*b*), σ_z (*c*), $\tau_{yz} = \tau_{zy}$ (*d*)

From the viewpoint of initiation of fatigue damage and fracture, the conclusions are the following. (1) Subsurface nucleation of fatigue cracks is due primarily to the existence of tangent damaged volumes $2S_\tau$. (2) Surface fracture (appearance of the pits of spalling) is governed in many respects by the ratio between the components V_x, V_y, V_z of the normal damaged volume. (3) The relation between the values of tangent and normal damaged volumes determines the real pattern of damage and fracture processes in rolling.

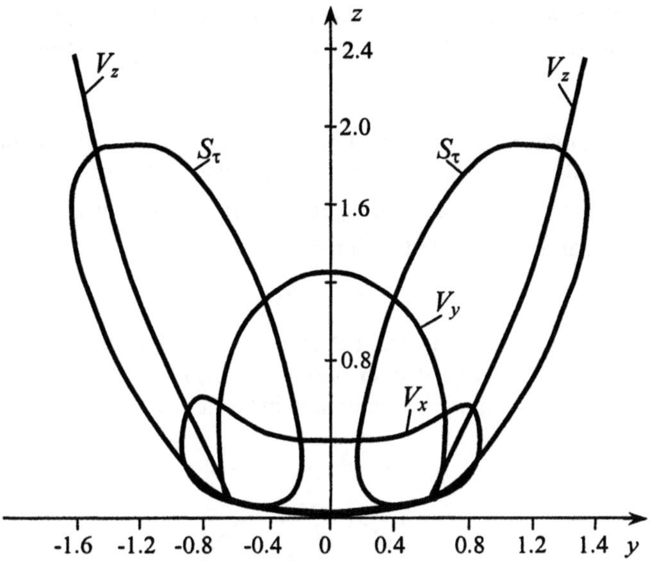

Fig. 2.24. Superposition of damaged volume components in the coordinate scale ($F_N = 5000$ N, $r_1 = 6$ mm, $r_2 = 50$ mm, $l = 1$ mm)

It should be verified that the analysis in Figs. 2.23 and 2.24 relates to the static contact between rollers. When the friction process evolves, a tangent damaged volume (2.23) appears on the surface of contact between two bodies due to frictional tangent stresses τ_W; its role in the processes of wear may be governing (see Sect. 1.4.3).

If the distributions of main stresses $\sigma_1 \geq \sigma_2 \geq \sigma_3$ around the contact site are known, it is possible to calculate *main damaged volumes* V_1, V_2, V_3.

Using formulas (1.105a) record the equations for the *surfaces of main damaged volumes*:

$$\left.\begin{aligned} \sigma_1 &= -\frac{p_0}{\sqrt{1+\left(\frac{z}{b}\right)^2}} = \sigma_*; \\ \sigma_2 &= -p_0\left[\frac{1+2\left(\frac{z}{b}\right)^2}{\sqrt{1+\left(\frac{z}{b}\right)^2}} - 2\frac{z}{b}\right] = \sigma_*; \\ \sigma_3 &= -2\mu p_0\left[\sqrt{1+\left(\frac{z}{b}\right)^2} - \frac{z}{b}\right] = \sigma_*. \end{aligned}\right\} \tag{2.29}$$

Since equations for surfaces (2.29) are determined with the condition that $y = 0$, the main damaged volumes V_1, V_2, V_3 lie in the plane xz. The combination

$$V_{123} = V_1 \cup V_2 \cup V_3 \tag{2.30}$$

is represented graphically for each of two rollers in Fig. 2.25 (shaded). In both rollers (with radii $r_2 >> r_1$) the values of the volumes (2.30) are equal: $V_{123}^{(r_1)} = V_{123}^{(r_2)}$. Since the main damaged volumes are shown graphically as rectangles, they can be calculated with the formulas

$$V_1 = lz_1, \quad V_2 = lz_2, \quad V_3 = lz_3, \tag{2.31}$$

where l – the thickness of the rollers (cf. Fig. 2.21), and z_1, z_2, z_3 – the coordinates of intersection of the curves of distribution of main stresses $\sigma_1(z)$, $\sigma_2(z)$, $\sigma_3(z)$ determined by the corresponding equation of system (2.29) with a straight line $\sigma_* = \text{const}$ (Fig. 2.26).

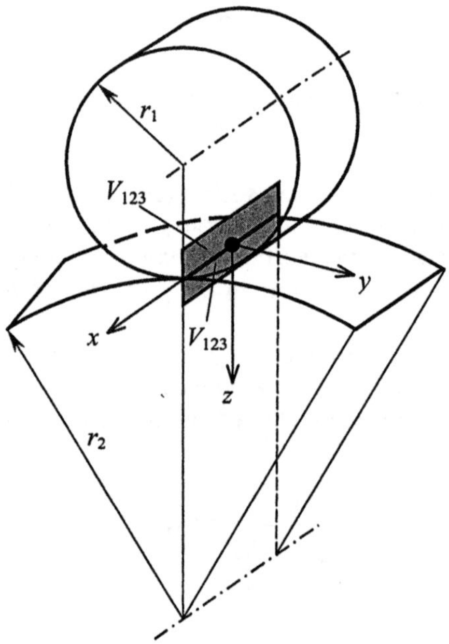

Fig. 2.25. Static contact between rollers and appearance of joins of main damaged volumes V_{123}

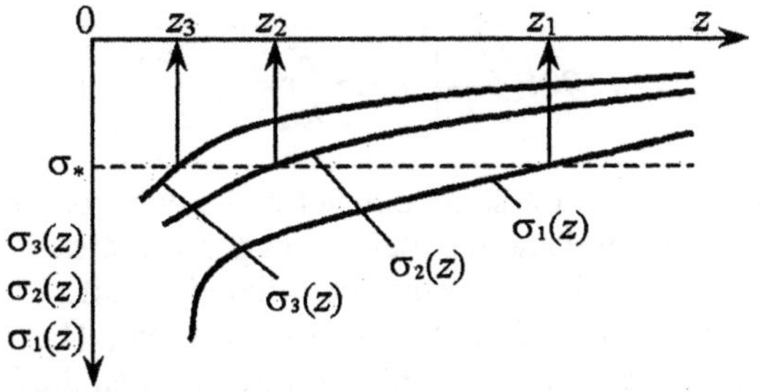

Fig. 2.26. To determination of coordinate z of bottom limits of main damaged volumes

Figure 2.27 combines the main damaged volumes V_1, V_2, V_3 determined with formulas (2.31) and shows the numerical values of their sizes calculated with the input parameters listed in Table 2.11. The rectangle z_1l provides a combination (2.30) of the damaged volumes; in this case we practically have $V_{123} = V_1$.

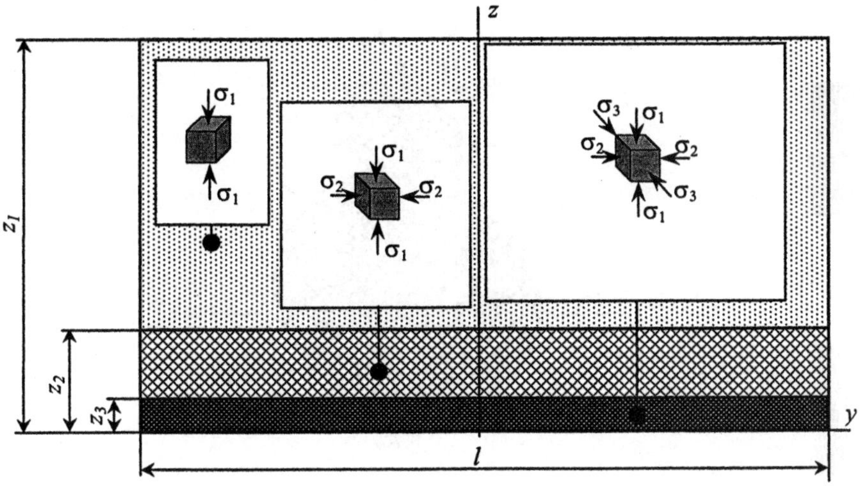

Fig. 2.27. Graphic images of main damaged volumes: $V_1 = 0.902$ mm^3; $V_2 = 0.17$ mm^3; $V_3 = 0.123$ mm^3

At a depth z_3 (cf. Fig. 2.27) there is an intersection $V_1 \cap V_2 \cap V_3$ of the main damaged volumes. They are all brought into coincidence (packed) within a single area with the dimensions z_3l within which a triaxial state of stress occurs ($\sigma_1 > \sigma_2 > \sigma_3$). At a depth ($z_2 - z_3$) there is an intersection of only two main damaged volumes ($V_2 \cap V_1$) and the state of stress is biaxial ($\sigma_1 > \sigma_2$).

The results of modelling of main damaged volumes lead to a conclusion that combination (2.30) of the main damaged volumes coincides with the volume V_1 within a quite broad range of loads. An analytical expression can be recorded relating its value to the input parameters, such as contact load, geometrical dimensions of rollers and constants of the material:

$$V_{123} = 1.522l\sqrt{\frac{F_N}{l}\left(0.175\frac{F_N}{l\sigma_{*1}^2} - E\frac{r_1 r_2}{r_1 + r_2}\right)} \ . \qquad (2.32)$$

Remember that in case of a static contact at a given compressive load F_N the damaged volumes in general and the main damaged volumes in particular have the same magnitudes in two rollers, though their diameters differ almost ten times. It is due to the fact that according to formulas (1.105), (1.106) and (2.29) the components of stresses are the same in both rollers. Then the bearing capacity of both rollers and their durability in rolling friction should be the same. It contradicts the experience. To resolve this contradiction the idea about *dynamic damaged volumes* is resorted to like in the case of mechanical fatigue (see Sect. 2.4.1).

Figure 2.28 shows a design diagram of the dynamic contact of the *roller / roller* pair: one body moves relatively to the other, the first roller having the angular speed ω_1, the second having the angular speed ω_2. In simple rolling $\omega_1 = \omega_2$.

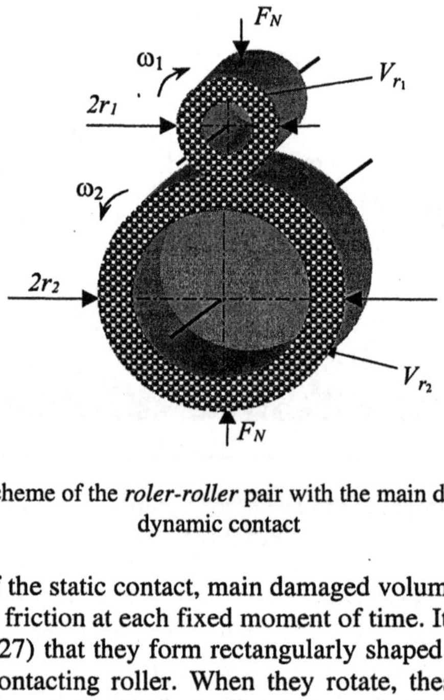

Fig. 2.28. Design scheme of the *roler-roller* pair with the main damaged volumes in dynamic contact

Like in the case of the static contact, main damaged volumes appear around the contact site in rolling friction at each fixed moment of time. It is established above (cf. Figs. 2.25 and 2.27) that they form rectangularly shaped sites arranged in the plane x, z of each contacting roller. When they rotate, these sites "sweep" the circularly shaped regions in the body that form the *main damaged volumes in the dynamic contact* (they are shaded in Fig. 2.28); let us designate them V_{r1} and V_{r2} for the first and second rollers, respectively. The following formulas enable to calculate the values of the main damaged volumes for each roller in dynamic contact:

$$\frac{V_{rj}}{V_{0j}} = 1 - \frac{K_{12}^2}{r_j^2}, \quad j = 1, 2, \tag{2.33}$$

where

$$K_{12}^2 = 0.41\left(\frac{F_N}{l}\right)^2 - \frac{F_N}{l}E\frac{r_1 r_2}{r_1 + r_2} = 0.41 q_l^2 (1 - 0.57 E_q R_{12}). \tag{2.34}$$

The linear load (N/m) is

$$q_l = F_N/l; \tag{2.35}$$

the reduced radius of curvature of the friction pair is

$$R_{12} = \frac{r_1 r_2}{r_1 + r_2}; \tag{2.36}$$

the linear rigidity (1/m) of the friction pair is

$$E_q = E/l; \tag{2.37}$$

V_{0j} – the geometrical volume of the roller for which the damaged volume is calculated.

Formula (2.33) for calculating a relative damaged volume when two cylindrical rollers roll has the structure similar to that of formula (2.18) for calculating damaged volumes in structural components in bending with torsion. Note that linear rigidity (2.37) of the friction pair in formula (2.34), hence, in (2.33), is associated with linear loading (2.35).

Summarizing we show how to solve the problem of assessment of damage of a given roller (with the radius r). If the epures of main stresses σ_i (i = 1, 2, 3) are determined, the *critical tensor of stresses* can be established for an isotropic material:

$$T_{\sigma*} = \begin{vmatrix} \sigma_1 = p_{f\,\min} & 0 & 0 \\ 0 & \sigma_2 = p_{f\,\min} & 0 \\ 0 & 0 & \sigma_3 = p_{f\,\min} \end{vmatrix}, \tag{2.38}$$

where $p_{f\,\min}$ – the lower boundary of scatter of contact fatigue limits p_f determined by the maximum pressuring the contact site center. If (2.38) is known, we record the *tensor of the damaged volumes* T_V and the *tensor of damage* T_ω:

$$T_V = \begin{vmatrix} V_{1r} & 0 & 0 \\ 0 & V_{2r} & 0 \\ 0 & 0 & V_{3r} \end{vmatrix}; \quad T_\omega = \begin{vmatrix} \omega_1 & 0 & 0 \\ 0 & \omega_2 & 0 \\ 0 & 0 & \omega_3 \end{vmatrix}, \tag{2.39}$$

where V_{ir} – the components of the dynamic damaged volume V_r dictated by the main stresses σ_i (i = 1, 2, 3). If the corresponding components of the static damaged volume are

$$V_i = \iiint\limits_{\sigma_i(x,y,z)>p_{f\min}} dxdydz, \tag{2.40}$$

then with some approximation

$$V_{ir} \approx 2\pi V_i r_c ,$$

where r_c – the radius of the centroid of a given component of the static damaged volume. A precise solution for V_{ir} is given, for example, by formula (2.33).

In the tensor T_ω (cf. (2.39)) the *main measures of damage*

$$0 \le \omega_i = \frac{V_{ir}}{V_k} \le 1 \tag{2.41}$$

are determined as a ratio between the components V_{ir} of the dynamic damaged volume excited in rolling friction and the working volume V_k of the roller ($V_k <<$ V_0, V_0 – its geometrical volume). For the case of contact in question (cf. Figs. 2.21 and 2.28) it is

$$V_k = \pi r^2 \left[1 - \left(1 - \frac{\Delta h_k}{r}\right)\right] l, \tag{2.42}$$

where Δh_k – the thickness of the worn surface layer of the roller when it reaches the limiting state.

Using the invariants of the tensor of damage (2.39)

$$\left.\begin{aligned} J_{\omega 1} &= \omega_1 + \omega_2 + \omega_3 ; \\ J_{\omega 2} &= -(\omega_1\omega_2 + \omega_2\omega_3 + \omega_3\omega_1); \\ J_{\omega 3} &= \omega_1\omega_2\omega_3, \end{aligned}\right\} \tag{2.43}$$

we determine the *tensorial measure of damage*

$$\omega_p^{(T)} = (J_{\omega 1} + J_{\omega 2} + J_{\omega 3})R_\omega = \left[1 - \prod_{i=1}^{3}(1-\omega_i)\right]R_\omega , \tag{2.44}$$

where R_ω – the parameter of interaction between damages(ω_i) set by formula (2.41). It follows from (2.44) that the main measures of damage (2.41) are in fact identical to particular probabilities of damage $0 \le P(\sigma_i) \le 1$ due to the main stresses σ_i in the sense that the algebra of events connected with the given test is applicable to both (see formulas (134), (134a) and (135)).

Note that the measures of damage (2.41) can be substituted for the *measures of continuity* opposite to them:

$$\psi_i = 1 - \omega_i = 1 - \frac{V_{ir}}{V_k}, \quad i = 1, 2, 3, \tag{2.45}$$

then, taking into account (2.45) in (2.44), we have the *tensorial measure of continuity*

$$\psi_p^{(T)} = \left(1 - \prod_{i=1}^{3} \psi_i\right) R_\psi . \tag{2.46}$$

The values ψ_i and $\psi_p^{(T)}$ vary within the interval

$$1 \geq \psi \geq 0. \tag{2.47}$$

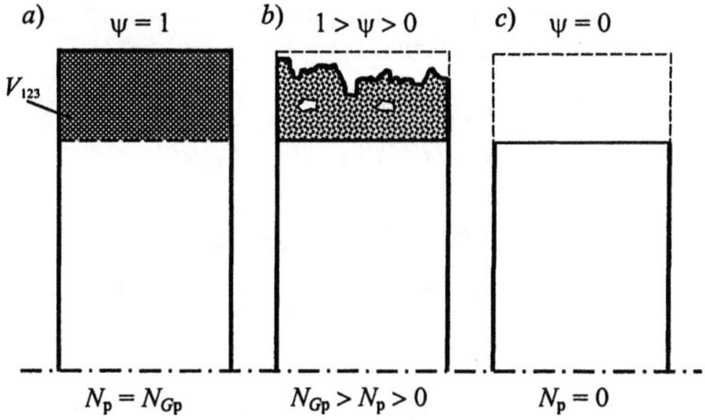

Fig. 2.29. Model of continuity loss

According to (2.45) and (2.46), the material is considered continuous at $\psi = 1$, i.e. undamaged, hence the durability N_p of the roller in case of contact fatigue under specified conditions remains unspent and equal, for example, to the value N_{Gp} (Fig. 2.29, *a*). It is the opposite case at $\psi = 0$, $N_p = 0$ because the material has lost its continuity within the working volume, it is destroyed, or worn (cf. Fig. 2.29, *c*). If $1 > \psi > 0$, it means that the durability is included into a corresponding interval $N_{Gp} > N_p > 0$ (Fig. 2.29, *b*). It is clear in this connection that the ideas of damage accumulation are physically justifiable in respect of the phenomenon of mechanical fatigue (see Sect. 1.3.2), while the idea of the loss of continuity by the surface layer of the object forms a model that better suits the wear processes (see Sect. 1.4.3).

2.4.3 Active system

Let us consider the active system shown in Fig. 2.1, *b*. It is apparent that all the solutions obtained in Sect. 2.4.1 for the structural component (namely, the shaft) and in Sect. 2.4.2 for the friction pair (*roller*/ *roller*) are the particular solutions for the given active system.

For the active system we will calculate the *complex damaged volume* according to the idea in (2.12), but we make an allowance that the dynamic damaged volume (2.33) is excited in friction pair; we have then

$$W_{P\gamma} = \varphi_{SV}(V_{P\gamma}, V_{rj}) \,. \tag{2.48}$$

An essential feature of the complex damaged volume (2.48) is that it is involved in forming two different *complex measures of damage* in response to the conditions of operation of the active system. If the direct effect occurs (hence, the effect of contact pressure on fatigue resistance is studied with due account of the parameter of interaction $R_{\sigma/p}$), then

$$\begin{aligned} \frac{W_{P\gamma}}{V_0} &= \frac{V_{P\gamma}}{V_0} \cup \frac{V_r}{V_0} = \left(\frac{V_{P\gamma}}{V_0} + \frac{V_r}{V_0} - \frac{V_{P\gamma}}{V_0}\frac{V_r}{V_0} \right) R_{\sigma/p} = \\ &= \omega_\sigma \left[1 + \omega_{\sigma p}(1 - \omega_\sigma) \right] \, R_{\sigma/p} \,, \end{aligned} \tag{2.49}$$

where the *particular measure of damage* $\omega_\sigma = \omega_{VP} = V_{P\gamma} / V_0$ (see 2.19), but

$$\omega_{\sigma p} = V_r / V_{P\gamma} \tag{2.50}$$

is the *integrated measure of damage* that indicates the proportion of the dynamic damaged volume as a function of contact load (V_r) in the damaged volume ($V_{P\gamma}$) as the function of the off-contact load and that dictates the direct effect.

Taking into account the tensorial measure of damage (2.44), we obtain a solution more general than (2.49):

$$\begin{aligned} \frac{W_{P\gamma}}{V_0} &= \omega_\sigma \cup \omega_p^{(T)} = \left(\omega_\sigma + \omega_p^{(T)} - \omega_\sigma \omega_p^{(T)} \right) R_{\sigma/p} = \\ &= \omega_\sigma \left[1 + \omega_{\sigma p}^{(T)} (1 - \omega_\sigma) \right] R_{\sigma/p} \,, \end{aligned} \tag{2.49a}$$

where

$$\omega_{\sigma p}^{(T)} = \frac{\omega_p^{(T)}}{\omega_\sigma} \,. \tag{2.50a}$$

If the back effect occurs (hence, the effect of cyclic stresses on the damage in rolling friction is studied with the account of the parameter of interaction $R_{p/\sigma}$), the complex measure of damage is

$$\frac{W_{P\gamma}}{V_k} = \frac{V_r}{V_k} \cup \frac{V_{P\gamma}}{V_k} = \left(\frac{V_r}{V_k} + \frac{V_{P\gamma}}{V_k} - \frac{V_r}{V_k} \frac{V_{P\gamma}}{V_k} \right) R_{p/\sigma} = \tag{2.51}$$

$$= \omega_\tau \left[1 + \omega_{p\sigma}(1 - \omega_p) \right] \ R_{p/\sigma} = \omega_p \left(1 + \omega_{p\sigma} \psi_p \right) R_{p/\sigma},$$

where, according to (2.41) and (2.45), the particular measures of damage $\omega_p = V_r/V_k$ and continuity $\psi_p = 1 - \omega_p$, and

$$\omega_{p\sigma} = V_{P\gamma} / V_r \tag{2.52}$$

are the total measure of damage indicating the proportion of the damaged volume as a function of the off-contact load ($V_{P\gamma}$) in the damaged volume as a function of contact load (V_r) dictating the back effect.

Taking into account the tensorial measures of damage (2.44) and continuity, (2.46), we obtain a solution more general than (2.51):

$$\frac{W_{P\gamma}}{V_k} = \omega_p^{(T)} \left(1 + \omega_{p\sigma}^{(T)} \psi_p^{(T)} \right) R_{p/\sigma}, \tag{2.51a}$$

where

$$\omega_{p\sigma}^{(T)} = \frac{\omega_\sigma}{\omega_p^{(T)}}. \tag{2.52}$$

It is essential to note that numerical values of the total measures of damage (2.50), (2.50□), (2.52) and (2.52□) can be larger, equal to or smaller than a unity:

$$\omega_{\sigma p}, \omega_{p\sigma} \gtrless 1,$$

and it makes them principally different from the measures of damage and continuity of type (2.19), (2.41), (2.45), in which the numerical values are included into a single interval (0; 1).

Figure 2.30 gives the analysis of formula (2.49) during a change of the parameter of damage $0.5 \le R_{\sigma/p} \le 1.33$. The following is apparent:

(1) the complex measure of damage ($W_{P\gamma}/V_0$) augments proportionally to the growth of $\omega_{\sigma p}$ and / or $R_{\sigma/p}$;

(2) ($W_{P\gamma}/V_0$) > ($V_{P\gamma}/V_0$) = ω_σ, always at $R_{\sigma/p} \ge 1$, i.e. the complex damaged volume during mechano-rolling fatigue cannot be less than the damaged volume during mechanical fatigue, providing damages *strongly* interact under the effect of contact and off-contact loads;

(3) a situation appears at $R_{\sigma/p} < 1$ when ($W_{P\gamma}/V_0$) < ω_σ, i.e. the complex damaged volume during mechano-rolling fatigue becomes smaller than during mechanical fatigue, providing damages interact *weakly* under the effect of contact and off-contact loads.

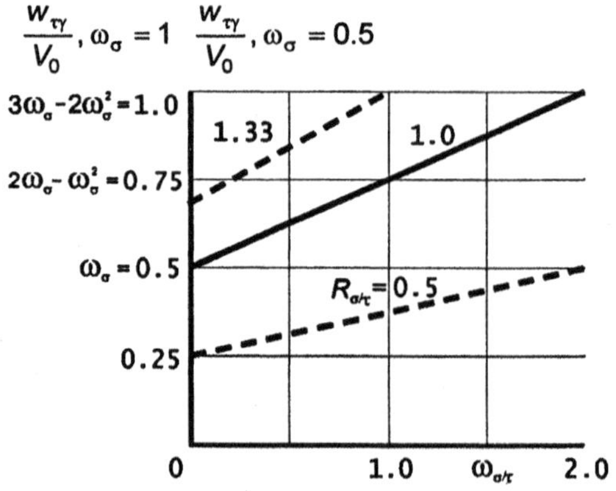

Fig. 2.30. To analysis of formula (2.49)

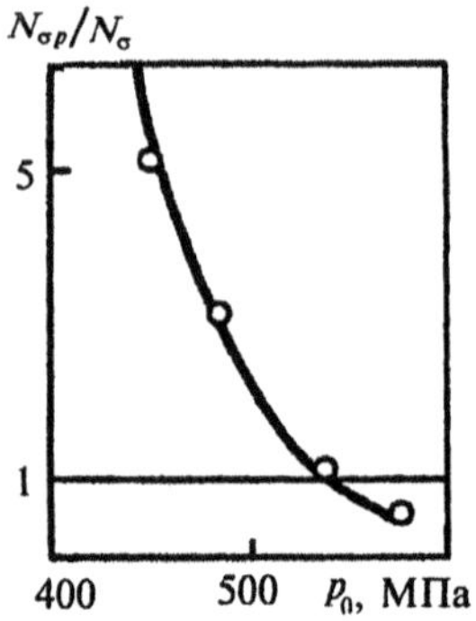

Fig. 2.31. Relation between durability of steel 40X during mechano-rolling fatigue and pressure in the contact site center

It is apparent that in case (2) strong interaction between damages results in the *softening*, meanwhile in case (3) weak interaction results on the opposite in the *hardening of the material.* Formula (2.51) leads to similar conclusions if the complex damaged volume during mechano-rolling fatigue is compared with the damaged volume during contact fatigue. It means that the following ratios of durabilities are possible during mechanical (N_σ), contact (N_p) and mechano-rolling ($N_{\sigma p}$ or $N_{p\sigma}$) fatigue:

$$N_{\sigma p} \gtrless N_\sigma,\ N_{p\sigma} \gtrless N_p. \tag{2.53}$$

Note that the subscript in $N_{\sigma p}$ designates the durability based on the criterion of mechanical fatigue (σ) with the account of the effect of contact load (p), i.e. the complex durability in case of the direct effect. The subscript in $N_{p\sigma}$ designates the durability based on the criterion of contact fatigue (p) with the allowance for cyclic stresses (σ), i.e. the complex durability in case of the direct effect.

Figure 2.31 provides an experimental confirmation of one of the conclusions (2.53). The durability at $p_0 \approx 550$ MPa during mechanical and mechano-rolling fatigue turns out approximately the same ($N_{\sigma p}/N_\sigma = 1$). If the contact pressure drops to ~450 Mpa, the durability during mechano-rolling fatigue is approximately 5 times larger than during mechanical fatigue ($N_{\sigma p}/N_\sigma = 5$), but if pressure grows to ~580 Mpa, the durability during mechanical fatigue becomes on the opposite two times larger than during mechano-rolling fatigue ($N_{\sigma p}/N_\sigma \approx 0.5$).

The complex damaged volume (2.48) is the absolute measure of damage of the active system; its relative value is

$$0 \le \frac{W_{P\gamma}}{V} \le 1 \tag{2.54}$$

similarly to measures (2.19), (2.20) and (2.26) during mechanical fatigue and friction, so the *condition of failure-free operation* of the system is

$$\frac{W_{P\gamma}}{V} = 0 \tag{2.55}$$

and the *condition of its limiting state* is

$$\frac{W_{P\gamma}}{V} = 1\,. \tag{2.56}$$

The *similarity criterion of the WFD* of active systems is the relation

$$\frac{W_{P\gamma}}{V_{st}} = \Pi_{WP}\,. \tag{2.57}$$

In formulas (2.54)–(2.57) $V = V_0$ if the direct effect occurs (see (2.49)) or $V = V_{kj}$ if the back effect occurs (see (2.51)).

It follows from the above that the *complex measures of damage* (2.49), (2.49a), (2.50), (2.51a) and (2.54) *estimate the degree of damage of the components of an active system as functions of contact and off-contact loads*; in other words, *they indirectly characterize the concentration of damage in dangerous regions (zones) of deformable solids*. In response to the conditions of interaction the concentration of damage may either grow (and then $R_{\sigma/p}$, $R_{p/\sigma} > 1$) or, vice versa, reduce (and then $R_{\sigma/p}$, $R_{p/\sigma} < 1$). Hence, the *parameter (or function) of interaction* $R_{\sigma/p}$, $R_{p/\sigma}$ *characterizes generally both the result and the direction of interaction between damages*. While the *complex damaged volume* in the components of an active system *describes a specific mechanical state* (see Sect. 1.2.1) *of the material or the state of its damage due to the fields of interacting stresses as functions of contact and off-contact loads*. The damaged volumes in structural components (see Sect. 2.4.1) or in friction pairs (see Sect. 2.4.2) describe similarly the state of damage in specific conditions of cyclic deformation or friction.

2.5 Interaction between damages

Interaction between dangerous volumes under the effect of contact and off-contact loads described in formulas (2.48), (2.49) and (2.51) makes the processes of damage and fracture in the active system strongly different from those in the friction pair or in the structural component. It can be illustrated by three examples.

(1) Figure 1.22 shows typical initial damage of the surface of the shaft in the process of mechanical fatigue (under the effect of off-contact load): they are extrusion and intrusion. If the shaft becomes one of the bodies of the friction pair in rolling, its initial surface damage is cardinally different. Figure 2.32, *a* shows the surface of the shaft after tests for contact fatigue. As a result of contact friction processes in rolling a specific granular structure appears and microcracks are clearly obvious against its background. Both grains and cracks are elongated and

unidirectional in respect to motion in friction. They result from deformation fragmentation and initial fracture of the original structure. Meanwhile, if the shaft becomes a component of the active system, its initial surface damage (Fig. 2.32, *b*) is principally different from the one observed during mechanical (cf. Fig. 1.22) and contact (cf. Fig. 2.32, *a*) fatigue. Now, during mechano-rolling fatigue, complex damage occurs: an intersecting system of multiple strips of sliding and submicrocracks or pores. It is the result of interaction between damages within the complex damaged volume.

Fig. 2.32. Half-tone images of the surface of the shaft from steel 45 obtained with atom force microscopy (scanning area 35×35 μm^2)

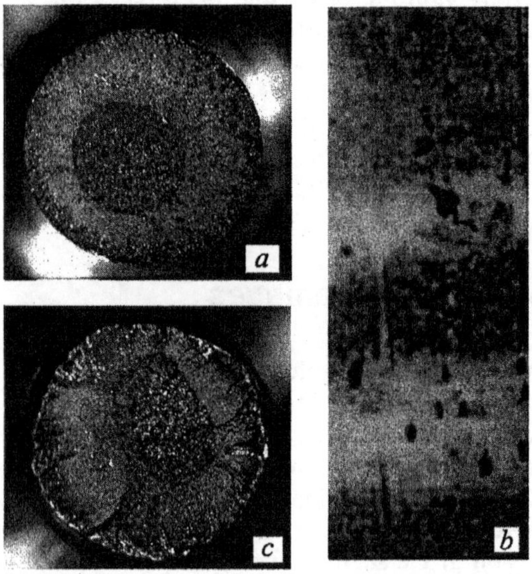

Fig. 2.33. Fracturing of the shaft during mechanical fatigue (*a*), pitting spots on its surface during contact fatigue (*b*) and fracturing during mechano-rolling fatigue (*c*)

(2) The limiting state of the shaft under cyclic loading when it snaps into two pieces due to the nucleation and development of the main fatigue crack. As a rule, the center of crack is detected in one "weak zone" located near the surface (Fig. 2.33, *a*, see also Fig. 1.10). The limiting state of the same shaft during contact fatigue can be reached when pitting spots of critical density appear on the surface of rolling (Fig. 2.33, *b*). In case of mechano-rolling fatigue the rupture of the shaft is principally different (Fig. 2.33, *c*): a large number of multiple fatigue cracks appear in the circular surface zone; the pattern of rupture can be determined as multicenter (multiblade). It is exactly the result of interaction between damages within the complex damaged volume located in the circular surface zone of the shaft. Smooth edges of a common fatigue crack and jagged edges of multiple contact fatigue cracks can be seen in Fig. 2.34, *a*, the cracks develop in a zigzag way from one to another (weak) groups of surface pitting spots (Fig. 2.34, *b*).

a) *b*)

Fig. 2.34. Edges of the main crack during mechanical (*a*) and mechano-rolling (*b*) fatigue

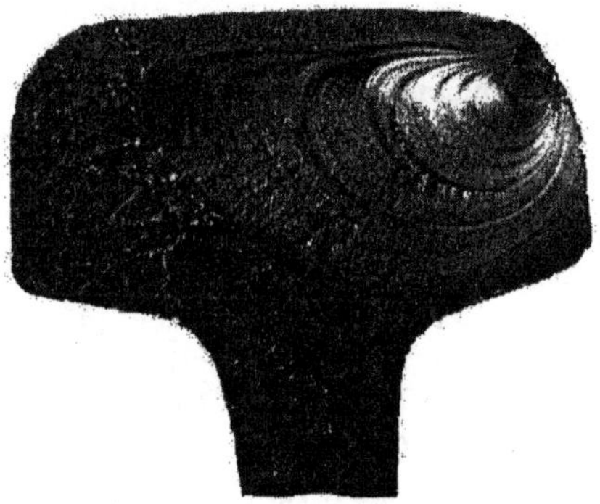

Fig. 2.35. Subsurface center of the main crack during mechano-rolling fatigue of the rail

(3) Small multiple initial subsurface cracks can nucleate during contact fatigue in response to the conditions of deformation. They can develop parallel to the contact site and lead to a peculiar surface damage – wear by delamination (see Sect. 1.4.6). A subsurface center of the transverse main fatigue crack is found

during mechano-rolling fatigue under definite conditions that develops and leads to disintegration of the object into pieces (Fig. 2.35). The conditions for this very cardinal change in the pattern of fracture results from interaction between damaged volumes. It is corroborated by a numerical example. Consider the active system of type 2.1, *b*, but let us additionally load it with torque M_K (Fig. 2.36).

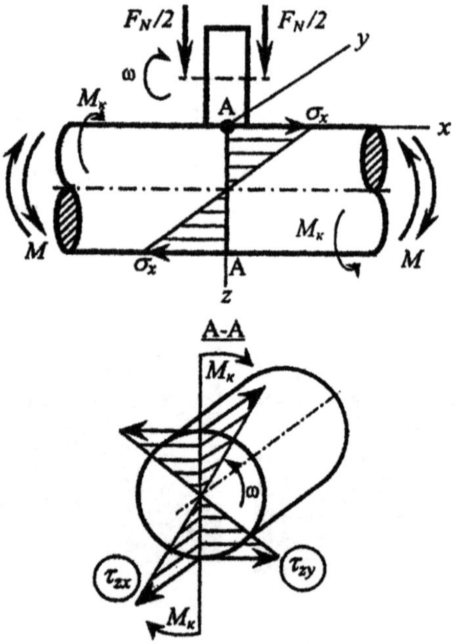

Fig. 2.36. Active system of the *shaft / roller* type loaded with bending moment and torque

It is understandable that the configuration and numerical values of damaged volumes as a function of contact load of both elements of this active system are the same like those of the friction pair. Figure 2.22 exemplifies the damaged volumes using the stress components σ_x, σ_y, σ_z, $\tau_{yz} = \tau_{zy}$. Hence, the problem of calculating the damaged volumes and assessing damage of one of the components of the system (namely, the roller) is already solved and discussed (see Sect. 2.4.2). Yet, damaged volumes under both off-contact (M and M_K) loads appear additionally in one of the components of the active system (namely, in the shaft). They are due to the distribution of normal and tangent stresses through the cross section of the shaft. Figure 2.19, *a* shows the configuration of the damaged volume due to bending stresses; based on Fig. 2.37 it is clear that the damaged volume produced by tangent stresses will have a similar configuration when it is due to torque.

Accepting the requirement that the system of coordinates in the active system should be the same lies in the friction pair (see Figs. 2.24 and 2.20), the distributions of normal and tangent stresses due to the bending M and torsion M_K moments are calculated with the formulas

$$\sigma_x = \frac{1}{r}\frac{M}{W}(z-r)\,;\;\; \tau_{zy} = \frac{1}{r}\frac{M_K}{W_p}(z-r)\,, \tag{2.58}$$

where W_p – the polar moment of resistance to torsion. Assume that the roller contacts the shaft in the zone of compression with the force $F_N = 4000$ N, assume that $\sigma_x = 200$ MPa at $z = 0$ and $\tau_{zy} = 200$ MPa at $z = 0$ in formulas (2.58). Figure 2.37 shows the corresponding damaged volumes.

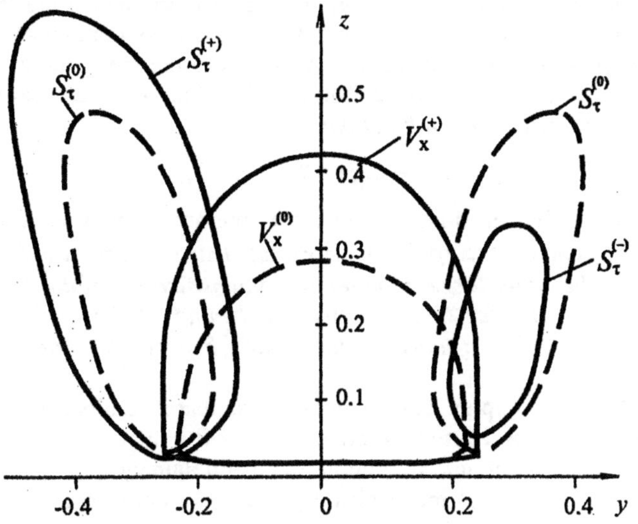

Fig. 2.37. Changes of damaged volumes in response to the contact and off-contact loads during one revolution of the shaft

Two combinations of damaged volume are shown that appear during one revolution of the shaft. Dotted lines show the damaged volumes $V_x^{(0)}$ and $S_\tau^{(0)}$ due exclusively to the contact load (the volumes $V_y^{(0)}$ and $V_z^{(0)}$ are not shown because their formation is independent on stresses σ_x determined from formula (2.58)). Full lines outline the damaged volumes $V_x^{(+)}$, $S_\tau^{(+)}$ and $S_\tau^{(-)}$ due to all the loads, the superscripts imply that during calculation of damaged volumes stresses due to contact and off-contact loads are summed up (+) if they have the same sign or deducted (–) if the signs are different. Figure 2.37 leads to two conclusions. (1) Subsurface nucleation of cracks (during contact fatigue) is possible in friction pair within the tangent damaged volumes $S_\tau^{(0)}$. These volumes arrange symmetrically on both sides of the plane *zx*, therefore, the first cracks are also detected in the similar regions under the contact site. (2) Subsurface development of cracks in the active system (during mechano-rolling fatigue) is expected in the tangent damaged volume $S_\tau^{(+)}$ because it is significantly larger than the volume $S_\tau^{(-)}$. The volume

$S_\tau^{(+)}$ is located on one side of the plane zx, hence the center of the main fatigue crack is always located on one side of the symmetry plane through the cross section of the rail, as it is obvious in Fig. 2.35.

Thus, *since diverse and innumerable events and effects of interactions between damages of many types cannot be described or predicted precisely, an idea is introduced about interaction between damaged volumes that contain a real complex of damages (defects) originating under the effect of corresponding fields of stresses (deformations). The damaged volume can serve an equivalent of the complex of damages* because its magnitude is proportional to the level of actual stresses, hence to the number (concentration) of defects (damages). The problem becomes essential in this connection how to determine the function of interaction in models (2.12), (2.44), (2.46), (2.48), (2.51). In fact, it is necessary to introduce and identify *a specific class of functions*

$$\Lambda_t(\omega_\sigma \leftrightarrow \omega_p) = R_{\sigma/p} \gtreqless 1, \tag{2.59}$$

that will help predict, in time t as well, the interaction between damages due to off-contact (σ) and contact (p) loads (stresses). *This interaction is dialectic, i.e. its result $R_{\sigma/p}$ can be either larger than a unity (intensification of the damaging effect, or softening), or smaller than a unity (weakening of the damaging effect, or hardening), or equal to a unity (a stable ratio between the hardening-softening processes).*

Thus, WFD is a complex damage in the sense that it results from the interaction between any damages caused by contact and off-contact loads that develop at different scale levels (submicro-, micro- and macrodamages).

From the standpoint of mechanics, a *full fatigue* curve *reflects basic possible types of volume fracture during alternating loading* (cf. Fig. 1.15): quasistatic (region I), low-cycle (II), multicycle (III) and high service life (IV) fatigue. It is to recall that they are due to the mechanics of deformation: the open loop of elasticoplastic hysteresis (I), the unclosed loop of elasticoplastic hysteresis (II), the closed loop of mechanical hysteresis (III), the degenerate loop of mechanical hysteresis (IV). A *similar curve in friction reflects the basic possible types of surface damage* (cf. Fig. 1.54). In the general case there are similar typical regions: quasistatic (I), low-cycle (II), multicycle (III) and high service life (IV) damages. They are due to the mechanics of surface deformation and illustrated by motion (with the speed ϑ) of a single irregularity imbedded into the plane: microcutting (region I), plastic pushing (II), elastic pushing (III), fracture of films (IV).

Comparison of Figs. 1.15 and 1.54 shows that interaction between damages due to contact and off-contact loads may be highly variable and intricate. So, if a component undergoes deformation in the multicycle region (III in Fig. 1.15), the processes of friction and wear may follow any mechanism (I, II, III or IV in Fig. 1.54). Yet, in each case there is the most essential feature relating to the dialectic interaction between damages due to contact and off-contact loads.

2.6 Stages of damage and fracture

2.6.1 General

It is noted in Sect. 1.3.4 that the process of mechanical fatigue in the general case evolves in two stages: stage *I* before the main crack nucleates rated as durability N_I, and stage *II* or the stage of survivability of a cracked structural component rated as durability N_{II}. Taking into acocunt the concept of damaged volume in the deformable solid (see Sect. 2.4.1), it should be made clear that *scattered damage* is observed at stage *I* and *localized fracture* at stage *II*.

Scattered damage within the damaged volume is typical both for the so-called smooth bodies and for the components with design stress concentrators. Figure 2.38 shows several microcracks in sharp (the radius is $r = 0.5$ mm, the theoretical coefficient of stress concentration is $\alpha_\sigma = 8$ – Fig. 2.38, *a*) and sloping ($r = 2$ mm, $\alpha_\sigma = 2.55$ – Fig. 2.38, *b*) notches and also two fatigue cracks spaced at a distance 25 mm one from another over the fillet portion from the crank pin to the crank web of the crankshaft ($r = 18$ mm, $\alpha_\sigma = 3.2$ – Fig. 2.38, *c*); the diameter of the crank pin is 360 mm.

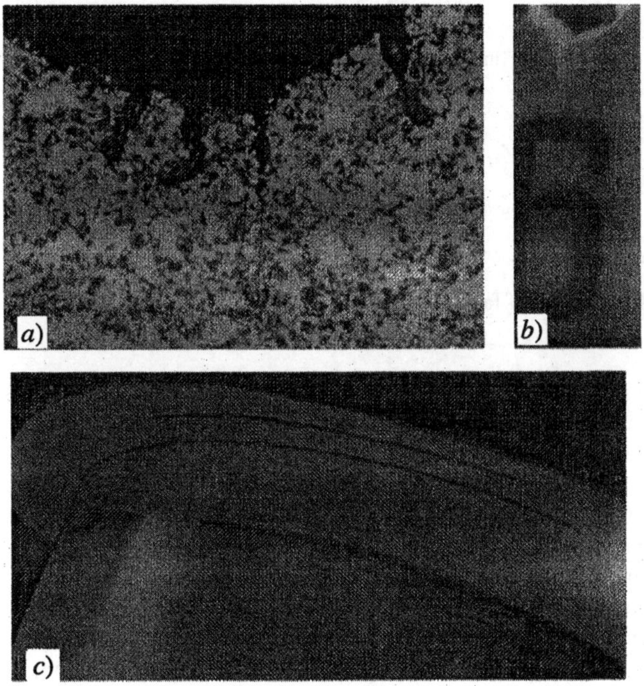

Fig. 2.38. Fatigue microcracks in the zones of concentration of stresses

Relation between the fatigue limits of metallic specimens and the theoretical coefficient of stress concentration is a proof that a damaged volume comparable with the geometrical volume is needed for fatigue fracture of a body (Fig. 2.39). Fatigue fracture due to the main crack occurs only in the region *C* determined by the critical point with the coordinates (α_σ^*, σ_{-1min}). If $\sigma < \sigma_{-1min}$ and $\alpha_\sigma > \alpha_\sigma^*$ there is region *B*: microscopic cracks in it are localized in the small volume and therefore they do not develop. Microcracks do not appear at all in region *A*. Hence, fatigue fracture is determined both by the level of actual stresses and the degree of their localization in the volume of the body. Primary damage in the form of submicro- and microcracks in the microvolumes of the body can grow into fatigue macrofracture providing there is a sufficiently large damaged volume in which the necessary conditions appear that favor interactions between numerous primary defects, their kinetic accumulation up to the critical concentration followed by aggregation of the most dangerous defects into the destroying main macrocrack.

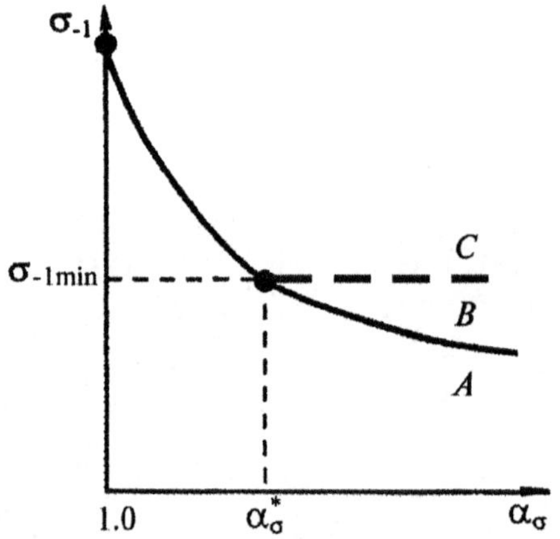

Fig. 2.39. Regions of fatigue damage and fracture due to the level of localization of actual stresses

Thus, the damage by *small (short) fatigue cracks* usually preceeds destruction by a (long) main crack. The crack is *long* when its typical linear size is one order of magnitude larger than a typical structural component (a grain) in the material. Short cracks have the length comparable with microstructural components of the material; they are cracks that have the dimensions from 0.001 mm for high-strength and to 0.1...1.0 mm for low-strength materials.

There is no distinct boundary between mechanical fatigue stages *I* and *II* since there is no unambigous division of cracks into short and long. However, the onset of stage *II* is attributed to the appearance of conditions of applicability of linear fracture mechanics to the analysis of the state of stress at the tip of the crack, hence, to the assessment of the coefficient of intensity of stresses *K* because its magnitude governs the rate of development of the main crack (see Sect. 1.3.4).

Studies have revealed that a specific region of plastic deformation appears ahead of the front of the developing crack under certain conditions (Fig. 2.40). A similar region of damage appears during mechanical fatigue and it in fact represents a peculiar damaged volume Ω_K. The cross sectional dimension of this volume can be approximately determined from the formula

$$d_* = \frac{1}{\alpha\pi}\left(\frac{K_{Ic}}{\sigma_*}\right)^2,$$

where α – the parameter depending on the type of the state of stress; K_{Ic} – the critical value of the coefficient of intensity of stresses during static loading corresponding to the onset of unsteady crack development; σ_* – the proportionality limit during cyclic loading.

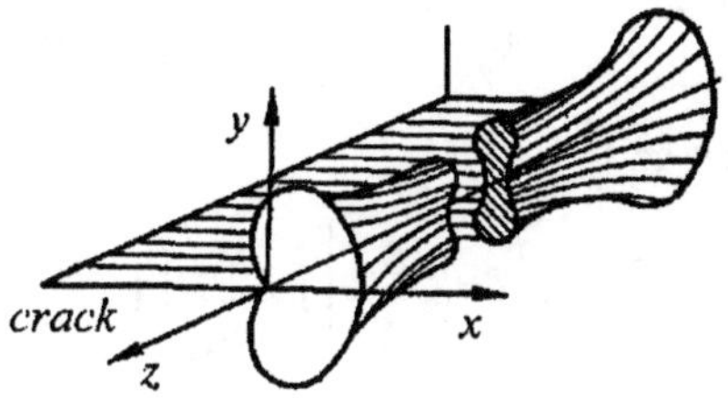

Fig. 2.40. Distribution of plastic deformation at the tip of the crack

The relative damaged volume during stage *II* of fatigue fracture (in case of deterministic approach) is

$$\omega_K = \frac{\Omega_K}{\Omega_0}, \tag{2.60}$$

where Ω_0 – the working volume determined by the area A_0 of dangerous cross section where the main crack develops; it is assumed that the area A_0 has a single thickness so that $\Omega_0 = 1 \cdot A_0$. Apparently (2.60) is the *measure of damage of the body with the main crack.* If the latter is flat, relation (1.48) between the area A_l the crack occupies and the working area A_0 of transverse cross section is used instead of (2.60).

So, the state of damage during mechanical fatigue in the general case is due to the level of cyclic stresses (normal and / or tangent) at stage *I* and the size of the damaged volume $V_{P\gamma}$ (see Sect. 2.4.1), and at stage *II* to the level of the coefficient of intensity of stresses (K_I, K_{II} and/or K_{III}) and the size of the damaged volume Ω_K.

The process of damage and fracture in friction (see Sect. 1.4.3) lacks stage *II* of main crack development because it is absent. Yet multiple (scattered) cracks growing within the damaged volume in friction and governing its wear (cf., for example, Fig. 2.29) are considered in some cases as *reduced single* inclined and subsurface cracks (cf. Fig. 1.79) to which the ideas of linear fracture mehanics are applicable. Hence, the state of damage of the material of a friction pair is believed to be due to contact stresses or the coefficient of intensity of stresses and to the size of the damaged volume V_r (see Sect. 2.4.2).

Fields of stresses excited by contact and off-contact loads interact and produce complex damaged volumes $W_{P\gamma}$ (see Sect. 2.4.3) in the active system. Both stages of damage and fracture evolve in case of direct effect when the limiting state is reached according to the criterion of mechanical fatigue and damage due to contact load is concomitant. Only the first stage of damage and fracture evolves in case of back effct when the limiting state is reached based on the criterion of wear and damage due to alternating loading is concomitant. In these cases the complex measures of damage are calculated with formulas (2.49), (2.49a) and (2.51), (2.51a), respectively, if the process of mechano-rolling fatigue is studied. Similarly in respect of mechano-sliding fatigue we have

$$\frac{W_{P\gamma}}{V_0} = \frac{V_{P\gamma}}{V_0} \cup \frac{S_{P\gamma}}{V_0} = \frac{V_{P\gamma}}{V_0}\left[1 + \frac{S_{P\gamma}}{V_{P\gamma}}\left(1 - \frac{V_{P\gamma}}{V_0}\right)\right] R_{\sigma/\tau} \tag{2.61}$$

in case of direct effect and

$$\frac{W_{P\gamma}}{V_k} = \frac{S_{P\gamma}}{S_k} \cup \frac{V_{P\gamma}}{S_k} = \frac{S_{P\gamma}}{S_k}\left[1 + \frac{V_{P\gamma}}{S_{P\gamma}}\left(1 - \frac{S_{P\gamma}}{S_k}\right)\right] R_{\tau/\sigma} \tag{2.62}$$

in case of back effect.

$R_{\sigma/\tau}$, $R_{\tau/\sigma}$ are relevant parameters (or functions) of interaction between damages due to contact and off-contact loads in formulas (2.61) and (2.62) .

Note that it is better to use the specific force of friction (frictional stresses) in case of sliding as the main damaging parameter, while contact stresses can be used for this purpose in case of rolling. If the process of rolling is accompanied by a tangent force, it can affect significantly the intensification of WFD.

2.6.2 Durability at stage I

We will show the solution of the problem of assessing the durability N_I of the active system at stage *I*. Assume for definiteness that the system operates under the conditions of mechano-sliding fatigue (cf. Fig. 2.1, *a*). Then both damaged volumes $V_{P\gamma}$, $S_{P\gamma}$ (shown schematiclly in Fig. 2.40) and the complex damaged volume $W_{P\gamma} = \varphi(V_{P\gamma}, S_{P\gamma})$ appear on the shaft's surface around the contact site during the first loading cycle.

Introduce the measure of WFD accumulated during n cycles of loading

$$\omega_{nT} = W_{nT} / W_{P\gamma}, \tag{2.63}$$

where

$$W_{nT} = \sum_1^q V_q ;$$

V_q – elementary volumes of scattered damage within the complex damaged volume $W_{P\gamma}$. Thus, , W_{nT} is a structurally damaged volume due to the number of loading cycles n with the unchanged level of cyclic and frictional stresses. Therefore, ω_{nT} is the measure of structural damage due to temperature – time and the state of stress of a component of the active system. It is clear that $W_{nT} \leq W_{P\gamma}$. The critical (or ultimate) state occurs in the damaged volume when the value W_{nT} attains the magnitude $W_{P\gamma}$; for example, it becomes entirely permeated with multiple (scattered) cracks in a critical concentration. Hence, at $\omega_{nT} = 1$ the component of the active system within the damaged volume becomes unable to resist effective loads. Practically it means that the dangerous volume is damaged with an initial main crack, like it is shown in Fig. 2.41. So, the main crack appears when

$$W_{nT} = W_{P\gamma},$$

or

$$\omega_{nT} = \omega_N = 1. \tag{2.64}$$

The process of accumulation of local damages and microfractures can be described with the curve of type *2* or *3* (cf. Fig. 2.41). In case there are no conditions for kinetic development of damages (see region *B* in Fig. 2.39), primary cracks remain underdeveloped (see dotted line *1* in Fig. 2.41).

Assume that the exponential function of cyclic σ and frictional τ_W stresses as well as temperature T determine the rate ϑ_I of accumuation of structural damages ω_{nT} scattered within the complex damaged volume at stage *I*. Then, with due regard of the basic postulates of the kinetic theory of strength of solids (see Sect. 1.3.2), we obtain

$$\vartheta_I = \frac{d\omega_{nT}}{dn} = C_U \exp\left(-\frac{U_0 - f(Q)}{kT}\right), \tag{2.65}$$

where the function of contact and off-contact loads is

$$f(Q) = \left[\gamma_\sigma(\sigma/\sigma_{-1\min}) + \gamma_\tau(\tau_W/\tau_{f\min})\right]\ \lambda\,(\sigma \leftrightarrow \tau_W); \tag{2.66}$$

γ_σ and γ_τ are structurally sensitive coefficients. It is taken into account that accumulation of damages ω_{nT} as the number of loading cycles n grows takes place only when $\sigma > \sigma_{-1\min}$ and $\tau_W > \tau_{f\min}$; if $\sigma < \sigma_{-1\min}$ and $\tau_W < \tau_{f\min}$, then it is believed that $\vartheta_I = 0$. Of course, assuming inverse relations

$$\gamma_\sigma = a_\gamma / \omega_{nT}, \quad \gamma_\tau = b_\gamma / \omega_{nT},$$

from (2.65) with the account of (2.64) we obtain

$$\int_0^{n=N_I} dn = \frac{1}{C_U} \int_{\omega_1}^{\omega_N=1} \frac{d\omega_{nT}}{\exp\left(-\dfrac{U_0}{kT} + \dfrac{1}{\omega_{nT}}\dfrac{f(Q)}{kT}\right)},$$

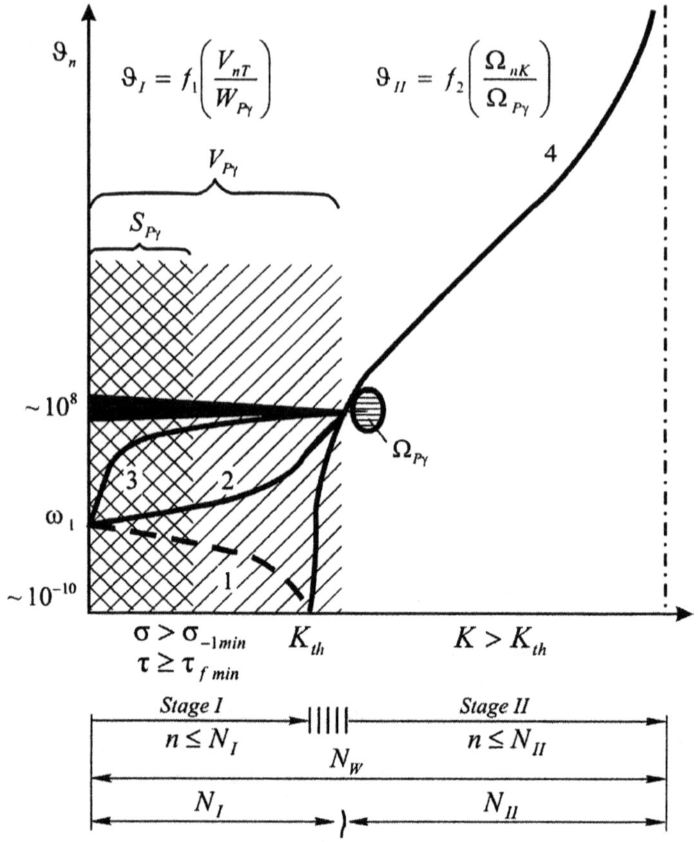

Fig. 2.41. Schematic representation of the kinetics of damages and fracture of the element of the active system

that after integration with some approximation yields

$$N_I = \frac{1}{C_U}\left[\exp\left(-\frac{U_0 - f(Q)}{kT}\right)\right]\varphi_1(\omega_1), \tag{2.67}$$

where the *function of damage* is

$$\varphi_1(\omega_1) = 1 - \omega_1 \exp\left[\frac{f(Q)}{kT}\left(1 - \frac{1}{\omega_1}\right)\right] \tag{2.68}$$

and the measure of damage during the 1-st loading cycle is

$$\omega_1 = W_{P\gamma}/V. \tag{2.69}$$

Measure (2.69) is determined from formulas (2.61), (2.62) during mechano-sliding fatigue or from formulas (2.49) and (2.51) during mechano-rolling fatigue.

Thus, formulas (2.67) and (2.68) have a rather general nature; they are applicable to the conditions of frictional, contact, mechanical, mechano-rolling and mechano-sliding fatigue, if the loading function $f(Q)$ is specified respectively, and measure of damage (2.69) is recorded with the allowance for the typical damaged volumes (Table 2.13). The methods of assessing them are disclosed above. It is assumed in Table 2.13 that $a_\gamma \approx b_\gamma$, the parameter C_U that makes sense of the initial rate of damage is designed with A with indexes corresponding to a given type of WFD.

Table 2.13. Formulas for determination of $f(Q)$ and ω_1

Types of wear-fatigue damage	$f(Q)$	ω_1	C_U
Mechanical fatigue	$a_{\gamma\sigma}(\sigma / \sigma_{-1\min})$	$\frac{V_{P\gamma}}{V_0}$	A_σ
Rolling fatigue	$a_{\gamma p}(p_0 / p_{f\min})$	$\frac{V_r}{V_k}$	A_p
Sliding fatigue	$a_{\gamma\tau}(\tau_W / \tau_{f\min})$	$\frac{S_{P\gamma}}{S_k}$	A_τ
Mechano-rolling fatigue:			
a) direct effect	$a_{\sigma p}\frac{\sigma}{\sigma_{-1\min}}\left(1+\frac{p_0/p_{f\min}}{\sigma/\sigma_{-1\min}}\right)^{R_{\sigma/p}}$	$\frac{W_{P\gamma}}{V_0}$	$A_{\sigma p}$
б) back effect	$a_{p\gamma}\frac{p_0}{p_{f\min}}\left(1+\frac{\sigma/\sigma_{-1\min}}{p_0/p_{f\min}}\right)^{R_{p/\sigma}}$	$\frac{W_{P\gamma}}{V_k}$	$A_{p\sigma}$
Mechano-sliding fatigue:			
a) direct effect	$a_{\sigma\tau}\frac{\sigma}{\sigma_{-1\min}}\left(1+\frac{\tau_W/\tau_{f\min}}{\sigma/\sigma_{-1\min}}\right)^{R_{\sigma/\tau}}$	$\frac{W_{P\gamma}}{V_0}$	$A_{\sigma\tau}$
б) back effect	$a_{\tau\sigma}\frac{\tau_W}{\tau_{f\min}}\left(1+\frac{\sigma/\sigma_{-1\min}}{\tau_W/\tau_{f\min}}\right)^{R_{\tau/\sigma}}$	$\frac{W_{P\gamma}}{S_k}$	$A_{\tau\sigma}$

Expression (2.67) is the *equation of the curve of limiting states of components of the active system based on cracking*. According to this equation the durability N_I lowers as loads (σ, τ_W, T) grow, hence the damaged volumes expand during the first loading cycle.

Note that loading function (2.66) determines in fact the mechanical energy $U_M(\sigma, \tau_W)$ due to cyclic and / or contact stresses and they exceed the corresponding endurance limits. Hence, $U_M(\sigma, \tau)$ in this case is the *damaging mechanical energy*.

Function of damage (2.68) is analyzed graphically in Fig. 2.42. It is done using the parameter

$$\rho_{M/T} = f(Q)/kT, \tag{2.70}$$

that determines the ratio between the mechanical (the index *M*) and thermodynamic (the index *T*) energies. Consequently, mechanical and thermodynamic loads affect the process of damage accumulation equally if $\rho_{M/T} = 1$, then $f(Q) = kT$. The processes of damage accumulation are predominantly due to mechanical stresses (contact and off-contact loads) if $f(Q) >> kT$, i.e. $\rho_{M/T} >> 1$. Thermodynamic factor dominates in the damage accumulation processes when $f(Q) << kT$, i.e. $\rho_{M/T} << 1$.

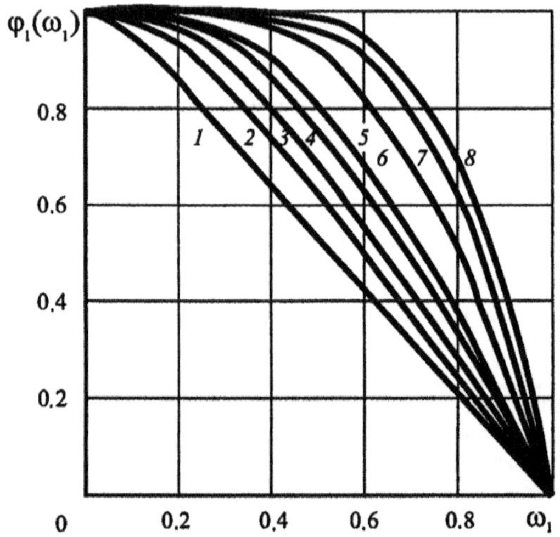

Fig. 2.42. Graphs of function (2.68) depending on the value ω_1 for different parameters $\rho_{M/T}$ (for curves *1*–*8* it is assumed that $\rho_{M/T}$ = 0.1; 0.3; 0.5; 0.8; 1.0; 2.0; 3.0 and 4.0, respectively)

It follows from Fig. 2.42 that, in case damage during the first loading cycle is ω_1 = const, the functions of damage (2.68) and therefore the durability N_T (according to (2.67)) depend significantly on the value $\rho_{M/T}$. The general regularity is the following: augmentation of $\rho_{M/T}$ and ω_1 leads to a correspong reduction of durability at stage *I*.

2.6.3 Durability at stage II

According to Fig. 2.41, if the curves of types *2* or *3* describe damage at stage *I*, then stage *II* follows of the survivability of the component of the active system with the main crack at the tip of which the damaged volume $\Omega_{P\gamma}$ appears. Curve *4* describes the kinetics of crack growth (see also Fig. 1.29).

The coefficient of stress intensity *K* controls the evolution of fracture at stage *II* (see Sect. 1.3.4), corresponding loads govern its magnitude. If an exponential relation is assumed between the rate of damage v_{II} at this stage and the coefficient of intensity of stresses, then it is possible to obtain an equation similar to (1.44):

$$N_{II} = \frac{1-\omega_K}{C_K \Delta K_\omega^{n_K} (n_K + 1)}, \tag{2.71}$$

where

$$\Delta K_\omega = \frac{K_{\max} - K_{th}}{1-\omega_K},$$

and ω_K is determined from (2.60) or (1.48).

Thus, according to (2.71) the durability both at stage *II* and at stage *I* is governed in many respects by the size of the relative damaged volume during the 1st loading cycle.

The equation of type (1.44a) is used for the case of an elasticoplastic crack.

Self-test questions

1. What is an active system? In what way does it differ from a friction pair?
2. How do you understand the term "wear-fatigue damage"? What basic types of WFD do you know?
3. Give the diagram of the simplest active system of the *shaft / sliding bearing* type. Give a characteristic of dynamic conditions of interaction between its components.
4. What is mechano-sliding fatigue? Give examples of active systems in which it occurs in operation.
5. Show the diagram of the simplest active system of the *shaft/ wheel (roller)* type. Describe dynamic conditions of interaction between its components.
6. What is mechano-rolling fatigue? Give examples of active systems in which it occurs in operation.
7. Give a scheme of the simplest active system of the *shaft / bush* type. Describe the dynamic conditions of interaction between its components.
8. What is fretting-fatigue? Give examples of active systems in which it occurs in operation.
9. What are the signs of the limiting state of *solid / solid* active systems? Briefly describe each sign you know.
10. What is the direct effect? What damages of an active system dominate in case the direct effect occurs? What are the sings of its limiting state?
11. What is the back effect? What damages of an active system dominate in case of the back effect? What are the indicators of its limiting state?
12. How is it possible to make the allowance for the complexity of the state of stress when analyzing interactions between components of an active system?

13. Do you discriminate interacting damages from interrelated damages? Show the schemes (examples) of the active systems of the *solid / solid* type that manifest such damage.
14. What is a combined active system? Give examples of such systems.
15. Give examples of active systems of the *solid / environment type*. Describe the conditions of interaction between the components of such systems.
16. What is mechano-corrosion fatigue? Give examples of active systems in which it occurs in operation.
17. What are the specific signs of the limiting state of an active system operating in the conditions of mechano-corrosion fatigue?
18. Give examples of active systems of the *solid / stream of particles* type. Describe the dynamic conditions of interaction between its components.
19. What is mechano-erosion fatigue? Give examples of active systems in which it occurs in operation.
20. Analyze the conditions of interactions between the components of the combined active system *wheel/ rail/sleeper/ roadbed*. What WFD types occur in this case?
21. Analyze the conditions of interactions between the components of the combined active system like the linear portion of the oil pipeline. What WFD types occur in this case?
22. Analyze the condition of operation of railway carriage axle. What WFD types are made evident in the conditions of operation?
23. Analyze briefly the advantages and disadvantages of two concepts of tests of active systems: *a*) bench (full-scale) tests; *b*) tests of small-size models. Do you believe that these concepts exclude one another or complement one another?
24. If mechanisms of motion of a fodder harvester are examined, what are the most common types of active systems (and WFD types)? What types are relatively few?
25. Analyze the main diagram of operability of the mechanical system *steel shaft/ polymeric sliding bearing* using the criteria of resistance to fatigue. Is the design diagram adequate to real conditions of operation of the given mechanical system? If not what is its idealization? What phenomena or factors are ignored?
26. Analyze the main diagram of operability of the mechanical system *steel shaft/ polymeric sliding bearing* using the criteria of tribology. Is the design diagram adequate to real conditions of operation of the given mechanical system? If not what is its idealization? What phenomena or factors are ignored?
27. Analyze the main diagram of operability of the mechanical system *steel shaft / polymeric sliding bearing* using the tribo-fatigue criteria. Is the design diagram adequate to real conditions of operation of the given mechanical system?
28. Describe tribo-fatigue as a complex scientific discipline. What objects does it study? What methods does it use?
29. The final practical task of the mechanics of fatigue fracture, tribology, reliability of mechanical systems and tribo-fatigue is common, it is to ensure the required durability of machinery. What are the principal differences of the methods and problems of tribo-fatigue from the methods and problems of related disciplines?
30. Analyze briefly the distinctive features of methods of tests for friction, methods of tests for mechanical fatigue and methods of wear-fatigue tests.

31. Briefly analyze the distinctive features of methods of theoretical studies of the bearing capacity and durability of structural elements (in the mechanics of fatigue fracture), friction pairs (in tribology) and active systems (in tribo-fatigue).
32. What is the essence of the model of a deformable solid with a damaged volume?
33. How does the damaged volume appear in the deformable solid under the effect of alternating (cyclic) loading? Indicate the main condition of limitation of this volume and show the general formula for its calculation.
34. How does the damaged volume appear in the body and the counterbody of a friction pair? Indicate the main condition of limitation of these volumes and show the general formula for their calculation.
35. How does the damaged volume appear in the components of the active systems? What are the conditions limiting this volume? How can the complex damaged volume be calculated in an active system?
36. Record generalized conditions of failure-free operation of structural components, friction pairs, and active systems. What is their essence?
37. How can the measure of damage of material or structural component be determined during mechanical fatigue? What is the interval of changes of its numerical values? What is the measure of damage when the limiting state occurs?
38. How can the measure of damage of a friction pair be determined? What is the interval of changes of its numerical values? What should the measure of damage be when the limiting state occurs?
39. How can the complex measure of damage of an active system be determined? What is the interval of changes of its numerical values? What is the measure of damage when the limiting state occurs?
40. What is the essence of the integral criterion of similarity of fatigue fracture? In what way does it differ from the measure of fatigue damage?
41. What is the essence of the integral criterion of similarity in respect to a friction pair? In what way does it differ from the measure of damage during friction and wear?
42. What is the essence of the integral criterion of similarity in respect to an active system? In what way does it differ from the measure of WFD?
43. Describe the method of calculating the damaged volume of the prismatic beam during transverse bending.
44. Describe the method of calculating the damaged volume during rolling friction of two rollers.
45. Describe the method of calculating the complex damaged volume in an active system.
46. Are the methods of calculating the complex damaged volume different in the case of the direct or back effect?
47. Record the tensor of main damaged volumes and explain the meaning of intersection and aggregation of damaged volumes.
48. Can you attribute any occurance of damaged volumes (normal, tangent, main) to the initiation of surface or subsurface damage in friction?
49. What is the geometric shape of an integration of main damaged volumes when rollers with parallel axes are compressed? What is its coordinate plane?

50. What is the difference between the dynamic damaged volume and static damaged volume? Why was the idea of the dynamic damaged volume necessary?
51. Write down three invariants of the main damaged volumes. Do they have any relation to the tensor of main stresses?
52. What is the difference between the invariants of damage and the invariants of main damaged volumes? What is the interval of changes of their numerical values?
53. Can you explain in what cases of damage (fracture) it is better to use the measure of continuity instead of the measure of damage?
54. Make a comparative interpretation of the measures of damage and continuity. What makes them common? What makes them different? Are the integral and tensor measures of damage different?
55. What do you mean when you say: interaction between damaged volumes? What are the main consequences of their interaction?
56. Describe typical aftereffects of interaction between damaged volumes you know in comparison – in a structural component, in a friction pair, in an active system.
57. If the dynamic damaged volumes in a friction pair (when two rollers with parallel axes roll) and in a structural component (the shaft bent with torsion) coincide in size, does it mean that the durability of these objects is the same? Corroborate your view of the problem.
58. What is the principal difference of interaction between phenomena from influence of factors? Give examples of such interactions and their aftereffects to prove your point.
59. What damaged volume can exactly be responsible for subsurface fatigue damage, for example, in the rail head?
60. What is the integrated measure of damage? What numerical values can it have?
61. Do you draw any difference between the total measure of damage in case of the direct and back effects?
62. How does the complex damaged volume depend on the total measure of damage and on the parameter of interaction between damages?
63. What general property should lambda-function of damage possess? Generally, what makes these functions predictable?
64. What is the difference between the measure of damage and the criterion of fracture similarity?
65. What is the sense of the parameter of interaction between damages? What is its role in rating the level of damage of components of an active system?
66. What is the ratio between durabilities during contact and mechano-rolling fatigue?
67. What is the ratio between durabilities during mechanical and mechano-rolling fatigue?
68. Write down the complex measures of damage during mechano-rolling fatigue when investigating both the direct and back effects. What do they have in common? What is their principal difference?
69. What numerical values can the parameter of interaction between damages have? For example, if $R_{\sigma p} = 1$, what does it signify?
70. Can the damaged volume be an equivalent of damage on different scale levels? Substantiate your view.

71. Analyze the commonness of and the difference between models of damage of a friction pair and an active system.
72. What geometrical shapes can the complex damaged volume acquire in the components of an active system?
73. If the numerical values of the complex damaged volume are known, is it possible to assess the level of concentration of damages of the components of an active system? If yes, how?
74. In what way does the state of damage differ from the mechanical state of a material? What parameters characterize the state of damage?
75. What is the principal difference between the stages of scattered damage and localized fracture?
76. Do you know the conditions of operation of active systems at which stage *II* of localized fracture does not occur? Are the conditions possible when stage *I* of scattered damage does not occur?
77. Do you know what cracks are called small (or short)? What cracks are called long? Under what conditions do fatigue cracks appear that do not grow?
78. If the direct effect occurs, what stages of damage and fracture are revealed? What are the stages of damage and fracture observed in case of the back effect?
79. How can the durability of an active system be estimated at stage *I*? What parameters govern it?
80. How can the survivability be assessed at stage *II*? What parameters govern it?

Tasks for research

1. Analyze the active systems of some machine that has the design known to you, such as a locomotive, a car, a tractor, a lifting crane, etc., as it is described in Sect. 2.2. Classify the identified active systems according to WFD types. This study may have substantial practical significance: you can make specific recommendations what methods are useful to design (or test models) all the active systems of this machine.
2. Study the material relating to the analysis of damage of the railway carriage axle or the *wheel/ rail* system or any other active system in operation. Identify what damages have the complex nature and relate to the WFD. Is it possible to obtain the statistics of failures for each WFD type that happen during operation of a given system? It enables to solve the problem of prioritization of the tasks of promoting the durable operation of an active system, provided, of course, extra material consumption (and cost) is taken into account.
3. The scale effect is analyzed in Sect. 2.4.1 using the results of fatigue tests. Try to find similar published experimental data regarding frictional or contact fatigue and plot the scale dependence of limiting stresses (or limiting pressures). Calculate the criterion of similarity of the objects based on the results of their tests.

4. Carry out the minimum program of tests at the laboratory to obtain the scale dependence of ultimate stresses (or durabilities) during rolling friction or sliding friction (using two–three points). Of course, the number of cycles should be limited in order to obtain the needed results during an acceptable testing time.
5. The formula for calculating the damaged volume when rollers with parallel axes roll is shown in Sect. 2.4.2. Try to derive a formula for calculating the damaged volume in the zone of the fillet passage applicable to step plates. Then it is easy to analyze the damage and the limiting state of the model of gearing shown in Fig. 2.2 (an active system with interrelated damages).
6. If research according to Sect. 5 is performed, damaged volume in full-scale cylindrical gearings can be calculated. Here is a rather hard task: if a gear wheel has 25 teeth, it has 25 pairs of interrelated damaged volumes. How should you rate the carrying capacity and durability of the gear wheel: should only one pair of interrelated damaged volumes or all 25 pairs of damaged volumes be considered? The following question provides a prompt how to solve the problem: does the durability of gear wheel depend on the number of teeth (with other equal conditions)?
7. Develop a method of assessing damaged volumes in case of the static contact between the wheel and the rail. It is better if 2–3 students are guided by the instructor to solve the problem.

3 METHODS OF WEAR-FATIGUE TESTS

No developed civilization can exist without testing materials.

George Gordon

3.1 Tasks

Special *methods of wear-fatigue tests* have been developed for experimental assessment of joint and combined effects of the processes of friction and mechanical fatigue on the serviceability of materials and models of active systems in complex conditions of loading.

As a rule, resistance to WFD is investigated in laboratory conditions by testing small size models of active systems. These tests are performed with special *machines for wear-fatigue tests*. Tests yield the quantitative characteristics of resistance to WFD. These characteristi cs are useful for:

- selecting structural materials for active systems and substantiating design and technological solutions;
- controlling quality of materials;
- calculating active systems at the stage of designing;
- issuing certificates of active systems based on the WFD criterion;
- designing and developing materials with specified mechanophysical behavior to ensure the required characteristics of resistance to WFD.

3.2 Methods

Combination of the existing methods of tests for mechanical fatigue with methods of tests for friction and wear is one of the ways of developing methods of complex (wear-fatigue) tests. Figure 3.1 shows an example of the principle of development when rotation bending serves as the basic method of tests for fatigue. Note that rotary motion is most common in modern machinery, hence, the methods in Fig. 3.1 are of practical significance.

This approach serves the purpose to make machines intended for wear-fatigue tests applicable to common tests for mechanical fatigue or friction and wear in definite conditions.

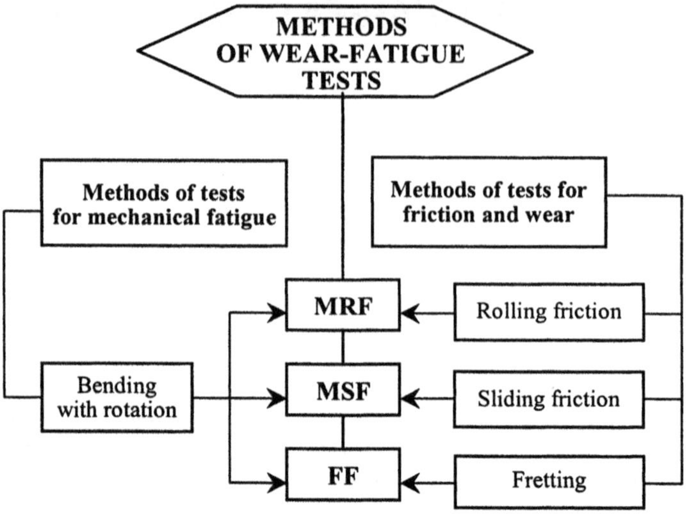

Fig. 3.1. Development of methods of wear-fatigue tests during main rotational movement: MRF – mechano-rolling fatigue; MSF mechano-sliding fatigue; FF – fretting fatigue

3.2.1 Basic schemes of tests

Tests for mechano-sliding fatigue (Fig. 3.2, *e*). One end of cylindrical specimen *1* is fixed in spindle *2* and rotates at the angular speed ω_1. Vertical bending (off-contact) load Q (upwards or downwards) is imposed to its other end. Non-rotating counterspecimen *3*, such as a plate or partial insert is pressed against its working zone $d = 10$ mm in diameter by contact load F_N. Maximum contact and bending stresses are thus created in the working zone of the specimen concurrently.

It is easy to see that in case the scheme of tests is configured according to Fig. 3.2, *e*, it is possible to perform

- wear-fatigue tests for mechano-sliding fatigue (Fig. 3.2, *e*) and vary the values F_N, Q and ω;
- tests for mechanical fatigue by bending and rotation (Fig. 3.2, *c*) and vary the values Q and ω. Counterspecimen *3* in this case is removed so that $F_N = 0$;
- tests for friction and wear in sliding (Fig. 3.2, *d*) and vary the values F_N and ω. In this case there is no bending load ($Q = 0$), specimen *1* is made shorter to save the material.

So, if to follow Fig. 3.1 and integrate (combine) the existing schemes of tests for mechanical fatigue and sliding friction, a scheme of tests for mechano-sliding fatigue will be like in Fig. 3.2, *e*.

Bending load Q may be constant (unchangeable in time t) during tests for mechanical fatigue (cf. Fig. 3.2, *c*), but the effective normal stresses in each point of the working cross section of specimen *1* change according to the symmetric cycle (Fig. 3.3) with the period T due to torsion of the specimen. If the maximum bending moment through the working cross section of the specimen is $M = Ql$, where l – the spacing between the dangerous cross section and the line of action of the load Q, the maximum normal stresses in this cross section are determined from formula (2.3).

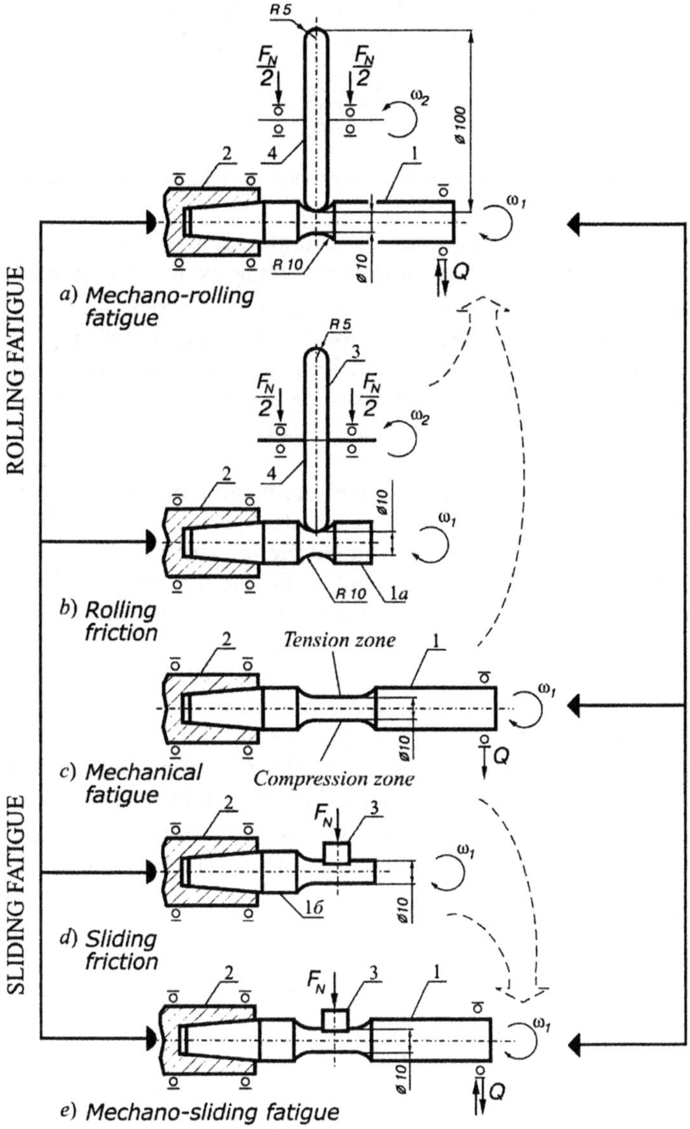

Fig. 3.2. Typical schemes of wear-fatigue tests: *1, 1a, 1b* – specimen; *2* – test machine spindle; *3, 4* – counterbody; Q – bending load; F_N – contact load; ω_1, ω_2 – rotational speed of specimen, counterbody

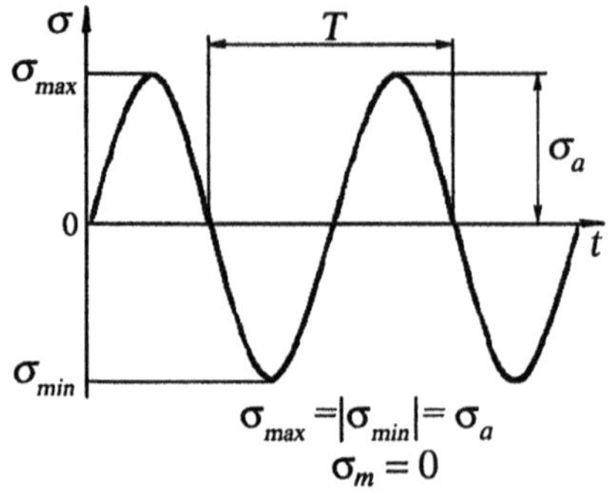

Fig. 3.3. Symmetrical cycle of stresses for mechanical fatigue tests

Maximum (σ_{max}), minimum ($|\sigma_{min}|$) and amplitude (σ_a) stresses are found to be equal numerically in case of a symmetric loading cycle; in such cases hereafter they will be called just cyclic stresses σ (= $\sigma_{max} = |\sigma_{min}| = \sigma_a$).

The contact load F_N during tests for sliding friction (cf. Fig. 3.2, *d*) can similarly be static, i.e. constant in time, yet the effective contact stresses are also cyclic (cf. Fig. 1.50). Hence, tests for sliding friction according to the scheme in Fig. 3.2, *d*, are, in fact, tests for *sliding fatigue* (during asymmetric tension-compression).

The conditions for the sliding fatigue to occur can integrally be described either by contact load F_N, or by mean (nominal) contact pressure (2.1), or by the specific force of friction by sliding that is also called *friction stress* (2.2).

Again about the tests for mechano-sliding fatigue (cf. Fig. 3.2, *e*), it is apparent that dynamic conditions of interactions between the specimen and counterspecimen can be characterized by two parameters: values of the cyclic stress (2.3) as a function of the off-contact (bending) load Q and friction stresses (2.2) (or mean pressure (2.1)) as a function of the contact load F_N.

Tests for mechano-rolling fatigue (Fig. 3.2, *a*). This scheme differs from the scheme of tests for mechano-sliding fatigue (cf. Fig. 3.2, *e*) because the counterspecimen fixed stationary is replaced with rotating roller *4*. In the general case the specimen and the roller can rotate with different angular speeds ω_1 and ω_2 and in different directions. When the scheme of tests in Fig. 3.2, *a*, it is possible to perform

- wear-fatigue tests for mechano-rolling fatigue (Fig. 3.2, *a*) and to vary the values F_N, Q, ω_1 and ω_2;
- tests for mechanical fatigue by bending and torsion (Fig. 3.2, *c*) and to vary the values Q and ω. Roller *3* is removed in this case, so that $F_N = 0$ and $\omega_2 = 0$;
- tests for rolling friction or tests for rolling friction with slip (Fig. 3.2, *b*) and to vary the values F_N, ω_1 and ω_2. In this case there is no bending load ($Q = 0$), specimen *1* is made shorter to save the material.

Therefore, according to Fig. 3.1, integration (combination) of the existing schemes of tests for mechanical fatigue with the tests for rolling friction results in a scheme of tests for mechano-rolling fatigue in Fig. 3.2, *a*.

Conditions for rolling friction (cf. Fig. 3.2, *a*, *b*) can be integrally described either by contact load F_N or by the maximum pressure in the center of the contact site determined from Hertzian formula (for elastic deformation)

$$p_0 = n_p F_N / A_p, \qquad (3.1)$$

or by the specific force of friction in rolling that is also called *friction stress* (2.4). A_0 is the contact area ($A_p = a^2$ for a round contact site with the radius a; $A_p = lb$ for contact along the strip with the dimensions $l \times b$; $A_p = ab$ for the elliptical contact site with the size $a \times b$; n_p – a coefficient (n_p = 0.478 for round and elliptical contact sites and n_p = 0.637 for strip contact).

The contact load F_N both during tests for rolling friction (cf. Fig. 3.2, *b*) and sliding friction can be static, i.e. constant in time, yet the effective contact stresses (for example, $p_0 = \sigma_{z\,max}$) are cyclic too (Fig. 3.4). Hence, tests for rolling friction according to the scheme in Fig. 3.2, *b*, are, in fact, tests for *rolling fatigue* of the surface layer of the material.

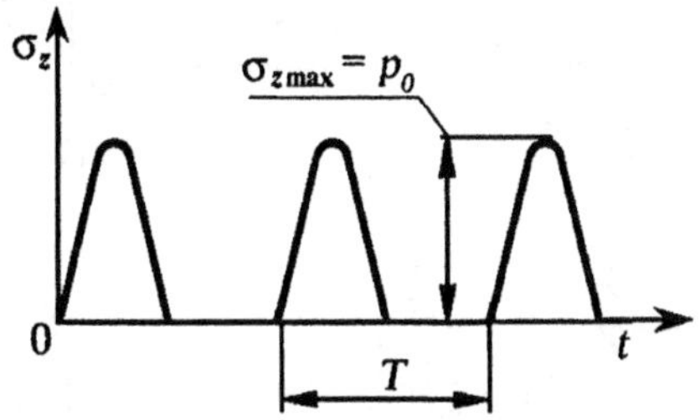

Fig. 3.4. Cycle of stresses during tests for rolling fatigue

Again about the tests of mechano-rolling fatigue (cf. Fig. 3.2, *a*), it is apparent that dynamic interactions between the specimen and the counterspecimen can be integrally characterized by two parameters: values of cyclic stresses (2.3) as a function of the off-contact (bending) load Q and friction stress (3.2) (or the maximum pressure (3.1) in the center of the contact site) as a function of the contact load F_N.

Tests for fretting fatigue (Fig. 3.5, *a*). Unlike the schemes of the tests for mechano-sliding fatigue (cf. Fig. 3.2, *e*) and mechano-rolling fatigue (cf. Fig. 3.2, *a*), two counterspecimens *3* called fretting bridges are pressed with the contact load F_N in this case (cf. Fig. 3.5, *a*) to the working zone of rotating cylindrical specimen *1* with bending load Q. The specimens can be imparted peripheral (with the speed v_1) or axial (with the speed v_2) low-amplitude oscillating movement or two motions can be excited simultaneously. When the scheme of tests in Fig. 3.5, *a*, it is possible to perform

- wear-fatigue tests for fretting fatigue (cf. Fig. 3.5, *a*) and to vary the values F_N, Q, ω, v_1 and v_2;
- tests for mechanical fatigue by bending and torsion (cf. Fig. 3.5, *b*) and to vary the values Q and ω. In this case the fretting bridges are not installed, so that $F_N = 0$, $v_1 = v_2 = 0$;

- tests for fretting during axial and / or peripheral slip (cf. Fig. 3.5, *c*) and to vary the values F_N, v_1 and v_2. In this case there is no bending load ($Q = 0$), specimen *1* is made shorter to save the material.

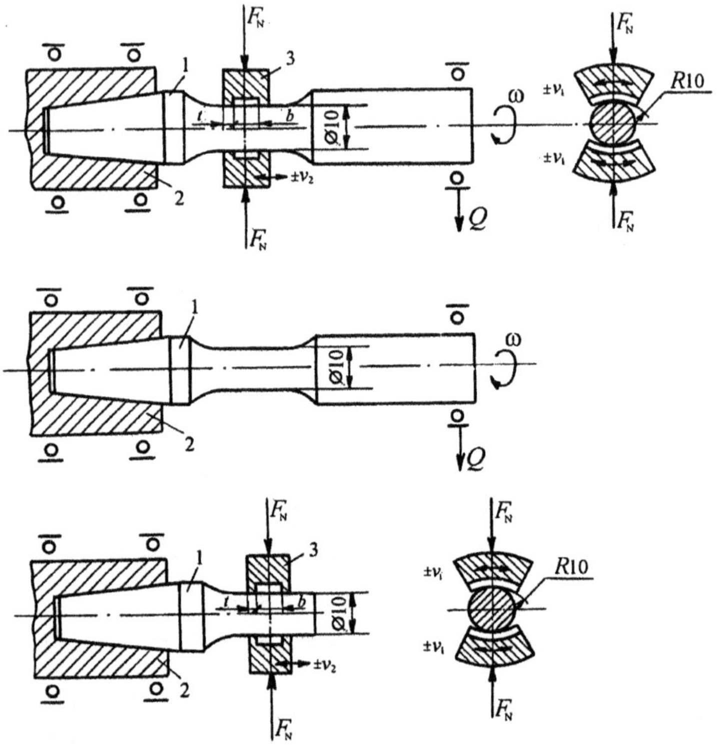

Fig. 3.5. Diagrams of tests for fretting fatigue (*a*), mechanical fatigue (*b*) and fretting (*c*)

The conditions of dynamic interactions between the specimen and the counterspecimen in fretting fatigue can be integrally characterized by cyclic stresses (2.3) and the nominal contact pressure

$$q = F_N / A_a$$

or friction stresses (2.5).

3.2.2 Basic characteristics of resistance to wear-fatigue damage

Basic characteristics of resistance to WFD results from the wear-fatigue tests of relevant objects.

During tests for mechanical fatigue the object is a *structural component*, for example, a *cylindrical specimen* of definite geometry (cf. Fig. 3.2, *c*). During tests for sliding or rolling friction the object is a friction pair (cf. Fig. 3.2, *b, d*) consisting of *specimen 1* and *counterspecimen 3* or *4*; they are also called the *body* and the *counterbody*. Note that the cylindrical structural component is always called the specimen (the body) and the partial insert or the roller is called the counterspecimen (the counterbody). Of course, these names can be changed on the opposite (like it is done in some publications). Finally, *active systems* consisting of two components *1* and *3* or *4* are objects of *wear-fatigue tests* (cf. Fig. 3.2, *a, e*).

The results of tests are used to plot the relevant *fatigue curve* serving to determine the main quantitative characteristics of resistance to fracture (see Figs. 1.16 and 1.57).

Figure 3.6 shows an example of four fatigue curves obtained experimentally, viz. the curve of mechanical fatigue $N(\sigma_a)$ plotted from the results of tests of carbon steel 45 (after normalization); the curve of rolling fatigue $N(p_0)$ plotted from the results of tests for rolling friction of the pair a *carbon steel 45 specimen / roller from steel 25XГТ* (after hardening and tempering) and also two curves of mechano-rolling fatigue plotted from the results of tests for wear-fatigue tests of the *steel 45 / steel 25XГТ* active system.

The limiting state criterion during tests for mechanical fatigue is disintegration of the specimen into pieces, that of tests for rolling fatigue is the critical density of the pits of spalling on the rolling surface. The limiting state during tests for mechano-rolling fatigue is determined by the criteria of damage and fracture typical for tests for mechanical and rolling fatigue.

Endurance limits (σ_{-1}, p_f, σ_{-1p}, $p_{f\sigma}$), parameters of the slope of the left branch of the fatigue curves (m_σ, m_p, $m_{\sigma p}$, $m_{p\sigma}$) and abscissas of the breakpoints of the fatigue curves ($N_{G\sigma}$, N_{Gp}, $N_{G\sigma p}$, $N_{Gp\sigma}$) are determined in all four cases. Note that the endurance limits during mechanical (σ_{-1}) and contact (p_f) fatigue are unambiguous and unique characteristics of the relevant objects of tests, meanwhile the endurance limits during mechano-rolling fatigue (σ_{-1p}, $p_{f\sigma}$) are not such characteristics. Similar fatigue curves can be plotted in any number – as many as the number of the values of the parameters p_0 = const or σ_a = const is set for wear-fatigue tests when regularities of direct and back effects are studied.

The effect of the friction and wear processes on changes of the characteristics of resistance to mechanical fatigue can be characterized by the *direct effect coefficient*

$$K_D = \sigma_{-1p} / \sigma_{-1}. \tag{3.2}$$

In fact, the coefficient K_D is a characteristic of strength. K_D = 256/165 = 1.62 during the tests that yielded results presented in Fig. 3.6.

The effect of the processes of mechanical fatigue on the changes in friction and wear characteristics can be characterized by the *back effect coefficient*

$$K_B = p_{f\sigma} / p_f. \tag{3.3}$$

Actually, the coefficient K_B is a tribological characteristic. K_B = 2200/1760 = 1.25 during the tests that yielded results presented in Fig. 3.6.

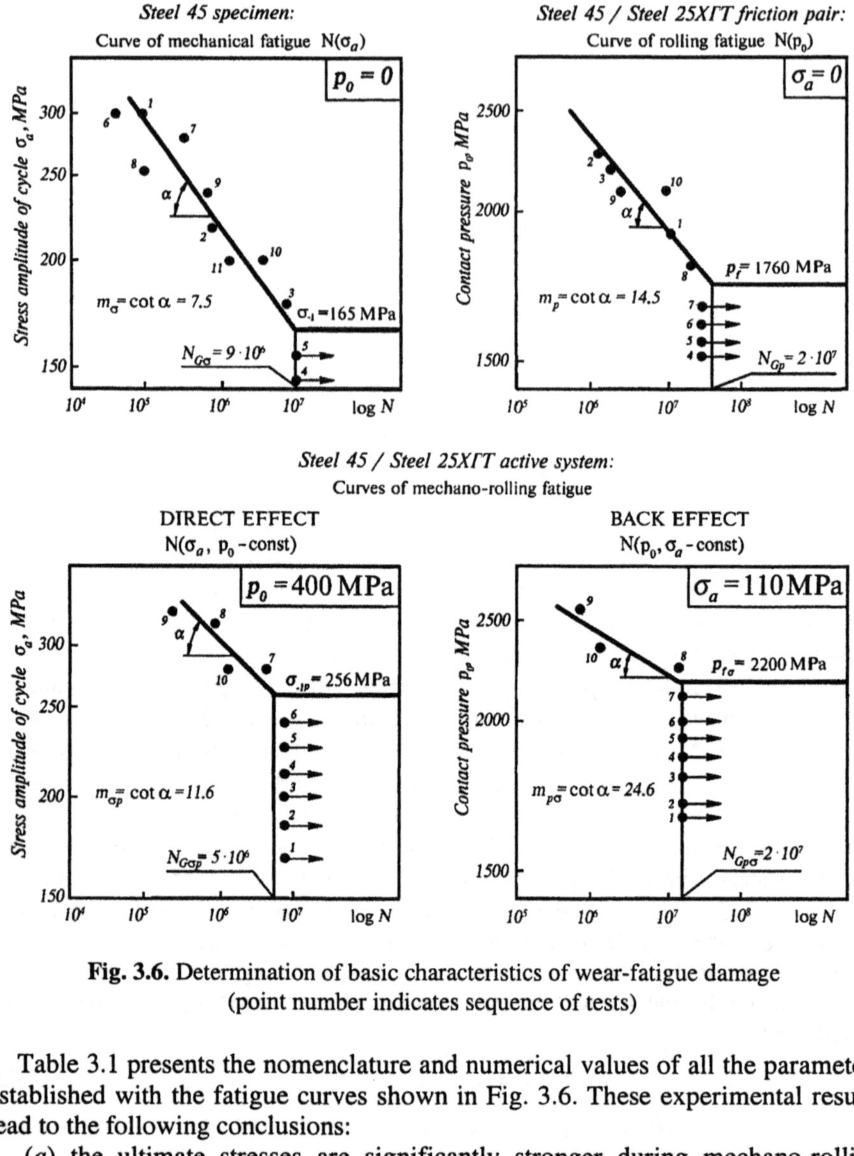

Fig. 3.6. Determination of basic characteristics of wear-fatigue damage (point number indicates sequence of tests)

Table 3.1 presents the nomenclature and numerical values of all the parameters established with the fatigue curves shown in Fig. 3.6. These experimental results lead to the following conclusions:

(*a*) the ultimate stresses are significantly stronger during mechano-rolling fatigue than during mechanical and rolling fatigue ($K_D > 1$, $K_B > 1$);

(*b*) the slope becomes steeper when switching over from the mechanical fatigue curve to the corresponding mechano-rolling fatigue curve ($m_{\sigma p} >> m_\sigma$); from the rolling fatigue curve to the corresponding mechano-rolling fatigue curve ($m_{p\sigma} >> m_p$).

Table 3.1. System of designations and numerical values of basic characteristics

Characteristics of properties	Mechanical fatigue curve	Rolling fatigue curve	Mechano-rolling fatigue curves	
	$N(\sigma_a)$	$N(p_0)$	$N(\sigma_a, p_0 = \text{const})$	$N(p_0, \sigma_a = \text{const})$
Endurance limit, MPa	$\sigma_{-1} = 165$	$p_f = 1760$	$\sigma_{-1p} = 256$	$p_{f\sigma} = 2200$
Abscissa of the break point of the fatigue curve, cycle	$N_{G\sigma} = 9 \cdot 10^6$	$N_{Gp} = 26 \cdot 10^7$	$N_{G\sigma p} = 5 \cdot 10^6$	$N_{G p\sigma} = 2 \cdot 10^7$
Fatigue curve slope indicator	$m_\sigma = 7.5$	$m_p = 14.5$	$m_{\sigma p} = 11.6$	$m_{ps} = 24.6$

In other words, these testing conditions lead to a stronger resistance to WFD than the resistance to mechanical or rolling fatigue. These regularities are explained below.

3.2.3 Determination of the fatigue curve parameters

A set of smooth-carbon steel specimens (12 pieces) was tested for mechanical fatigue at a frequency 50 Hz.

The experimental results of the tests are shown graphically in double logarithmic coordinates log σ – log N (Fig. 3.7).

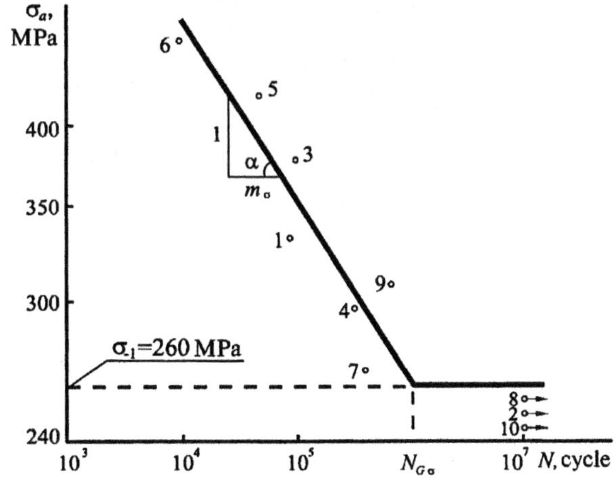

Fig. 3.7. Experimental curve of mechanical fatigue (steel 45)

There is a linear relation between the values $x = \log N$ and $y = \log \sigma$

$$y = ax + b. \tag{3.4}$$

According to the method of least squares the coefficients a and b of equation (3.4) are found from the system

$$\left.\begin{aligned} a\sum_{i=1}^{n} x_i^2 + b\sum_{i=1}^{n} x_i = \sum_{i=1}^{n} x_i y_i; \\ a\sum_{i=1}^{n} x_i + nb = \sum_{i=1}^{n} y_i, \end{aligned}\right\} \tag{3.5}$$

where $i = 1, 2, 3, \ldots, n$ – the order number of a specimen.

The solution of the system is

$$b = \frac{\sum_{i=1}^{n} y_i \cdot \sum_{i=1}^{n} x_i^2 - \sum_{i=1}^{n} x_i y_i \cdot \sum_{i=1}^{n} x_i}{n\sum_{i=1}^{n} x_i^2 - \left(\sum_{i=1}^{n} x_i\right)^2}; \tag{3.6}$$

$$a = \frac{n\sum_{i=1}^{n} x_i y_i - \sum_{i=1}^{n} x_i \cdot \sum_{i=1}^{n} y_i}{n\sum_{i=1}^{n} x_i^2 - \left(\sum_{i=1}^{n} x_i\right)^2}. \tag{3.7}$$

Table 3.2 presents the numerical results of tests of seven failed specimens. The Table indicates also the method of experimental data processing.
Using the data from Table 3.2 according to expressions (3.6) and (3.7), we obtain

$$b = \frac{17.743 \cdot 181.3143 - 89.3715 \cdot 35.3753}{7 \cdot 181.3143 - (35.3753)^2} = 3.12;$$

$$a = \frac{7 \cdot 89.3715 - 35.3753 \cdot 17.743}{7 \cdot 181.3143 - (35.3753)^2} = -0{,}116.$$

Using fatigue curve equation (3.4) with the account of the obtained coefficients a and b, we obtain

$$\log \sigma = 3.12 - 0.116 \log N. \tag{3.8}$$

The fatigue curve slope indicator is

$$m_\sigma = 1/|a| = 1/0.116 \approx 8.6.$$

Hence, the equation of type (1.26) for the steel in question is

$$\sigma^{8.6} N = \text{Const},$$

where, in accordance with (1.28), const = $260^{8.6} \cdot 10^6$, $N_{G\sigma} = 10^6$ cycles.

Table 3.2. Results of tests and their processing

Specimen number	σ_a, MPa	N, cycle	$y_i = \log \sigma_a$	$x_i = \log N$	x_i^2	$x_i y_i$
1	330	$7.22 \cdot 10^4$	2.5185	4.8585	23.6050	12.2361
3	371	$8.5 \cdot 10^4$	2.5697	4.9294	24.2989	12.6671
4	295	$3.33 \cdot 10^5$	2.4698	5.5224	30.4969	13.6392
5	413	$4.46 \cdot 10^4$	2.6154	4.6493	21.6159	12.1598
6	454	$9.3 \cdot 10^3$	2.6568	3.9685	15.7489	10.5435
7	268	$4.0 \cdot 10^5$	2.4281	5.6021	31.3835	13.6025
9	305	$7.0 \cdot 10^5$	2.4847	5.8451	34.1652	14.5233
$n = 7$	Sum Σ		17.7430	35.3753	181.3143	89.3715

Equation (3.8) is applicable to calculated estimate of the durability N_σ of specimens at given stresses σ.

The parameters of the curves of rolling and mechano-rolling fatigue (cf. Fig. 3.6) are determined similarly (like the corresponding parameters of the sliding and mechano-sliding fatigue curves).

3.2.4 Methods of studies of wear-fatigue damages

WFD is due to three groups of factors.

(1) A group of basic factors relating to the *conditions of the alternating loading process*: 1) type of the state of stress (homogenous, non-homogeneous, linear, flat, volume); 2) level (magnitude) of stresses; 3) pattern of the stress cycle (symmetric, pulsing, etc.); 4) frequency of loading, etc.

(2) A group of basic factors relating to the conditions of friction: 1) type of the friction process (sliding, rolling, rolling with slip, slip); 2) level of contact load; 3) amplitude of slippage (in fretting), an extent of slip (in rolling friction); 4) friction rate (in rolling, in sliding), frequency (in fretting), etc.

(3) A group of basic factors relating to the conditions of contact interactions between the components of the system: 1) materials of the body and the counterbody, their composition and condition; 2) design features of the system, in particular, the pattern of distribution of contact pressure (the shape of the contact site like a strip, a circle, an ellipse), etc.; 3) technological features of the process of fabrication of the components of the system, in particular, the structure of

contacting surfaces (roughness, waviness, etc.); 4) conditions of lubrication and composition of the lubricant; 5) the environment; 6) temperature in the contact zone; 7) duration (number of cycles) of contact interactions (loading), etc.

The task of studying the pattern and regularities of WFD with the allowance of a large variety of factors seems highly intricate and should be executed using the theory of experiment design. Yet, the analysis is somewhat simplified if the research methods include studies of the direct and back effects.

Direct effect can be studied experimentally with two methods.

(*D1*) *Method of complex tests* implying that the processes of friction, wear and (mechanical) fatigue occur simultaneously in integrity throughout the tests. Therefore, the effect of the conditions of the friction process on the changes of resistance characteristics of one component of the system is investigated (the endurance limit, fatigue durability, etc.);

(*D2*) *Method of successive tests* comprises two stages, that is why it may be also called a two-stage method. Tests for friction and wear at the first stage (factors 2) at specified conditions of contact interactions between the components of a friction unit are carried out during some fixed time (factors *3*). One of the components of the friction unit is subjected to fatigue tests at the second stage (factors*1*) in order to determine its characteristics of fatigue resistance.

The effect of preliminary damage in friction on the resistance to fatigue of a component of the system is investigated in this manner.

The back effect can also be studied experimentally with two methods.

(*B1*) *Method of complex tests* implies that the processes of friction, wear and (mechanical) fatigue occur simultaneously in integrity throughout the tests. The effect of the conditions of alternating loading of a component of the system (factors *1*) is studied to determine how it changes the characteristics of friction and wear of the unit and its both components (the friction coefficient, the wearing intensity, their durability in wear, etc.).

(*B2*) *Method of successive* (or *two-stage*) *tests*. At the first stage one component of the system is tested for (mechanical) fatigue at specified conditions (factors *1*) during a fixed number of loading cycles (without fatigue fracture), the unit is tested at the second stage for friction at specified conditions of contact interactions (factors 2 and *3*) in order to determine the characteristics of friction and wear resistance of both individual components and the unit as a whole (the intensity of wear of the components, the friction coefficient, etc.). That is how the effect of preliminary fatigue damage of one of the components of the system on the wear resistance of the unit friction is studied.

3.3 Testing machines

3.3.1 Technical characteristics

The following testing machines for wear-fatigue tests are manufactured (at the TRIBOFATIGUE Research and Production Group, Belarus) with their designs based on a number of inventions and upon customers' specifications:

- SI-01 machine (tests for mechano-sliding fatigue);
- SI-02 machine (for mechano-rolling fatigue);
- SI-03 full-set machine (tests for mechano-sliding fatigue and mechano-rolling fatigue).

All these machines support tests for fretting fatigue too.

Table 3.3 lists main technical characteristics of machines SI-01, SI-02, SI-03, Fig. 3.8 shows their general view.

Table 3.3. Technical characteristics of module machines of SI series

Indicators	SI-01	SI-02	SI-03
	Friction pair		
	Cylinder–terminal block	Cylinder–roller	Cylinder–terminal block Cylinder–roller
Specimen working portion diameter, mm	10	10	10
Dimensions of counterspecimen, mm	10×10×11.5	Ø100	10×10×11.5 Ø100
Frequency range of rotation of specimen, min^{-1}	40...4000	3000	600...6000
Frequency range of rotation of counterspecimen, min^{-1}	–	50...500	50...500
Range of bending loads, N	70...700	70...700	10...800
Range of contact loads, N	10...500	50...1000	10...2000
Range of measurement of total wear of specimen and counterspecimen, μm	10...3000	10–3000	10...4000
Range of measurement of friction torque, N · m:			
– during sliding friction	0.01...1.2	–	0.01...1.2
– during rolling friction	–	0.2...20	0.2...20

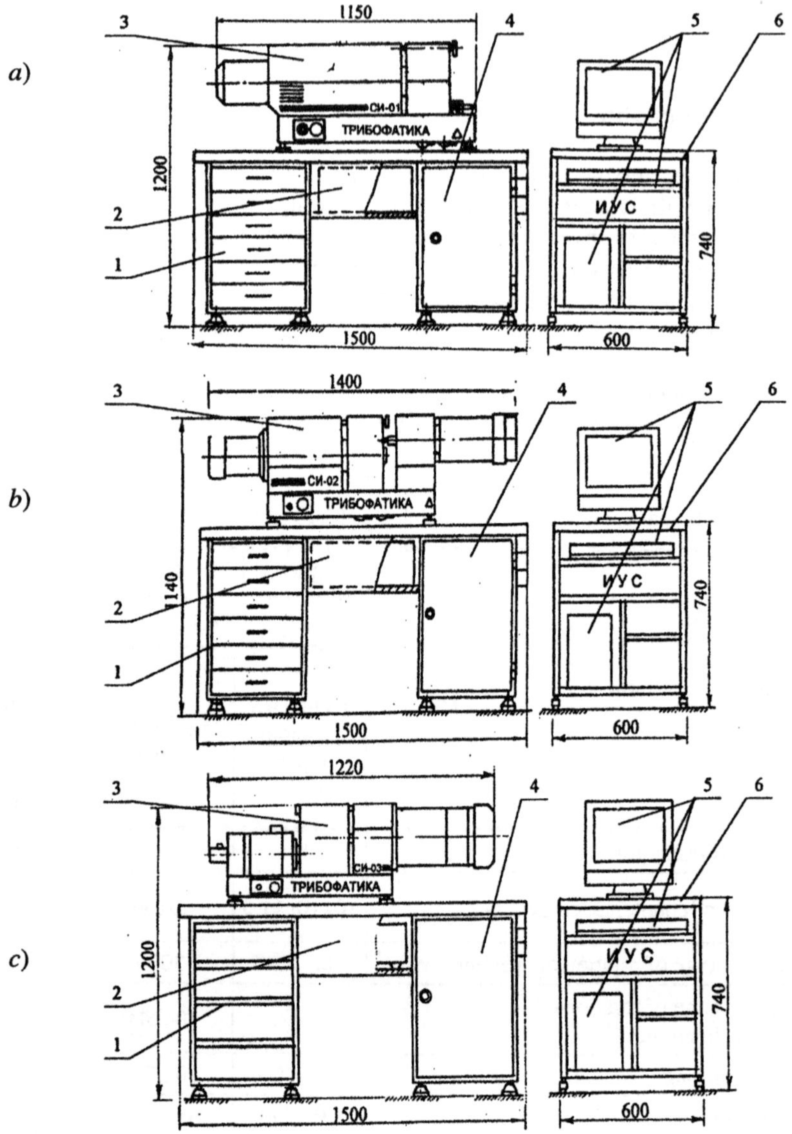

Fig. 3.8. General view of module SI series machines: *a* – SI-01; *b* – SI-02; *c* – SI-03

SI series machines are manufactured in accord with the requirements of the Interstate Standard GOST 30755–2001 "Tribo-fatigue. Wear-fatigue tests machines. General technical requirements".

3.3.2 Design features

Machines of the SI series comprise the following modules (cf. Fig. 3.8):

- test installation *3* containing units and mechanisms necessary to secure specimens of models of active systems;
- special tables *1* and *6*;
- electrical cabinet *4* built into the table's pedestal and containing power starting and control equipment, electronic drive controls, controls of drives for specimens, counterspecimens and loaders ;
- data control system (DCS) 2 including primary sensors of revolutions, rotation frequency, loading, temperature, vibration, linear wear, etc., amplifiers and analog-to-digital converter to convert signals from sensors and emergency signals into digital combinations to send to the PC, a digital-to-analog converter to control drives rotating specimens, counterspecimens and loaders;
- A PC with accessories *5* and software.

Scheme in Fig. 3.9 shows the configuration of the components of the SI-03 testing machine.

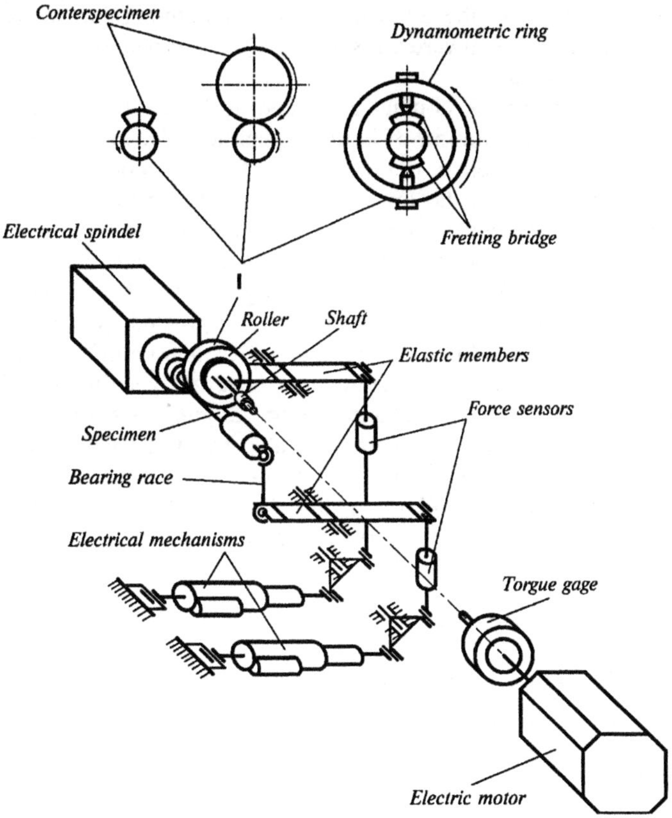

Fig. 3.9. Scheme of test installation of SI-03 machine

The specimen electric drive spindle rotates the shaft to which the tested specimen is attached.

The electric motor of the counterspecimen drive through a flexible shaft rotates the shaft to which the counterspecimen (a roller) is attached. In this case the machine creates rolling friction. The DCS controls a D.C. motor with the help of a thyristor control unit changing within a broad range and high accuracy the frequency of rotation of the counterspecimen maintaining a preset speed of slippage of a friction pair.

An electrical mechanism through a system of levers presses the counterspecimen against the working surface of the specimen creating the required contact load.

Instead of the rotating roller as a counterspecimen the lever can carry a holder with a fixed counterspecimen or a dynamometric ring with fretting bridges. Sliding friction or fretting is realized in case an active (*specimen / counterspecimen*) system is tested.

The electrical mechanism creates bending stresses in the specimen through a system of levers and the race with a bearing mounted on the shank of the rotating specimen.

The site of friction in the zone of tension or compression moves when the direction (upwards or downwards, respectively) of the bending force Q affecting the specimen is changed.

Force transducers check contact and bending loads. An optoelectric sensor reads the frequency of rotation of counterspecimen. An inductive pickup reads linear wear or convergence of the axes of the friction pair, a vibration accelerometer mounted on the lever in the zone where the counterspecimen is fixed reads the parameters of vibration (they are omitted on the diagram). A torque gage mounted on the shaft of the electric motor reads the friction torque.

The SI-01 machine for mechano-sliding fatigue tests (cf. Fig. 3.8) differs from the SI-03 machine because the tested specimen is rotated by a D.C. motor with stepless frequency control (then r.p.m. range is 40...4000 min^{-1}). The machine has no counterspecimen drive as it can simulate exceptionally sliding friction.

The SI-02 machine for mechano-rolling fatigue tests differs from SI-03 machine because the tested specimen is rotated by an asynchronous A.C. motor with the nominal r.p.m. 3000 min^{-1}. By changing the rotation frequency of the counterspecimen drive (a roller), a broad range of speeds of slippage is maintained. The counterspecimen drive is similar to that in the SI-03 machine.

3.3.3 Data control systems

Structure. PC-supported data control systems (DCS) are used in the modular machines SI-01, SI-02 and SI-03. The DCS are based on the principle: the test installation – the control/measurement system – the PC. Figure 3.10 shows the DCS structure.

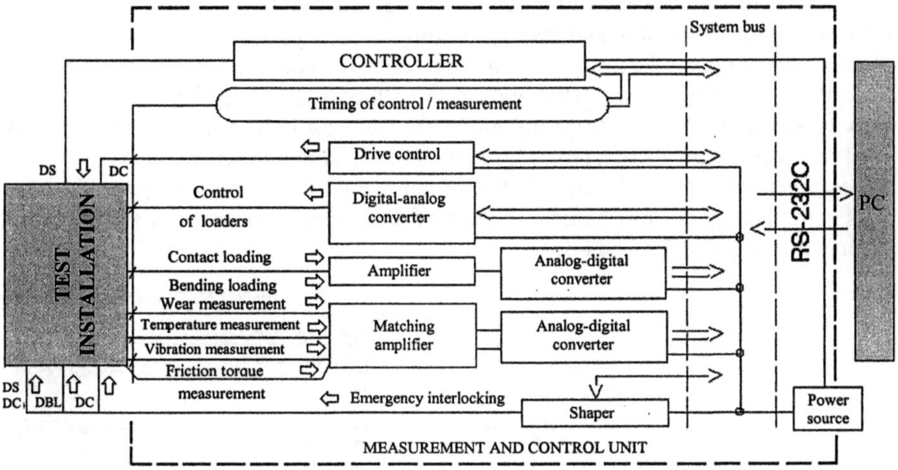

Fig. 3.10. Data control system of module machines: DS, DC, DC, DBL are drives of specimen, counterspecimen, contact and bending loading, respectively

Figure 3.10 shows that the DCS of the modular machines consists of two main parts: the controlling PC and the measuring and control unit interfaced with the PC through a standard cable. The necessary measuring and converting instruments are built into the measuring and control unit as electronic boards and modules.

The DCS has 4 channels to control the devices of the machine. They serve to control the speed of rotation of the specimen, the speed of rotation of the counterspecimen, the contact load, and the bending load.

The channels of registration of analog signals number up to 16. These channels serve to measure signals from the outputs of gages measuring the contact and bending load, gages measuring the friction torque, temperature sensors (thermopairs) of the chromel-capel type, sensors of wear, vibration accelerometers.

The measuring and control unit includes the following main functional units: a controller, a counter of revolutions of the specimen, control signal shapers, a unit of analog-to-digital signal converter, a transducer amplifier, matching amplifiers, an interlocking unit, a power supply unit.

The DCS maintains the following modes of operation: tests planning; calibration of measuring channels; performance of tests; examination of kinetic experimental data; processing of test results. The DCS carries out measurements and registers parameters throughout tests. A special program sends control parameters from the PC to the controller where they are converted into control signals for execution units and proper execution of the set task is monitored by controllers of the r.p.m. of the specimen and the counterspecimen, drives of loaders. The controller sends back to the PC the results of measurements. The PC monitor displays graphically the process of testing.

The software of the test process is a dialog executable code selecting the task from the menu of modes and testing conditions (the algorithm of test control); it controls the output of control actions (the control kinetics), collects primary data from the system of sensors (the algorithm of measurement), performs secondary data processing (the algorithm of processing of results), makes presentation of final results (test protocols, tables, graphs, limiting state curves, etc.).

Management. We will explain the principles of managing the parameters of tests using the electromechanical scheme of arrangement of sensors and drives of the machine SI-03 (Fig. 3.11).

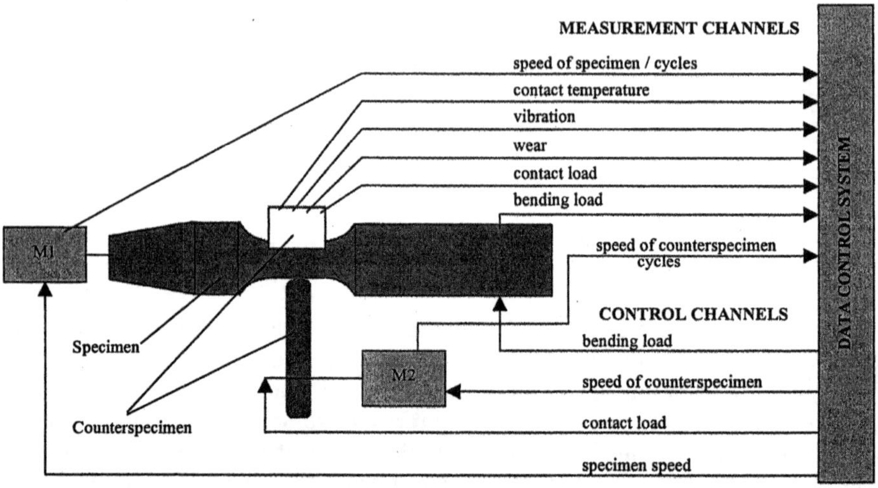

Fig. 3.11. Electromechanical scheme of arrangement of sensors in machine SI-03: M1, M2 – drives of specimen and counterspecimen, respectively

The electrical spindle sets the speed of rotation of the specimen and the electric motor of the roller sets its speed of rotation (in tests for rolling and mechano-rolling fatigue). Special frequency transducers control both the electrical spindle and the counterspecimen electric motor. Optoelectronic sensors mounted on the shafts of the motors read the frequencies of rotation of the specimen and roller counterspecimen, respectively. They output pulses of variable frequency proportional to the speed of rotation.

Strain gages mounted on loading springs (equal resistance beams) provide the DCS with the information about current contact and bending loads. Also information is collected about the temperature in the zone of contact between the specimen and the counterspecimen, parameters of vibration of the active system during tests. Specially devised instruments measure the friction torque in sliding (and mechano-sliding fatigue) and rolling (and mechano-rolling fatigue), the principle of functioning of the instruments is validated when laboratory operations are performed.

A special sensor shapes a discrete emergency signal when the specimen fails received by the DCS with immediate stops of the test installation.

Measurements. Figure 3.12 shows the diagram of measuring and registering two basic WFD parameters, viz. *wear* and *displacement of axes of the friction pair*.

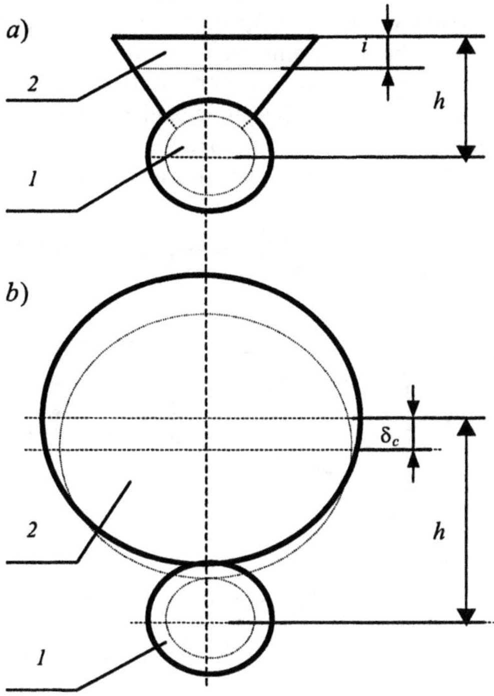

Fig. 3.12. Schemes explaining measurements of total wear in sliding friction (*a*) and displacement of axes in rolling friction (*b*): *1* – specimen; *2* – counterspecimen (full lines show contours of components of friction pair before testing, dotted lines show after or during testing process)

Wear i is the thickness of the removed layer of the material as a result of contact interactions between contacting specimen and counterspecimen during sliding friction and mechano-sliding fatigue. Displacement of the axes of the friction pair δ_c is the result of damage of the components surface of the active system during rolling friction and mechano-rolling fatigue. Displacement of the axes of the friction pair is due to wear, residual deformation and vibromovements. Figure 3.12 shows how theses parameters form.

An inductive pick-up of microdisplacements measures the total wear i (Fig. 3.12, *a* during tests for sliding friction and mechano-sliding fatigue) and the displacement of axes δ_c of the friction pair (Fig. 3.12, *b* during tests for rolling friction and mechano-rolling fatigue) in the SI machines.

Wear i (displacement of axes δ_c) can be measured with two methods – *integral* and *discrete*. According to the integral method the value i (or δ_c) is measured with some small enough intervals of time. Each measured i (or δ_c) corresponds to a random point over the perimeter of the dangerous cross section of the specimen and/or on the working surface of the counterspecimen.

Figure 3.13 shows the results of measurements of *discrete wear* of the *steel 45 / steel 45* active system during tests for mechano-sliding fatigue (cf. Fig. 3.2, *d*). The method implies that the maximum cyclic stresses are excited in the dangerous cross section of the specimen and concurrently sliding friction occurs; eight points (1), (2), ..., (8) are marked along the circle and local wear is measured highly precisely during one revolution of the specimen. These measurements can be naturally performed at any moment of tests (during the time t).

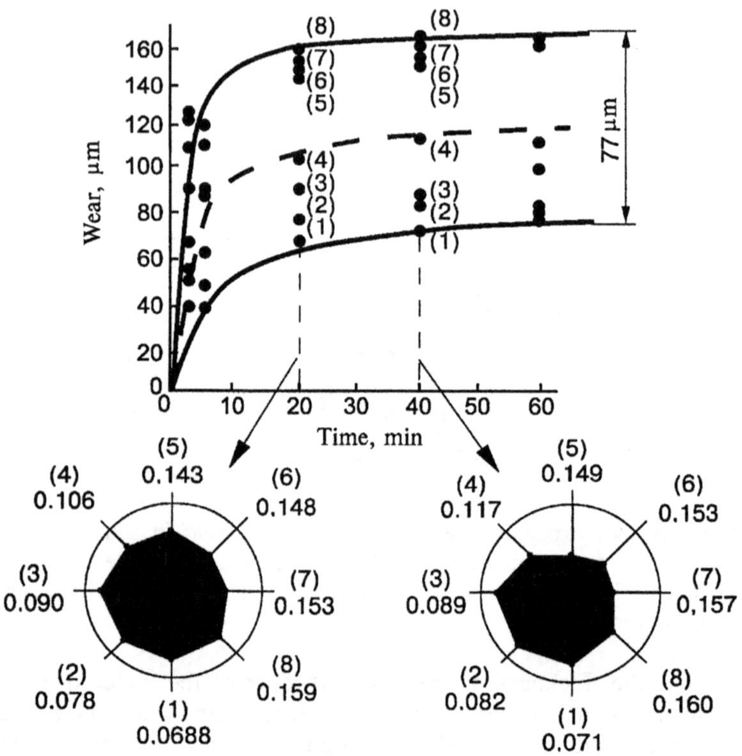

Fig. 3.13. Kinetic curves and circles of wear during wear-fatigue tests of metal-to-metal *steel 45 / steel 45* system

Processing of results. The obtained test results can be interpreted in two ways. The first interpretation is in the form of 8 *kinetic curves* of wear changes in time t (cf. Fig. 3.13, top). Each cross section has eight experimental points that in combination provide the scatter of wear through one cross section of the specimen at a given moment of time. Eight events of *random wear process* are obtained at once in this way. Full lines limit the scatter strip from the top and bottom and a dashed line shows variations of mean wear in Fig. 3.13. In fact, this mean can be identified with the integral wear that is commonly measured. It is quite obvious how strongly the local wear pattern differs quantitatively from the integral wear. The scatter strip reaches 77 μm with the mean wear being of the order of 110 μm. In other words, the "amplitude" of wear in respect of its mean value reaches ±35 μm (±30 %) during tests.

Another interpretation is in the form of *wear circles* obtained at specified moments of time. These circles represent corresponding cross sections of the kinetic processes of local wear. Straight lines connect conventionally the experimental points over the wear circles. It is quite obvious how the pattern of real (locally measured) wear in the points of the cross section of the specimen differs qualitatively from the integral wear during one revolution. Though loads (bending and contact) stay unchanged within one revolution, the surface layers of the metal respond strongly differently in different local zones of the friction path. It seems natural since the mechano-physical properties of the metal surface layers are substantially different too (in both measurements). Hence, local wear over such surface portions of the specimen should be different as these local portions differently resist fracture. Thus, the *anisotropy of local properties of the material* generates the *anisotropy of discrete* (*local*) *wear*.

PC-aided control of the SI machines enables to select schemes of tests to be performed under specified loading conditions with highly precise measurements and validity of results.

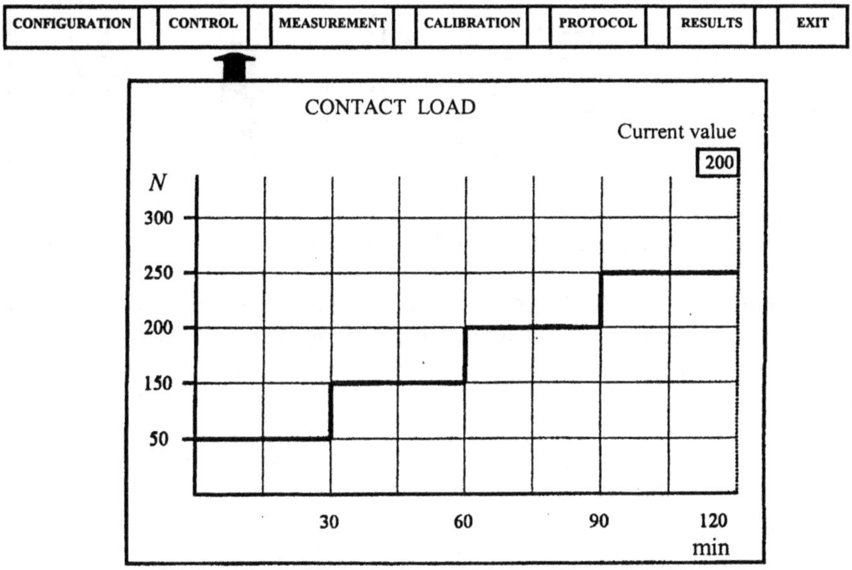

Fig. 3.14. Example of setting contact load in "CONTROL" mode

Dialogue with PC. The operator communicates with the testing machine by maintaining a dialogue with the PC. After the executive program is started, a menu line appears in the top of the display (Figs. 3.14 and 3.15). The experiment can be planned by selecting "CONFIGURATION" in the menu and then enter the duration of test, periodicity of saving results, periodicity of registration of test parameters, file name and other parameters. By selecting "CONTROL" in the menu it is possible to enter contact and bending loads, speed of rotation of the specimen and counterspecimen, the slippage factor. Figure 3.15 shows a fragment of entering the contact load in steps totally lasting 120 minutes. Note that contact and bending loads can be changed with any regularity in time.

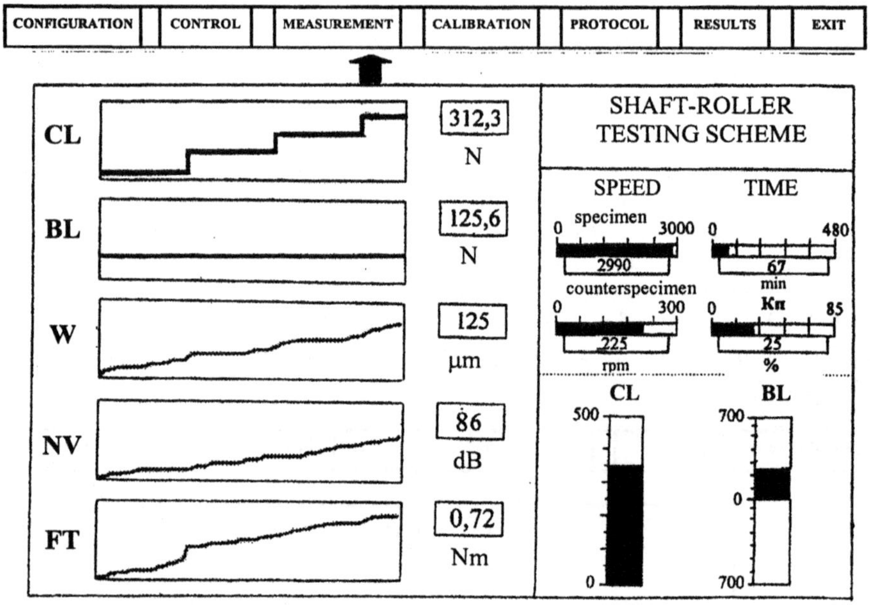

Fig. 3.15. Displayed tests of *shaft / roller* system in "MEASUREMENT" mode: CL – contact load; BL – bending load; W – wear; NV – noise and vibration; FT – friction torque

The item "MEASUREMENT" in the menu serves to monitor current parameters of loading and damage of the components of the active system. A testing program is started and the monitor displays the testing scheme, the speed parameters (the speed of the specimen and the counterspecimen), the current testing time and the slippage factor (K_{II}) – in the upper right-hand corner of the display; the parameters of loading of the specimen (nomographs of preset contact (CL) and bending (BL) loads) – in the bottom right-hand corner of the display; graphs of measured parameters (loads (CL, BL), wear (W), vibration and noise (VN), friction torque (FT)) with the indication of the current mean value of each parameter in the left portion of the display (cf. Fig. 3.14).

The item "CALIBRATION" in the menu serves to adjust the measuring system of the testing performed by a specialist.

Tests may be recorded as standard protocols by selecting "PROTOCOL" in the menu.

The item "RESULTS" in the menu serves displaying graphically the monitored characteristics registered during the test (loading, wear, vibration acceleration, temperature, the friction torque, etc.).

The operator during tests can watch on the display the preset conditions and measured (registered) parameters that together give a full idea about the experiment on the whole. After tests the operator scans through the accumulated data, analyzes them and draws up a test protocol.

3.3.4 Auxiliary devices

Optionally the SI machine can be equipped with the following devices: chambers for tests at elevated and negative temperatures; chambers for tests in various environments, liquid or gas; devices for performing tests for mechano-erosion fatigue; devices for testing effects of laser irradiation.

A pilot high-speed SI-011S machine was fabricated in 2000 with the speed of rotation up to 17,000 min^{-1}; a similar super high-speed machine is being developed with the speed of rotation up to 50,000 min^{-1}.

The main advantages of SI machines:

(*a*) fully computerized tests and data processing ;

(*b*) a block-module principle of configuration of a number of desktop special-purpose machines;

(*c*) unique opportunities of investigating processes of wear-fatigue damage;

(*d*) high precision of measurements;

(*e*) combination of a variety of tests performed by a single testing machine making tests highly economical;

(*f*) a possibility to perform a variety of tests with specimens of unified shape and dimension making the results easily comparable.

Self-test questions

1. Describe the purpose of the methods of wear-fatigue tests.
2. What purposes can the results of wear-fatigue tests serve?
3. Say about the principles of developing methods of wear-fatigue tests during main rotary motion. Name the objects of tests for mechanical, sliding, rolling fatigue and the objects for wear-fatigue tests.
4. Describe the schemes of mechano-rolling fatigue tests. How do you divide the schemes into tests for mechanical and rolling fatigue?
5. Describe the schemes of tests for mechano-sliding fatigue tests. How do you divide the schemes into tests for mechanical and sliding fatigue?
6. Describe the schemes of tests for fretting fatigue. How do you divide the schemes into tests for mechanical fatigue and fretting?
7. Describe in comparison the conditions of dynamic interactions between the specimen and counterspecimen during rolling and mechano-rolling fatigue.
8. Describe in comparison the conditions of dynamic interactions between the specimen and counterspecimen during sliding and mechano-sliding fatigue.
9. Why is damage during sliding friction considered to be fatigue damage?

10. Why is damage during rolling friction considered to be fatigue damage?
11. What is the fatigue curve (in principle)? What coordinates are used to plot it? What is its simplest equation? Name the basic parameters of the fatigue curve as the most essential characteristics of fatigue resistance.
12. What is the sense of notion "endurance limit" and "fatigue limit"? How can the numerical values of these extreme stresses be established experimentally?
13. Describe the procedure of estimating the fatigue curve parameters using the method of least squares.
14. How many curves can be plotted for these objects of tests: *a*) mechanical fatigue; *b*) sliding fatigue; *c*) rolling friction; *d*) mechano-sliding fatigue; *e*) mechano-rolling fatigue?
15. What is the nomenclature of fatigue curves obtained with different schemes of wear-fatigue tests?
16. What is direct effect? How is the direct effect coefficient determined?
17. What is back effect? How is the back effect coefficient determined?
18. Are the coefficients of direct and back effects constant values? Do they change in response to the dynamic conditions of tests or not?
19. Describe the nomenclature of endurance limits during mechanical, rolling and mechano-rolling fatigue. What ultimate stresses characterize the direct and back effects?
20. Describe the nomenclature of indicators of the slope of curves during mechanical, rolling and mechano-rolling fatigue.
21. What groups of factors cause WFD?
22. List the basic factors relating to the conditions of cyclic loading.
23. List the basic factors relating to the conditions of friction.
24. List the basic factors relating to the conditions of interactions between the components of the active system.
25. Describe briefly the methods of complex wear-fatigue tests.
26. Describe briefly the methods of successive (double-stage) tests.
27. What tests are to be performed to investigate the regularities of direct effect?
28. What tests are to be performed to investigate the regularities of back effect?
29. What modifications of the SI machines do you know for wear-fatigue tests?
30. What types of tests can be performed with SI-01, SI-02, SI-03 machines?
31. What main blocks do the SI machines include?
32. How are contact loads set and adjusted to the SI machines?
33. How are bending loads set and adjusted to the SI machines?
34. What purpose does the data control system serve? What are its basic functions?
35. List basic parameters measured and monitored by the SI machines.
36. How are two WFD characteristics discriminated: *a*) total wear; *b*) convergence of the axes of the friction pair? Illustrate these notions with relevant schemes.
37. What is integral wear and what is discrete wear? How are they measured?
38. What possibilities are there to study WFD using the method of discrete wear measurement?
39. What is the "wear circle"? How can one plot it?
40. Describe the displayed representation of the *shaft / roller* system in the "measurement" mode.
41. Table 3.1 shows the nomenclature of characteristics for tests for rolling and mechano-rolling fatigue. Can you compose a similar nomenclature of characteristics for tests for sliding and mechano-sliding fatigue?
42. Note that the fatigue curves in the upper half and in the bottom half of Fig. 3.6 are plotted using different methods. Try to analyze these two methods and establish advantages and disadvantages of each. In case of a problem read the relevant publications or ask the instructor to help.

Tasks for research

1. Using Fig. 3.1 try to propose several schemes of wear-fatigue tests during main reciprocating motion. Analyze them with the instructor. May it happen that you invent anything new?
2. If you read publications relating to the methods of wear-fatigue tests, you will see an amazing variety of the methods. Propose your own methods of tests that model or simulate the operation of active systems well known to you. May it turn out that you propose something unique?
3. The scheme shown in Fig. 3.2, *a* is in principle applicable to the *wheel / rail* system. Give a critical view of this scheme? What does it ignore in operation of the real system? What conditions of operation should be additionally taken into account and reflected in the scheme of tests? Briefly, try to think up your own (other) schemes of tests for mechano-sliding fatigue that would reflect the operations of the *wheel / rail* system more reasonably. You will doubtlessly generate a new solution or may be an invention.
4. The SI machines use a lever mechanism to create contact and bending loads. Propose other methods, such as electrohydraulic, electromagnetic, etc. Compare the alternatives, yours may be better. But remember that loading both should be applied and controlled.
5. The methods of measuring basic parameters were and remain the most vulnerable factors in any testing machine. A small error is tolerable for a broad range of variations of values to be measured with stable sustained performance. Try to find a non-trivial solution for any measuring problem (that you like more).
6. When working at laboratory pay attention to the design of the SI machines. It is far from perfect, though it should be admitted that the design of the SI-03M machine seems to be rather good (Fig. 3.16). Calm down your imagination and stay sensible! Propose your alternative of the design.

Fig. 3. 16. SI-03M machine

7. Let a small team of students with the help of the instructor devise an acceptable machine to perform tests for fretting fatigue like the one shown in Fig. 3.5. The main hurdle is that there should be drive for oscillatory motion (along the generatrix of the

specimen and the periphery of its cross section) of the order of about 10...100 μm and frequency about 50 Hz or any other reliable means of such motion instead of the drive.

8. D.C. synchronous and A.C. asynchronous motors are used in the SI machines. Do you have any idea about ac electronic motors? If yes, find their characteristics (manufacturers) and answer the question (after a comparative analysis) whether these ac electronic motors are promising. Make an exhaustive list of their advantages and disadvantages for such solution.
9. Design and execute the following experiments: *a*) measure the friction torque during sliding friction with a specified contact load and then calculate the friction coefficient; *b*) measure again the friction torque during wear-fatigue tests with the same contact load (and assume cyclic stresses at the level of the endurance limit) and then calculate the friction coefficient. Compare the obtained results. What is the difference between them? What role do cyclic stresses play in governing the force (and the coefficient) of friction?
   It would be interesting to carry out experiments with metal-to-polymer and metal-to-metal friction pair and an active system. Prepare a report for students' scientific conference.
10. Design and execute experimental studies similar to 3.9, but now measure wear (during fixed time) and / or temperature within the contact site. You may expect as interesting results.

4 DIRECT AND BACK EFFECTS

He who poses questions, he gets answers.
But he should pose sensible questions.

William Ramsay

4.1 General

Any practical analysis of the processes of mechanical fatigue, the processes of friction and wear are based on the main idea that the characteristics of the above processes are affected (usually damaging) by numerous factors, all their variety being classified into four groups, viz. factors of design, metallurgy, production process and operation (see Sect. 1.6). The effect of a factor, for example, on changes in the fatigue limit is, as a rule, unidirectional, i.e. it is the same qualitatively irrespective of the variations of the parameter characterizing this factor.

Yet, the experience of operation and tests of active systems as specific objects has revealed that mechanical fatigue, on the one hand, and friction (with wear), on the other hand, cannot be considered as the factors affecting, being complex phenomena, strength and durability, in a definite and independent manner. These phenomena evolve concurrently in a single zone of the active system within a complex dangerous volume and dialectically interact. The result of such interactions may in principle be of double kind: (1) accelerated development of damage because softening becomes dominating and leads to sharp loss of durability by the active system; (2) on the opposite, delayed development of damage due to dominating hardening strongly increases its durability. While result (1) seems trivial (clear and explainable, see, for example, Sect. 1.6), result (2) needs proof and explanation: load growth, from the standpoint of the mechanics of deformation and fracture, always leads to the corresponding loss of the bearing capacity and durability of the material.

So, we proceed from the fact that it is sufficient and effective to analyze the governing factors (the *factor* analysis) to estimate the bearing capacity and durability of a component of a design or a friction couple, while interactions between phenomena should be analyzed dialectically to estimate the bearing capacity and durability of active systems (the phenomena analysis).

4.2 Mechano-sliding fatigue

4.2.1 Direct effect

Since direct effect is determined as changes of characteristics of resistance to fatigue due to the processes of friction and wear, its basic regularities are studied experimentally, from the standpoint of fatigue fracture mechanics (see Sects. 1.3 and 3.3).

Let us design the simplest experiment observing the following basic principles:

- a metal-to-polymer system is to be tested, i.e. the specimen (metal) and the counterspecimen (polymer) are made from the materials with contrasting mechanophysical properties of unlike origin;
- the process of sliding friction occurs without any lubricating material, so the effect of the latter is ignored;
- the metallic specimen experiences practically no wear during tests, hence only one component in the pair, viz. the counterspecimen, undergoes wear;
- a linear state of stress appears in the working zone of the specimen during cyclic bending, i.e. they are the simplest conditions of tests for fatigue;
- the process of friction locates in the zone of tension of the specimen in bending.

The following pieces were prepared for the tests:

- cylindrical specimens from high-chrome steel 40X (the ultimate strength in tension is 970 MPa);
- counterspecimens from glass-filled (≈25%) polyamide "Durethane" BKV-30H (the ultimate strength in compression is 170 MPa).

The configuration of the experiment causes some doubt. It seems apparent that the relatively soft counterbody cannot significantly affect the resistance of the quite hard steel to fatigue because no physical wear of the specimen is expected.

Figure 4.1 shows the results of the experiments. From Fig. 4.1, *a* it follows that in the *process of wear-fatigue tests of the metal-to-polymer system the durability of the steel specimen at the amplitude of stresses* σ_a *= 200 MPa and contact pressure* p_a *= 8.5 MPa reduces ten times and the fatigue limit reduces by 32%* (compared with common fatigue). If the amplitude of stresses diminishes to 150...160 MPa, the durability during wear-fatigue tests is approximately 10^6 cycles, meanwhile the specimens do not fail at all during usual fatigue tests.

Thus, the processes of friction affect considerably the changes of characteristics of fatigue resistance (direct effect). Since no physical wear of the metallic specimen occurs in these test conditions, it may not be responsible for the above effect. In this case it is due to a complex of chemophysical phenomena in the friction zone. In particular, products of tribodestruction are known to possess the properties of surfactants. They accumulate in the contact zone and facilitate migration and multiplication of dislocations on the metallic friction surface (Rebinder effect, see Sect. 1.4.3). It causes acceleration of the surface fatigue damage. Also, as the contact pressure and duration of tests increase, the average temperature in the friction zone grows (to 70 °C in the conditions of the test) inducing thermal activation of many chemophysical processes so that the resistance of the specimens to fatigue becomes still less.

Figure 4.1, *b* shows the fatigue limits as functions of contact pressure. The curve is the relation between σ_{-1p} and p_a, so that according to this relation pressure rise leads to loss of fatigue resistance. The horizontal dotted line is the fatigue limit during mechanical fatigue that is definitely independent of the contact pressure.

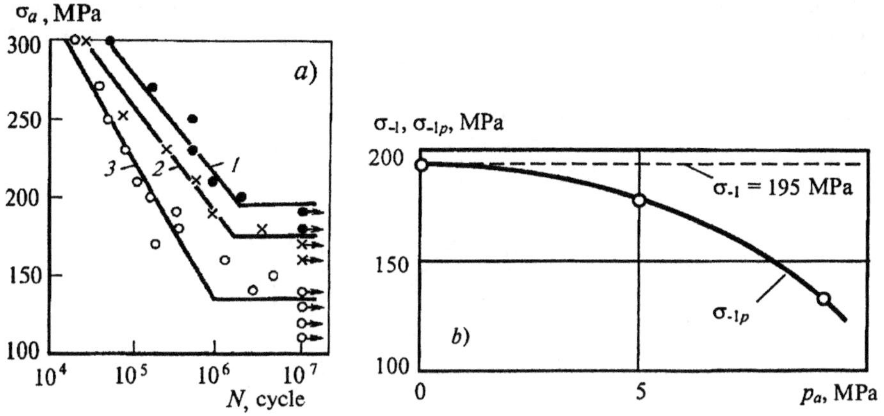

Fig. 4.1. Results of wear-fatigue tests of *steel 40X / polyamide "Durethane" BKV-30H* system: *a* – fatigue curves (*1* – mechanical fatigue curve; *2*, *3* – mechano-sliding fatigue curves at $p_a = 5$ and 8.5 MPa, respectively); *b* – fatigue limit as function of contact pressure (L A Sosnovskiy)

The theoretical analysis yielded the following equation that satisfactorily describes the results of the tests (see the curve in Fig. 4.1, *b*):

$$\sigma_{-1p} = \sigma_{-1T}(1-\varphi_p)^{1/m_V}, \tag{4.1}$$

where

$$\varphi_p = b_S \frac{S_{0.5\gamma}}{S_k}\left(\frac{p_{tf}^{(1)}\Delta T}{p_d - p_a}\right)^{m_S}. \tag{4.2}$$

According to Eq. (4.1) the mean fatigue limit σ_{-1p} of steel specimens during mechano-sliding fatigue of the metal-to-polymer active system is governed both by the conditions of tests and by the complex of mechanophysical properties of the metal and the polymer. The fatigue limit with the account of temperature effect (σ_{-1T}) and the parameter of isotropy of steel (m_V) characterize integrally the conditions of testing for fatigue and mechanophysical properties. Nominal contact pressure (p_a), temperature variations of the polymer (ΔT), the scheme of contact interactions between the components of the system (b_S), relative damaged volume in friction ($S_{0.5\gamma}/S_k$) describe the conditons of tests in sliding friction. Meanwhile, the mechanophysical properties of the polymer are rated by the *destruction limit* ($p_d = U_0/\gamma_p$, where U_0 – the energy of breaking of interatomic bonds, γ_p – the structurally sensitive coefficient), the parameter of the number of defects m_S, a *single thermofluctuation stress* $p_{tf}^{(1)} = k/\gamma_p$ (k – the coefficient of Boltzman).

From Eq. (4.1) it follows that the effect of friction processes (described integrally by the function φ_p) on fatigue resistance of the steel specimen is a damaging effect under these test conditions (it is predicted that $\sigma_{-1} \leq \sigma_{-1T}$ always) and it is really due to the complex of mechanical and chemophysical phenomena described by the corresponding parameters and coefficients (see function (4.2) for φ_p).

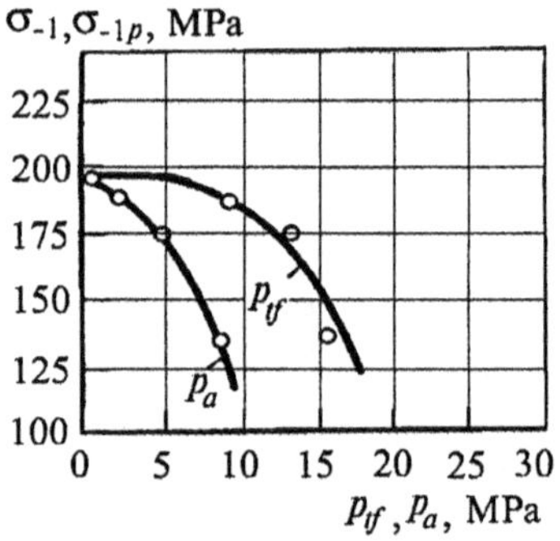

Fig. 4.2. Role of thermofluctuation stresses in wear-fatigue damage processes (L A Sosnovskiy)

Figure 4.2 illustrates the role of thermoactivating phenomena in the processes of mechano-sliding fatigue of the metal-to-polymer active system. The curve $\sigma_{-1p}(p_a)$ is borrowed from Fig. 4.1, *b* and represents the relation between ultimate stresses σ_{-1p} and nominal contact pressure p_a. If *thermofluctuation stresses* are estimated in the polymer

$$p_{tf} = p_{tf}^{(1)} \Delta T = \frac{k}{\gamma_p} \Delta T , \qquad (4.3)$$

that occured under the conditions of the experiment and the relation between the ultimate stresses σ_{-1p}, and the value p_{tf} is plotted, then it turns out that thermofluctuation stresses exceed substantially (nearly two times) contact pressures. Hence, the effect of thermofluctuation processes both on the surface damage (or wear) of the polymer and generation of resistance to fatigue of steel specimens is governing during the tests. In other words, it is a convincing proof of the conclusion above that mechano-sliding fatigue of the active system in question is actually due to the chemophysical phenomena in the contact zone.

Let us consider again the results of tests of the metal-to-metal active system. Unlike the metal-to-polymer system the main distinction of the metal-to-metal system is that both components, i.e. the specimen and the counterspecimen, undergo physical wear in the process of fatigue tests. Curve in Fig. 4.3 depicts the relation of the fatigue limit of steel specimens and contact pressure in the *steel 45 / iron* active system (friction and lubrication with oil CY), the horizontal dashed line represents the fatigue limit of the steel specimens during common fatigue tests (naturally, the limit does not depend on pressure). Comparison of the curves in Fig. 4.2, *b* and 4.3 enables to establish their principal difference, viz. the *ultimate stress during wear-fatigue tests of the metal-to-metal system is higher within a relatively broad range of variations of contact pressure* (from ≈1.05 MPa in Fig. 4.3) *than the ultimate stress during common fatigue tests*. In other words, these conditions of the processes of friction and wear do not cause damage, they lead, on the opposite, to hardening.

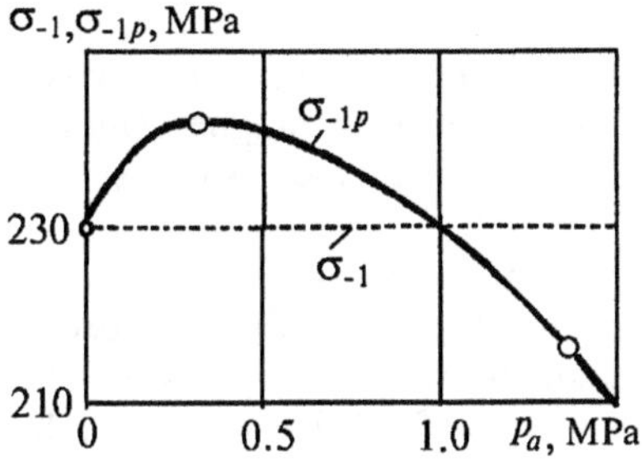

Fig. 4.3. Ultimate stresses as function of contact pressures in *steel 45 / pig iron* active system (V I Pokhmursky, et al.)

A similar abnormal behavior of the function $\sigma_{-1}(p_a)$ is explained by the ratio between the processes of hardening-softening and removal of surface impurities by friction.

The curve in Fig. 4.3 is satisfactorily described by the equation

$$\frac{\sigma_{-1p}}{\sigma_{-1}} = \left(1 + \frac{p_a}{p_f}\right)\left(1 - \mu_\varepsilon \frac{p_a}{p_f}\right)^2, \tag{4.4}$$

where p_f – the ultimate value of pressure during sliding friction (the sliding fatigue limit); μ_ε – the parameter of strain hardening.

From the above it follows that it is not *justifiable to consider friction and wear as some factors that are necessarily harmful for the active system. It is more opportune to imply some complex processes and results of interactions between two damaging phenomena, viz. mechanical fatigue and friction (including attendant wear)*. These interactions can lead to ambiguous consequences (while the effect of this or that factor is usually unambiguous. *The fatigue limit of the specimen can either grow or fall or remain unchanged in response to the conditions of wear-fatigue tests and origin of contacting materials.*

These are the basic regularities the direct effect established during the experiments performed, as it is noted above, from the standpoint of fatigue fracture mechanics. In fact, their understanding helps overcome the traditional limits of fatigue fracture mechanics and familiarize with tribo-fatigue as it has become clear and proved that friction and wear are the phenomena capable of mechano-physico-chemical interactions with the phenomenon of fatigue rather than the factors producing a simple effect on the resistance of materials to fatigue. It is the result of these interactions that complex wear-fatigue damage of the material occurs. Since it is complex, it is not just a simple sum of individual (particular) damages plus fatigue damages and damages due to friction and wear.

4.2.2 Back effect

Since the back effect is defined as changes of the characteristics of friction and wear under the effect of the processes of fatigue damage, its basic regularities are studied by designing and performing experimental studies from the standpoint of tribology (see Sects. 1.4 and 3.3).

The principles of experiment designing remained the same. A metal-to-polymer system *steel 40X* (the *specimen*) / *formaldehyde copolymer* (the *counterspecimen*) was subjected to wear-fatigue tests: the ultimate strength at compression 56 MPa) at a constant contact pressure $p_a = 5.7$ MPa. However, this time the linear wear of the polymeric counterbody was measured in the process of tests. The result served to calculate the volume intensity of wear using the formula

$$I_V = \Delta V / 2\pi r n,$$

where ΔV – the volume of the worn polymer; r – the steel specimen radius; n – the number of loading cycles.

Figure 4.4 shows the relation between the wear intensity increment ΔI_σ of the polymer and the amplitude of stresses σ_a in the steel specimen. The value ΔI_σ at a given contact pressure p_a = const was calculated using the results of measurements in the following manner:

$$\Delta I_\sigma = \frac{I_\sigma(n, \sigma_a) \ - I_V(n)}{I_V(n)},$$

where $I_\sigma(n, \sigma_a)$ – the wear intensity of the counterbody in the active system in which $\sigma_a > 0$, i.e. during the wear-fatigue tests; $I_V(n)$ – the wear intensity of the counterbody in a usual friction couple in which there are no cyclic stresses ($\sigma_a = 0$).

From the data in Fig. 4.4 it follows that the amplitude of stresses in the steel specimen significantly affects the wear intensity of the polymeric counterbody. If *cyclic stresses grow from 160 to 300 MPa, the wear intensity increment due to these stresses changes from 110 to 180%* (versus the wear intensity in a common friction couple when $\sigma_a = 0$). Hence, the *durability of metal-to-polymer active system based on the wear criterion is governed by the back effect in many respects.*

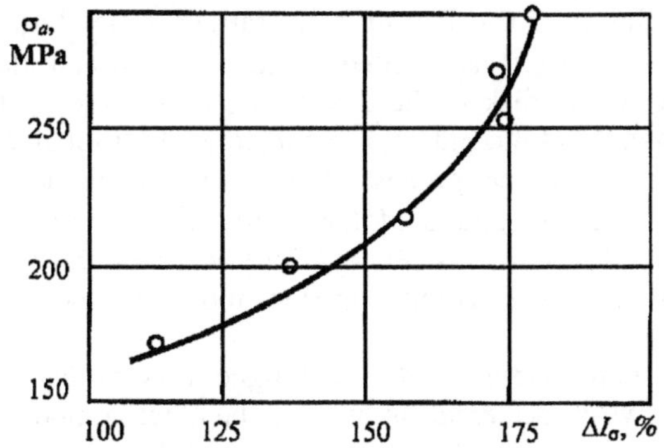

Fig. 4.4. Incrementations of wear intensity of polymer as function of amplitude of cyclic stresses (alloyed steel *40X / formaldehyde copolymer*) (L A Sosnovskiy)

The described back effect in the metal-to-polymer active system is due to additional intensification of kinetic processes of breakup of polymeric molecules by cyclic stresses in the actual contact spots. This breaking is much due to the phenomenon of thermodestruction of the polymer because of intensive heat emission in the contact. It intensifies due to non-elastic cyclic deformation of the surface layer on the steel specimen during tests for fatigue. Effective transfer of the polymer to steel observed visually during tests is an indirect proof of this assumption.

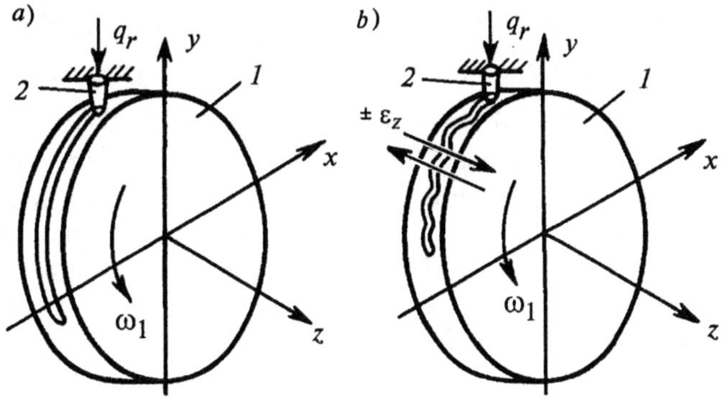

Fig. 4.5. To analysis for back effect

Let us examine Fig. 4.5 in order to answer the question, from the standpoint of mechanics, why the wear of the polymeric counterbody strongly intensifies when cyclic stresses are excited in the conjugated steel body. The body is shown as rotating disk *1* with smooth working surface and the counterbody as fixed single indentor *2*. During usual tests for friction (Fig. 4.5, *a*) only the contact load q_r is operative, indentor *2* statically bends (in the direction opposite to rotation ω_1), so the deformable zone on the working surface of the disk looks like a strip (a friction path). During wear-fatigue tests (Fig. 4.5, *b*) additional cyclic deformation $\pm\varepsilon_z(\sigma)$ is excited in the disk. Small deformation of the working surface of the disk in the direction z makes the friction path over the surface look zigzag and the indentor is subjected additionally to cyclic bending (in the direction z). The wear process of both components naturally intensifies in accordance with the magnitude of the cyclic stresses $\pm\sigma_z$. If the indentor is polymeric, while the disk is steel, only the wear of the polymer as a softer material intensifies. If the indentor is steel too, the wear of both components may intensify.

Thus, during such conditions of wear-fatigue tests the back effect may lead to two phenomena: wear accelerates in one and the other component under the effect of cyclic stresses excited in only one component of the active system.

Theoretical analysis has yielded the following formula to estimate the wear intensity $I_V(\sigma)$ of the polymeric counterbody with the allowance for both the amplitude of stresses σ_a and the complex of the mechanophysical properties of steel and the conditions of tests for fatigue:

$$I_V(\sigma) = \frac{I_V}{1-\varphi_\sigma}; \tag{4.5}$$

$$\varphi_\sigma = b_v \frac{V_{0.5\gamma}}{V_0}\left[\left(\frac{T_M}{T_0}\right)^{m_T}\left(\frac{\sigma_a - \sigma_{-1\min}}{\sigma_w}\right)\right]^{m_V}, \tag{4.6}$$

where I_V – the wear intensity of the polymer without any cyclic stresses in the steel specimen (i.e. in the common friction couple); b_v – the coefficient known a priori and making allowance for the conditions of fatigue tests; $V_{0.5\gamma}/V_0$ – a relative dangerous volume of the cyclically deformable steel specimen; T_M/T_0 – relative temperature of the metallic specimen in the friction zone; m_T – the parameter of temperature activation of the processes of fatigue damage; $\sigma_{-1\min}$ and σ_W – the parameters of the function of distribution of fatigue limits of steel specimens in the form of the law of Waybull; m_V – the parameter of steel isotropy.

Equation (4.5) predicts the damaging effect of cyclic stresses: the value $(1-\varphi_\sigma)<1$ and $\varphi_\sigma<1$, hence, $I_V(\sigma) > I_V$.

Experimental studies thus show (cf. Fig. 4.4) that the *value I_σ depends strongly non-linearly on the conditions of cyclic deformation. It means that the resulting wear in the active system is not just a simple sum of usual wear (in a friction couple) and additional wear due to cyclic stresses.* Therefore, it is another confirmation of the main assumption that wear-fatigue damage results from an intricate interaction between the phenomena of fatigue and friction (together with wear). Since the effect is complex, it is represented by the non-linear function (4.5) of individual (particular) damages.

The principal distinction between two notions – *wear-fatigue damage* and *fatigue wear – should be outlined now.* In this connection it is opportune to quote the corresponding definitions in the Standards:

- *wear-fatigue damage* is damage due to kinetic interactions between the phenomena of fatigue, friction in any its manifestations, wear and (or) erosion (GOST 30638–99);
- *fatigue wear* is mechanical wear resulting from fatigue fracture when microvolumes of the material in the surface layer undergo re-deformation (GOST 27674–88).

The following formula has been obtained for estimating the wear intensity of the shaft in the metal-to-metal system *crankpin* / connecting-rod end (*with a sliding bearing*):

$$I_h(\sigma) = I_h \Phi^{m_\tau}, \tag{4.7}$$

where I_h – linear wear intensity in the common friction couple (when $\sigma_a = 0$) and the function Φ depends on the size of the crankpin (the ratio between length and radius: L/R), the design of the unit (the ratio between half-width of the contact strip and radius: a/R), the coefficient of sliding friction f, the Poisson coefficient ν and the elasticity moduli of contacting metals:

$$\Phi = \left\{ 1 + \frac{2\frac{L}{R}\chi(1-\mu) + \chi^2\left[2\left(\frac{L}{R}\right)^2 + 3\left(\frac{a}{R}\frac{1+2\mu}{1+\mu}\right)^2\right]}{1-\mu-\mu^2} \right\}^{1/2}. \tag{4.8}$$

The loading parameter $\chi = \sigma_a/(fp_a)$ makes allowance for both cyclic (σ_a) and contact (p_a) stresses in the friction zone. Since, according to formula (4.8), $\Phi > 1$ always, Eq. (4.7) predicts like (4.5) the damaging effect of cyclic stresses, i.e. $I_h(\sigma) > I_h$. Equation (4.7) describes the wear of the shaft as one manifestation of the back effect.

Understanding of the back effect enables to go beyond the traditional frame (this time the frame of tribology) and come close thus to tribo-fatigue, on the other hand. In fact, it turns out that the *wear intensity can be controlled non-traditionally by exciting cyclic stresses in one component of the friction couple.* This control is highly effective: the wear intensity can change tens or even hundreds of per cent. If it is borne in mind that according to the experimental data, certain wear can exceed significantly the reliability of an active system, it becomes clear that we go beyond the common approach to ensuring the reliability of mechanical systems based on individual criteria of fatigue or wear resistance. We approach the *complex problem of control over the reliability of active systems in modern machinery based on the criterion of wear-fatigue damage.* In other words, it has become clear that tribo-fatigue should be created on the verge of tribology and fatigue fracture mechanics.

4.3 Mechano-rolling fatigue

4.3.1 Direct and back effects

Another experiment was designed from the standpoint of tribo-fatigue. Its purpose was to investigate the direct and back effects in the metal-to-metal active system during mechano-rolling fatigue, this time friction was created once again in the zone of tension of a bending test specimen (cf. Fig. 3.2, *a*). Figure 4.6 illustrates the results of tests of the *carbon steel 45* (the *specimen*) / *alloyed steel 25XTT* (the *roller*).

The diagram *ABCD* is plotted in the following coordinates: pressure p_0 in the center of the contact site (the abscissa axis) – the amplitude σ_a of cyclic stresses in bending (the ordinate axis).

The point *A* is the fatigue limit σ_{-1} of steel 45 specimens determined by common tests for mechanical fatigue following the scheme in Fig. 3.2, *c*. The limiting state criterion is when the specimen breaks into two pieces because of the main fatigue crack in its vulnerable cross section. Hence, this point implies the mechanics of fatigue fracture. In general, the *ordinate axis* σ_a is the *strength scale*: this scale should accommodate results of fatigue tests of any components of structures made from any materials.

The point *D* is critical pressure p_f in rolling friction without slip, it was determined by common tests for friction (following the scheme in Fig. 3, *b*). The limiting state criterion is the appearance of pits of spalling of critical density along the rolling path. Hence, this point implies tribology. In general, the *abscissa axis* p_0 is the *tribological scale*: this scale should accommodate test results of any friction pairs the components of which are made from any materials.

Curves *ABCD* are a *diagram of limiting states of the active system during mechano-rolling fatigue*, it as plotted from the results of wear-fatigue tests (following the scheme in Fig. 3, *a*). Hence, it implies tribo-fatigue.

The limiting state along the portion *AB* is predominantly due to the development of the main fatigue crack, meanwhile the processes of appearance of pits of spalling are attendant. Therefore, the direct effect occurs satisfactorily described by the equation

$$\sigma_{-1p} = \sigma_{-1}\left[1 - \frac{p_0}{p_f}\left(1 - \frac{p_0}{p_f}\right)\ln(1 - \mu_p)\right], \tag{4.9}$$

where μ_p – the *parameter of rolling hardening*; it is $\mu_p = 0.92$ in the conditions of the experiment.

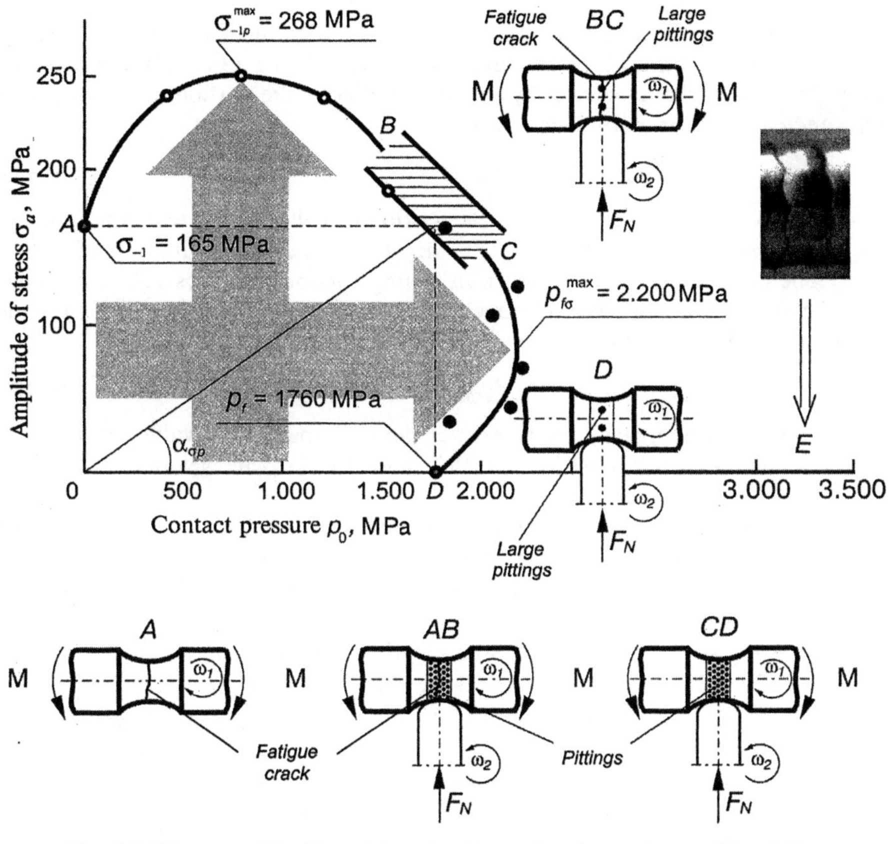

Fig. 4.6. Diagram of limiting states of active system in mechano-rolling fatigue (L A Sosnovskiy, A V Bogdanovich, S A Tyurin)

On the opposite, the limiting state along the portion *CD* is governed by the critical concentration of pits of spalling, meanwhile the development of mechanical fatigue microcracks is an attendant damage. It is the back effect satisfactorily described by the equation

$$p_{f\sigma} = p_f \left[1 - \frac{\sigma_a}{\sigma_{-1}} \left(1 - \frac{\sigma_a}{\sigma_{-1}} \right) \ln(1 - \mu_\sigma) \right], \tag{4.10}$$

where μ_σ – the *parameter of cyclic hardening*; it is $\mu_\sigma = 0.65$ in the conditions of the experiment.

The portion *BC* is transient; the kinetic processes of interactions between the phenomena of friction (with wear) and mechanical fatigue evolve at larger parameters of loading σ_a and p_0 close (or equal) to critical (σ_{-1}, p_f). The limiting state under these conditions of tests can be reached concurrently based on two criteria.

Examination of the *ABCD* diagram leads to the following basic conclusions.

(1) The fatigue limit of the specimen increases 1.5...1.6 times if the process of rolling friction occurs concurrently (the direct effect – the portion *AB*). The *direct effect factor* advanced in tribo-fatigue (3.2)

$$K_D = \sigma_{-1p}/\sigma_{-1} \tag{4.11}$$

is, in fact, a strength characteristic; its maximum value in the test conditions is $K_{D\,\max} = 268/165 = 1.62$. Factor (4.11) is incorporated, naturally, into Eq. (4.9).

(2) The critical (ultimate) pressure in rolling friction increases 1.2...1.25 times if cyclic stresses are concurrently excited in the specimen (the back effect – the portion *BC*). The *back effect factor* advanced in tribo-fatigue (3.3)

$$K_B = p_{f\sigma}/p_f \tag{4.12}$$

is, in fact, a tribological characteristic too; its maximum value in the test conditions is $K_{B\,\max} = 2200/1760 = 1.25$. Factor (4.12) is incorporated, naturally, into Eq. (4.10).

(3) The process of wear in rolling within the optimum range of contact pressures ($p_0 \approx$ 400...1300 MPa) significantly increases the reliability of the system based on the criterion of fatigue resistance so that a tendency to wearless friction is unjustifiable in this case.

(4) Tensile stresses during cyclic loading in the optimum conditions ($\sigma_a \approx$ 50...100 MPa) are favorable because they lead to a significant rise of the reliability of the system based on the criterion of resistance to rolling friction.

Improvement of the limiting state characteristics σ_{-1p} and $p_{f\sigma}$ in the process of wear-fatigue tests versus the characteristics during rolling friction (p_f) and mechanical fatigue (σ_{-1}) can be explained from the viewpoint of mechanics by the following reasons:

- addition of stresses with opposite signs (contact and bending) causing the shift of the mean stress cycle towards negative values and thus to the reduction of the maximum stress cycle;
- hardening of the working portion of the specimen by surface plastic deformation;
- appearance of favorable residual compressive stresses;
- healing of the primary fatigue cracks by elastoplastic deformation in the process of rolling friction.

The *controlling parameter of wear-fatigue damage* (cf. Fig. 4.6)

$$0 \le \psi_{\sigma p} = \tan \alpha_{\sigma p} = \sigma_a / p_0 \le \infty \tag{4.13}$$

has the *critical value*

$$\psi_{-1f} = \sigma_{-1}/p_f = 165/1760 = 0.094. \qquad (4.14)$$

This critical value separates the regions of direct and back effects on the diagram of limiting states of the active system. If $\psi_{\sigma p} < \psi_{-1f}$, we obtain the curve *CD*. If $\psi_{\sigma p} > \psi_{-1f}$, we obtain the curve *AB*. The value $\psi_{\sigma p} = \infty$ (pure mechanical fatigue) corresponds to the point *A*, the value $\psi_{\sigma p} = 0$ (pure rolling friction) corresponds to the point *D*.

Application of fine experimental methods of research enables to study and get insight into the specifics of complex wear-fatigue damage. Figure 4.7 exemplifies the results of studies (with the method of atom force microscopy) of the processes of cracking on steel 45 specimens during rolling friction and wear-fatigue tests as a function of the level of contact pressure p_0 and the value of the amplitude of cyclic stresses σ_a. Figure (their dimension is ~35×35 μm^2) shows the morphology of cracks typical for the corresponding conditions of tests. The histogram shows the relation between the critical depth *h* of the damaged layer and the level of cyclic stresses (at unchanged contact pressure p_0 = 2130 MPa). These experimental data lead to the following conclusions.

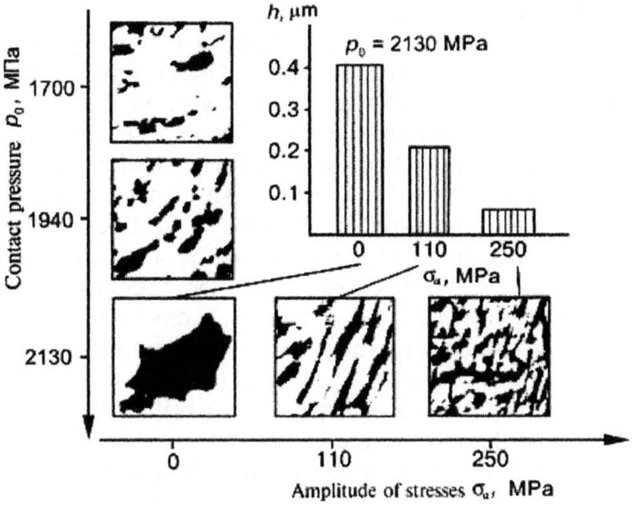

Fig. 4.7. Microtopography of surface damage during rolling friction (vertical column of figures) and during wear-fatigue tests (remaining figures) (L A Sosnovskiy, S A Chizhik, et al.)

Any rise of contact pressure during pure rolling friction intensifies plastic deformation, hence it leads to deformation fragmentation of grains, initially to the appearance of discrete pores and cracks which later form chains. The system of the deformed grains, chains, pores and cracks is unidirectional and it is oriented along the rolling direction. This process leads to the formation of relatively large discrete pits of spalling. Delamination and spalling are two main types of wear. The critical damage depth of the layer is estimated at ~0.4...0.5 μm.

During wear-fatigue tests similarly deformation fragmentation of grains, appearance of pores and cracks are observed. Yet the pattern of damage changes significantly. As the amplitude of cyclic stresses grows, the process of appearance of the second system of cracks accelerates, now they are transverse in respect of the rolling direction. That is why damage scatters and an almost regular net of intersecting pores and cracks appears, that fringes with finely dispersed particles (fragments of grains) of the material. The higher the cyclic stresses the denser is the net of pores and cracks and the finer and thinner are the separating particles. The critical depth of the damaged layer grows smaller to 0.05 μm. It prevents the appearance of larger and deeper pits of spalling, and they are not observed under these conditions. *Surface crushing* is the dominating wear process in this case. It is characterized by separation of finely dispersed particles from the working surface that result from multiple microshearing over intersecting planes and generation of a huge number of scattered microscopic pores and cracks and fine crushing of grains. This mechanism of complex surface damage is called the *scattered effect of multiple microshearing* (SEMMS).

The above results enable to identify additional causes why wear-fatigue damage in certain conditions is less menacing than the damage in friction (at a similar contact pressure).

1. Superposition of the fields of contact and bending stresses leads to dissipation of more applied energy in a finer surface layer of the material and localization of the processes of cracking and wear in the layer. Deformation energy is expended faster for finer crushing of grain fragments and their multiple separation than for penetration of damage into the depth of the material.

2 Wear of the surface layer damaged by a net of pores and cracks exposes a new relatively sound surface highly resistant to fracture. The appearance of relatively larger (in response to the loading conditions) pits of spalling is thus delayed in time or even prevented entirely at the bottom of which dangerous micrconcentration of stresses and a dangerous main crack develop.

3 Approximately tenfold rejuvenation of the working surface is required by fragmentation, crushing and separation of metal particles during wear-fatigue tests for the damage to reach the same depth like in rolling friction, providing the contact pressure is similar in both cases.

In this way, it has been established experimentally that wear-fatigue damage is a *specific and peculiar type of surface damage* of the main component of the active system. Its specific feature in these conditions is the surface crushing because of SEMMS over the intersecting planes of sliding. Its peculiarity is that the process does cause damage, but it is useful for it boosts significantly the reliability and durability of the active system. It is evident that in case of an optimal combination of the loading parameters σ_a and p_0 the active system reaches the state when its bearing capacity is maintained spontaneously (or controlled automatically) for a long time by the wear and removal of a fine damage surface layer in the friction zone. Summarizing it should be mentioned that the active system is a peculiar dynamic system, its behavior can and should be controlled, for example, by non-traditional method of the wear intensity control.

It should be remarked that the diagram of limiting states of the active system (cf. Fig. 4.6) differs cardinally from ultimate double-parametric diagrams known in mechanics (for example, $\sigma_a - \sigma_m$, cf. Fig. 1.17). As a rule, the diagrams of the limiting states of components of structures and friction pairs are plotted using a single criterion of damage (fracture), for example, the appearance of main crack of a definite length (for a structural component) or the appearance of the critical concentration of pits of spalling (for a friction pair). Meanwhile, the diagram of the limiting states of the active system shown in Fig. 4.6, is based on three criteria: fatigue fracture over the portion *AB* (direct effect), ultimate wear over the portion *CD* (back effect) and the critical state based on both criteria concurrently over the portion *BC*. It means that a single equation cannot describe analytically the full diagram of limiting states of the active system; there should be separate equations for the portion *AB* and *CD*. Of course, these equations may be similar (cf., for example, (4.9) and (4.10)), but their parameters should be specific (like they are in Eqs. (4.9) and (4.10)).

4.3.2 Translimiting state

Another experiment served the purpose of studying the manifestations of the back effect when contact pressure increases in *multiple steps* within a broad range of variations (Fig. 4.8, steps *I*, *II*, ..., *XII*). In the process of tests the convergence δ_c of axes of the pair components of the system comprising the *specimen from soft steel / roller from high-strength steel* was measured during rolling friction (cf. Fig. 3.12, *b*) (when $\sigma_a = 0$) and during mechano-rolling fatigue (at $\sigma_a = 0.8\sigma_{-1}$ and $\sigma_a = 1.0\sigma_{-1}$). It is visible that (cf. Fig. 4.8) the process of accumulation of wear-fatigue damage decelerates substantially compared with the process of damage during rolling friction, the range of normal friction based on the contact pressure expands by approximately 14%. We will explain the difference between the process of addition and interaction between damages using these experimental data (see also Sect. 2.5).

Assume that during the time t_1 damages due to contact (ω_p) and off-contact (ω_σ) loads accumulate as Fig. 4.9, *a* shows it: none of these criteria leads to the limiting state ($\omega_p \ll 1.0$; $\omega_\sigma \ll 1.0$). If damages are added up ($\omega_p + \omega_\sigma = \Sigma\omega$), then in case of wear-fatigue tests the limiting state ($\Sigma\omega = 1.0$) is reached during the time $t_2 < t_1$. Yet, evidently this prediction turns out to be untrue for the experimental data shown in Fig. 4.8. If it is considered that damages due to contact and off-contact loads interact

$$(\omega_p + \omega_\sigma)R_{\sigma/p} = \omega_\Sigma,$$

then the scheme adequately reflecting the experimental data in Fig. 4.8 looks like Fig. 4.9, *b* shows it. The limiting state during rolling friction is reached within the time t_2, while during mechanical fatigue it does not occur even at $t_1 \gg t_2$. During wear-fatigue tests the durability (t_1) turns out larger than during rolling friction (t_2). Whence a general conclusion follows: during wear-fatigue damage the *deformation energy due to contact (U_p) and off-contact (U_σ) loads do not add, they interact dialectically*:

$$(U_\sigma + U_p)\,\lambda\,(\sigma \leftrightarrow p) = U_\Sigma. \qquad (4.15)$$

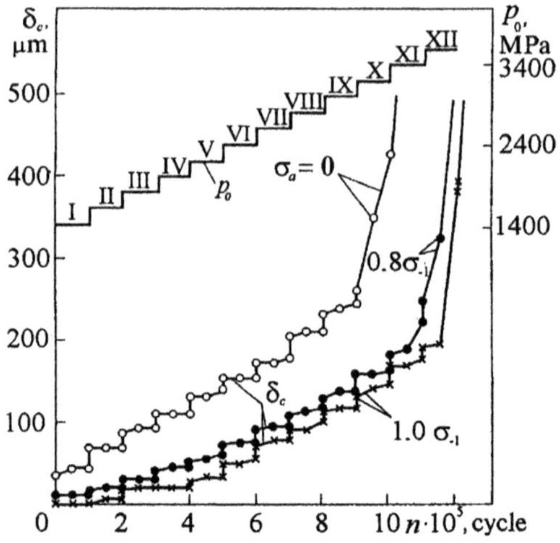

Fig. 4.8. Variations of δ_c during step-by-step contact pressure rise (L A Sosnovskiy, S A Tyurin)

The result of such interactions is determined both by the loading conditions and the direction of the processes of hardening-softening (see Sect. 2.5). It follows from (4.15) that a particular case of interactions between damages occurs, or their addition, at $\lambda\,(\sigma \leftrightarrow p) = 1$ (the sign of equality is assumed in condition (4.15)).

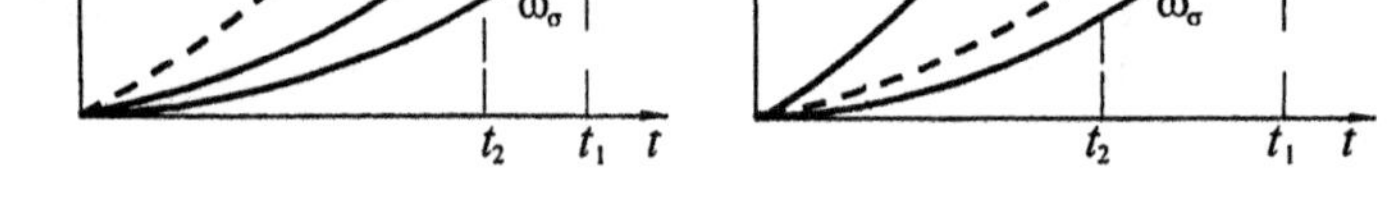

Fig. 4.9. Diagrams explaining summation of (*a*) and interactions between (*b*) damages

An *unexpected phenomenon* was discovered during tests (cf. Fig. 4.8): *residual undulatory damages or immovable irregular plasticity waves along the rolling path on the soft steel specimen* (see the photo in the upper right-hand corner of Fig. 4.6). Meanwhile the shape of the high-strength steel roller remains unchanged in the contact zone, i.e. geometrically undistorted. Figure 4.6 shows that in case of regular loading the ultimate pressure $p_{f\sigma}$ did not exceed $p_{f\sigma}^{\max} = 2200$ MPa, residual wave-like damages during multistage loading appear under much higher pressure (see zone E in Fig. 4.6) usually exceeding 3000 MPa (pressure rise by approximately 30...40%). Hence, a *translimiting state instead of the limiting state* (more precisely, one of the possible forms of the translimiting state) was reached during multistage loading.

Figure 4.10, *a* shows the scan of the rolling path with several irregular (congealed) waves of plastic surface deformation that appeared under these conditions of tests. Each wave is a combination of two peculiar semipunctiform craters and a lintel with the tip resembling a wavy ridge. Figure 4.10, *a* shows the typical dimensions of the craters and lintels that lead to the following conclusions.

a)

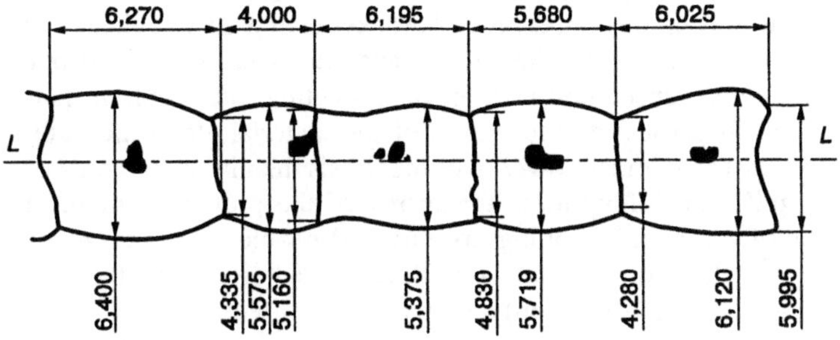

b)

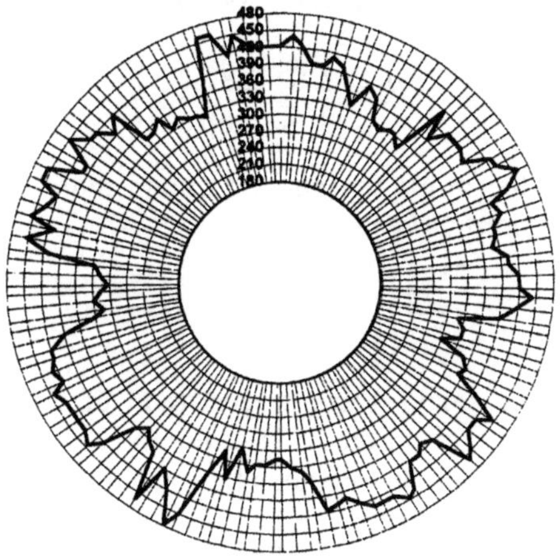

Fig. 4.10. Specific type of limiting state: surface undulatory damages (pits of spalling are shaded) (*a*) and distribution of microhardness along length L of rolling path (*b*) (L A Sosnovskiy, S A Tyurin, V A Yakovlev)

None of the congealed deformation waves repeats: each crater and lintel has its own dimensions different from others. The step between craters is also variable. The relative plastic deformation in the radial direction is 4...8%, while it reaches

50...70% in the axial direction. Hence, the appearance of residual surface undulatory damage is due to the non-stationary process of elastoplastic deformation. The anisotropy of the mechanophysical properties of the material in local zones of the rolling path can be assumed to be responsible for the deformation anisotropy in these zones leading to the formation of discrete pits of spalling as sources of the nonsteady state. The stronger the deformation anisotropy and the larger the pits of spalling, the stronger is the dynamic force excited during local collisions of the roller with the specimen. Thus, the form of the translimiting state described for this case is due to the nonsteady impact fatigue processes.

The method of microhardness was applied to corroborate the conclusion about the anisotropic properties of the friction surface. Figure 4.10, *b* shows the distribution of the microhardness of the material over the circumference of the specimen passing through the centers of the craters (see line *L* in Fig. 4.10, *a*). It is apparent, that, on the one hand, microhardness changes periodically according to the step of the craters. On the other hand, the pattern of distribution of the microhardness is substantially irregular reflecting the random nature of the anisotropy of properties of local zones of the material along the path of rolling. Hardness is, as a rule, much lower over the lintels than in the bottoms of craters.

So, a significant deformation anisotropy of the properties of the material in local zones of the path of rolling appears and develops in the process of wear-fatigue damage. It becomes manifest in three typical directions: the circumference, depth (the radius of the specimen) and in the axial direction.

It seems that it dictates introduction of special characteristics of the local wear-fatigue damage process: the *coefficient of asymmetry*

$$R_a = \frac{1}{4}\sum_{i=1}^{4}\frac{r_{\min(i)}}{r_{\max(i)}}, \tag{4.16}$$

where $r_{\min}$ and $r_{\max}$ – the minimum and maximum radii of one diameter of the specimen, and the *coefficient of irregularity*

$$\eta_a = \frac{r_{\rm sma}}{r_{\rm lar}}, \tag{4.17}$$

where $r_{\rm sma}$ and $r_{\rm lar}$ – the smallest and largest radii of the specimen during one revolution. Figure 4.11 (cf. also Fig. 3.13) shows the conventional designations of the radii of the specimen, also the relations between the coefficients R_a and η_a and the level of cyclic stresses during the tests of the active system *steel 45/steel 25XTT* for mechano-rolling fatigue by changing the bending loads in steps under the contact pressure $p_0 = 0.7p_f =$ const. It is apparent that the degree of irregularity (or anisotropy) of local wear-fatigue damage grows according to the augmentation of cyclic stresses. Note that the smaller are the values R_a and η_a the larger is the anisotropy of wear-fatigue damage.

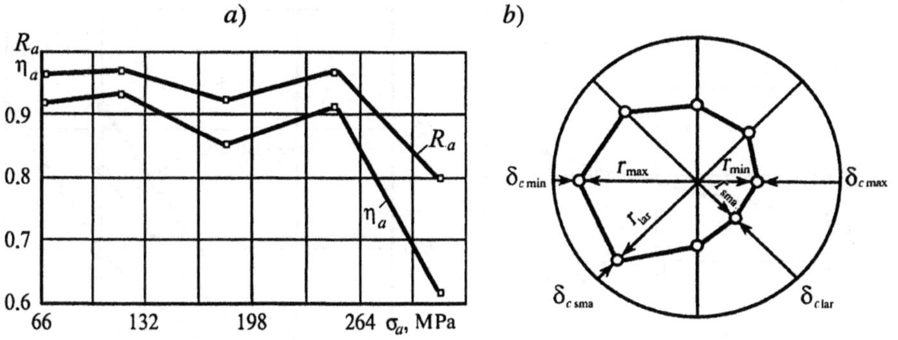

Fig. 4.11. Dependence of asymmetry coefficients and irregularity of wear-fatigue damage during tests for mechano-rolling fatigue of *steel 45 / steel 25XTT* active system (L A Sosnovskiy, S A Tyurin)

The procedure of different determination of the asymmetry and irregularity coefficients can be used, viz. they can be recorded in the order of magnitudes δ_c:

$$R_\delta = \frac{1}{4}\sum \frac{\delta_{c\,\min(i)}}{\delta_{c\,\max(i)}}, \quad \eta_\delta = \frac{\delta_{c\,\mathrm{sma}}}{\delta_{c\,\mathrm{lar}}}. \tag{4.18}$$

It is quite apparent that the coefficients determined from formulas (4.18) are unequal to the corresponding coefficients determined from formulas (4.16) and (4.17). Selection of the type of presentation of the coefficients R and η is dictated by the purpose of a specific analysis.

4.4. Effect of conditions of interactions

Since the conditions of damage interactions from contact and off-contact loading are highly diversified (see Sect. 2.5), it should be expected that the regularities of the direct and back effect may change correspondingly.

Really, let us study, for example, the results of tests of the metal-to-polymer active system for mechano-sliding fatigue. If the friction process evolves in the zone of tension of the cyclically bending specimen (cf. Fig. 4.1, *b*), growth of contact pressure reduces its resistance to fatigue. On the contrary, the fatigue limit follows the rise of contact loading F_N if friction occurs in the zone of compression (Fig. 4.12, *a*). Measurements of wear in relation to the effect of cyclic stresses show (Fig. 4.12, *b*) that the wear process is more intensive in the zone of tension (σ = +330 MPa) than in the zone of compression (σ = –330 MPa), i. e. the cyclical tensile stresses intensify wear stronger than the cyclic compressive stresses of the same level at

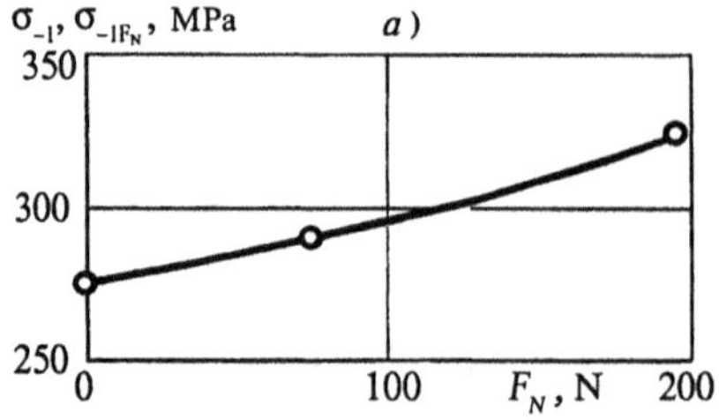

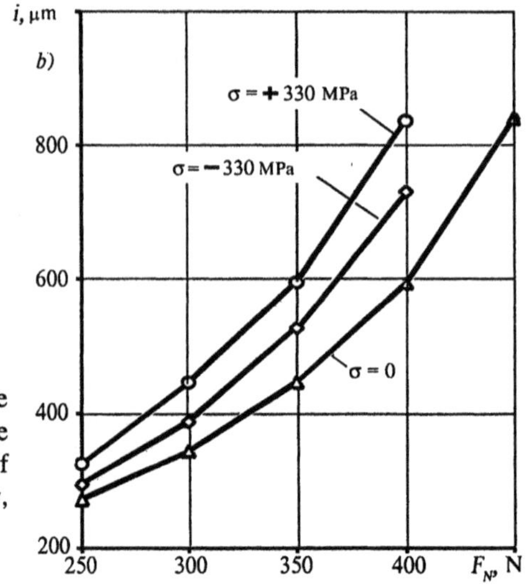

Fig. 4.12. Results of wear-fatigue damage tests of *steel 45 / polymer Φ4-BM* in case friction occurs in zone of compression of specimen being bended (L A Sosnovskiy, V V Vorobyev)

these testing conditions. As a rule, the polymer in the friction couple wears less ($\sigma = 0$) than it wears during mechano-sliding fatigue in the corresponding active system ($\sigma = \pm\ 300$ MPa).

Now let us weigh the results of tests for mechano-sliding fatigue of metal-to-metal active systems.

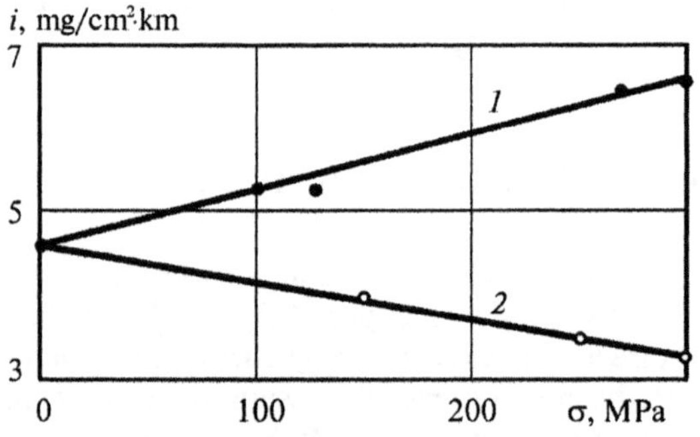

Fig. 4.13. Effect of cyclic compressive (*1*) and tensile (*2*) stresses on wear of steel 45 specimens (V T Sharai)

Figure 4.13 shows the results of wear-fatigue tests of the *steel 45 / steel 45* system (without lubrication) within a broad range of variations of cyclic stresses $\sigma < \sigma_{-1} = 320$ MPa. Wear regularities were different from those of the metal-to-polymer active system (compare Figs. 4.12, *b* and 4.13). A specific feature of the results of tests of the metal-to-metal system during oxidation wear is that cyclic stresses intensify wear in the zone of compression and its rise is up to 40% (in the test conditions), meanwhile it slows down in the zone of tension (and reduces to 32.5%) compared with the wear in the friction couple (when $\sigma = 0$). It is because the friction surface in the zone of tension is coated with oxides that protect it against fracture (the effect of Roscoe, see Sect. 1.4.3). The friction surface in the zone of compression shows just traces of oxides and its fracture naturally intensifies in this case.

Figure 4.14 demonstrates the role of lubrication in ensuring the durability of the active system.

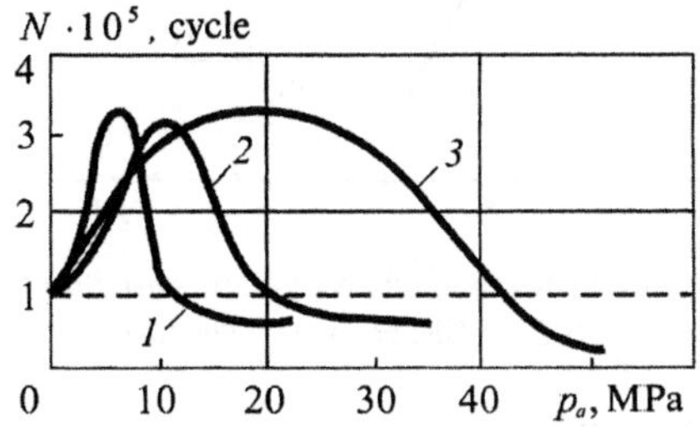

Fig. 4.14. Effect of pressure during friction with lubricating material on fatigue durability of steel 45 specimens at $\sigma_a = 400$ MPa: *1* – oil MC-20 + P; *2* – oil MC-20 (without additive); *3* – oil MC-20 + ДФ (I G Nosovsky, et al.)

For example, oil MC-20 with various additives does not affect practically the fatigue durability of the specimens (at $p_a = 0$). Yet, during wear-fatigue tests the ratio $N(p_a)$ has a bell-shaped pattern. The durability during such tests and within a broad range of variations of contact pressure is much (nearly 3.5 times) higher than during common fatigue tests (when $p_a = 0$). The higher the load the stronger the durability is; this is the main regularity in this case. Meanwhile the maximum durability is practically the same with all three lubricants, yet it is reached at strongly different pressures. The range of pressures within which the maximum durability is maintained depends on the additive type: it somewhat reduces with the oil MC-20 + P and strongly increases with the oil MC -20 + ДФ versus the case when the oil MC-20 is used without additives.

If the metal-to-metal active system is tested by tough loading (when the deformation range ε is assigned instead of the range of stresses σ) in the low-cycle region, the durability (Fig. 4.15) during wear-fatigue tests (when $F_N > 0$) is less than in case of common fatigue ($F_N = 0$) just at relatively small deformation; all three fatigue curves practically merge at $\varepsilon \approx 0.5...0.6\%$.

Note in conclusion that the experiments described in this Chapter can be divided into two groups based on the author's formulation: (1) the data that resulted from the factor analysis (cf. Figs. 4.3, 4.13, 4.14), and (2) the data that resulted from the phenomena analysis (see, for example, Figs. 4.1, 4.4, 4.6–4.8 etc.). It has taken several decades that the data that resulted from the factor analysis were interpreted on the phenomena analysis, and therefore it has become possible to conceive them as fundamental for tribo-fatigue to come into being. Moreover, extensive experimental results from studies of fretting fatigue and mechano-corrosion fatigue accomplished during the last decades on the basis of the factor analysis can and should be similarly interpreted on the basis of the phenomena analysis.

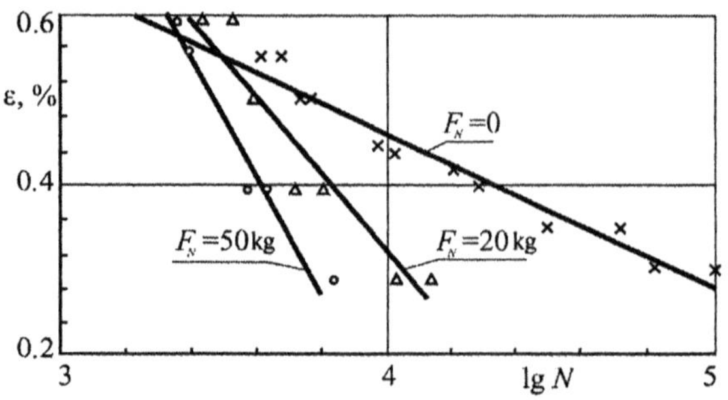

Fig. 4.15. Results of tests for low-cycle fatigue of *steel 30XГСА / hard alloy P6M5* (Zh M Blednova, A N Shauro)

Modern ideas (and methods) of *physical mesomechanics of materials* will definitely add to knowledge of new regularities of wear-fatigue damage. Deformation carriers principally different from dislocations are considered at the mesolevel, they are three-dimensional structural elements (*mesovolumes*), translation-rotation motion of which leads to the appearance of deformational dissipative *mesostructures* in the loaded material. The nature of the latter too (the type, dimensions of components, kinetics of appearance and subsequent development) governs wear-fatigue damage in many respects. Figure 4.16 shows the pattern of vectors of displacements in the mesovolume during fretting fatigue obtained for the first time.

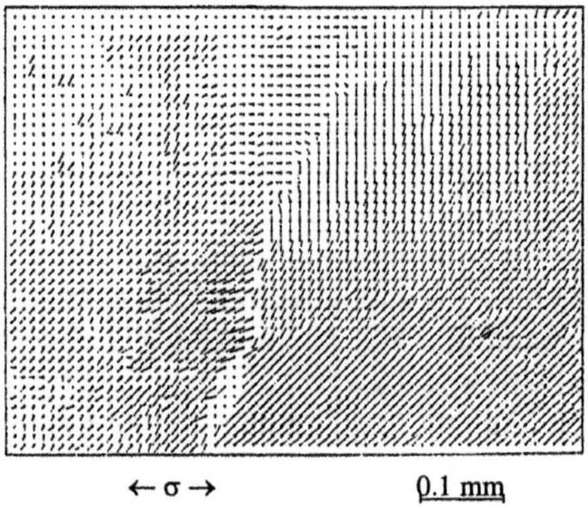

Fig. 4.16. Field of vectors of displacements ahead of the front of fatigue crack front on friction surface (alloy Д16AT, $N = 5.5 \cdot 10^4$ cycles) (V E Panin, V S Pleshanov, V V Kibitkin)

Three stages of wear-fatigue damage have been identified at the mesolevel: (1) appearance of stochastically distributed zones of plastic shear and centers of fretting damage on contacting surfaces; (2) nucleation and quasibrittle growth of fatigue cracks activated by fretting-processes; (3) brittle-plastic growth of cracks preceded by the appearance of the deformation small-domain *mesosubstructure* with discrete disorientations ahead of the front of the main crack (cf. Fig. 4.16). No systematic studies in the sphere of mesomechanics of wear-fatigue damage have yet been accomplished.

Self-test questions

1. What is the direct effect? Is it possible to study the regularities of the direct effect from the standpoint of tribology? Corroborate your view.
2. What is the back effect? Is it possible to study the regularities of the back effect from the standpoint of mechanical fatigue? Corroborate your view.
3. Formulate the main experimentally established regularities of the direct effect during mechano-sliding fatigue of the metal-to-polymer active system.
4. Formulate the main theoretically predictable regularities of the direct effect during mechano-sliding fatigue of the metal-to-polymer active system.
5. Describe the role of thermofluctuation stresses in the development of wear-fatigue damage of the polymer during mechano-sliding fatigue. How is a singe thermofluctuation stress calculated?
6. What is the principal difference between the direct effect during mechano-sliding fatigue of the metal-to-polymer and that of metal-to-metal active systems?
7. Formulate basic experimentally established and theoretically predictable regularities of the direct effect during mechano-sliding fatigue of the metal-to-metal active systems.
8. What is the basic experimentally established regularity of the back effect in the metal-to-polymer active system during mechano-sliding fatigue.
9. Formulate basic theoretically predictable regularities of the back effect in the metal-to-polymer active system during mechano-sliding fatigue
10. Explain (from the standpoint of the mechanics of deformation) why the wear of the polymeric counterbody strongly intensifies when cyclic stresses are excited in the conjugated metallic body under pressure?
11. Do you know two manifestations of the back effect during mechano-sliding fatigue of the metal-to-metal active systems? Describe them.
12. Do you discriminate the notions *wear-fatigue damage* and *fatigue wear*? What do they have in common? What is their principal difference? What is its essence?
13. How do the basic regularities of the direct and back effect change in response to the level of contact and off-contact loads?
14. What do you think about the role of physical wear in the active system, whether it is positive or negative? Or in any other way? Corroborate your view.
15. What will the result be if two damaging phenomena, like friction (including wear with friction) and mechanical fatigue develop in one and the same zone of the active system's components? Does the result depend on the conditions of loading of the active system? If yes, how?

16. Describe the basic regularities of the direct effect in the metal-to-metal active system during mechano-rolling fatigue if the specimen is made from soft steel and the roller from high-strength steel. According to what criterion is the limiting state reached? What damages are attendant?
17. Describe the basic regularities of the back effect in the metal-to-metal active system during mechano-rolling fatigue if the specimen is made from soft steel and the roller from high-strength steel. According to what criterion is the limiting state reached? What damages are attendant?
18. When are undulatory residual surface damages in the metal-to-metal active system possible? Can you indicate the causes of their appearance?
19. What is the direct effect factor? What numerical values may it have during (*a*) mechano-rolling fatigue? (*b*) mechano-sliding fatigue?
20. What is the back effect factor? What numerical values may it have during (*a*) mechano-rolling fatigue? (*b*) mechano-sliding fatigue?
21. Can you explain why the active system may have stronger durability than a similar friction pair (when contact loads are the same)? How can an "additional" cyclic load in the active system lead to stronger and not to weaker durability?
22. The popular idea is that when the energy of deformation excited in the deformable solid body augments, its durability (bearing capacity) diminishes correspondingly. Is this idea always true in respect of active systems? If not, why? What is the role of interactions between damages due to contact and off-contact loads?
23. Describe the wear of the metal-to-metal active system by crushing during mechano-rolling fatigue. When does it occur?
24. What is the scattered effect of multiple microshearing? What are its symptoms? What is its role in the appearance of wear-fatigue damage?
25. How is the governing parameter of the wear-fatigue damage calculated? What are its numerical values? Can they help identify the effect, direct or back, that occurs in these conditions?
26. How is it possible to determine the critical value of the wear-fatigue damage governing parameter and what does it imply?
27. After you have got an idea about the regularities of the direct and back effects is it clear now what tribo-fatigue is? In what way is it different from tribology, fracture fatigue mechanics, other disciplines studying the problems of strength, surface and volume fracture (strength of materials, machinery, structural mechanics)?
28. What is the principal difference of two-parametric diagrams of the limiting state of objects during mechanical fatigue and during friction from the diagram of limiting states of the active system during mechano-rolling fatigue?
29. What phenomenon is called the translimiting state during mechano-rolling fatigue of the *soft steel / high-strong steel (roller) active system*? Describe this state.
30. What are possible causes (and mechanisms) of appearance of residual undulatory damage along the rolling path?

31. Do you discriminate between the processes of addition and interaction between damages due to contact and off-contact loads? What results of these two processes are possible? Can the addition be considered as a particular case of interaction between damages?
32. What coefficients of anisotropy of wear-fatigue damage do you know? How is it possible to calculate them using the results of relevant measurements?
33. Describe the relation between the coefficient of asymmetry and the coefficient of irregularity of wear-fatigue damage and the level of cyclic stresses during mechano-rolling fatigue.
34. What do you know about the differences between the processes of wear of active system components if friction occurs either in the zone of compression or in the zone of tension?
35. Compare the wear processes in the friction pair and in the similar active system. What regularities can you outline? How does the pattern of friction in the zone of tension and in the zone of compression affect wear? What stresses – compressive or tensile – are more hazardous?

Tasks for research

1. Carry out the following experimental study (if you have learned to determine the characteristics of surface roughness in some other discipline).
   a) Test a metallic specimen for fatigue at stresses $\sigma > \sigma_{-1}$ during 10–15 minutes.
   b) Test a metal-to-polymer friction pair at pressure $p_a > p_f$ during the same time.
   c) Test a metal-to-polymer active system with the same loading parameters assumed in a) and b).

   Every time obtain a profilogram of the working surface before and after tests. Use it to determine basic characteristics of the surface purity. Then make an exhaustive comparison: (1) how the condition of the surface changes after each of three tests? 2) if the surface characteristics are different depending on the type of tests? if the difference is only quantitative or qualitative? You must realize that to establish qualitative difference is more essential than quantitative. Note that the roughness of surfaces of both tested bodies should be measured and analyzed.
2. Your results may be incorporated into your presentation at the students' conference.
3. Using PC make a graphic analysis of functions (4.1) and (4.4). Compare the obtained graphs. Does your analysis contain at least one conclusion that is not described in the manual?
4. Using PC make a graphic analysis of functions (4.5) and (4.7). Does your analysis contain any conclusions not described in the manual? If yes, do you intend to carry on research?
5. Functions (4.9) and (4.10) look identical but their content is different, of course. Find out one common feature of these two functions when predicting both the direct and back effects. Will this feature still exist if tests for mechano-rolling fatigue are carried out during rolling friction with slip and not just during pure rolling friction? Corroborate you view.

6. Figure 4.6 shows the diagram of limiting states of the active system during mechano-rolling fatigue. In your view, what will a similar diagram look like during mechano-sliding fatigue? Try to make one, for example, for a metal-to-metal (or metal-to-polymer) active system. Do not imagine it is an easy task.

5 METHODS OF CALCULATION OF ACTIVE SYSTEMS

Practice should always be constructed on a good theory...

Leonardo da Vinci

5.1 Limiting state

5.1.1 General

Energy approach is the most common approach to solving individual problems of strength and wear resistance of structural components because energy criteria are universal and they characterize integrally the stress-strain state. The above criteria are used to analyze the processes of static, long-term and cyclic (volume and surface) fracture of materials and friction units.

Let us try to apply the energy approach to any active system (cf. Fig. 2.1). WFD in the active systems is due, in the most general case, (see Sect. 2.1) to the following effects: a) contact load – it is characterized in the first approximation by the specific force of friction $\tau_W = fp_a$, where p_a – the maximum contact pressure, f – the friction coefficient; b) alternating (off-contact) loading – it is characterized in the first approximation by cyclic stresses σ; c) thermodynamic loading – it is characterized integrally by temperature T_Σ generated by all heat sources; d) electrochemical loading – indirectly it is characterized by the corrosion parameter (D), note that corrosion under stress (D_σ), corrosion in friction (D_τ) and thermal corrosion (D_T) should be discriminated. The case in question is called *general* in the sense that practically an entire complex of damaging phenomena occur in the active system. On the other hand, it is quite apparent that two simplifying assumptions are made: not any spatial system of contact and cyclic stresses but just their linear equivalents. However, this schematization of loads remains *principal* because all loads determining damage of the active system are taken into account.

Let us formulate *basic notions* of such active systems that can serve as a basis of the *theory of limiting states*.

(1) Origination and evolution of complex WFD is governed, in the first place by four particular phenomena: mechanical fatigue, friction and wear, thermal and electrochemical (corrosion) processes. These phenomena are called particular in

the sense that each can occur as an independent and separate event and it results in the corresponding limiting state based on particular (individual) criteria.

(2) All these particular phenomena and processes evolve in the active system simultaneously within the same zone; therefore, the system's limiting state is due to the combined (integral) effect of these phenomena and not to a single phenomenon all producing the WFD of a critical value.

(3) The WFD kinetics is not determined by all the total energy U the system receives but it is determined only by its effective (dangerous) portion $U^{eff} << U$ being expended for damage.

(4) The condition of attainment of critical value U_0 by the effective energy U^{eff} within some region of limited dimensions of a component of the active system in its dangerous volume is the criterion of the limiting state.

(5) Energy U_0 is considered a fundamental constant for a given substance; it should be independent of the conditions of tests, types of input energy and mechanisms of damage.

(6) Effective energy U^{eff} can be represented, in the general case, by the function of four components: thermal U_T^{eff}, force U_σ^{eff}, frictional U_τ^{eff} and electrochemical U_{ch}^{eff} energies

$$U^{eff} = R(U_T^{eff},\ U_\sigma^{eff},\ U_\tau^{eff},\ U_{ch}^{eff}), \tag{5.1}$$

where R allows for kinetic interactions between particular damaging phenomena in the complex WFD process.

(7) The processes of electrochemical (corrosion) damage can be taken into account as thermal corrosion ($D_{T(ch)}$), corrosion under stress ($D_{\sigma(ch)}$) and friction corrosion ($D_{\tau(ch)}$), so that function (5.1) becomes

$$U^{eff} = R(U_{T(ch)}^{eff}, U_{\sigma(ch)}^{eff}, U_{\tau(ch)}^{eff})\,. \tag{5.2}$$

(8) WFD accumulates in time t non-linearly in the general case.

(9) The limiting state of the active system appears if at least one (any) particular damaging phenomenon occurs, while the remaining damaging phenomena are attendant.

(10) The limiting state of the active system can appear with any two, three or all four criteria concurrently.

(11) If the conditions of operation or tests of the active system are such that the direct effect occurs, the limiting state appears due to the volume fracture criteria.

(12) If the conditions of operation or tests of the active system are such that the back effect occurs, the limiting state appears due to the criteria of surface damage (fracture).

5.1.2 Energy criterion

Assume that the active system operates in the environment with the temperature T, with one of its components being in the linear state of stress under the effect of cyclic stresses σ, while the field of contact stresses is described by the mean frictional stress τ_W (cf. Fig. 2.1). Then the full input energy is

$$U = U_T + U_\sigma + U_\tau , \tag{5.3}$$

where U_T – thermal; U_σ – force and U_τ – frictional components of the full energy.

Values U_σ and U_τ are easily calculated with allowance for the known relations between stresses and deformations (see Sect. 1.2). Yet, there is no point in using (5.3) for practical calculations because most of the input energy dissipates in the system and in the environment without damaging the material.

Let us introduce the notion *effective energy* $U^{eff} << U$, i.e. the portion of the full energy directly spent for generation and accumulation of damage in the active system. It is clear that the total effective energy also includes thermal U_T^{eff}, force U_σ^{eff} and frictional U_τ^{eff} components that (like values U_T, U_σ, U_τ) should be proportional to corresponding parameters:

$$\left.\begin{aligned} U_T^{eff} &\sim T; \\ U_\sigma^{eff} &\sim \sigma^2; \\ U_\tau^{eff} &\sim \tau_W^2. \end{aligned}\right\}$$

It can be assumed that

$$\left.\begin{aligned} U_T^{eff} &= a_T T; \\ U_\sigma^{eff} &= a_\sigma \sigma^2; \\ U_\tau^{eff} &= a_\tau \tau_W^2, \end{aligned}\right\} \tag{5.4}$$

where the coefficients $a << 1$ isolate the effective portions U^{eff} from the full thermal and mechanical energies.

Note that here and further we deal with specific quantities of the total effective energy and its components (for example, energy values attributed to the unit of quantity of the matter).

As values σ and (or) τ_W and (or) T and (or) time (the number of loading cycles) grow, respectively, so does the total effective energy (5.1) until it reaches a critical (limit) value U_0. Then the *limiting state of the active system* should occur characterized, for example, by the appearance of a fatigue crack of critical dimensions or the tolerable wear limit of the system is reached or by the occurrence of both these states simultaneously.

An assumption that the limiting state of the active system occurs when a simple algebraic sum of effective energies reaches the critical value is untrue, in the general case, (see Sects. 2.5 and 4.3.2). In fact, if the mechanism of damage is due to the kinetics of accumulation of broken interatomic bonds, as the thermofluctuation concept of strength of solids treats it (see Sect. 1.3.2), a possible process of their recombination should be taken into account. Or if the appearance and accumulation of dislocations (or vacancies) govern the mechanism of damage, as the dislocation (or vacancy) theories treat it (see Sect. 1.3.2), a possible process of their disappearance should be taken into account. If interactions between various damages are taken into account integrally, then the numerical values of $R \gtrless 1$ (see Sect. 2.6). Hence, the *energy criterion* of the *limiting state of the active system* can be recorded as:

$$U^{eff} = R_{T/M}\left[a_T T_\Sigma + R_{\sigma/\tau}\left(a_\sigma \sigma^2 + a_\tau \tau_W^2\right)\right] = U_0 . \tag{5.5}$$

Here $R_{\sigma/\tau}$ allows for interactions between effective portions of the mechanical energy due to normal σ and frictional τ_W stresses, $R_{T/M}$ – the interaction between thermal and mechanical components of the effective energy; moreover, the values R allow for the processes of "healing" damage, whatever their mechanism is. Note also that the effective portion of thermal energy in expression (5.5) is determined by the change in total temperature $T_\Sigma = T_2 - T_1$ in the dynamic contact zone due to all heat sources, including the heat generated during mechanical (volume and surface) deformation, structural transformations, etc.

Criterion (5.5) has a very general nature. It is free of unjustifiable coefficients and is independent, for example, of the manner with which the system is loaded (static, protracted, cyclic loading), or of the mechanisms of accumulation of damage and fracture.

It is easy to obtain from general criterion (5.5) a number of essential particular cases. Therefore, the conditions of *purely thermal* (or *thermodynamic*) *fracture* (when $\sigma = 0$ and $\tau_W = 0$) or *purely mechanical fracture* (when $T_\Sigma = 0$) are, respectively, the following:

$$a_T T_\Sigma = U_0 ; \tag{5.6}$$

$$R_{\sigma/\tau}\left(a_\sigma \sigma^2 + a_\tau \tau_W^2\right) = U_0 . \tag{5.7}$$

In case of the *isothermal mechanical fatigue* (when $\tau_W = 0$) we have

$$R_{T/M}\left(a_T T_\Sigma + a_\sigma \sigma^2\right) = U_0 , \tag{5.8}$$

and for the isothermal *frictional fatigue* (when $\sigma = 0$) we similarly obtain

$$R_{T/M}\left(a_T T_\Sigma + a_\tau \tau_W^2\right) = U_0 . \tag{5.9}$$

In order to make the method of calculation of energy (5.2) more specific, a mode should be indicated how an allowance is made for the effect of electrochemical processes on the damage of the active system. Introduce the parameter $0 \le D \le 1$ and assign the following sense to it: its growth should be equivalent to the growth of effective (spent for the WFD appearance and accumulation) energy in the active system due to the evolution of electrochemical damage. This effect is easy to describe by changing correspondingly the values of the parameters a in criterion (5.5). In fact, the value a is reduced $(1 - D)$ times, i.e. if the expression $a/(1 - D)$ is introduced into criterion (5.5), we obtain that the growth D means a corresponding augmentation of a. Then criterion (5.5) in the generalized form can be recorded as

$$R_{T/M}\left[\frac{a_T}{1-D_T}T_\Sigma + R_{\sigma/\tau}\left(\frac{a_\sigma}{1-D_\sigma}\sigma^2 + \frac{a_\tau}{1-D_\tau}\tau_W^2\right)\right] = U_0\,. \tag{5.10}$$

Introduce *relative measures* ω *of thermodynamic* (index T), *force* (index σ) and *frictional* (index τ) *damages* with the allowance for *corrosion* $(1 - D)$ (index ch):

$$\omega_{T(ch)} = \frac{a_T T_\Sigma}{U_0(1-D_T)};\quad \omega_{\sigma(ch)} = \frac{a_\sigma \sigma^2}{U_0(1-D_\sigma)};\quad \omega_{\tau(ch)} = \frac{a_\tau \tau_W^2}{U_0(1-D_\tau)}\,. \tag{5.11}$$

Then criterion (5.10) acquires the form

$$R_{T/M}\left[\omega_{T(ch)} + R_{\sigma/\tau}\left(\omega_{\sigma(ch)} + \omega_{\tau(ch)}\right)\right] = 1\,, \tag{5.10a}$$

or

$$\omega_\Sigma = 1, \tag{5.10b}$$

where the *measure of complex* WFD is

$$\omega_\Sigma = R_{T/M}\left[\omega_{T(ch)} + R_{\sigma/\tau}\left(\omega_{\sigma(ch)} + \omega_{\tau(ch)}\right)\right]. \tag{5.12}$$

Criterion (5.10) reads that the limiting state of the active system occurs when the sum of interactive effective energy components due to force, frictional and thermal effects (with the allowance for the processes of corrosion under stress, thermal and tribochemical corrosion) reaches the critical value U_0. Criterion (5.10) in the form (5.10a) or (5.10b) is convenient because all the measures of damage are dimensionless and have a single interval $(0 \le \omega \le 1)$ of changes of values.

If the concept of damaged volumes of the deformable solid (see Sect. 2.4) is used during cyclic loading ($V_{P\gamma}$), friction ($S_{P\gamma}$) and thermodynamic loading ($V_{T\gamma}$), the damage measures (5.11) can be determined in the following manner:

$$\omega_{\sigma(ch)} = \frac{V_{P\gamma}}{V_0(1-D_\sigma)};\quad \omega_{\tau(ch)} = \frac{S_{P\gamma}}{S_k(1-D_\tau)};\quad \omega_{T(ch)} = \frac{V_{T\gamma}}{V_0(1-D_T)}, \tag{5.13}$$

where V_0, S_k – working volumes. Then criterion (5.10a) with the account of (5.13) becomes the following:

$$R_{T/M}\left[\frac{V_{T\gamma}}{V_0(1-D_T)} + R_{\sigma/\tau}\left(\frac{V_{P\gamma}}{V_0(1-D_\sigma)} + \frac{S_{P\gamma}}{S_k(1-D_\tau)}\right)\right] = 1\,. \tag{5.14}$$

The advantage of criterion (5.10) in the form (5.14) is that here an allowance is made for the effect of the complex of designing, technological and metallurgical factors because they govern relative damaged volumes $V_{P\gamma}/V_0$, $S_{P\gamma}/S_k$, $V_{T\gamma}/V_0$.

Note that no limitations for values $T_\Sigma > 0$, $\tau_W > 0$, $\sigma > 0$ are made in criteria (5.5), (5.10) and (5.14). Therefore, they can describe the attainment of the limiting state during both the complex wear-fatigue damage and particular loading conditions, for example, during pure thermal or pure mechanical fracture, as it is noted above. If normal stresses σ are replaced with stress intensity $\sigma_{int} = \varphi(\sigma_i, \tau_{ij})$, where σ_i, τ_{ij} – stress components, these criteria are also applicable to the conditions of arbitrary complex state of stress of the system's cyclicly deformable component (see Sect. 1.2.1). The integral parameter, viz. the specific force of friction $\tau_W = fp_a$ (see Sect. 2.1) in these criteria allows for the complex state of stress in the contact problem too. It has been shown that the value τ_W is proportional to the equivalent stress determined from the known theories of strength. Moreover, the effect of lubrication on the damage of the system can be taken into account if it is assumed that $f = f_{lub}$, where f_{lub} – the friction coefficient with lubrication. Finally, it should be assumed that criteria (5.5), (5.10) and (5.14) describe both brittle (elastic) and plastic fracture (see Sect. 1.1.7) if the known law $\sigma_{int} = E'\varepsilon_{int}$, is used, where E' – the secant modulus of deformation (see Sect. 1.2.1), ε_{int} – deformation intensity determined both by contact and off-contact loads.

These ample capabilities of criteria (5.5), (5.10) and (5.14) relate to the fact that they are based on the most general energy ideas about the conditions of damage and fracture of solids. A general analysis of these criteria leads to three basic conclusions.

(1) Growth of loading parameters (σ, τ_W, T_Σ, D) leads to corresponding acceleration of the limiting state.

(2) The active system can reach its limiting state also if only one (any) loading parameter increases (while the remaining parameters stay unchanged).

(3) If $R > 1$, the active system degrades quicker, while at $R < 1$ degradation slows down compared with the damage due to the combined effect of the loading parameters solely.

5.1.3 Parameters

Criteria (5.5), (5.10) and (5.14) should have validated methods how to determine the values U_0, a, R, D for practical use.

Above it is noted that the parameter U_0 has fundamental nature. In the thermofluctuation theory of strength (see Sect. 1.2.3) U_0 is interpreted as the *initial energy of activation of the fracture process*. It has been shown that the value U_0 coincides approximately with the heat of sublimation of metals and crystals with ion bonds and the energy of activation of thermodestruction for polymers:

$$U_0 \approx U_T.$$

On the other hand, the value U_0 is interpreted as the energy of activation of mechanical fracture:

$$U_0 \approx U_M.$$

Hence, energy U_0 can be considered as the constant of matter:

$$U_0 \approx U_M \approx U_T = \text{const.} \tag{5.15}$$

Taking into account the mechanophysical and thermodynamic ideas about the fracture processes, we record (5.15) as

$$U_M = s_k \frac{\sigma_{th}}{E} \frac{C_a}{\alpha_V} = U_0 = kT_S \ln \frac{k\theta_D}{h} = U_T, \tag{5.15a}$$

where s_k – the coefficient of reduction; σ_{th} – theoretical strength; E – the elasticity modulus; C_a – atomic heat capacity; α_V – the coefficient of the thermal expansion of volume; k – the Boltzmann constant; T_S – the melting temperature; θ_D – the temperature of Debye; h – Plank's constant. According to (5.15a), it can be approximately assumed that

$$U_0 \approx \varepsilon_* \frac{C_a}{\alpha_V}, \tag{5.15b}$$

where $\varepsilon_* \approx 0.6$ – the ultimate deformation of the atomic bond.

From equality (5.15a), it follows that U_0 is the energy of activation of the substance equal in the order of values to 1…10 eV per particle, atom or molecule ($\sim 10^2 \ldots 10^3$ kJ/ mole), i.e. the *value close to the energy of breaking of atomic bonds in the solid. Its level does not depend on the way the fracture is reached, whether it is mechanical, thermal or through a combined effect.* Methods of experimental determination U_0 are available. The numerical values U_0 for materials of various classes are listed in Table 5.1.

Using (5.15a), the formula for estimation of the *theoretical strength* is recorded as

$$\sigma_{th} = \frac{E\alpha_V kT_S}{C_a} \ln \frac{k\theta_D}{h}.$$

This formula yields the thermomechanical constant of the material:

$$\frac{\sigma_{th}}{T_S} = E \frac{\alpha_V k}{C_a} \ln \frac{k\theta_D}{h} = \theta_\sigma,$$

that characterizes the loss of strength per 1 K.

The coefficients a in Eq. (5.5) are determined from the following boundary conditions:

$$\left.\begin{aligned} T = 0, \tau_W = 0: \quad & a_\sigma \sigma_d^2 = U_0, \quad a_\sigma = U_0 / \sigma_d^2; \\ T = 0, \sigma = 0: \quad & a_\tau \tau_d^2 = U_0, \quad a_\tau = U_0 / \tau_d^2; \\ \sigma = 0, \tau_W = 0: \quad & a_T T_d = U_0, \quad a_T = U_0 / T_d, \end{aligned}\right\} \tag{5.16}$$

where σ_d, τ_d – normal and frictional ultimate stresses at $T \to 0$ called *limits of* (mechanical) *destruction*; T_d – the temperature of destruction (at $\sigma = 0$, $\tau_W = 0$) or the *limit of thermodestruction.*

Table 5.1. Values of U_0 for some materials

Material	U_0, kJ/mole
METALS	
Aluminium	222
Iron	419
Cadmium	117
Copper	339
Niobium	629
Platinum	503
Titanium	503
Zinc	507
POLYMERS	
Kapron	188
Polymethyl metacrylate	750
Polypropylene	235
Polystyrene	130
Polyvinylchloride	147
ION CRYSTALS	
Rock salt	285
Lithium chloride	302
Silver chloride	126

In order to determine a_σ using the first of formulas (5.16), the material should be tested, for example, statically for tension at a temperature close to absolute zero. Then practically pure mechanical fracture occurs by normal separation at stresses $\sigma = \sigma_d$. To estimate a_τ using the second of formulas (5.16), for example, static tests for simple shear or torsion (of thin-walled tube) should be performed at $T \to 0$. Then practically pure mechanical fracture by shearing occurs under tangent stress $\tau_W = \tau_d$. The coefficient a_T from the third of formulas (5.16) can be determined if fracture of the material is achieved by thermal method only (in this case the temperature is $T = T_d$). Thus, the values σ_d, τ_d, T_d are *physical constants of the material* determined with the corresponding tests. It means the coefficients a should be the constants of the material too under given conditions of loading.

Regarding the parameters R, it follows from (5.5) with the account of (5.4) that their values depend on the ratio between effective energies under the assigned conditions of tests of a specific active system. Thus, the parameter $R_{T/M} = f_1 \left[\left(U_\sigma^{eff} + U_\tau^{eff}\right)/U_T^{eff}\right]$ depends on the ratio between mechanical and thermal effective energies, while the parameter $R_{\sigma/\tau} = f_2\left[U_\tau^{eff}/U_\sigma^{eff}\right]$ depends on the ratio between the frictional and force portions of mechanical energy.

In the first approximation the values $R_{T/M}$, $R_{\sigma/\tau}$ can be determined from the results of two experiments that are used to construct a system of two equations of the type (5.5). If, for example, $T = 0$, then $R_{T/M} = 1$ and (5.7) yields

$$R_{\sigma/\tau} = \frac{U_0}{a_\sigma \sigma^2 + a_\tau \tau_W^2}. \tag{5.17}$$

Then, at known $R_{\sigma/\tau}$ and a given temperature $T > 0$, we obtain

$$R_{T/M} = \frac{U_0}{a_T T_\Sigma + R_{\sigma/\tau}\left(a_\sigma \sigma^2 + a_\tau \tau_W^2\right)}. \tag{5.18}$$

The contribution of corrosion processes into the WFD of the active system can be determined as

$$\left.\begin{aligned} 1 - D_T = b_{e(T)}\left(\frac{v_{ch}}{v_{ch(T)}}\right)^{m_{v(T)}}; \quad 1 - D_\sigma = b_{e(\sigma)}\left(\frac{v_{ch}}{v_{ch(\sigma)}}\right)^{m_{v(\sigma)}}; \\ 1 - D_\tau = b_{e(\tau)}\left(\frac{v_{ch}}{v_{ch(\tau)}}\right)^{m_{v(\tau)}}, \end{aligned}\right\} \tag{5.19}$$

where v_{ch} – the rate of corrosion in a given environment; $v_{ch(T)}$, $v_{ch(\sigma)}$, $v_{ch(\tau)}$ – the rate of corrosion in the same environment due to thermal, force and frictional effects, respectively; b_e – the coefficients that allow for the processes of corrosive erosion; m_v – the parameters determining the electrochemical activity of materials during force (index σ), frictional (index τ) and thermodynamic (index T) loading.

The parameters D can also be calculated with corresponding ultimate stresses:

$$1 - D_T = (T_{tch} / T_t)^{n_T} g_T; \quad 1 - D_\sigma = (\sigma_{-1ch} / \sigma_{-1})^{n_\sigma} g_\sigma; \qquad (5.19a)$$
$$1 - D_\tau = (\tau_{fch} / \tau_f)^{n_\tau} g_\tau,$$

where T_t, σ_{-1}, τ_f – the limits of thermal, mechanical, friction fatigue in the air, respectively, while T_{tch}, σ_{-1ch}, τ_{fch} – the same values in a given corrosive environment. The coefficients g have the same sense with the coefficients b in (5.19). The parameters n are similar to the parameters m_v in (5.19).

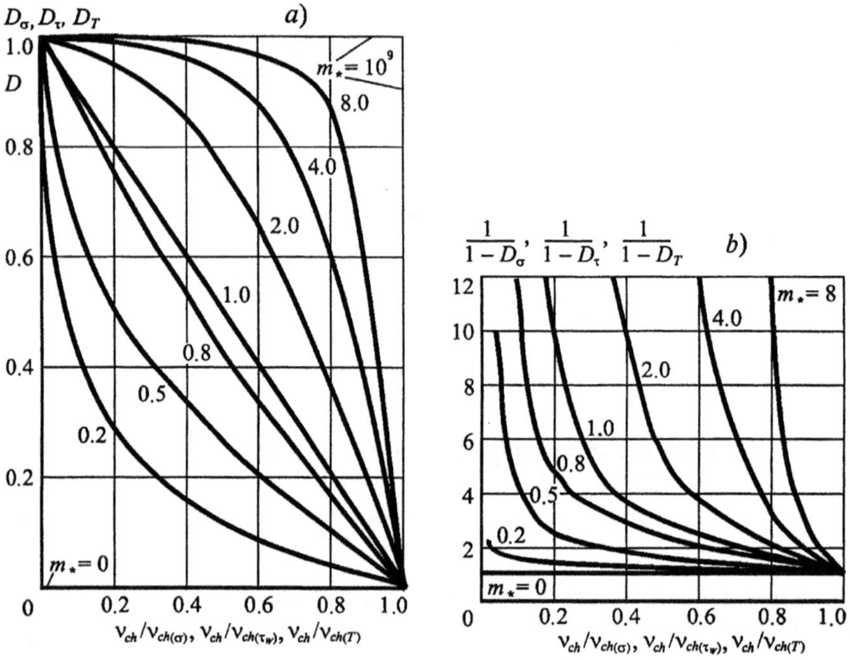

Fig. 5.1. Graphs of functions (5.19) (*a*) and values $1/(1 - D)$ (*b*) due to change of values of parameter m_v

Figure 5.1 gives the general analysis of the role of electrochemical damage (the parameter D) in the development of the limiting state in the active system. When formulas (5.10) and (5.19) are examined together with Fig. 5.1, the following conclusions can be made.

(*a*) If the parameter D grows (cf. Fig. 5.1, *a*), then $(1 - D)$ reduces respectively. Hence, the slower the relative rate of damage $v_{ch}/v_{ch(*)}$ (cf. Fig. 5.1, *b*) the more the value $1/(1 - D)$ augments. In other words, the higher the value of the parameter D and/or the rate $v_{ch(*)}$ of thermal corrosion, friction corrosion and corrosion under stress, the stronger the electrochemical damage boosts the development of the limiting state in the active system.

(*b*) The larger the parameter m_v the stronger its effect on the WFD of the system (cf. Fig. 5.1). An essential feature of this effect is that a given environment is very sensitive to any excitation of mechanical stresses in the active system and to any temperature rise if its parameter is $m_v > 5$ (cf. Fig. 5.1, *b*). In other words, the *translimiting state* can occur in such a case when the measure of damage exceeds a unity ($\omega_\Sigma > 1$), then, according to (5.10a), it is enough to have $\omega_\Sigma = 1$ to reach the limiting state.

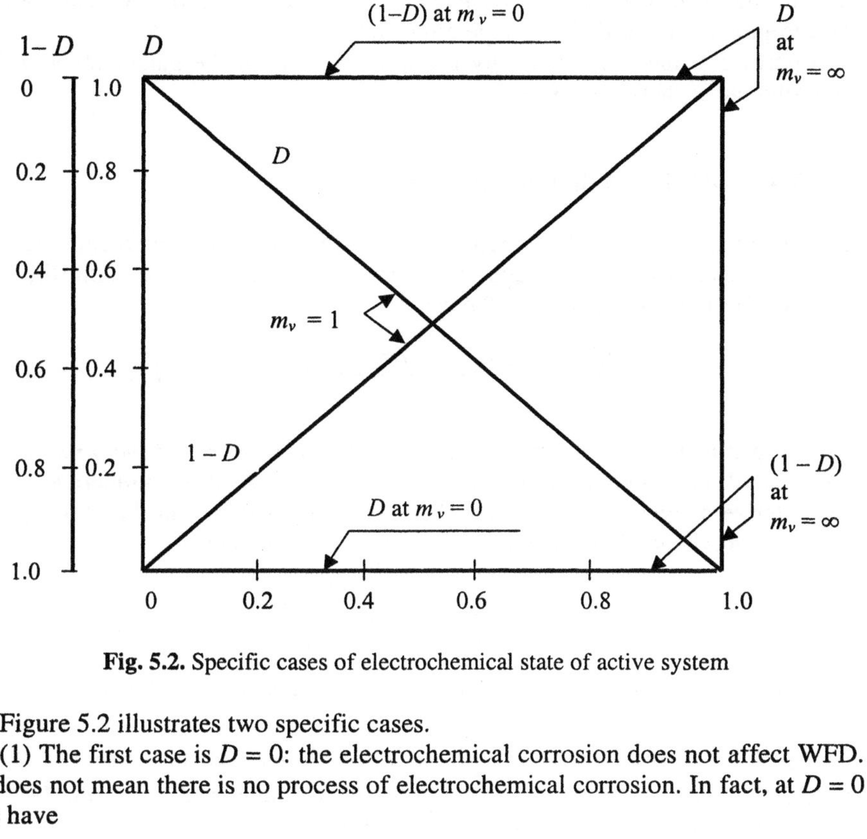

Fig. 5.2. Specific cases of electrochemical state of active system

Figure 5.2 illustrates two specific cases.

(1) The first case is $D = 0$: the electrochemical corrosion does not affect WFD. It does not mean there is no process of electrochemical corrosion. In fact, at $D = 0$ we have

$$1 - \frac{v_{ch}}{v_{ch(*)}} b_* = 0 ,$$

whence it follows that there should be $b_* = 1$ and $v_{ch}/v_{ch(*)} = 1$, i.e. the rate of corrosion is insensitive to this factor (or some value of mechanical or frictional stress or a certain temperature). It means that there are some thresholds of σ^0, τ_W^0 and T^0 for a given environment. The rate of corrosion does not change in this environment at $\sigma \le \sigma^0$, $\tau_W \le \tau_W^0$ and $T \le T_0$.

(2) The second case is $D = 1$, i.e. $(1 - D) = 0$ and $1/(1 - D) \to \infty$, i.e. an *explosive damage occurs in the system* because $\omega_\Sigma \to \infty$. In this case it should be

$$\frac{v_{ch}}{v_{ch(*)}} b_* = 0 .$$

If $v_{ch} = 0$, it is an impossible event, then it remains to assume that $v_{ch(*)} \to \infty$. It is the condition for chemical explosion in the active system. Explosion is due not just to the effect of the environment, it is the effect of the environment catastrophically amplified by temperature and mechanical stresses.

Figures 5.1 and 5.2 graphically illustrate the role of the parameter m_v of the environment in these two cases.

5.1.4 Asymmetry of damage processes

It is noted above that the parameter $R_{T/M}$ allowing for the interaction between effective energies (thermal and mechanical) should correlate with the ratio between these energies, i.e. with the value

$$\rho_{T/M} = \frac{U_M^{eff}}{U_T^{eff}} = \frac{R_{\sigma/\tau}(a_\sigma \sigma^2 + a_\tau \tau_W^2)}{a_T T_\Sigma} = R_{\sigma/\tau}(\rho_{\sigma/T} + \rho_{\tau/T}) . \tag{5.20}$$

The parameter $\rho_{T/M}$ is the measure of the *asymmetry of the processes of mechanical and thermal damage of the active system*; its numerical values characterize comparatively the contributions of thermal and mechanical effective energies into the total damage. In expression (5.20) the values

$$\rho_{\sigma/T} = \frac{a_\sigma \sigma^2}{a_T T}, \quad \rho_{\tau/T} = \frac{a_\tau \tau_W^2}{a_T T} \tag{5.21}$$

serve as *measures of asymmetry of either force or frictional and thermal damage*, respectively.

Figure 5.3 shows schematically how U_T^{eff} and U_M^{eff} depend on $\rho_{T/M}$ according to (5.20). If $\rho_{T/M} = 1$, the effect of T (on the one hand) and σ, τ_W (on the other hand) on the damage of a system is equal. If $\rho_{T/M} >> 1$, it means that under given conditions of operation mechanical mechanisms dominate in the fracture process (the right hand portion of the diagram in Fig. 5.3) and in the extreme case (i.e. at $T \to 0$) $\rho_{T/M} \to \infty$ fracture may be due exceptionally to the mechanical energy, i.e. it is determined by the deformation statistics of structural damages. Yet, if $\rho_{T/M} < 1$ (the left-hand portion of the diagram in Fig. 5.3), thermoactivation mechanisms become predominant in the process of fracture; fracture may evolve only thermally and in the extreme case at $\rho_{T/M} = 0$ (i.e. at $\sigma = 0$, $\tau_W = 0$). Hence, the thermofluctuation statistics of broken atomic bonds determine it. Thus, in the general case (under the effect of σ, τ_W, T) the fracture process results from the statistical stream of (interacting) local structural damage and microfractures caused by various mechanisms.

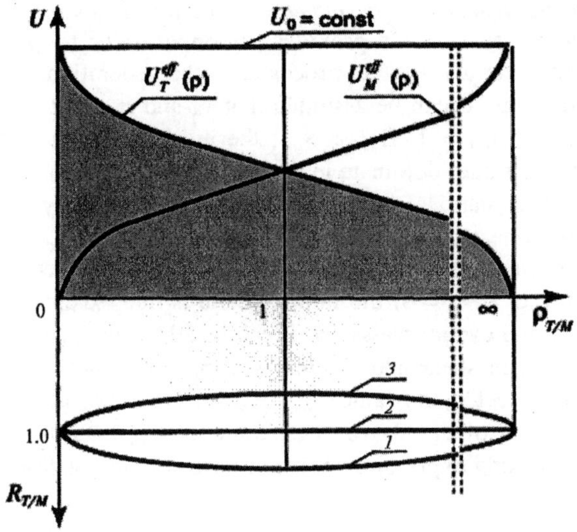

Fig. 5.3. Energy diagrams of ultimate states of active system

Figure 5.3 shows conventionally several types of relations between $R_{T/M}$ and $\rho_{T/M}$ that may occur in real active systems. Curve *3* predicts the weakening of the damaging effect when U_T^{eff} and U_M^{eff} are combined, i.e. the full effective energy corresponding to the moment of fracture should be smaller than a simple sum of $U_T^{eff} + U_M^{eff}$ that it includes throughout the range of possible changes of $\rho_{T/M}$. Line *2* characterizes the conditions when the interaction between the energies U_T^{eff} and U_M^{eff} is not apparent externally and does not influence the durability of the active systems. Curve *1* predicts intensification of the damaging effect when thermal and mechanical energies combine throughout the range of changes of the parameter $\rho_{T/M}$, i.e. the full effective energy at the moment of fracture in this model should be greater than a simple sum of $U_T^{eff} + U_M^{eff}$. In other words, it is either the process of softening in the given active system under specific conditions of its operation (curve *1*) or the processes of hardening of some active systems under given conditions of operation (curve *3*). As far as dependence 2 is concerned, it is true for the cases when the processes of hardening and softening become mutually equal and the ratio between thermal and mechanical energies throughout the range of its changes does not affect the system's durability. Of course, dependencies *1–3* are models of just some possible occurrence of limiting states in active systems. Actual patterns may be more complicated.

Introduce also into consideration the parameter

$$\rho_{\tau/\sigma} = \frac{U_\tau^{eff}}{U_\sigma^{eff}} = \frac{a_\tau \tau_W^2}{a_\sigma \sigma^2} = \frac{\rho_{\tau/T}}{\rho_{\sigma/T}}, \tag{5.22}$$

that serves as the *measure of asymmetry of the processes of frictional and force damage in active systems.* It provides an opportunity to predict a priori the contributions of shear and rupture processes in the generation and accumulation of wear-fatigue damage. It can be assumed that τ_W and σ affect the damage of active systems equally if $\rho_{\tau/\sigma} = 1$. If $\rho_{\tau/\sigma} > 1$, the mechanisms of surface fracture (for example, friction fatigue) dominate in active systems, while in case $\rho_{\tau/\sigma} \to \infty$ (i.e. at $\sigma = 0$) the limiting state is due exceptionally to the energy of friction, i.e. the statistics of microshear structural damage determine it. If $\rho_{\tau/\sigma} < 1$, mechanisms of volume fracture (mechanical fatigue) are predominant in the active systems; in case when $\rho_{\sigma/\tau} \to 0$ (i.e. at $\tau_W = 0$) the limiting state is due exceptionally to the energy produced by normal cyclic stresses σ. It can be added that, since the ratio between the strength limits in shear and rupture is $\tau_b/\sigma_b \approx 0.5$, hence according to (5.16) $a_\tau/a_\sigma \approx 4$ can be expected, the value $\rho_{\tau/\sigma} = \rho_{\sigma/\tau} \approx 1$ is reached at $\tau_W = 0.5\sigma$.

With the account of the above, the parameter $R_{\sigma/\tau}$ from the experimental data can be estimated using expression (5.17) or based on the relation $R_{\sigma/\tau}(\rho_{\tau/\sigma})$.

5.1.5 Multicriterial diagram

If it is assumed that $\sigma = \sigma_{-1\tau}$ in (5.10), then *normal ultimate stresses* are calculated with the account of the processes of friction, wear and corrosion at a given temperature (the direct effect):

$$\sigma_{-1\tau} = \left[\frac{U_0 \dfrac{1}{R_{T/M} R_{\sigma/\tau}} - \dfrac{1}{R_{\sigma/\tau}} \dfrac{a_T}{1 - D_T} T_\Sigma - \dfrac{a_\tau}{1 - D_\tau} \tau_W^2}{a_\sigma / (1 - D_\sigma)} \right]^{1/2} . \tag{5.23}$$

If it is assumed that $\tau_W = \tau_{f\sigma}$ in (5.10), then *ultimate friction stresses* are calculated with the account of the effect of cyclic stresses and processes of corrosion at a given temperature (the back effect):

$$\tau_{f\sigma} = \left[\frac{U_0 \dfrac{1}{R_{T/M} R_{\sigma/\tau}} - \dfrac{1}{R_{\sigma/\tau}} \dfrac{a_T}{1 - D_T} T_\Sigma - \dfrac{a_\sigma}{1 - D_\sigma} \sigma^2}{a_\tau / (1 - D_\tau)} \right]^{1/2} . \tag{5.24}$$

Figure 5.4 represents Eqs. (5.10), (5.23) and (5.24) graphically as *multicriterial diagrams 1–5* of *limiting states* of various *active systems.*

The ordinate axis is a strength scale and the abscissa axis is a tribological scale.

Common tests for fatigue (no friction, so that $\tau_W = 0$) yield the fatigue limit of the shaft σ_{-1} (cf. Fig. 5.4). During wear-fatigue tests of the active system its value changes due to the effect of the processes of friction and wear (it is designated by $\sigma_{-1\tau}$ in Fig. 5.4, *a*). This change determines the basic regularities of the direct effect. They can be described by typical curves *1–5* (cf. Fig. 5.4, *a*) depending on the type of an active system and conditions of its operation (the contact load level, the temperature, properties of the environment, etc.). Curves *1* and *2* are typical for mechano-rolling fatigue, curves *2*, *3* and *4* for mechano-sliding fatigue, curves *3*, *4* and *5* for fretting fatigue during various conditions of tests (temperature, environment, etc.).

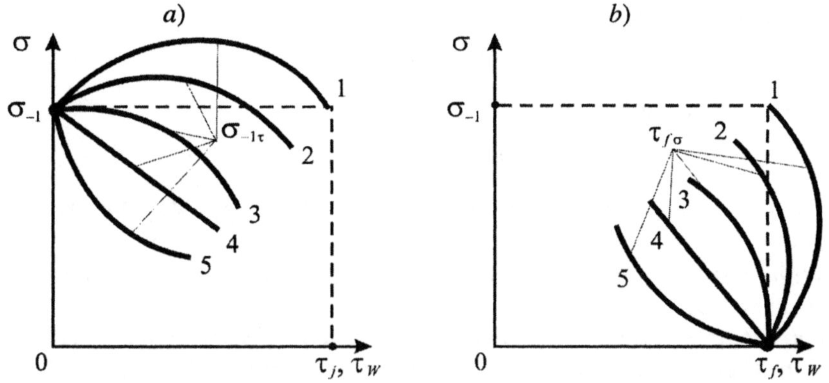

Fig. 5.4. Diagrams of ultimate states of various active systems

Common tests of friction pairs (no cyclic stresses, i.e. $\sigma = 0$) yield the ultimate value of friction stress τ_f, which is also called the *friction fatigue limit* (or the ultimate value of contact pressure p_f that corresponds to the value τ_f) (cf. Fig. 5.4). During wear fatigue tests of the active system its value changes due to the effect of the level of cyclic stresses (it is designated by $\tau_{f\sigma}$ in Fig. 5.4, *b*). This change determines the basic regularities of the back effect. They can be similarly described by typical curves *1–5* (cf. Fig. 5.4, *b*) depending on the type of an active system and conditions of its operation (the cyclic load level, temperature, properties of the environment, etc.). Curves *1–5* have the same sense with curves *1–5* in Fig. 5.4, *a*. Significant difference is that in case of the direct effect, as it is noted above, the limiting state of the system follows the criteria of resistance to mechanical fatigue, while in case of the back effect it follows the criteria of friction and wear.

When analyzing Eqs. (5.10), (5.23), (5.24) and Fig. 5.4 the following most essential conclusions can be made.

1) In response to the conditions of appearance the processes of friction and wear can both significantly reduce (cf. curves *3*, *4* and *5* in Fig. 5.4, *a*) and significantly enhance (cf. curves *1* and 2 in Fig. 5.4, *a*) resistance of the active system to fatigue. It means that friction and wear are beneficial in definite conditions of operation. In addition, the processes of wear-fatigue damage can be effectively controlled by suitably varying the conditions of friction and wear in a specific active system.

2) Depending on the conditions of tests the cyclic stresses can both significantly reduce (cf. curves *3*, *4* and *5* in Fig. 5.4, *b*) and significantly enhance (cf. curves *1* and *2* in Fig. 5.4, *b*) the wear resistance of the active system. It means that cyclic stresses are beneficial in definite conditions of operation. In addition, the processes of wear-fatigue damage can be effectively controlled by suitably varying the conditions of cyclic loading in a specific active system.

In both cases during the direct and back effects the WFD *controlling parameter* is the relation of type (4.11)

$$\Psi = \sigma / \tau_W = \tan\alpha , \tag{5.25}$$

that has a critical value of type (4.12)

$$\Psi_{-1f} = \sigma_{-1} / \tau_f = \tan\alpha_{-1f} . \tag{5.26}$$

When $\Psi > \Psi_{-1f}$ the direct effect occurs and when $\Psi < \Psi_{-1f}$ the back effect occurs.

With (5.10), (5.23) and (5.24) it is easy to consider and analyze a number of particular cases.

For example, assume that no corrosion damage occurs in the active system ($D = 0$). Then the energy criterion has the form (5.5). If there is no friction ($\tau_W = 0$), we obtain from (5.5) the *criterion of the limiting state during isothermal mechanical fatigue* (5.8). On the opposite, if there are no cyclic stresses ($\sigma = 0$), we obtain from (5.5) the *criterion of isothermal friction fatigue* (5.9).

Criteria (5.8) and (5.9) yield the formulas for ultimate stresses during isothermal mechanical fatigue

$$\sigma_{-1} = \left[\frac{U_0 / R_{T/M} - a_T T}{a_\sigma} \right]^{1/2} \tag{5.27}$$

and during isothermal friction fatigue

$$\tau_f = \left[\frac{U_0 / R_{T/M} - a_T T}{a_\tau} \right]^{1/2} . \tag{5.28}$$

Criterion (5.8) and formula (5.27) are useful in those cases when the limiting state of the structural component is due to fatigue fracture. Criterion (5.9) and formula (5.28) are true when the limiting state of the friction pair appears following the criteria of wear resistance (ultimate wear, critical density or depth of pits of spalling, intolerable noise or vibration, etc.). If the limiting state of the active system is analyzed according to (5.5), it is required to consider two cases when direct or back effects occur.

When investigating the direct effect from (5.5) or from (5.23) at $\sigma > 0$, $T_\Sigma > 0$, $\tau_W > 0$, $D = 0$, we obtain a formula for ultimate stresses

$$\sigma_{-1\tau} = \left[\frac{(U_0 / R_{T/M} - a_T T) / R_{\sigma/\tau} - a_\tau \tau_W^2}{a_\sigma} \right]^{1/2}, \tag{5.29}$$

that can be transformed with the account of (5.27) and (5.28) into

$$\sigma_{-1\tau} = \sigma_{-1} \sqrt{\frac{1}{R_{\sigma/\tau}} - \frac{\tau_W^2}{\tau_f^2}} = \sigma_{-1} \varphi_\sigma(\tau_W), \tag{5.29a}$$

where there should be

$$\varphi_\sigma(\tau_W) = \frac{1}{R_{\sigma/\tau}} - \frac{\tau_W^2}{\tau_f^2} \geq 0. \tag{5.30}$$

When investigating the back effect, we similarly from (5.5) or (5.24) obtain

$$\tau_{f\sigma} = \tau_f \sqrt{\frac{1}{R_{\sigma/\tau}} - \frac{\sigma^2}{\sigma_{-1}^2}} = \tau_f \varphi_\tau(\sigma), \tag{5.31}$$

where there should be

$$\varphi_\tau(\sigma) = \frac{1}{R_{\sigma/\tau}} - \frac{\sigma^2}{\sigma_{-1}^2} \geq 0. \tag{5.32}$$

Now let us plot the *diagrams of limiting states* for typical active systems following any criterion of damage and/or fracture using relations (5.29a) and (5.31) and taking into account formulas (5.27) and (5.28). Unlike Fig. 5.4, the diagrams of limiting states are constructed in relative coordinates $\sigma_a/\sigma_{-1} - \tau_W/\tau_{f\sigma}$ (Fig. 5.5, *a*). The limiting state appears every time when the equalities $\sigma_a = \sigma_{-1\tau}$ or $\tau_W = \tau_{f\sigma}$ occur. Additionally we plot the functional relations between the coefficient of interaction $R_{\sigma/\tau}$ and the relative parameter (Fig. 5.5, *b*) in double logarithmic coordinates:

$$\overline{\rho} = \left(\frac{\tau_W}{\tau_f}\right)^2 \left(\frac{\sigma_{-1}}{\sigma}\right)^2 = \left(\frac{\tau_{f\sigma}}{\tau_f}\right)^2 \left(\frac{\sigma_{-1}}{\sigma_{-1\tau}}\right)^2. \tag{5.33}$$

The parameter $\overline{\rho}$ can be recorded with the *coefficients of direct* (K_D) and *back* (K_B) *effects* (cf. (4.11) and (4.12)). If we introduce the *particular margin coefficients*

$$K_\sigma = \sigma_{-1} / \sigma, \quad K_\tau = \tau_f / \tau_W, \tag{5.34}$$

the parameter $\overline{\rho}$ can be related then to these coefficients. So,

$$\overline{\rho} = (K_\sigma / K_\tau)^2 = (K_D / K_B)^2. \tag{5.35}$$

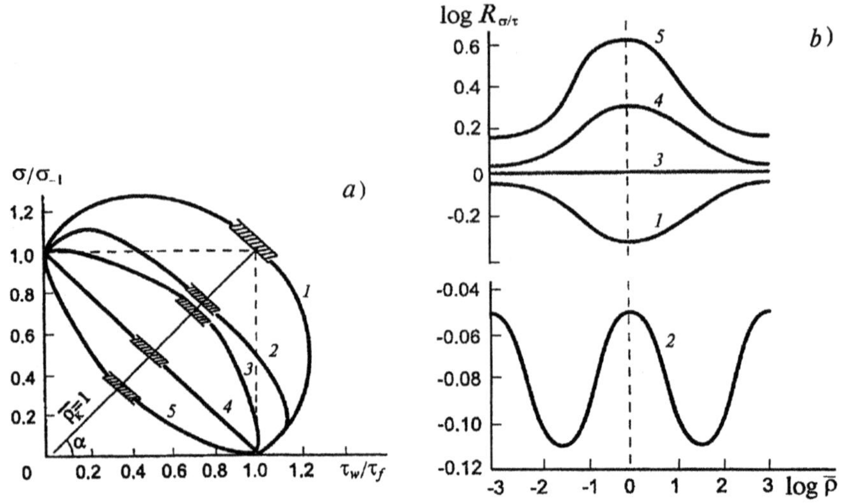

Fig. 5.5. Schemes: *a* – five possible (*1–5*) diagrams of ultimate states of active systems; *b* – corresponding five types (*1–5*) of dependencies $R_{\sigma/\tau}$ ($\bar{\rho}$)

The geometrical sense of the parameter $\bar{\rho}$ ensues from the diagram of limiting states:

$$\bar{\rho} = \cot^2 \alpha .$$

The critical value $\bar{\rho}_\kappa$ is reached at $\alpha = 45°$; in this case

$$\bar{\rho}_\kappa = K_\sigma / K_\tau = K_D / K_B = 1.$$

The parameter $\bar{\rho}$ in general has an energy sense; it is determined by the following ratio between the effective components of wear-fatigue damage energy:

$$\bar{\rho} = \frac{a_\tau \tau_W^2}{a_\sigma \sigma^2} \frac{a_\sigma \sigma_{-1}^2}{a_\tau \tau_f^2} .$$

Comparison of this formula with (5.33) shows that they are identical. Combined examination of Figs. 5.5, *a* and 5.5, *b* leads to the conclusion that the *value* $R_{\sigma/\tau}$ and related *parameter* $\bar{\rho}$ control the *processes of hardening and softening in the active system.* Curve *1* differs on the diagram (cf. Fig. 5.5, *a*) by the fact that hardening processes dominate in the active system ($\varphi_\sigma(\tau_W) > 1$, $\varphi_\tau(\sigma) > 1$). Note that for the above relation it is typical that $R_{\sigma/\tau} \leq 1 =$ var throughout the range $0 \leq \tau_W \leq \tau_f$ of variations of frictional τ_W and throughout the range $0 \leq \sigma \leq \sigma_{-1}$ of variations of volume σ stresses, while the relation $R_{\sigma/\tau}$ (ρ) in double logarithmic coordinates (cf. Fig. 5.5, *b*) is an exponential function with one minimum (at $\bar{\rho} = \rho_k = 1$); here $R_{\sigma/\tau} = 1$ when $\bar{\rho} = 0$ or $\bar{\rho} = \infty$. Curve *2* on diagram (cf. Fig. 5.5, *a*) characterizes the WFD of the active systems and the conditions of their operation when the processes of hardening ($\varphi_\sigma(\tau_W) > 1$, $\varphi_\tau(\sigma) > 1$) prevail over one portion of the given interval of changes of σ or τ_W, while the processes of softening ($\varphi_\sigma(\tau_W) < 1$, $\varphi_\tau(\sigma) << 1$) prevail over the other. In this example the former occur when τ_W grows from 0 to ~$0.45\tau_f$ and σ grows from 0 to ~$0.45\sigma_{-1}$. It is also noteworthy that the relation of $R_{\sigma/\tau}$ ($\bar{\rho}$) (cf. Fig. 5.5, *b*) in the case in question is a complex exponential function with one maximum (at $\bar{\rho} = \bar{\rho}_k = 1$) and two minima corresponding to the maximum possible values max $\sigma_{-1\tau}$ and max $\tau_{f\sigma}$ of ultimate stresses. The value $R_{\sigma/\tau} \leq 1$ persists throughout the set range of variations of stresses, where $R_{\sigma/\tau} = 1.0$ at $\bar{\rho} = 0$ and $\bar{\rho} = \infty$.

Lines *3–5* correspond to the active systems and to the conditions of their operation that have the limits of mechano-rolling fatigue $\sigma_{-1\tau}$ and $\tau_{f\sigma}$ declining throughout the range of variations of stresses, i.e. $\varphi_\sigma(\tau_W) < 1$, $\varphi_\tau(\sigma) < 1$, while the processes of softening prevail during WFD. For curve *3* on the diagram the values of the functions are

$$\varphi_\sigma(\tau_w) = \sqrt{1 - \frac{\tau_W^2}{\tau_f^2}}, \quad \varphi_\tau(\sigma) = \sqrt{1 - \frac{\sigma^2}{\sigma_{-1}^2}},$$

i.e. $R_{\sigma/\tau} = 1.0 =$ const within given ranges of variations of τ_W and σ. In this case the wear-fatigue damage is due to the independent combined action of forces of friction and cyclic (volume) stresses. Hence, there is no link between $R_{\sigma/\tau}$ and $\bar{\rho}$. Curve *5* in Fig. 5.5, *a* characterizes a high degree of softening of the active system during wear-fatigue damage when ultimate stresses $\sigma_{-1\tau}$ and $\tau_{f\sigma}$ drop down sharply as τ_W and σ increase respectively. Therefore, the dependence $R_{\sigma/\tau}(\bar{\rho})$ corresponding to it (curve *5* in Fig. 5.5, *b*) is described by the function inverse to the dependence $R_{\sigma/\tau}(\bar{\rho})$ in the active systems in which the processes of hardening occur (cf. curve *1* in Fig. 5.5). A similar function describes the link $R_{\sigma/\tau}$ ($\bar{\rho}$) in the systems with inversely proportional linear dependence *4* on the diagram (cf. Fig. 5.5, *a*); it also has $R_{\sigma/\tau} \geq 1 =$ var at $0 \leq \tau_W \leq \tau_f$ and $0 \leq \sigma \leq \sigma_{-1}$ (cf. Fig. 5.5, *b*).

It is essential to emphasize the multicriterial pattern of the diagram of limiting states (cf. Fig. 5.5, *a*). Really, the point with the coordinates ($\tau_W/\tau_f = 0$; $\sigma_{-1\tau}/\sigma_{-1} = 1$) corresponds to the fatigue limit σ_{-1} of a component of the active system determined by performing standard tests for common mechanical fatigue, (the limiting state criterion is the appearance of the main crack). Hence, this point implies the mechanics of fatigue fracture. The point with the coordinates ($\tau_{f\sigma}/\tau_f = 1$; $\sigma/\sigma_{-1} = 0$) corresponds to the friction fatigue limit τ_f of a component of the system determined by performing standard tests for friction. Hence, this point implies tribology. Lines *1–5* on the diagram are the curves of limiting states of components of the active system during complex wear-fatigue damage that are plotted using the results of wear-fatigue tests. It means that these curves imply tribo-fatigue.

It should be added that the relative parameter $\overline{\rho}$ of the asymmetry of processes of friction and force damage varies within the range $0 \leq \overline{\rho} \leq \infty$.

In fact, the parameter $\overline{\rho}$ provides an opportunity to estimate contribution of shear and rupture processes into the complex damage of an active system. It can be assumed that if $\overline{\rho} = 1$, then σ and τ_W affect the damage of the system equally. Figure 5.5, *b* shows a dotted line corresponding to $\overline{\rho} = 1$ ($\log \overline{\rho} = 0$); a straight line corresponds to it plotted from the point of the origin of the coordinates at an angle $\alpha = 45°$ on the diagram (cf. Fig. 5.5, *a*). If $\overline{\rho} > 1$, mechanisms of surface fracture dominate in the active system (wear, for example, during sliding friction); in the ultimate case when $\overline{\rho} \rightarrow \infty$ (at $\sigma = 0$) the system may fail due exceptionally to the energy of friction (following the criterion of wear limit attainment). If $\overline{\rho} < 1$, mechanisms of volume fracture (mechanical fatigue) become predominant and in the extreme case of $\overline{\rho} = 0$ (at $\tau_W = 0$) fracture may be due exclusively to the mechanical energy induced by normal cyclic stresses σ (following the criterion of main crack appearance). Note that the limiting state of the system in the transitional zone $\overline{\rho} \rightarrow 1$ (each such zone for curves *1–5* is shaded in Fig. 5.5, *a*) may occur following the criterion of main crack appearance and the criterion of attainment of ultimate wear simultaneously (or almost simultaneously).

The parameter of interaction $R_{\sigma/\tau}$ can be calculated with the following expressions:

a) for active systems with dominating processes of softening (curves *3–5* in Fig. 5.5),

$$R_{\sigma/\tau} = 1 + C_s z\,; \tag{5.36a}$$

b) for active systems with dominating processes of hardening (ascending branches of curves *1*, *2* in Fig. 5.5),

$$R_{\sigma/\tau} = 1 - C_h z\,, \tag{5.36b}$$

where z – the exponential function determined from the formula

$$z = \overline{\rho}\exp(-\overline{\rho})\,. \tag{5.37}$$

C_s, C_h are constant in formulas (5.36a), (5.36b), here $(-C_h)$ has the sense of the coefficient of hardening and $(+C_s)$ that of the coefficient of softening. In fact, the coefficients K_D and K_B (cf. formulas (4.11) and (4.12)) of the direct and back effects are also indicators of deformation hardening (softening) of active systems.

According to (5.36) a linear relation between $R_{\sigma/\tau}$ and z is predicted. Therefore, we have corresponding straight lines starting at the point ($R_{\sigma/\tau} = 1$; $z = 0$) (Fig. 5.6) in uniform Cartesian coordinates $R_{\sigma/\tau} - z$ for dependencies *1–5* (cf. Fig. 5.5, *a*, *b*). Note that broken straight lines in Fig. 5.6 describe dependencies *1*, *2* because the ascending branches (within the range of variations of σ/σ_{-1} from 1 to $(\sigma/\sigma_{-1})_{max}$ during the direct effect and within the range of variations of τ_W/τ_f from 1 to $(\tau_W/\tau_f)_{max}$ during the back effect) obey the law of hardening, while the descending branches obey the law of softening. The model for the relation $R_{\sigma/\tau}(z)$ is selected accordingly.

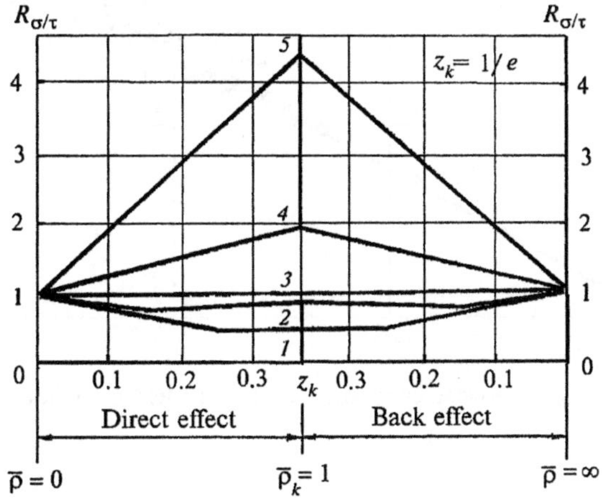

Fig. 5.6. Dependencies $R_{\sigma/\tau}(z)$ corresponding to five types of diagrams of ultimate states of active systems (cf. Fig. 5.4)

5.1.6 Isothermal fatigue: interactions between damages

Let us consider one of the particular cases of criterion (5.5), viz. *isothermal mechanical fatigue*. Formula (5.27 follows from (5.8) that we represent as

$$\log \sigma_{-1} = \frac{1}{2} \log C_T; \; C_T = \frac{U_0 / R_{T/M} - a_T T}{a_\sigma}. \tag{5.38}$$

According to (5.38) the relation between ultimate stresses and the *parameter of thermal force resistance* C_T in double logarithmic coordinates should be a straight line with the slope ratio 1/2. Figure 5.7 convincingly confirms this relation for various classes of materials tested under various conditions. It is apparent that values of C_T change by more than two orders of magnitude, i.e. 100 times and more; values of the fatigue limit σ_{-1} by more than an order of magnitude, i.e. 10 times and more, when the temperature of tests varied within the range from that from the helium temperature to $0.8T_s$. The general regularity is that the larger the value C_T the higher is σ_{-1}.

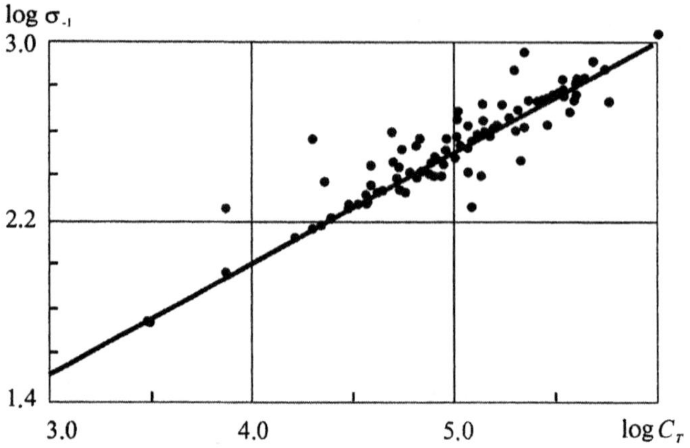

Fig. 5.7. Relation between endurance limits of materials and the parameter C_T

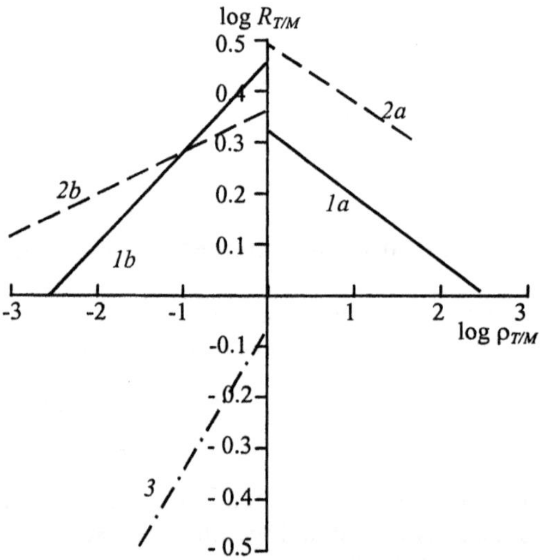

Equation	Coefficient of correlation	Curve
$\log R_{T/M} = 0.35 - 0.16 \log \rho_{T/M}$	$r = 0.876$	*1a*
$\log R_{T/M} = 0.414 + 0.16 \log \rho_{T/M}$	$r = 0.862$	*1b*
$\log R_{T/M} = 0.48 - 0.14 \log \rho_{T/M}$	$r = 0.917$	*2a*
$\log R_{T/M} = 0.346 + 0.072 \log \rho_{T/M}$	$r = 0.999$	*2b*
$\log R_{T/M} = -0.075 + 0.273 \log \rho_{T/M}$	$r = 0.939$	3

Fig. 5.8. Logarithmic relations $R_{T/M}$ ($\rho_{T/M}$)

Figure 5.8 analyzes the function of interaction $R_{T/M}$ in response to the parameter $\rho_{T/M}$ of asymmetry of thermomechanical damage of steels, aluminium alloys and nickel (based on the results of extensive experimentation). The coefficient of correlation was quite high: from $r = 0.862$ to $r = 0.999$. As a rule, the dependencies $R_{T/M}(\rho_{T/M})$ show a leap (a discontinuity) at $\log \rho_{T/M} = 0$, i.e. when $\rho_{T/M} = 1$ and thermal and force damage equalize. A proportional relation between $R_{T/M}$ and $\rho_{T/M}$, appears for steels (curves *1*) and nickel (curves *2*) at $\rho_{T/M} < 1$, it becomes inverse at $\rho_{T/M} >> 1$. The relation $R_{T/M}(\rho_{T/M})$ is inversely proportional for aluminium alloys (curve *3*) if $\rho_{T/M} < 1$. Therefore, the parameter of interaction $R_{T/M}$ is sensitive both to the ratio between effective (thermal and force) energies and to the structure and composition of metallic materials.

From (5.9) we obtain for the case of *isothermal sliding fatigue* formula (5.28) that we represent as

$$\log \tau_f = \frac{1}{2}\log C_T\,, \quad C_T = \frac{U_0 / R_{T/M} - a_T T}{a_\tau}, \tag{5.38a}$$

that is similar to Eq. (5.38). If an allowance is to be made for the effect of corrosive damage, the parameter C_T in formulas (5.38) and (5.38a) is respectively

$$C_{T(ch)} = \frac{U_0 / R_{T/M} - a_T T/(1-D_T)}{a_\sigma /(1-D_\sigma)}, \quad C_{T(ch)} = \frac{U_0 / R_{T/M} - a_T T/(1-D_T)}{a_\tau /(1-D_\tau)}.$$

In this case *it implies isothermal mechano-corrosion and isothermal corrosion friction fatigue.*

5.1.7 Calculations based on the limiting state

If the direct effect occurs, the condition of strength is based on the criterion of resistance to fatigue with the allowance for the processes of friction and wear

$$\sigma \le [\sigma] = \sigma_{-1\tau} / n_{\sigma\tau}\,, \tag{5.39}$$

where $[\sigma]$ and $n_{\sigma\tau}$ – allowable stress and the margin of safety of a component of the active system bearing both contact and off-contact loads.

Condition (5.39) of unattainability of the limiting state, like the similar condition of strength (1.16), enables to solve three problems: (1) validation of strength, (2) determination of the cross section of a structural component and (3) selection of a material for its fabrication.

A special experiment is carried out or formula (5.23) or its modifications (5.29), (5.29a) and others are used to estimate the ultimate stress $\sigma_{-1\tau}$. If we accept (5.29a), we find, for example, the required moment of resistance W_{TF} of the shaft when it is bent with the moment M:

$$W_{TF} = \frac{M n_{\sigma\tau}}{\sigma_{-1\tau}} = \frac{M n_{\sigma\tau}}{\sigma_{-1}\sqrt{\dfrac{1}{R_{\sigma/\tau}} - \dfrac{\tau_W^2}{\tau_f^2}}}. \tag{5.40}$$

From the practical point of view, an essential feature of formula (5.40) is that it is enough to have just four particular fatigue limits (σ_{-1}, τ_f) in order to select the dimensions of the component of the active system based on the criterion of tribo-fatigue (W_{TF}). The structure of (5.40) is such that it contains also the usual moment of resistance (W) determined from the condition of strength recorded on the ultimate stress σ_{-1},

$$W = W_F = \frac{Mn_\sigma}{\sigma_{-1}} = \frac{Mn_{\sigma\tau}}{\sigma_{-1}},$$

if it is assumed that strength margins of safety coincide in both cases: $n_\sigma = n_{\sigma\tau}$. Then formula (5.40) with designations in (5.30) becomes

$$\frac{W_{TF}}{W_F} = \frac{1}{\sqrt{\dfrac{1}{R_{\sigma/\tau}} - \dfrac{\tau_W^2}{\tau_f^2}}} = \frac{1}{\varphi_\sigma(\tau_W)}. \tag{5.41}$$

According to (5.41) the ratio between the dimensions of the shaft from the criterion of tribo-fatigue and the criterion of mechanical fatigue is inversely proportional to the function $\varphi_\sigma(\tau_W)$. Figure 5.9 provides a graphic analysis of formula (5.41) with the condition that $W_F = 1$.

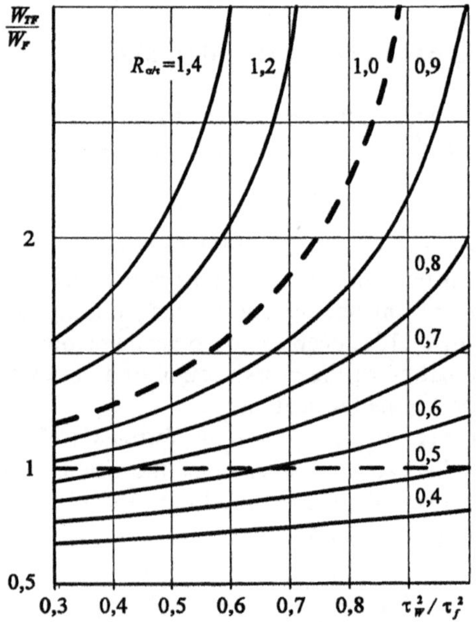

Fig. 5.9. Analysis of relation (5.41)

A horizontal dotted line corresponds to the case when $W_{TF}/W_F = 1$, i.e. when the shaft is calculated only from the particular criterion of mechanical fatigue ignoring the effect of wear and friction processes. Another dotted line corresponds to the case when this effect is taken into account according to (5.41) at $R_{\sigma/\tau} = 1$. Full lines characterize the role of interactions between damages due to contact and off-contact loads ($R_{\sigma/\tau} \neq 1$) with the account of their direction ($R_{\sigma/\tau} > 1$ or $R_{\sigma/\tau} < 1$). The following conclusions can be made from Fig. 5.9.

(1) If $R_{\sigma/\tau} = 1$, then $W_{TF} > W_F$ already at $\tau_W^2/\tau_f^2 \approx 0{,}3$, while at $\tau_W^2/\tau_f^2 \approx 0.7$ the moment of resistance W_{TF} should be assumed exceeding the value W_F ~2...3 times.

(2) If $R_{\sigma/\tau} < 1$, the moment of resistance $W_{TF} < W_F$ is usually required.

(3) Finally, if $R_{\sigma/\tau} > 1$, the value W_{TF} grows rapidly (compared with the value W_F) and its growth accelerates as the ratio τ_W^2/τ_f^2 increases. For example, the moment of resistance W_{TF} should exceed W_F approximately 3 times at $R_{\sigma/\tau} = 1.2$ and $\tau_W^2/\tau_f^2 \approx 0.7$.

If the back effect occurs, the condition of strength is based on the criterion of sliding fatigue with the allowance for cyclic stresses:

$$\tau_W = fp_a \leq [\tau] = \tau_{f\sigma}/n_{\tau\sigma}. \tag{5.42}$$

Condition (5.42) in terms of contact pressure is

$$p_a \leq [p] = p_{f\sigma}/n_{p\sigma}. \tag{5.42a}$$

Conditions of unattainability of the limiting state (5.42), (5.42a), like similar conditions of strength (1.113a, b, c) enable too to solve three problems: (1) verification of strength, (2) determination of the dimension of a structural component of an active system and (3) selection of a material for its fabrication.

Special experiment is carried out or formula (5.24) or its modifications, for example (5.31) are used to estimate the ultimate stress $\tau_{f\sigma}$. If (5.31) is used, (5.42) yields the required nominal contact area A_{TF} of active system components under the effect of contact load F_N:

$$A_{TF} = \frac{fF_N n_{\tau\sigma}}{\tau_{f\sigma}} = \frac{fF_N n_{\tau\sigma}}{\tau_f \sqrt{\dfrac{1}{R_{\sigma/\tau}} - \dfrac{\sigma^2}{\sigma_{-1}^2}}}. \tag{5.43}$$

Formula (5.43) can be analyzed as it is done with (5.40).

If it is needed to estimate the *state* of the active system as an integrity, measure (5.12) of complex WFD is analyzed correspondingly. The criterion of the limiting state attainability is set by equality (5.10). The condition of *unattainability of the limiting state*

$$0 < \omega_\Sigma < 1 \tag{5.44}$$

serves simultaneously as the *criterion* for assessing the *state of damage*: the larger the value ω_Σ, the more severe is the damage. Finally, the *condition of translimiting states* is

$$\omega_{\Sigma} > 1. \tag{5.45}$$

The theory of translimiting states of active systems has not been developed so far.

5.2 Reliability

5.2.1 Metal-to-polymer active system

General

Set out the solution of the problem of reliability of the metal-to-polymer active system in which the metallic body (the shaft) bears and transmits cyclic bending loads, and the polymeric counterbody (a sliding bearing) makes a force sliding friction contact with it without lubrication (cf. Fig. 2.1, *a*). Formulation of this problem is generally *limited*: only the case of regular loading of the body in the linear state of stress is considered. The effect of lubrication on the wear processes is ignored; a quantitative assessment of wear of rubbing surfaces is made easier because the metallic body wears negligibly little compared to the polymer, so it can be ignored. Yet, it is apparent that this problem is *basic* (mechanical fatigue and sliding friction occur in mating under study in their simple form) and *principal* (the materials of unlike origin with strongly varied mechanophysical properties are in force contact). Therefore, this problem has a sufficiently *general* nature.

The basis of the problem's solution is the theory of strength of the weakest link in its modern interpretation; the conception of dangerous regions of the deformable solid during friction and fatigue; the kinetic theory of strength of solids; the fatigue theory of wear; the molecular mechanical concept of friction of solids (see Sects. 1.3 and 1.4). The basic assumptions in these theories and conceptions have been thoroughly validated experimentally and are considered universal in modern science, therefore their application is justifiable.

Assume that regular cyclic stresses σ appear in one of its elements in the metallic body – in the process of operation of the metal-to-polymer active system, while the polymeric counterbody is forced into contact with a body by pressure p_α. Hence, the specific force of friction (the friction stress) is $\tau_W = fp_\alpha$. Let us assume that the probability $P(\sigma, \tau_W)$ of failure after preset time of operation under combined effect of normal σ and frictional τ_W stresses is the main characteristic of reliability of the metal-to-polymer active system. Let us treat $P(\sigma, \tau_W)$ as the probability of attainment by the metallic body or polymeric counterbody or by both simultaneously of the limiting state in the process of operation or wear-fatigue tests. Assume that the limiting state occurs when a fatigue crack of critical dimensions appears in the body, and the limiting state of the counterbody occurs when wear reaches a critical extent. It is clear that in both cases the metal-to-polymer active system fails; a combination of both also means failure.

Three approaches to calculating failure probability $P(\sigma, \tau_W)$ are possible.

(1) The functions of distribution $\varphi(\sigma_{-1})$ and $\varphi(\tau_f)$ of ultimate stresses σ_{-1} and τ_f in the body and the counterbody as well the functions of distribution $\varphi(\sigma)$ and $\varphi(\tau)$ of the effective stresses σ and τ_W are known.

(2) The functions of distribution $\varphi(\sigma_{-1})$ and $\varphi(\tau_f)$ are known and the values of effective stresses σ and τ_W are considered deterministic.

(3) The functions of distribution $\varphi(\sigma)$ and $\varphi(\tau)$ are known and the values of ultimate stresses σ_{-1} and τ_f are considered deterministic.

We use approach (2). In order to apply it, a two-dimensional function of distribution $P(\sigma_{-1}, \tau_f)$ of ultimate stresses σ_{-1} and τ_f is needed.

The two-dimensional function of distribution of ultimate stresses

Assume σ_{-1} and $\tau_f = fp_f$ are independent random values. Indeed, they are determined experimentally from the results of independent tests for friction (τ_f and/or p_f) and mechanical fatigue (σ_{-1}). The task is to find a two-dimensional function of distribution $P(\sigma_{-1}, \tau_f)$.

According to the statistical theory of strength of the weakest link, the integral function of distribution $P(\sigma_{-1})$ of fatigue limit σ_{-1} of the metallic body (a shaft, for example) is

$$P(\sigma_{-1}) = 1 - [1 - F(\sigma_{-1})]^{n_1 V_\sigma}. \tag{5.46}$$

Here n_1 – the number of dangerous defects per unit of volume of the metallic body, V_σ – its working volume, so $n_1 V_\sigma$ – the number of defects in the volume V_σ.

Assume that the function of distribution $P(\tau_f)$ of the friction fatigue limits τ_f of the polymeric counterbody (for example, an insert of the sliding bearing) is similar to (5.46). It is justifiable because statistical regularities of resistance to the same phenomenon – fatigue damage and fracture (including the fatigue fracture of the polymer) are meant in both cases. If values σ_{-1} and τ_f are both independent and compatible, the two-dimensional function of their distribution is

$$P(\sigma_{-1}, \tau_f) = 1 - [1 - F(\sigma_{-1})]^{n_1 V_\sigma} [1 - F(\tau_f)]^{n_2 S_\tau}, \tag{5.47}$$

where n_2 – the number of dangerous defects per unit of volume of the polymeric counterbody, S_τ – its working volume, so $n_2 S_\tau$ – the number of defects in the volume S_τ.

Since $n_1 V_\sigma$ and $n_2 S_\tau$ are sufficiently large, function (5.47) can be substituted with its asymptotic approximation

$$P(\sigma_{-1}, \tau_f) = 1 - \exp[-c_1 n_1 V_\sigma F(\sigma_{-1}) - c_2 n_2 V_\tau F(\tau_f)], \tag{5.48}$$

where c_1, c_2 – the coefficients of proportionality ensuring the identity of (5.47) and (5.48).

Concrete definition of expressions that are under the exponent in (5.48) leads to a *two-dimensional S-function of distribution* of ultimate stresses.

$$P(\sigma_{-1},\tau_f)=$$

$$=1-\exp\left\{-C_V\frac{V_{P\gamma}}{V_0}\left[\left(\frac{T_M}{T_0}\right)^{m_T}\frac{\sigma_{-1}-\sigma_{-1\min}}{\sigma_W}\right]^{m_V}-C_S\frac{S_{P\gamma}}{S_k}\left(\frac{\tau_{tf}^{(1)}\Delta T}{\tau_d-\tau_f}\right)^{m_S}\right\}, \tag{5.49}$$

that has six parameters ($\sigma_{-1\min}$, σ_W, m_V, $\tau_{tf}^{(1)}$, τ_d, m_S). Here C_V – the coefficient determining the body's shape and the pattern of its cyclic deformation, C_S – the coefficient determining the counterbody's shape and the pattern of its contact interaction with the body in the process of friction. The thermomechanical function of ultimate stresses that (5.46)–(5.49) contain

$$F(\sigma_{-1})=\left[\left(\frac{T_M}{T_0}\right)^{m_T}\frac{\sigma_{-1}-\sigma_{-1\min}}{\sigma_W}\right]^{m_V}$$

corresponds to the function in the theory of strength of the weakest link, but it is recorded here with the allowance for the thermodynamic state of the metal: T_M/T_0 – its relative temperature, T_0 – initial, for example, room temperature, m_T – the parameter of thermal activation of fatigue damage.

The molecular mechanical function that (5.47)–(5.49) contain

$$F(\tau_f)=\left(\frac{\tau_{tf}^{(1)}\Delta T}{\tau_d-\tau_f}\right)^{m_S}$$

corresponds to the function in the kinetic (thermofluctuation) theory of strength of solids

$$F(\tau_f)=\left(\frac{R\Delta T}{U_0-\gamma_\tau\tau_f}\right)^{m_S}=\left(\frac{R\Delta T}{\gamma_\tau\left(\dfrac{U_0}{\gamma_\tau}-\tau_f\right)}\right)^{m_S},$$

but here it is recorded in (5.49) is in terms of frictional stresses

$$\tau_d=U_0/\gamma_\tau; \tag{5.50}$$

$$\tau_{tf}^{(1)}=R/\gamma_\tau; \tag{5.51}$$

$$\tau_{tf}=\tau_{tf}^{(1)}\Delta T, \tag{5.52}$$

where γ_τ – the structurally sensitive coefficient; R – gas constant; $\Delta T = T_2 - T_1$ – increment (in the process of operation or tests) in the temperature of polymer from T_1 to T_2.

The structure of (5.49) is such that it allows to single out both the strength (fatigue) function

$$\varphi_\sigma = C_V \frac{V_{P\gamma}}{V_0}\left[\left(\frac{T_M}{T_0}\right)^{m_T} \frac{\sigma_{-1} - \sigma_{-1\min}}{\sigma_W}\right]^{m_V}, \tag{5.53}$$

and the tribological function

$$\varphi_\tau = C_S \frac{S_{P\gamma}}{S_k}\left(\frac{\tau_{tf}^{(1)} \Delta T}{\tau_d - \tau_f}\right)^{m_S}. \tag{5.54}$$

If we introduce the designations

$$\eta_\sigma = C_V \frac{V_{P\gamma}}{V_0}\left(\frac{T_M}{T_0}\right)^{m_T m_V}; \tag{5.55}$$

$$\eta_\tau = C_S \frac{S_{P\gamma}}{S_k}, \tag{5.56}$$

distribution (5.49) then becomes

$$P(\sigma_{-1}, \tau_f) = 1 - \exp\left[-\eta_\sigma \left(\frac{\sigma_{-1} - \sigma_{-1\min}}{\sigma_W}\right)^{m_V} - \eta_\tau \left(\frac{\tau_{tf}}{\tau_d - \tau_f}\right)^{m_S}\right]. \tag{5.57}$$

Thus, we have obtained an integral function of distribution of ultimate stresses (5.49) and its modification (5.57). The specific feature of these functions is that both random values have their limits

$$\sigma_{-1} > \sigma_{-1\min}; \quad \tau_W < \tau_d.$$

In other words, the distribution of fatigue limits σ_{-1} has its bottom threshold of scatter $\sigma_{-1\min}$, such that the probability is

$$P\,(\sigma_{-1} < \sigma_{-1\min}) = 0.$$

The distribution of ultimate stresses τ_f has, on the opposite, the upper threshold of scatter τ_d, such that the probability is

$$P\,(\tau_f > \tau_d) = 1.$$

It means that fatigue function (5.53) $\varphi_\sigma = 0$ if $\sigma_{-1} \le \sigma_{-1\min}$. It is apparent it should be $V_{P\gamma} = 0$ in this case. Tribological function (5.54) $\varphi_\tau = 0$ if $\tau_f = 0$, hence, $S_{P\gamma} = 0$.

Determination of parameters

Methods of estimating parameters contained in (5.49) are established by analyzing particular cases.

If no friction (hence, no contact interaction between the body and the counterbody) occurs, we obtain from (5.49) at $S_{P\gamma} = 0$ a particular function of distribution of fatigue limits of the metallic body (the shaft)

$$P(\sigma_{-1}) = 1 - \exp\left\{ -C_V \frac{V_{P\gamma}}{V_0} \left[\left(\frac{T_M}{T_0} \right)^{m_T} \frac{\sigma_{-1} - \sigma_{-1\min}}{\sigma_W} \right]^{m_V} \right\}. \quad (5.58)$$

By differentiating (5.58) with the variable σ_{-1}, we obtain with the account of (5.55) the density of probabilities

$$\varphi(\sigma_{-1}) = \eta_\sigma \frac{m_V}{\sigma_W} \left(\frac{\sigma_{-1} - \sigma_{-1\min}}{\sigma_W} \right)^{m_V - 1} \exp\left[-\eta_\sigma \left(\frac{\sigma_{-1} - \sigma_{-1\min}}{\sigma_W} \right)^{m_V} \right]. \quad (5.59)$$

Note that (5.58) and (5.59) represent the function of distribution of Waybull. The distribution of Waybull is known to be three-parameter ($\sigma_{-1\min}$, σ_W, m_V) or two-parameter if it is assumed that $\sigma_{-1\min} = 0$; an additional parameter η_σ is introduced into function (5.59)

GOST 11.007 specifies the methods of estimate determination and confidence limits of the parameters of distribution of Waybull irrespective of the physico-mechanical content of analyzed random values, here the parameter σ_W is called the parameter of scale, m_V – the parameter of shape and $\sigma_{-1\min}$ – the parameter of shear. The methods of determination of basic parameters described below differ from the standard methods, still they are informative and experimentally validated in respect of the fatigue resistance (problems of strength) of materials.

It is established experimentally that

$$\sigma_{-1\min} \approx (0.4...0.6)\overline{\sigma}_{-1d=10}, \quad (5.60)$$

where $\overline{\sigma}_{-1d=10}$ – the mean value of fatigue limits determined during statistical fatigue tests of specimens with the diameter $d = 10$ mm under conditions of bending and torsion. If the results of such tests are unavailable, it can be assumed with some approximation that

$$\sigma_{-1\min} \approx 0.5\sigma_{-1}, \quad (5.61)$$

where σ_{-1} – the deterministic value of fatigue limit of specimens of the same diameter.

The value $\sigma_{-1\min}$ determined by the described method turns out to be stable: it is insensitive both to the scale effect and to the level of concentration of stresses. That is the reason that the value $\sigma_{-1\min}$ is accepted for limitation of the damaged volume (2.13).

The parameter σ_W can be determined from function (5.58) if the designation $\eta_\sigma = 1$ is used (cf. formula (5.55)). Let us introduce some magnitude σ_0 of the random value σ_{-1} so that $\sigma_0 = \sigma_{-1\min} + \sigma_W$. Then from (5.58) with the account of $\eta_\sigma = 1$ we obtain

$$P(\sigma_0 = \sigma_{-1\min} + \sigma_W) = 1 - e^{-1} \approx 0.63 .$$

Hence, $\sigma_W = \sigma_0 - \sigma_{-1\min}$, where σ_0 – the fatigue limit with the probability $P = 0.63$ (Fig. 5.10).

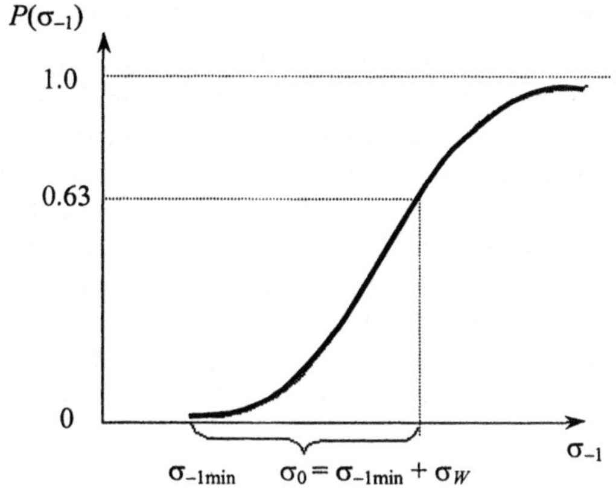

Fig. 5.10. Determination of parameter σ_W

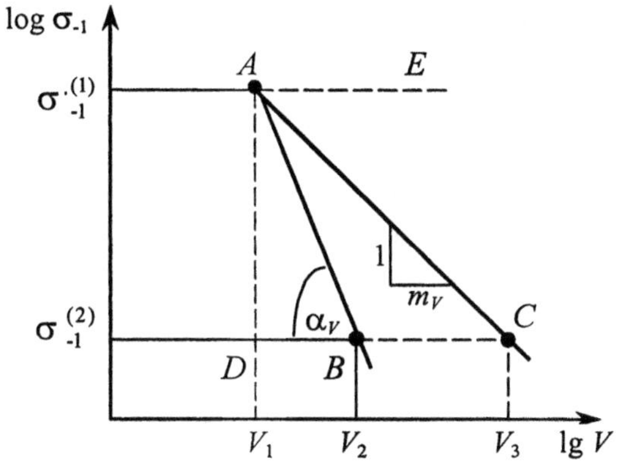

Fig. 5.11. Determination of parameter m_V of mechanical non-homogeneity of material

The parameter m_V of mechanical non-homogeneity (isotropy) describes integrally, in fact, the extent of presence of defects of a material. Since it is established that the extent of presence of defects augments as a logarithmic function as the sizes of an article become larger, the relation of the fatigue limit σ_{-1} (or strength limit σ_b) and the body's volume V (or the diameter d, or the area A_0 of its cross section) is plotted in double logarithmic coordinates (Fig. 5.11). The value m_V is determined, for example, from the coordinates of two points (A and B) as a cotangent of the slope angle of the scale dependence of ultimate stresses to the abscissa axis:

$$\cot \alpha_V = \frac{\log V_2 - \log V_1}{\log \sigma_{-1}^{(1)} - \log \sigma_{-1}^{(2)}} = m_V . \tag{5.62}$$

Thus, the parameter m_V characterizes the intensity (or the rate) of drop of the fatigue limit as the body increases in size. If the denominator in relation (5.62) is made equal to a unit, it becomes clear that the parameter m_V characterizes quantitatively a growth of the body's volume to such an extent that the logarithm of the fatigue limit drops by a unity (cf. curve AC in Fig. 5.11). If $\alpha_V = 0$, $m_V = \infty$ and it corresponds to a defect-free (ideally isotropic) material (the dotted line AE in Fig. 5.11) for which the scale effect does not exist. If $\alpha_V = 90°$, $m_V = 0$ and it corresponds to an ultimately defective metal (the dotted line AD) that is incapable to bear any load (it can be imagined that the body's volume is made up by metallic powder). Table 5.2 lists the numerical values of the parameter m_V for typical technical materials. It follows from Fig. 5.11 (compare lines AB and AC) that the larger the parameter m_V is (or the smaller the angle α_V), the better is the quality of the material regarding the presence of defects, hence the stronger is its resistance to fracture.

Table 5.2. Parameter m_V for various classes of materials

Materials	m_V
Structural steels	11.2...19.6
Polymers	3.1...9.8
SiC-based materials	5.90...12.14
Cermets	8.32...8.71
Pig iron	5.97...18.00

The parameter m_T of thermal activation of fatigue damage is determined from the temperature dependence of the fatigue limits that becomes a straight line (Fig. 5.12) in double logarithmic coordinates (for a limited range of elevated working temperatures):

$$\cot \alpha_T = \frac{\log T_2 - \log T_1}{\log \sigma_{-1}^{(1)} - \log \sigma_{-1}^{(2)}} = m_T . \tag{5.63}$$

Table 5.3 lists numerical values of the parameter for typical classes of materials. In fact, the parameter m_T, similarly with the parameter m_V characterizes quantitatively such temperature rise that leads to the reduction of the fatigue limit logarithm by unity. The larger the parameter m_T (or the smaller the angle α_T), the stronger is the resistance of the material to thermodynamic loads.

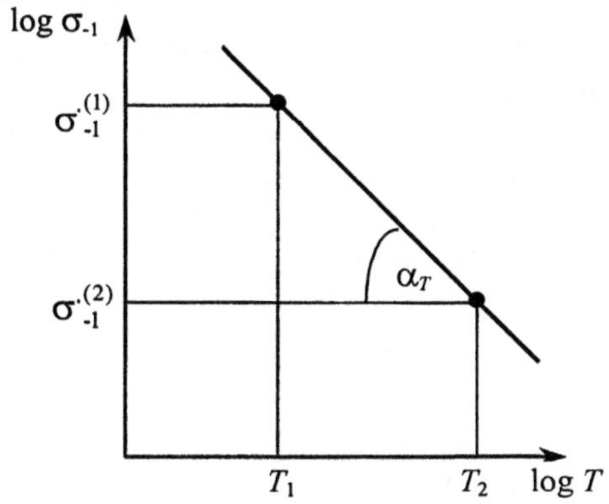

Fig. 5.12. Determination of parameter m_T of thermal activation of fatigue fracture

Table 5.3. Parameter m_T for various classes of materials

Materials	m_T
Structural steels	0.82...2.60
High-temperature steels	2.09...4.65
Nickel alloys	3.00...5.47
Deformable aluminium alloys	1.99...3.31

In order to determine the tribological parameters, let us consider a particular case of the function of distribution (5.49). If no cyclic deformation of the metallic body occurs, we obtain the function of distribution of limits of friction fatigue of the polymeric counterbody from (5.49) at $V_{P\gamma} = 0$:

$$P(\tau_f) = 1 - \exp\left\{ -C_S \frac{S_{P\gamma}}{S_k} \left(\frac{\tau_{tf}^{(1)} \Delta T}{\tau_d - \tau_f} \right)^{m_S} \right\}. \tag{5.64}$$

By differentiating (5.64) with the variable τ_f, we have the following density of probabilities:

$$\varphi(\tau_f) = \eta_\tau \frac{m_S}{\tau_{tf}} \left(\frac{\tau_{tf}}{\tau_d - \tau_f} \right)^{m_S+1} \exp\left[-\eta_\tau \left(\frac{\tau_{tf}}{\tau_d - \tau_f} \right)^{m_S} \right]. \quad (5.65)$$

Note that (5.64) and (5.65) are the function of distribution in the form of Freshét. The distribution of Freshét has two parameters; this distribution is three-parameter ($\tau_{tf}^{(1)}$, τ_d, m_S). There are no standard methods of assessing the parameters of distribution of Freshét.

The essence and methods of determination of some parameters of distribution have been validated experimentally for the phenomenon of friction fatigue.

It is established that the value τ_d can be estimated from the results of tests of a polymeric specimen for compression. If its strength limit during compression σ_b^{com} is determined, then

$$\tau_d \approx f \, \sigma_b^{com}, \quad (5.66)$$

where f – the usual friction coefficient.

Another method to determine the parameter τ_d is to use formula (5.50). By comparing it with formula (5.16), it is easy to determine that

$$\gamma_\tau = \sqrt{a_\tau U_0}, \quad (5.67)$$

in order to identify $\tau_d \equiv \tau_Д$. So, τ_d is the limit of destruction of the polymer during friction loading.

The lower limit $\tau_{f\min}$ of scatter of friction fatigue limits τ_f should be set for functions (5.64) and (5.65) because the value $\tau_f > 0$. It has been revealed experimentally that the probability $P = 0.001$ approximately corresponds to the value $\tau_{f\min}$. Hence,

$$\tau_{f\min} = \tau_{fP=0.001}, \quad (5.68)$$

i.e. $\tau_{f\min}$ is the friction fatigue limit with the probability of appearance equal to $P = 0.1\%$. Thus the criterion of limitation of the damaged volume is set (2.23).

The parameter $\tau_{tf}^{(1)}$ is determined from formula (5.51). It is a single thermofluctuation stress, i.e. the stress that appears in the polymer when its temperature changes by 1 K. If the temperature increment is $T_2 - T_1 = \Delta T$ K, formula (5.52) serves to calculate the thermofluctuation stress

$$\tau_{tf} = \tau_{tf}^{(1)} \Delta T,$$

that can be introduced into functions (5.57), (5.49) and others.

The parameter of mechanical non-homogeneity (the extent of defects) of the polymer m_S is determined by the same method like the similar parameter m_V for the metallic body, yet the allowance is made for the fact that the polymer operates in the conditions of friction loading. Hence,

$$\cot \alpha_S = \frac{\log S_2 - \log S_1}{\log \tau_f^{(1)} - \log \tau_f^{(2)}} = m_S, \tag{5.69}$$

where $S_2 > S_1$ – the working volumes of the polymer.

Note that the functions of distribution in question contain probabilistic measures of damage of the metal and the polymer

$$\omega_{VP} = \frac{V_{P\gamma}}{V_0}; \quad \omega_{SP} = \frac{S_{P\gamma}}{S_k},$$

that were discussed earlier (see Sects. 2.4.1 and 2.4.2). It determines a direct link of dangerous volumes or the extent of damage of the active system's components with the probability of its failure.

The notional generality of all the parameters of the material m (m_V, m_S, m_T, etc.) should be highlighted: they are all defined as a cotangent of the angle of slope of the corresponding dependence to the abscissa axis. As a rule, it is the relation "stress–determining factor" which is a straight line in double logarithmic coordinates. Thus, these logarithmic functions play a specific role in describing the properties of materials: namely, they characterize the rate (the intensity) of changes of the characteristics of mechanophysical properties as the real combination of operation conditions (or tests) of an active system (m_T, m_σ, m_τ, etc.) and design and technological parameters of its components (m_V, m_S, etc.) vary.

Probability of failures

Function (5.49) serves to estimate the failure probability $P(\sigma, \tau_W)$ of the metal-to-polymer active system under simultaneous (joint) effect of contact (τ_W) and off-contact (σ) loads. The ultimate stresses are substituted in the function for effective stresses: $\sigma_{-1} = \sigma$ and $\tau_f = \tau_W$. Yet, since the failure probability of the active system is governed by interaction between two damaging phenomena (mechanical fatigue and friction (including wear)), an additional function of interaction $R_{T/M}$ should be introduced into (5.49) determined, for example, from (5.18). The result is the following:

$$P(\sigma, \tau_W) =$$
$$= \left[1 - \exp\left\{-C_V \frac{V_{P\gamma}}{V_0}\left[\left(\frac{T_M}{T_0}\right)^{m_T} \frac{\sigma - \sigma_{-1\min}}{\sigma_W}\right]^{m_V} - C_S \frac{S_{P\gamma}}{S_k}\left(\frac{\tau_{tf}^{(1)} \Delta T}{\tau_d - \tau_W}\right)^{m_S}\right\}\right] R_{T/M}. \tag{5.70}$$

According to this equation:

(*a*) the failure probability of the metal-to-polymer active system grows together with the growth of

- effective cyclic ($\sigma > \sigma_{-1\min}$) and contact ($\tau_W < \tau_d$) stresses;
- defects in the metal (m_V) and the polymer (m_S);

- dimensions of dangerous volumes during cyclic loading ($V_{P\gamma}$) and friction ($S_{P\gamma}$);
- the temperature of the metal (T_M) and the polymer (ΔT);
- the thermodynamic state of the metal m_T and the polymer $\tau_{tf}^{(1)}$;

(*b*) the probability of failure of the metal-to-polymer active system is the function of

- sizes and shapes of its components and the pattern of their cyclic deformation and contact interactions (C_V/V_0, C_S/S_k);
- a complex of mechanophysical properties of the polymer (τ_d, $\tau_{tf}^{(1)}$, m_S, U_0, γ_τ);
- a complex of the mechanophysical properties of the metal ($\sigma_{-1\min}$, σ_W, m_V, m_T);
- the pattern and direction of interactions between damaging phenomena ($R_{T/M}$).

So, Eq. (5.70) takes into account a whole number of phenomena and factors that shape decisively the regularities of WFD of the active system in general and its individual components.

Pay attention to three specific features of Eq. (5.70).

(1) It is not the value (σ) of cyclic stresses that determines the failure probability, it is the level to which the effective stresses exceed the lower threshold of scatter of fatigue limits. Indeed:

$$\sigma - \sigma_{-1\min} = \sigma_{-1\min}\left(\frac{\sigma}{\sigma_{-1\min}} - 1\right),$$

where the *overload factor*

$$\frac{\sigma}{\sigma_{-1\min}} = k_\sigma > 1. \tag{5.71}$$

It is of basic significance. For example, the probability of failure of the shaft from high-strength steel under the effect of the stress σ = 600 MPa and of the same shaft from carbon steel under the effect of the stress σ = 300 MPa contributes similarly providing the overload factors for both steels are the same (of course, with other equal conditions).

(2) It is not the value (τ_W) of friction stresses that determines the failure probability, it is the remoteness of the effective stress from the limit of destruction of the polymer. Indeed:

$$\tau_d - \tau_W = \tau_d\left(1 - \frac{\tau_W}{\tau_d}\right),$$

where the remoteness factor is

$$\frac{\tau_W}{\tau_d} = k_\tau < 1. \tag{5.72}$$

It is of basic significance: in order to determine the probability of failure, it is not so much essential how large is the absolute value of friction stress, it is essential how much it is less than (or how far it is remote from) the limit of destruction.

(3) In fact, the failure probability of the active system is corrected by two parameters of interaction ($R_{T/M}$, $R_{\sigma/\tau}$), each of them may be equal to, exceed or be less than a unity as a function of the complex of the real conditions of operation (or tests).

It is also basically significant that two (and more) damaging effects turn out to interact statistically in the dialectical sense. This interaction may improve the situation (reduce the failure probability) or aggravate (increase the failure probability) it. Yet, calculation with (5.70) yields $P(\sigma, \tau_W) > 1$, so it is assumed that $P(\sigma, \tau_W) = 1$. So, there are three limits set for formula (5.70): $\sigma > \sigma_{-1\min}$, $\tau_W < \tau_d$, $P(\sigma, \tau_W) \leq 1$.

Note that the allowance for time in (5.70) is made implicitly. Since the characteristics of resistance to damage are established on a definite time basis (for example, the fatigue limit is determined on the base of 10^7 loading cycles), the failure probability predicted by Eq. (5.70) relates to the same duration of operation (or tests).

In case there is no contact interaction between the body and the counterbody (hence, there is no friction), a particular formula follows from (5.70) for calculating the failure probability of the metallic body (the shaft) following the criterion of resistance to mechanical fatigue:

$$P(\sigma) = 1 - \exp\left\{ -C_V \frac{V_{P\gamma}}{V_0} \left[\left(\frac{T_M}{T_0} \right)^{m_T} \frac{\sigma - \sigma_{-1\min}}{\sigma_W} \right]^{m_V} \right\}. \tag{5.73}$$

In case there is no cyclic deformation of the body, a particular formula follows from (5.70) for estimating the failure probability of the polymeric counterbody (the sliding bearing) following the criterion of resistance to friction fatigue:

$$P(\tau_W) = 1 - \exp\left\{ -C_S \frac{S_{P\gamma}}{S_k} \left(\frac{\tau_{tf}^{(1)} \Delta T}{\tau_d - \tau_W} \right)^{m_S} \right\}. \tag{5.74}$$

When (5.73) and (5.74) are derived from (5.70), it should be naturally assumed that $R_{T/M} = 1$.

Record function (5.70) for the specific active system consisting of the *shaft/* the *sliding bearing* (cf. Fig. 2.1). Assume the shaft is made from normalized steel 45. The parameters of this grade of carbon steel are $\sigma_{-1\min} = 150$ MPa, $\sigma_W = 140$ MPa, $m_V = 16.4$. Now assume the insert of the sliding bearing is made from the polymer BKV-30H. The characteristics of the polymer are $\tau_d = 49.5$ MPa, $\tau_{tf}^{(1)} = 0.21$ MPa, $m_S = 4.6$, its thermodynamic condition in the specified conditions of operation (or tests) are characterized by the temperature increment $\Delta T = 60 - 20 = 40$ °C, so that $\tau_{tf} = 8.4$ MPa. The estimates of two other parameters are $\eta_\sigma = 0.016$, $\eta_\tau = 0.12$. Record (5.70) with the account of the specified parameters:

$$P(\sigma,\tau_W) = 1 - \exp\left[-0.016\left(\frac{\sigma - 150}{160}\right)^{16.4} - 0.12\left(\frac{8.4}{49.5 - \tau_W}\right)^{4.6}\right]. \tag{5.75}$$

Assigning $\sigma > \sigma_{-1\min} = 150$ MPa and $\tau_W < \tau_d = 49.5$ MPa, we plot surface (5.75) (Fig. 5.13). Figure 5.14 shows two views of the surface: τ_W along the arrow (cf. Fig. 5.14, *a*) that indicates growth of friction stresses within the range from 38 to 46 MPa and along the arrow σ that indicates growth of cyclic stresses within the range from 280 to 360 MPa.

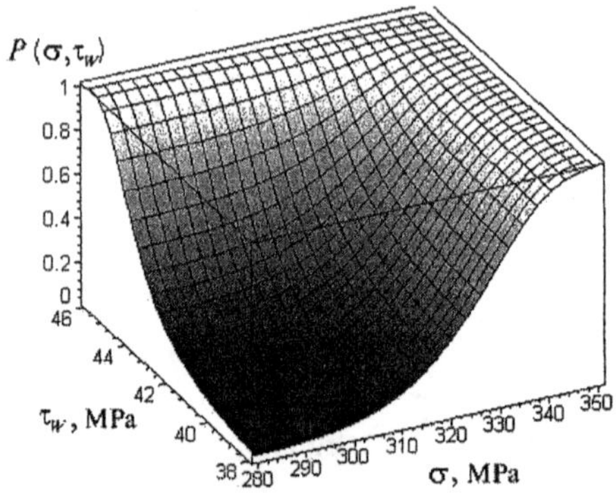

Fig. 5.13. Two-dimensional function of distribution $P(\sigma, \tau_W)$ according to (5.75)

Compile a PC program and calculate the probability of failure of the metal-to-polymer active system using formula (5.75) when σ varies from 200 to 326.3 MPa and τ_W from 28 to 46 MPa providing that $R_{T/M} = 1$ (Table 5.4). Also calculate the probability of failures $P(\sigma)$ and $P(\tau_W)$ using formulas (5.73) and (5.74).

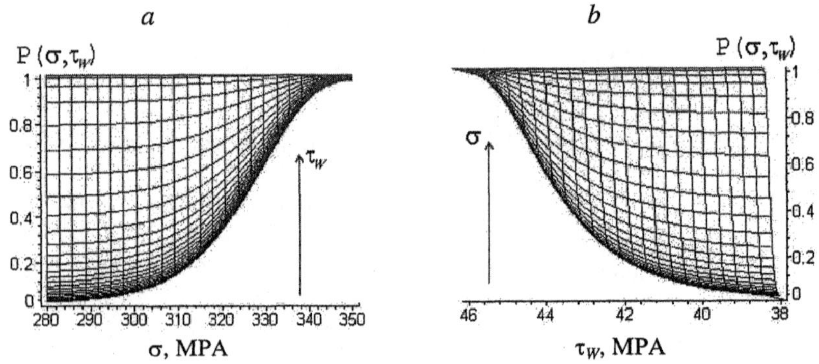

Fig. 5.14. Effect of value of friction stresses τ_W on changes of function $P(\sigma, \tau_W)$ during variations of cyclic stresses within range from 280 to 350 MPa (*a*) and effect of cyclic stresses σ on changes of function $P(\sigma, \tau_W)$ when friction stresses vary within range from 38 to 46 MPa (*b*)

Table 5.4. Probability of failures $P(\sigma)$, $P(\tau_W)$ and $P(\sigma,\ \tau_W)$ during independent development of damages due to contact and cyclic (off-contact) loads

τ_W, MPa	σ, MPa						
	0	200	240	270	300	326,3	
0		$7.0\cdot10^{-10}$	$1.14\cdot10^{-5}$	$1.28\cdot10^{-3}$	$4.84\cdot10^{-2}$	0.504	$P(\sigma)$
28	$1.58\cdot10^{-3}$	$1.58\cdot10^{-3}$	$1.60\cdot10^{-3}$	$2.86\cdot10^{-3}$	$4.99\cdot10^{-2}$	0.505	⇑ $P(\tau_W,\ \sigma = \text{const})$
32	$4.09\cdot10^{-3}$	$4.09\cdot10^{-3}$	$4.10\cdot10^{-3}$	$5.36\cdot10^{-3}$	$5.23\cdot10^{-2}$	0.506	
35	$9.69\cdot10^{-3}$	$9.69\cdot10^{-3}$	$9.70\cdot10^{-3}$	$1.10\cdot10^{-2}$	$5.76\cdot10^{-2}$	0.509	
38	$2.78\cdot10^{-2}$	$2.78\cdot10^{-2}$	$2.79\cdot10^{-2}$	$2.91\cdot10^{-2}$	$7.49\cdot10^{-2}$	0.518	
42	0.182	0.182	0.183	0.184	0.223	0.595	
46	0.99880	0.99880	0.99880	0.99880	0.99887	0.99940	
	$P(\tau_W)$	$P(\sigma,\ \tau_W = \text{const})$ ⇐					$P(\sigma,\ \tau_W)$

From Fig. 5.14 and Table 5.4 three basic conclusions follow.

(1) If friction stresses grow, the relation between $P(\sigma,\ \tau_W)$ and the value of cyclic stresses σ weakens, and in the extreme case when $\tau_W \to \tau_d$, the role of cyclic stresses in generating the failure probability loses its significance. For example, if $\tau_W = 46$ MPa, the probability of failure of the system changes little at any value of cyclic stresses.

(2) If friction stresses grow, the relation between $P(\sigma,\ \tau_W)$ and the value of friction stresses τ_W weakens, and in the extreme case when $\sigma > 350$ MPa, the role of friction stresses in generating the failure probability loses its significance. For example, if $\sigma = 326.3$ MPa, the growth of friction stresses from 28 to 40 MPa affects little the value of the system's failure probability.

(3) In any one-dimensional case (i.e. either during mechanical or during friction fatigue) the failure probability is less than in the two-dimensional case at similar stresses and $R_{T/M} = 1$, of course.

Note that this analysis yields the values of mean stresses $\overline{\sigma} = \overline{\sigma}_{-1} \approx 326$ MPa and $\overline{\tau}_W = \overline{\tau}_f \approx 43.8$ MPa.

Now examine the effect of interactions between damages on the failure probability of the metal-to-polymer active system. We proceed from an assumption that the ambient temperature is unchangeable. Then instead of the general function of interactions $R_{T/M}$ it is enough to consider only the function $R_{\sigma/\tau}$, that allows for the interaction between damages during friction and cyclic deformation.

Record an expression of type (5.36) for the function of interaction at $C = \pm 1$ with the allowance for the mean ultimate stresses ($\overline{\sigma}_{-1}, \overline{\tau}_f$):

$$R_{\sigma/\tau} = 1 \pm \frac{\overline{\sigma}_{-1}}{\overline{\tau}_f}\frac{\tau_W}{\sigma}\exp\left[-\frac{\overline{\sigma}_{-1}}{\overline{\tau}_f}\frac{\tau_W}{\sigma}\right], \qquad (5.76)$$

and introduce into it numerical values of the parameters

$$R_{\sigma/\tau} = 1 \pm \frac{326}{43.8}\frac{\tau_W}{\sigma}\exp\left[-\frac{326}{43.8}\frac{\tau_W}{\sigma}\right]. \tag{5.76a}$$

Figure 5.15 presents a graphic image of the function of interaction.

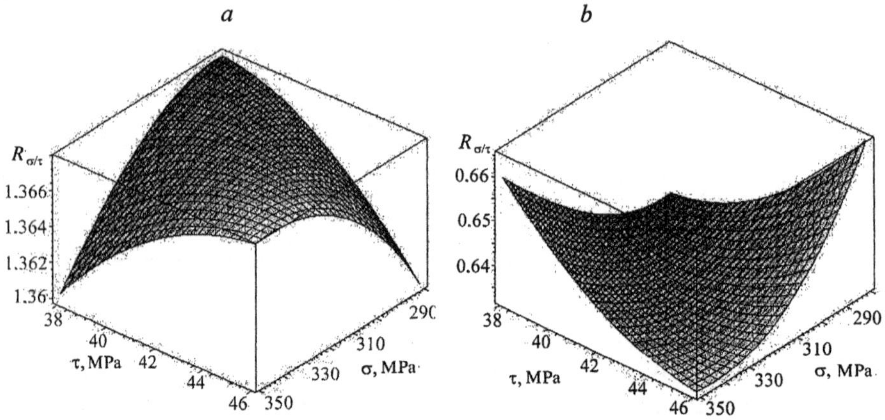

Fig. 5.15. Graphic representation of function (5.76a)

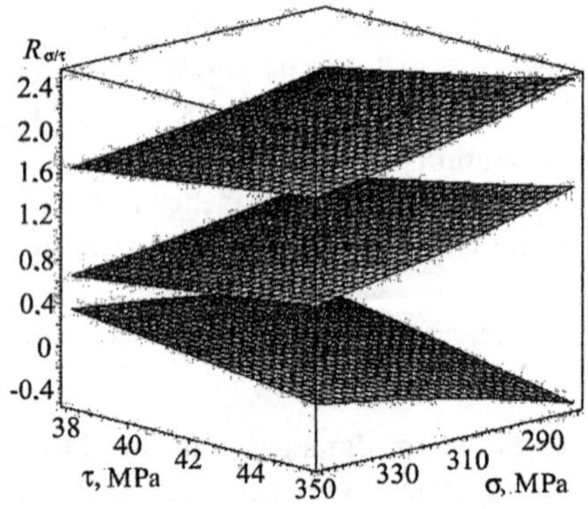

Fig. 5.16. Graphic representation of function (5.36) at $C = 1$

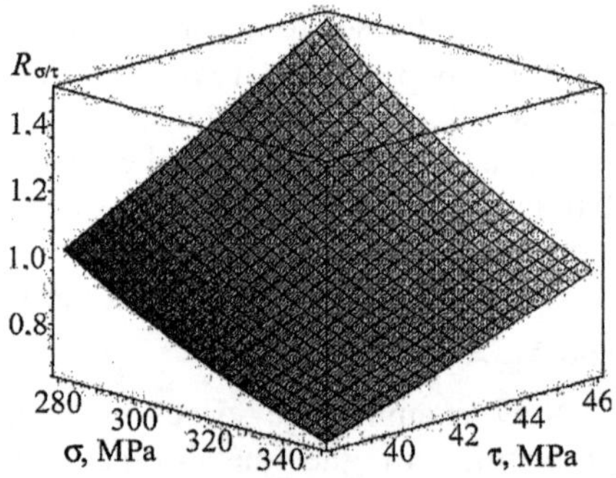

Fig. 5.17. Graphic representation of function (5.33)

Figures 5.16 and 5.17 present graphically their two other modifications; they all describe possible cases of interactions between damages: (1) $R_{\sigma/\tau} < 1$ (cf. Figs. 5.15, *a*; 5.16 – the upper surface), (2) $R_{\sigma/\tau} < 1$ (cf. Figs. 5.15, *b*; 5.16 – the lower surface) and (3) $R_{\sigma/\tau} \gtrless 1$ (cf. Fig. 5.16 – the middle surface; 5.17). It is apparent that they satisfy fully the requirements to λ-functions of interactions between damages (see Sect. 2.5).

5.2.2 Metal-to-metal active system

It is apparent that solutions disclosed above can be obtained for the metal-to-metal active system like for the metal-to-polymer system providing tribological function (5.54) in the corresponding equations is substituted for another function more suitable for describing resistance to friction fatigue (and friction damage) of the metallic counterbody (the insert of the sliding bearing). This function can be constructed in terms of the coefficient of intensity of stresses K (see Sect. 1.3.4) in the following manner:

$$\varphi_K = C_\Omega \frac{\Omega_{P\gamma}}{\Omega_k}\left(\frac{K_{\max} - K_{th}}{K_{fc} - K_{\min}}\right)^{m_\Omega}. \tag{5.77}$$

Here $K_{\max}$, $K_{\min}$ – the maximum and minimum values of the cycle of changes of the value K in time; K_{th}, K_{fc} – the threshold and ultimate values of the coefficient of intensity of stresses; m_Ω – the parameter of mechanical homogeneity (isotropy) of the material determined in the local zone – damaged volume $\Omega_{P\gamma}$ ahead of the front of the fatigue crack; Ω_k – the working volume of the insert ($\Omega_k = S_k$ can be assumed).

Without repeating the full analysis made above for the metal-to-polymer active system, we just provide the determining equation for calculating the probability of failure of the metal-to-metal active system using the given time base:

$$P(\sigma, K) = \left(1 - \exp\left\{-C_V \frac{V_{P\gamma}}{V_0}\left[\left(\frac{T_0}{T_M}\right)^{m_T} \frac{\sigma - \sigma_{-1\min}}{\sigma_W}\right]^{m_V} - C_\Omega \frac{\Omega_{P\gamma}}{\Omega_0}\left(\frac{K_{\max} - K_{th}}{K_{fc} - K_{\min}}\right)^{m_\Omega}\right\}\right) R_{T/M}. \tag{5.78}$$

Function (5.78) can be analyzed similarly as it was done in respect of the metal-to-polymer active system.

5.2.3 System of reliability conditions

A generalized system of reliability conditions is based on the statistical model of a deformable solid with a dangerous volume (see Sects. 2.3 and 2.4).

The system is based (Table 5.5) on the fundamental idea that the damaged volume is equal to zero providing the field of effective stresses does not induce damage, hence the failure probability is $P = 0$. On the contrary, if the failure probability appears only when the condition occurs that this or that component of the active system with some probability shows a corresponding dangerous volume ($V_{P\gamma}, S_{P\gamma}$); it is equivalent to the appearance of nonzero damage ($\omega > 0$). The complex dangerous volume $W_{P\gamma}$ during WFD is determined as a corresponding function of particular dangerous volumes $V_{P\gamma}$ and $S_{P\gamma}$ with the allowance for interactions between damaging phenomena (lambda-function (2.59)).

Table 5.5. System of reliability conditions

Damage	Dangerous volume	Condition of failure-free operation	Measure of damage	Conditions of damage and fracture
Mechanical fatigue	$V_{P\gamma} = \iiint\limits_{\sigma(x,y,z)>\sigma_{-1\min}} dxdydz$	$V_{P\gamma} = 0$	$\omega_{VP} = \dfrac{V_{P\gamma}}{V_0}$	$0 < \omega_{VP} \le 1$
Friction and wear	$S_{P\gamma} = \iiint\limits_{\tau_W(x,y,z)>\tau_{f\min}} dxdydz$	$S_{P\gamma} = 0$	$\omega_{SP} = \dfrac{S_{P\gamma}}{S_k}$	$0 < \omega_{SP} \le 1$
Wear-fatigue damage	$W_{P\gamma} = (V_{P\gamma} \cup S_{P\gamma})\varphi_{SV}$	$W_{P\gamma} = 0$	$\omega_{WP} = \dfrac{W_{P\gamma}}{V}$	$0 < \omega_{WP} \le 1$

Based on the model of the deformable body with a dangerous volume the probabilistic indicators of similarity Π_{VP}, Π_{SP} and Π_{WP} are established that serve to model active systems (cf. formulas (2.22), (2.27) and (2.57)). The general sense of the indicators Π is the following: if a given indicator has an identical numerical value for the objects of different standard sizes, these objects are similar, i.e. they similarly resist damage (fracture) following this criterion.

The reliability condition (exemplified by the metal-to-polymer active system) is

$$P(\sigma, \tau_W) \le [P] \tag{5.79}$$

or

$$Q(\sigma, \tau_W) = 1 - P(\sigma, \tau_W) \ge [Q], \tag{5.80}$$

where $[P]$ – the normative failure probability and $[Q]$ – the normative probability of failure-free operation $Q(\sigma, \tau_W)$.

Note that with the function of failure probability of type (5.70) it is possible to solve the *problem of optimization*:

$$\left.\begin{aligned} &P(\sigma,\tau_W) \Rightarrow \min;\\ &C_0(\sigma_{-1\min}, \sigma_W, m_V; \tau_d, \tau_{tf}^{(1)}, m_S) \Rightarrow \min, \end{aligned}\right\} \tag{5.81}$$

where $C_0(\sigma_{-1\min}, \sigma_W, m_V; \tau_d, \tau_{tf}^{(1)}, m_S)$ – total cost of labor, means and materials in the sphere of its production and operation. Of course, target function (5.81) should be supplemented with corresponding restrictions of all the parameters and values used to formulate the optimization problem.

5.3 Service life

5.3.1 Regular loading

Technical service life is the indicator of durability implying the time of operation of the active system since it is commissioned until it reaches the limiting state. The number of cycles (or blocks) of loading or duration of operating time can be the measure (the characteristic) of service life.

In case of regular loading the service life can be estimated if the equation of corresponding fatigue curves is known. In the general case two types of fatigue curves can be plotted for the active system (see Sect. 3.2.2): 1) curve $\sigma - N_{\sigma\tau}$ based on the parameter $\tau_W = \text{const} > 0$ (Fig. 5.18, *a*) and 2) curve $\tau_W - N_{\tau\sigma}$ based on the parameter $\sigma = \text{const} > 0$ (Fig. 5.18, *b*). The former is plotted when examining the direct effect, the latter is plotted for the back effect.

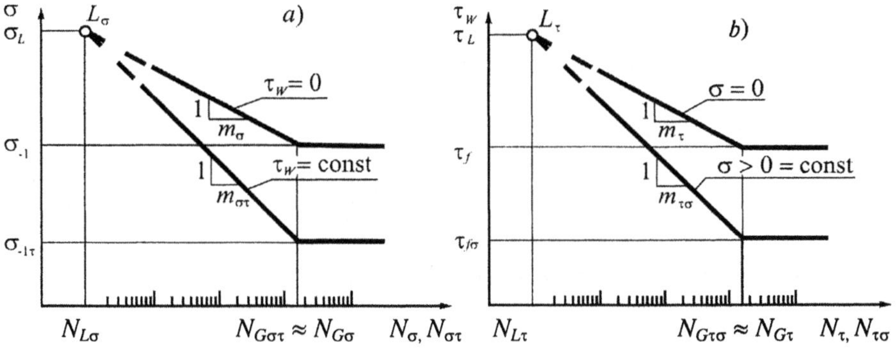

Fig. 5.18. Schemes of mutual arrangement of curves of sliding, mechanical and mechano-sliding fatigue

The problem is formulated in the following way: assume that the common mechanical fatigue curves are known as $\sigma - N_\sigma$ (at $\tau_W = 0$ – cf. Fig. 5.18, *a*) and sliding fatigue curve is known as $\tau_W - N_\tau$ (at $\sigma = 0$ – cf. Fig. 5.18, *b*). It means that their basic parameters are known: σ_{-1}, $N_{G\sigma}$, m_σ and τ_f, $N_{G\tau}$, m_τ respectively. It is required to find similar parameters of the curves of mechano-sliding fatigue $\sigma_{-1\tau}$, $N_{G\sigma\tau}$, $m_{\sigma\tau}$ and $\tau_{f\sigma}$, $N_{G\tau\sigma}$, $m_{\tau\sigma}$ (cf. Fig. 5.18, *a*, *b*). Let us make three assumptions in order to solve the problem. The first assumption is that the corresponding fatigue curves in the low-cycle region converge in a single point $L_\sigma(\sigma_L, N_{L\sigma})$ and $L_\tau(\tau_L, N_{L\tau})$. The second assumption is that the abscissas of the breakpoint of the corresponding fatigue curves are identical: $N_{G\sigma\tau} = N_{G\sigma}$ and $N_{G\tau\sigma} = N_{G\tau}$ (cf. Fig. 5.18). Finally, it is assumed that all the fatigue curves can be justifiably described by an exponential equation.

Assume the direct effect (σ = var, τ_W = const) is studied when the limiting state of the system appears following the criterion of appearance of a main fatigue crack in a component of the system. Equation of the fatigue curve in this case (cf. Fig. 5.18, *a*) is

$$N_{\sigma\tau} = \left(\frac{\sigma_{-1\tau}}{\sigma}\right)^{m_{\sigma\tau}} N_{G\sigma\tau}\,. \tag{5.82}$$

Two parameters are to be determined in Eq. (5.82). The first is ultimate stress σ_{-1} determined from formulas (5.23) or (5.29a). The second unknown parameter ($m_{\sigma\tau}$) can be estimated with the following relation:

$$m_{\sigma\tau} = m_\sigma \varphi_m(\tau_W)\,, \tag{5.83}$$

combining the indicator of the slope $m_{\sigma\tau}$ of mechano-sliding fatigue curve with the indicator of the slope m_σ of the mechanical fatigue curve. The scheme of mutual arrangement of the curves of mechanical fatigue and WFD (Fig. 5.18, *a*) is used to determine the function $\varphi_m(\tau_W)$ of effect of friction stresses. For the point L_σ the following equality is true:

$$\left(\frac{\sigma_{-1}}{\sigma_L}\right)^{m_\sigma} N_{G\sigma} = \left(\frac{\sigma_{-1\tau}}{\sigma_L}\right)^{m_{\sigma\tau}} N_{G\sigma\tau}\,, \tag{5.84}$$

when at $_{G\sigma\tau} = N_{G\sigma}$, we obtain

$$m_{\sigma\tau} = m_\sigma \frac{\log(\sigma_{-1}/\sigma_L)}{\log(\sigma_{-1\tau}/\sigma_L)}\,, \tag{5.85}$$

i.e., according to (5.83), (5.85), the required function is

$$\varphi_m(\tau_W) = \frac{\log(\sigma_{-1}/\sigma_L)}{\log(\sigma_{-1}\varphi_\sigma(\tau_W)/\sigma_L)}\,. \tag{5.86}$$

In practical calculations it can be assumed that $\sigma_L \approx 0.8\sigma_b$.

The analysis of (5.86) reveals that at $\sigma_{-1\tau} < \sigma_{-1}$ the function $\varphi_m(\tau_W) < 1$, it means that according to (5.83) $m_{\sigma\tau} < m_\sigma$, i.e. the slope of the left branch of the mechano-sliding fatigue curve is steeper than that of the left branch of the mechanical fatigue curve. If $\sigma_{-1\tau} > \sigma_{-1}$, then from (5.86) it follows that $\varphi_m(\tau_W) > 1$, and according to (5.83), $m_{\sigma\tau} > m_\sigma$, i.e. the slope of the left branch of the mechano-sliding fatigue curve is less, on the contrary, than the slope of the same branch of the mechanical fatigue curve .

Hence, equation of durability (5.82) can be rewritten with the account of (5.86) and (5.29a) in the following form:

$$N_{\sigma\tau} = \left(\frac{\sigma_{-1}\varphi_\sigma(\tau_W)}{\sigma}\right)^{m_\sigma \cdot \varphi_m(\tau_W)} N_{G\sigma}. \tag{5.87}$$

Expression (5.87) enables to calculate the durability $N_{\sigma\tau}$ of a component of the active system during mechano-sliding fatigue using the characteristics of resistance to mechanical (σ_{-1}, m_σ, $N_{G\sigma}$) and sliding ($\varphi_\sigma(\tau_W)$, $\varphi_m(\tau_W)$) fatigue with the account of the level (σ, τ_W) of its loading, including the conditions of friction too, because $\tau_W = f_S\, p_a$.

Assume the back effect is studied during WFD (τ_W = var, σ = const) when the limiting state of the active system occurs following the criterion of attainment of ultimate wear. The exponential equation of the curve of complex damage is recorded as follows:

$$N_{\tau\sigma} = \left(\frac{\tau_{f\sigma}}{\tau_W}\right)^{m_{\tau\sigma}} N_{G\tau\sigma}. \tag{5.88}$$

The procedure of rapid assessment of the values $m_{\tau\sigma}$, $\tau_{f\sigma}$ is in principle not different from the above procedure of assessment of $m_{\sigma\tau}$, $\sigma_{-1\tau}$ for the direct effect. The value $\tau_{f\sigma}$ can be determined with the help of formulas (5.24) or (5.31) providing the sliding fatigue limit τ_f and the function $\varphi_\tau(\sigma)$ are known.

For determining the parameter of slope $m_{\tau\sigma}$ of the WFD curve the relation similar to (5.83) can be used:

$$m_{\tau\sigma} = m_\tau \varphi_m(\sigma), \tag{5.89}$$

where m_τ – the indicator of slope of the sliding fatigue curve. In order to determine the function $\varphi_m(\sigma)$ of effect of cyclic stresses, let us assume that the scheme of mutual arrangement of fatigue curves during the direct effect (cf. Fig. 5.18, *a*) remains true during the back effect (cf. Fig. 5.18, *b*). Then for $\varphi_m(\sigma)$ we obtain

$$\varphi_m(\sigma) = \frac{\log(\tau_f / \tau_L)}{\log(\tau_f \varphi_\tau(\sigma) / \tau_L)}. \tag{5.90}$$

As it follows from (5.90), at $\tau_{f\sigma} < \tau_f$ the function $\varphi_m(\sigma) < 1$, hence from (5.89) $m_{\tau\sigma} < m_\tau$, i.e. the slope of the left branch of the WFD curve is steeper than that of the left branch of the sliding fatigue curve. If $\tau_{f\sigma} > \tau_f$, then $\varphi_m(\sigma) > 1$, it means from (5.89) that $m_{\tau\sigma} > m_\tau$, i.e. the slope of the left branch of the WFD curve is, on the contrary, less than the slope of the same branch of the sliding fatigue curve.

Record equation of durability (5.88) with the account of (5.89) and (5.31) as

$$N_{\tau\sigma} = \left(\frac{\tau_f\, \varphi_\tau(\sigma)}{\tau_W}\right)^{m_\tau \varphi_m(\sigma)} N_{G\tau}. \tag{5.91}$$

Hence, Eq. (5.91) enables to calculate the durability $N_{\tau\sigma}$ of a component of the active system during WFD based on the characteristics of resistance to frictional (τ_f, m_τ, $N_{G\tau}$) and mechanical ($\varphi_\tau(\sigma)$, $\varphi_m(\sigma)$) fatigue with the account of the level (σ, τ_W) of its loading.

5.3.2 Block loading

Similarly (1.56), introduce the kinetic (time) function of WFD accumulation in the active system

$$\omega_{\Sigma n} = \left[1 - \left(1 - \frac{n}{N}\right)^h\right]^q = \omega(\sigma, \tau_W), \tag{5.92}$$

where $n = n\,(\sigma, \tau_W)$ – the number of loading cycles of the active system, $N = N\,(\sigma, \tau_W)$ – the number of cycles before the limiting state shows any sign, i.e. the durability of the active system; $h \geq 1$, $q \geq 1$ – the *parameters of softening – hardening of the material*. Function (5.92) predicts a linear law of WFD accumulation at $h = 1$, $q = 1$; non-linear softening at $q = 1$, $h \geq 1$; non-linear hardening at $q > 1$, $h = 1$; complex processes of hardening – softening at $h > 1$, $q > 1$.

Let us consider two modes of loading of the active system. Mode *I* (Fig. 5.19, *a*): block loading with cyclic stresses σ_i ($i = 1, 2, 3, ..., s$ – the number of loading steps in a block), while the frictional stress remains unchanged ($\tau_W = \text{const} > 0$). It is the problem of direct effect: it is required to estimate the fatigue durability of the system $N_{\sigma\tau\Sigma}$ with the allowance for the effect of frictional stresses τ_W. Mode *II* (Fig. 5.19, *b*): block loading with frictional stresses τ_{Wj} ($j = 1, 2, 3, ..., r$ – the number of loading steps in a block), while the cyclic stresses remain regular ($\sigma = \text{const} > 0$). It is the problem of back effect: it is required to estimate the durability of the system $N_{\tau\sigma\Sigma}$ based on the wear resistance criteria with the allowance for the effect of cyclic stresses σ.

Function (5.92) is transformed with the account of (5.87), (5.91) as follows:

(*a*) when studying the direct effect (the criterion of the limiting state is nucleation of the main crack) –

$$\omega_{\Sigma n}(\sigma, \tau_W = \text{const}) = \left[1 - \left(1 - \frac{n}{N_{\sigma\tau}}\right)^{h_\sigma}\right]^{q_\sigma} =$$
$$= \left[1 - \left(1 - \left(\frac{\sigma}{\sigma_{-1}\,\varphi_\sigma(\tau_W)}\right)^{m_\sigma \varphi_m(\tau_W)} \frac{n}{N_{G\sigma}}\right)^{h_\sigma}\right]^{q_\sigma}; \tag{5.93}$$

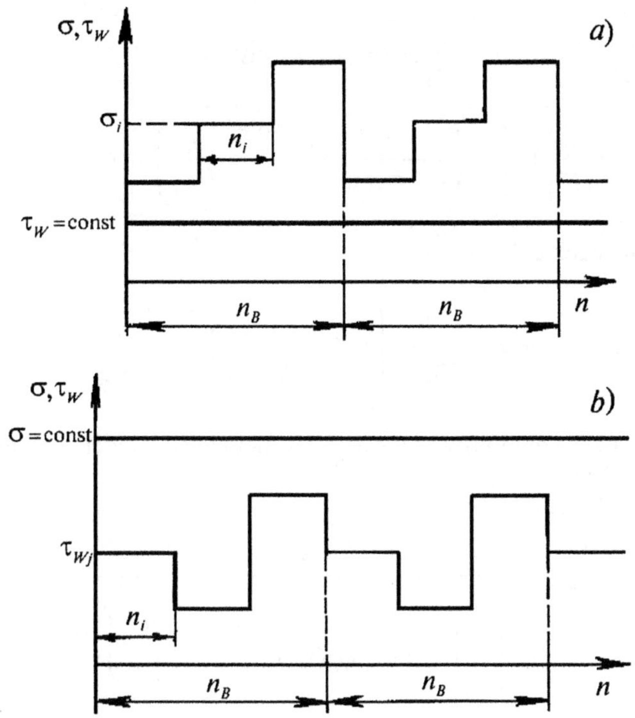

Fig. 5.19. Schemes of block loading: *a* – mode *I*; *b* – mode *II*

(*b*) when studying the back effect (the criterion of the limiting state is the appearance of the ultimate wear) –

$$\omega_{\Sigma n}(\tau_W, \sigma = \text{const}) = \left[1 - \left(1 - \frac{n}{N_{\tau\sigma}}\right)^{h_\tau}\right]^{q_\tau} =$$

$$= \left[1 - \left(1 - \left(\frac{\tau_W}{\tau_f \ \varphi_\tau(\sigma)}\right)^{m_{\tau T}\varphi_m(\sigma)} \frac{n}{N_{G\tau}}\right)^{h_\tau}\right]^{q_\tau}. \tag{5.94}$$

The conditions of appearance of the limiting state during block loading in modes I and II (cf. Fig. 5.19) are the following with the account of functions (5.93) and (5.94):

(*a*) for the direct effect –

$$\sum_{i=1}^{s}\omega_{\Sigma}(\sigma_i,\tau_W=\text{const})=$$
$$=\sum_{i=1}^{s}\left[1-\left(1-\left(\frac{\sigma_i}{\sigma_{-1}\,\varphi_\sigma(\tau_W)}\right)^{m_\sigma\varphi_m(\tau_W)}\frac{n_i}{N_{G\sigma}}\right)^{h_\sigma}\right]^{q_\sigma}=1\,; \tag{5.95}$$

(*b*) for the back effect –

$$\sum_{j=1}^{r}\omega_{\Sigma}(\tau_{Wj},\sigma=\text{const})=\sum_{j=1}^{r}\left[1-\left(1-\left(\frac{\tau_{Wj}}{\tau_f\,\varphi_\tau(\sigma)}\right)^{m_\tau\varphi_m(\sigma)}\frac{n_j}{N_{G\tau}}\right)^{h_\tau}\right]^{q_\tau}=1\,. \tag{5.96}$$

Formulas of the durability of the active system during block loading ensue from (5.95) and (5.96):

(*a*) for the direct effect –

$$N_{\sigma\tau\Sigma}=\frac{n_B}{\sum_{i=1}^{s}\left[1-\left(1-\left(\frac{\sigma_i}{\sigma_{-1}\,\varphi_\sigma(\tau_W)}\right)^{m_\sigma\varphi_m(\tau_W)}\frac{n_i}{N_{G\sigma}}\right)^{h_\sigma}\right]^{q_\sigma}}\,; \tag{5.97}$$

(*b*) for the back effect –

$$N_{\tau\sigma\Sigma}=\frac{n_B}{\sum_{j=1}^{r}\left[1-\left(1-\left(\frac{\tau_{Wj}}{\tau_f\,\varphi_\tau(\sigma)}\right)^{m_\tau\varphi_m(\sigma)}\frac{n_j}{N_{\tau G}}\right)^{h_\tau}\right]^{q_\tau}}\,. \tag{5.98}$$

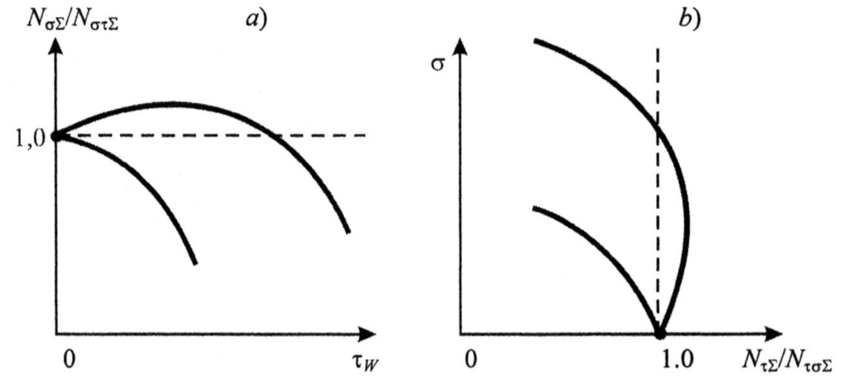

Fig. 5.20. Analysis of durability of active systems during block loading

If it is assumed in Eq. (5.98) that $\varphi_m(\sigma) = 1$ and $\varphi_\tau(\sigma) = 1$, the equation yields (pure) wear durability $N_{\tau\Sigma}$ during block loading. If it is similarly assumed in Eq. (5.97) that $\varphi_\sigma(\tau_W) = 1$ and $\varphi_m(\tau_W) = 1$, the equation yields (pure) fatigue durability $N_{\sigma\Sigma}$ during block loading. The analysis shows that in response to the conditions of tests: *a*) during direct effect the effect of frictional stresses is contradictory: $N_{\sigma\tau\Sigma}$ can be smaller or larger than $N_{\sigma\Sigma}$ (Fig. 5.20, *a*); *b*) during back effect the effect of cyclic stresses is also contradictory: $N_{\tau\sigma\Sigma}$ can be smaller or larger than $N_{\tau\Sigma}$ (Fig. 5.20, *b*).

5.3.3 Random loading

Let us calculate the durability of the active system with the loading as shown in Fig. 5.21 that is similar to Fig. 5.19. Assuming $h = q = 1$, the following expression for mean durability is obtained:

(*a*) during direct effect (Fig. 5.21, *a* – the criterion of the limiting state is appearance of the main crack) –

$$N_{\sigma\tau\Sigma} = [\sigma_{-1}\varphi_\sigma(\tau_W)]^{m_\sigma \varphi_m(\tau_W)} N_{G\sigma} \left\{ \int\limits_{\sigma_{-1}\varphi(\tau_W)}^{\sigma_{\max}} \overline{\sigma}^{m_\sigma \varphi_m(\tau_W)} \varphi(\sigma) d\sigma \right\}^{-1} ; \quad (5.99)$$

(*b*) during back effect (Fig. 5.21, *b* – the criterion of the limiting state is ultimate wear in the active system) –

$$N_{\tau\sigma\Sigma} = [\tau_f \varphi_\tau(\sigma)]^{m_\tau \varphi_m(\sigma)} N_{G\tau} \left\{ \int\limits_{\tau_f \varphi_\tau(\sigma)}^{\tau_{W\max}} \overline{\tau}^{m_\tau \varphi_m(\sigma)} \varphi(\tau_W) d\tau_W \right\}^{-1} , \quad (5.100)$$

where σ and τ_W – mean values of effective stresses.

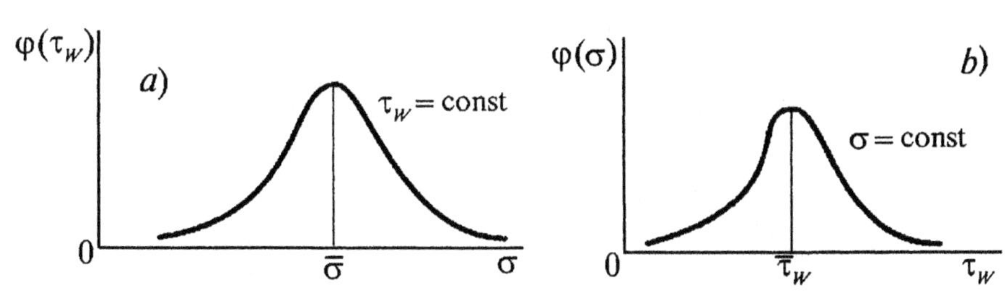

Fig. 5.21. Schemes of random loading of active system during direct (*a*) and back (*b*) effects

If the functions $\varphi(\sigma)$ and $\varphi(\tau_W)$ are approximated with some theoretical laws of distribution, the integrals in formulas (5.99) and (5.100) can be expressed through the tabulated function of the integral of probabilities χ^2:

$$P(\chi, n) = \frac{1}{2^{\frac{n}{2}} \Gamma\left(\frac{n}{2}\right)} \int_{x}^{\infty} y^{\frac{n}{2}-1} e^{-\frac{y}{2}} dy .$$

The routines of calculating the above integrals using the function $P(\chi, n)$ for various theoretical laws of distribution are known.

5.4 Force and coefficient of friction

The frictional force in an active system can be treated as a function of the common force of friction in sliding F_S, that appears in the peripheral direction (the object – the friction pair) and the cyclic component F_C of the frictional force that appears additionally due to excitation of cyclic stresses (deformation) in the axial direction (cf. Fig. 2.1, *a*).

Friction in the active system in accordance with Fig. 2.1, *a* can be described by four individual (independent) characteristics:

– the force of friction in sliding

$$F_S = f_S F_N = \tau_W A_a = \gamma_S G A_a ; \tag{5.101}$$

– the friction coefficient in sliding

$$f_S = \frac{F_S}{F_N} = \frac{\tau_W A_a}{F_N} = \frac{\tau_W}{p_a} = \frac{\gamma_S G}{p_a} ; \tag{5.102}$$

– the cyclic force of friction

$$\pm F_C(t) = f_C(t) F_N = \sigma(t) A_a = \varepsilon(t) E A_a = M(t) \frac{A_a}{W} ; \tag{5.103}$$

– the cyclic friction coefficient

$$\pm f_C(t) = \frac{F_C(t)}{F_N} = \frac{\sigma(t) A_a}{F_N} = \frac{\varepsilon(t)}{p_a} = \frac{M(t)}{F_N} \cdot \frac{A_a}{W} . \tag{5.104}$$

The nomenclature in formulas (5.101)–(5.104) is the following: f_S – the friction coefficient; F_N – contact load; p_a – contact pressure; $\tau_W = f_S p_a$ – the specific force of friction or frictional stress; A_a – the contact area; γ_S – relative shear; G – the shear modulus; $\sigma = M/W$ – the maximum cyclic stress; ε – relative deformation (tension – compression); E – the modulus of normal elasticity; M – the bending moment; W – the moment of resistance to bending.

These four characteristics (5.101)–(5.104) can be reduced to two relative characteristics of friction in the active system

$$\pm F_C / F_S = \pm F_{C/S} \text{ and } \pm f_C / f_S = \pm f_{C/S} ,$$

that are quite apparently have the identical numerical value

$$\pm F_{C/S} = \pm f_{C/S} = \pm\sigma / \tau_W .$$

So, while friction in the active system is a more intricate phenomenon than friction in (simple) sliding, it is nevertheless can be described by a single characteristic

$$f_F = \pm\sigma / \tau_W = \sigma / f_S p_a = \psi_{\sigma\tau} , \tag{5.105}$$

that is the *complex friction indicator* (in the active system); it is called *friction index (in the active system)*. In fact, the friction index f_F is the governing parameter of WFD (5.25).

The friction index has the simplest form but complex nature. It allows simultaneously for the "normal effect" (through p_a), the "friction effect" (through τ_W or f_S) and the "cyclic effect" (through $\pm\sigma$) in the processes of friction in the active system.

The value inverse to the friction index (5.105) is called the *parameter of friction in the active system:*

$$\pm f_{\tau/\sigma} = f_S \frac{p_a}{\sigma} . \tag{5.106}$$

It is easy to show that parameter (5.106) is linked with the parameter $\overline{\rho}$ of asymmetry of WFD (5.33) and critical value (5.26) of the controlling parameter ψ_k:

$$f^2_{\tau/\sigma} = \frac{\overline{\rho}}{\psi^2_{-1f}} . \tag{5.107}$$

As it follows from (5.106), the friction parameter can be more, less or equal to the common friction coefficient:

$$\pm f_{\sigma/\tau} \gtreqless f_S ,$$

because the ratio between contact and cyclic stresses is $fp_a / \sigma \gtrless 1$. It should be emphasized that the friction parameter does not depend on the level of fp_a and σ, it depends only on the ratio between them.

Thus, a general conclusion can be made: the friction index (or parameter) in the active system reflects the level (or extent) of complex WFD of components of the active system and therefore the pattern of their wear.

Note that friction index (5.105) can apparently be applicable to the processes of elastoplastic deformation too. The laws of plasticity in tension (compression) and shear yield

$$f_F = \varepsilon E_p / \gamma_S G_p ,$$

where E_p, G_p – the moduli of plasticity.

Now let us consider friction in the active system in case of direct and back effects.

We will use the methods of the theory of elasticity for engineering analysis of the stress-strain state in the point located close to the surface of the shaft. We cut out an elementary parallelepiped in the vicinity of this point and assume with some approximation that the triaxial state of stress is characterized by stresses σ_0, σ_n, σ_θ , as Fig. 5.22 shows it. Cyclic stresses ($\pm\sigma_0$) due to the bending moment are directed along the axis O_1O of the shaft and are applied to the sites perpendicular to this axis. Compressive stresses ($-\sigma_n$) are identified with the mean contact pressure p_a, acting normally (n) to the contact surface. Tensile stresses ($+\sigma_\theta$) active under the surface are detected peripherally; in some cases their value is close to the value of frictional stresses τ_W. Our task is to disclose the *procedure of analyzing the direct and back effects during friction in the active system* rather than to obtain precise estimates of theses stresses.

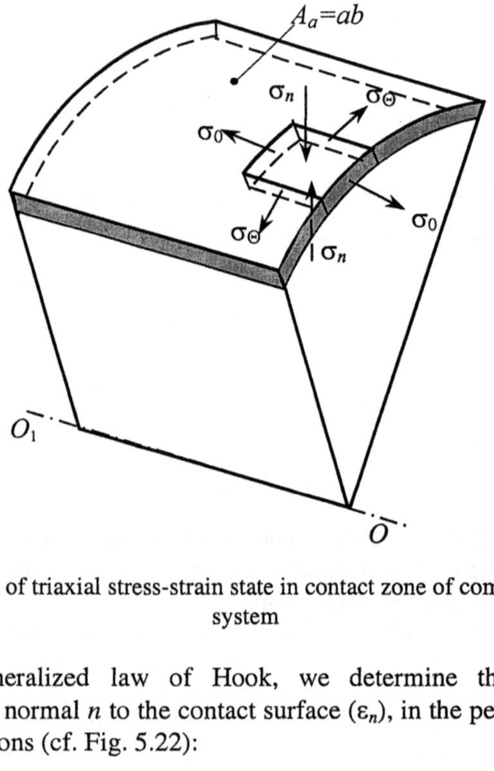

Fig. 5.22. Scheme of triaxial stress-strain state in contact zone of components of active system

Using the generalized law of Hook, we determine the relative linear deformation along normal n to the contact surface (ε_n), in the peripheral θ (ε_θ) and axial O (ε_0) directions (cf. Fig. 5.22):

$$\left.\begin{aligned} \varepsilon_n &= \frac{1}{E}\left[-\sigma_n - \mu(\pm\sigma_0 + \sigma_\theta)\right] + \alpha T_\Sigma; \\ \varepsilon_\theta &= \frac{1}{E}\left[\sigma_\theta - \mu(\pm\sigma_0 - \sigma_n)\right] + \alpha T_\Sigma; \\ \pm\varepsilon_0 &= \frac{1}{E}\left[\pm\sigma_0 - \mu(\sigma_\theta - \sigma_n)\right] + \alpha T_\Sigma, \end{aligned}\right\} \tag{5.108}$$

where α – the coefficient of linear expansion of the shaft's material; T_Σ – total temperature in the contact zone generated by all heat sources.

Rewrite (5.108) for the stresses over the same sites perpendicular to the directions n, θ, O (cf. Fig. 5.22):

$$\left.\begin{aligned} \sigma_n &= 2G\left[-\varepsilon_n + \frac{3\mu}{1-2\mu}\varepsilon_m - \frac{1+\mu}{1-2\mu}\alpha T_\Sigma\right]; \\ \sigma_\theta &= 2G\left[\varepsilon_\theta + \frac{3\mu}{1-2\mu}\varepsilon_m - \frac{1+\mu}{1-2\mu}\alpha T_\Sigma\right]; \\ \pm\sigma_0 &= 2G\left[\pm\varepsilon_0 + \frac{3\mu}{1-2\mu}\varepsilon_m - \frac{1+\mu}{1-2\mu}\alpha T_\Sigma\right]; \end{aligned}\right\} \tag{5.109}$$

$$G = \frac{E}{2(1+\mu)}, \quad \varepsilon_m = \frac{1}{3}(-\varepsilon_n + \varepsilon_\theta \pm \varepsilon_0). \tag{5.110}$$

Assume that the stress σ_θ is proportional to the peripheral force of friction F_θ in the active system. Then (5.109) and (5.110) yield

$$F_\theta \approx 2GA_a\left[\varepsilon_\theta + \frac{\mu}{1-2\mu}(-\varepsilon_n + \varepsilon_\theta \pm \varepsilon_0) - \frac{1+\mu}{1-2\mu}\alpha T_\Sigma\right]. \tag{5.111}$$

According to (5.111), the peripheral force F_θ is determined both by static ($\varepsilon_\theta - \varepsilon_n$) and cyclic deformations $\pm\varepsilon_0$; hence, formula (5.111) describes the back effect (the effect of cyclic stresses on changes of friction characteristics). In this connection it is appropriate to call the value F_θ as *quasistatic force of friction* (in the active system).

Now assume that the stress σ_0 is proportional to the axial force of friction F_0 in the active system. Then (5.109) and (5.110) yield

$$F_0 \approx 2GA_a\left[\pm\varepsilon_0 + \frac{\mu}{1-2\mu}(-\varepsilon_n + \varepsilon_\theta \pm \varepsilon_0) - \frac{1+\mu}{1-2\mu}\alpha T_\Sigma\right]. \tag{5.112}$$

According to (5.112), the axial force F_0 is determined both by cyclic deformations $\pm\varepsilon_0$ and by static deformations ($-\varepsilon_n + \varepsilon_\theta$); hence, formula (5.112) describes the direct effect (the effect of friction conditions on changes of the fatigue resistance characteristics). In this connection the value F_0 is called *quasicyclic force of friction* (in the active system).

Note that we indicate in a number of formulas the signs (±) for cyclic forces and stresses, though in the general case these values can change in time and follow the cycle of constant sign, i.e. they can have only one sign (plus or minus).

The advantage of formulas (5.111), (5.102) is that they allow for physical interactions between three (component) deformations – in the normal (radial), axial and peripheral directions; these deformations are due both to contact and off-contact loads. In other words, formulas (5.111) and (5.112) and the systems of Eqs. (5.108) and (5.109) describe quite fully what is termed in tribology *mechano-friction effect*, yet from a different standpoint by additional allowance for the thermodynamic state in the contact zone. Hence, we mean the *mechano-thermo-friction effect* (the MTF-effect) *in the active system.*

Two problems appear here: 1) whether the force of friction F_S calculated with formula (5.101) is identical with the quasistatic force of friction F_θ calculated with formula (5.111) as soon as they act in one and the same (peripheral) direction, and 2) whether the maximum cyclic force of friction F_C determined with formula (5.103) is identical with the quasicyclic force of friction F_0 determined with formula (5.112) as soon as they act in one (axial) direction? Here is one answer: these forces are not identical. First, formulas (5.111) and (5.112) allow for thermal effects, while formulas (5.101) and (5.103) ignore them. Second, the total effect of both contact and off-contact loads together induce forces F_0 and F_θ, while the force F_S is determined by single (only) contact load and the force F_C is determined by single (only) off-contact load.

From the above reasoning, it follows that the force of friction in the active system measured in peripheral direction may be larger, smaller or equal to the traditional force of friction measured in a similar friction pair in response to the loading conditions.

This analysis is based on the assumption that the material of the components of the active system operates in the contact zone within perfect elasticity range; the problems of non-elastic behavior and elastoplastic interactions are omitted here. Still note that a similar analysis is possible based on any theory of elasticity and/or theory of creep.

5.5 Damage intensity

Relations (1.92) provide a general idea about the *surface damage intensity* called *wear intensity*. Let us determine it as *linear wear i per unit of length of friction path per one loading* (one revolution of the shaft) in the *shaft/ sliding bearing* friction pair. If N_τ is the number of cycles before the limiting state occurs, then the linear wear intensity is

$$I_h = \frac{\Delta h}{L_f} = \frac{i}{\pi d N_\tau}, \qquad (5.113)$$

where d – the diameter of the shaft. The value $I_h(n)$ can be calculated from (5.113) at any number of loading cycles $n(\tau_W) \leq N_\tau$ if the wear $i(n)$ is known corresponding to this operating time n (cf. Fig. 1.67)

Write formula (1.92a) for calculating the wear intensity of the friction pair to allow for the effect of temperature and corrosion processes using relations (5.19):

$$I_h = K_0 \left(\frac{fp_a}{(1 - D_{\tau/T})\tau_f} \right)^{1+\beta_0 m_\tau} \frac{K_m}{N_\tau} K_{RR} \,. \tag{5.114}$$

The parameters τ_f and m_τ of the friction fatigue curve are assumed in accordance with Fig. 1.57, the parameter $D_{\tau/T}$ is complex, unlike its representations in (5.19), because it allows simultaneously for the effect of the values τ_W and T on the intensity of corrosion.

Similarly to (5.113), introduce the idea about the *volume damage intensity – intensity of fatigue damage* of a structural component. It is determined as non-elastic (residual) deformation $\Delta\varepsilon_\sigma$, per unit of length of the shaft per revolution (per loading cycle). If N_σ is fatigue durability, then

$$I_\varepsilon = \frac{\Delta\varepsilon_\sigma}{\pi d N_\sigma} \,. \tag{5.115}$$

The value $I_\varepsilon(n)$ is calculated from (5.115) at any number of loading cycles $n(\sigma) \leq N_\sigma$ if non-elastic deformation $\Delta\varepsilon_\sigma(n)$ is measured corresponding to this operation time n (cf. Fig. 1.26). The fatigue damage intensity can be determined from (5.115) using other characteristics (parameters) of damage, for example, flexure (in bending), the length and area of a crack (when studying its development), the damping decrement of vibrations (in the region of weak stresses $\sigma \leq \sigma_{-1}$), a relative dangerous volume $\omega_\sigma = V_{P\gamma} / V_0$, etc.

Similarly to (5.114), we record the formula for calculating the fatigue damage intensity

$$I_\varepsilon = K_\varepsilon \left(\frac{\sigma}{(1 - D_{\sigma/T})\sigma_{-1}} \right)^{m_\sigma} \frac{1}{N_\sigma} \,. \tag{5.116}$$

The parameters σ_{-1} and m_σ of the mechanical fatigue curve are assumed here in accordance with Fig. 1.16, the parameter $D_{\sigma/T}$ unlike its representations in (5.19), is complex because it allows simultaneously for the effect of the values σ and T on the intensity of corrosion.

Based on (5.19) we obtain

$$1 - D_{\tau/T} = \left(\frac{v_{ch}}{v_{ch(\tau,T)}} \right)^{m_{\tau/T}} b_e(\tau_W, T);$$

$$1 - D_{\sigma/T} = \left(\frac{v_{ch}}{v_{ch(\sigma,T)}} \right)^{m_{\sigma/T}} b_e(\sigma, T).$$

Here v_{ch} – the rate of corrosion in the given environment; $v_{ch(\tau,T)}$ and $v_{ch(\sigma,T)}$ – the rate of corrosion in the same environment during isothermal fatigue (sliding fatigue – the index τ, T and mechanical fatigue – the index σ, T), the parameters b_e allow for the effect of erosion processes during thermodynamic (σ, T) and thermofriction (τ_W, T) loading.

Using (5.114), (5.116) and lambda-function (2.71) of interactions between damages, we obtain an equation for estimating the *damage intensity of the active system* (cf. Fig. 2.1, *a*)

$$I_W = \lambda(\omega_\sigma \leftrightarrow \omega_\tau) \times$$

$$\times \left[K_0 \left(\frac{fp_a}{(1 - D_{\tau/T})\tau_f} \right)^{1+\beta_0 m_\tau} \frac{K_m}{N_\tau} K_{RR} + K_\varepsilon \left(\frac{\sigma}{(1 - D_{\sigma/T})\sigma_{-1}} \right)^{m_\sigma} \frac{1}{N_\sigma} \right]. \quad (5.117)$$

After simple transformations we have an approximate formula for calculating *the wear intensity of the active system with the allowance for the effect of processes of corrosion and mechanical isothermal fatigue (the back effect)*

$$I_h(\sigma) = I_h \left[1 + \left(\frac{\sigma}{\tau_W} \frac{\tau_f}{\sigma_{-1}} \frac{1 - D_{\tau/T}}{1 - D_{\sigma/T}} \right)^{m_\sigma} \frac{k_0}{K_{RR}} \frac{N_\tau}{N_\sigma} \right] R_{\tau/\varepsilon}, \quad (5.118)$$

where $R_{\tau/\varepsilon}$ – the parameter of interactions between damage during back effect, and a formula for calculating *the fatigue damage intensity of the active system with the allowance for the effect of the processes of friction and corrosion isothermal fatigue (the direct effect)*

$$I_\varepsilon(\tau_W) = I_\varepsilon \left[1 + \left(\frac{\tau_W}{\sigma} \frac{\sigma_{-1}}{\tau_f} \frac{1 - D_{\sigma/T}}{1 - D_{\tau/T}} \right)^{1+\beta_0 m_\tau} \frac{K_{RR}}{k_0} \frac{N_\sigma}{N_\tau} \right] R_{\varepsilon/\tau}, \quad (5.119)$$

where $R_{\varepsilon/\tau}$ – the parameter of interactions between damage during direct effect, and

$$k_0 = \frac{K_\varepsilon}{K_0} \frac{1}{K_m}.$$

In accordance with formulae (5.118) and (5.119) the *process of friction in the zone of compression and the process of friction in the zone of tension* can be discriminated if to assume $\pm\sigma$, respectively. It turns out that the processes of damage intensify during tensile cyclic stresses and slow down during compressive cyclic stresses (cf. also Sect. 4.4).

According to Eqs. (5.118) and (5.119) the damage intensity of the active system depends on:

- the ratio between the effective cyclic and contact stresses(σ/τ_W or τ_W/σ);
- the ratio between the ultimate stresses during mechanical and sliding fatigue (σ_{-1}/τ_f or τ_f/σ_{-1});
- the ratio between the rates of corrosion and erosion processes during isothermal sliding and mechanical fatigue $(1 - D_{\tau/T})/(1 - D_{\sigma/T})$ or $(1 - D_{\sigma/T})/(1 - D_{\tau/T})$;
- electrochemical activity of the tested materials during dynamic ($m_{\sigma/T}$) and frictional ($m_{\tau/T}$) loading.

Finally, it is apparent from (5.117)–(5.119) that the damage intensity of the active system is directly due to the parameter controlling the WFD (5.25) and to the parameter of damage asymmetry like (5.33) and also to the friction index (5.106).

This analysis is performed with the account of schematization of the mechanical and sliding fatigue curves represented as two rectilinear segments (cf. Figs. 1.16 and 1.57). If the analysis is based on the full mechanical and sliding fatigue curves (cf. Figs. 1.15 and 1.54), solutions can be obtained in a similar way with the account of the whole variety of damage mechanisms and their interactions (see Sect. 2.6).

GOST 30638–99 defines (Table 5.6) the groups of wear resistance of active systems similarly as it is proposed for friction pairs (cf. Table 1.3). It is clear that the *classes of resistance to fatigue damage of active systems* can be represented in a similar manner.

Table 5.6. Groups of wear resistance of active systems

$I_h(\sigma)$	$< 10^{-7}$	$10^{-7} \ldots 10^{-5}$	$> 10^{-5}$
Group of wear resistance	A	B	C
	> 7	$7 \ldots 5$	< 5

The wear intensity in friction can also be determined using the dangerous volume. For example, the product $\sqrt{(h/r)} \times (A_r/A_a)$ in formula (1.95) characterizes the deformable volume in friction, that is why it should be proportional to the relative dangerous volume in sliding $S_{P\gamma}/S_k$. Taking this into account, formula (1.95) becomes

$$I_h = \frac{S_{P\gamma}}{S_k} \eta_S \frac{K_m}{N_\tau} K_* ,$$

where η_S is the coefficient of proportionality.

If energy approach is used to estimate the wear intensity in formulae (1.98) and (1.99), it is apparent that the values ($\Delta V/V_D$) and ($S_{P\gamma}/S_k$) are proportional, so that formula (1.98) becomes

$$e_R^* = \frac{N_\tau e_{\mathrm{Re}}}{(S_{P\gamma}/S_k)\eta_V},$$

where η_V is the coefficient of proportionality. Taking (1.100) into account, we get

$$I_h = \frac{\tau_W}{e_R^*} = \frac{\tau_W}{N_\tau}\frac{S_{P\gamma}}{S_k}\frac{\eta_V}{e_{\mathrm{Re}}}.$$

Table 5.7. Parameters of wear and their relation to linear wear intensity

Wear parameters and factors	Formula	Relation with I_h
Abrasion coefficient	$\vartheta^* = \Delta V / W_T$	$\vartheta^* = I_h / fp_a = I_h / \tau_W$
	$\vartheta_f^* = \Delta V / W_T$	$\vartheta_f^* = I_h / p_a$
	$\gamma = \Delta V / FL$	$\gamma = I_h / p_a$
"Imaginary" energy density	$e_R^* = M_T / V$	$e_R^* = fp_a / I_h = \tau_W / I_h$
Wear coefficient	$K = h / p_a L$	$K = I_h / p_a$
	$K = h / p_a Vt$	$K = I_h / p_a$
Specific wear volume	$W = \Delta V / L$	$W = I_h / p_a$
Energy wear intensity	$K = h / Fvt$	$K = I_h / fp_a = I_h / \tau_W$
Specific wear intensity	$i_h = I_h (p_r / p_a)$	$I_h = i_h (p_a / p_r)$
Wear factors:		$\Phi_{eu} = I_h^{m_{eu}} / p_a \Delta^{n_{eu}}$
a) unrunning-in elastic contact	Φ_{eu}	$m_{eu} = 0.5...0.8$;
		$n_{eu} = 0.6...1.3$;
		$\Phi_{er} = I_h / p_a$
b) running-in elastic contact	Φ_{er}	$\Phi_{pl} = I_h^{m_{pl}} / p_a \Delta^{n_{pl}}$
c) plastic contact		$m_{pl} = 0.6...0.7$;
	Φ_{pl}	$n_{pl} = 0.9...1.0$;
Microshearing	Φ_m	$\Phi_m = I_h / p_a$ HB
Note: v – sliding velocity; L – friction path; Δ – surface roughness parameter; F – friction area; p_r and p_a – actual and nominal pressure on contact site.		

It should be pointed out that a lot of various wear parameters are used in books on tribology but all of them are expressed in terms of linear wear intensity I_h (Table 5.7). A knowledge of relations given in Table 5.7 makes possible comparing and analysing experimental data in practice received by different authors. They can be used to analyse wear intensity in active systems (see Formula (5.118)).

5.6 Quality, risk, safety

The state-of-the-art and competitiveness of a modern machine is determined by the integrity of interrelated safety, economic and ergonomic indicators (Fig. 5.23).

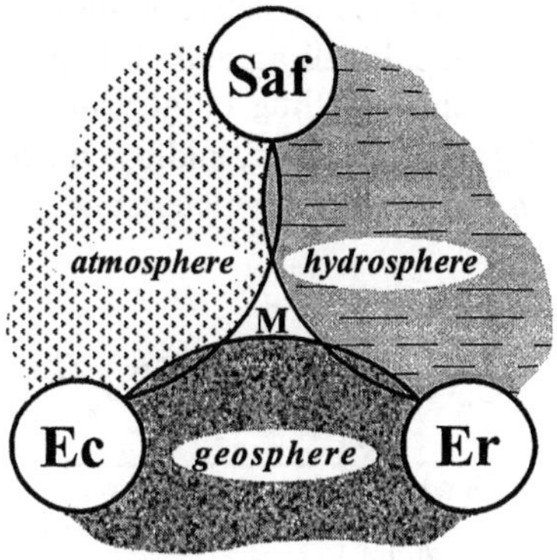

Fig. 5.23. Concept of SafEcEr: safety, economy, ergonomics (M – machine operating in environment)

Ergonomics is considered satisfied if:

- Labor conditions ⇒ opt;
- Convenience ⇒ max;
- Aesthetics ⇒ opt.

Economics is acceptable if:

- Production costs ⇒ opt;
- Operation costs ⇒ min;
- Maintenance costs ⇒ min.

Safety is considered ensured if:

- System quality ⇒ max;
- Hazard (for personnel and environment) ⇒ min;
- reliability of systems ⇒ opt.

Here we briefly consider the problem of ensuring safety of active systems using the *QRR approach* (*quality – risk – reliability*). The algorithm for solution can be constructed as follows:

$$\left.\begin{array}{c} \dfrac{D(x_i)}{\Pi(x_i)} = \rho(x_i) = \lambda_x \dfrac{P(t)}{Q(t)} \\ \Updownarrow \\ 1 - \dfrac{\rho(x_i)}{\lambda_x} = R_\rho = 1 - \dfrac{P(t)}{Q(t)} \end{array}\right\}; \tag{5.120}$$

$$\frac{P(t)}{Q(t)} = \rho(t) = \frac{1}{\lambda_x}\rho(x_i). \tag{5.121}$$

Only a limited number x_i, $i = 1, 2, ..., n$ is selected out of a large number of characteristics of mechanical properties of materials and resistance to WFD to analyze the quality and risk of use of active systems, for example: σ_{-1}, p_f, τ_f, – fatigue limits during mechanical, rolling and sliding fatigue; σ_{-1p}, $p_{f\sigma}$, $\sigma_{-1\tau}$, $\tau_{f\sigma}$ – ultimate stresses during mechano-rolling fatigue and mechano-sliding fatigue; I_p, I_τ, $I_{\sigma p}$, $I_{\sigma\tau}$ – wear intensity during rolling, sliding, mechano-rolling and mechano-sliding fatigue; σ_T, σ_b, δ, ψ – basic characteristics of properties of materials in tension, etc.

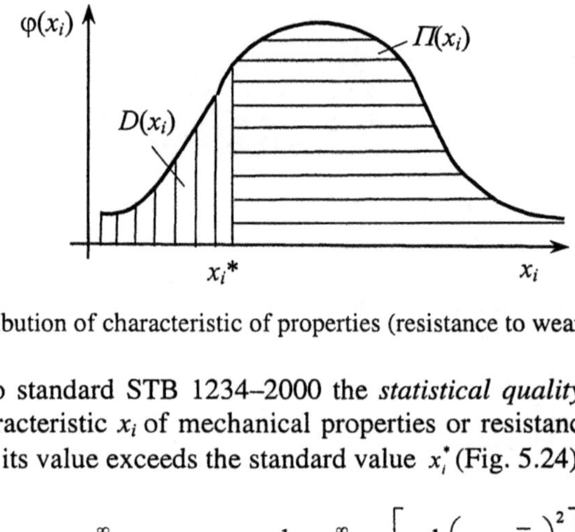

Fig. 5.24. Distribution of characteristic of properties (resistance to wear-fatigue damage)

According to standard STB 1234–2000 the *statistical quality indicator* based on a given characteristic x_i of mechanical properties or resistance to WFD is the probability that its value exceeds the standard value x_i^* (Fig. 5.24):

$$\Pi(x_i) = P(x_i \geq x_i^*) = \int_{x_i^*}^{\infty} p(x_i)dx_i = \frac{1}{\sqrt{2\pi}S_{\bar{x}_i}} \int_{x_i^*}^{\infty} \exp\left[-\frac{1}{2}\left(\frac{x_i - \bar{x}_i}{S_{\bar{x}_i}}\right)^2\right] dx_i\,. \tag{5.122}$$

Then *the statistical indicator of quality loss* (cf. Fig. 5.24)

$$D(x_i) = \int_{-\infty}^{x_i^*} p(x_i)dx_i = \frac{1}{\sqrt{2\pi}S_{\bar{x}_i}} \int_{-\infty}^{x_i^*} \exp\left[-\frac{1}{2}\left(\frac{x_i - \bar{x}_i}{S_{\bar{x}_i}}\right)^2\right] dx_i = 1 - \Pi(x_i)\,. \tag{5.123}$$

The risk indicator is determined as expectation of quality loss:

$$0 \le \rho(x_i) = \frac{D(x_i)}{\Pi(x_i)} \le 1. \tag{5.124}$$

The correlation between the risk indicator and the quality loss indicator is given by the formula

$$\rho(x_i) = \frac{1}{\dfrac{1}{D(x_i)} - 1}, \tag{5.125}$$

and the correlation with the quality indicator is given by the formula

$$\rho(x_i) = \frac{1}{\Pi(x_i)} - 1. \tag{5.126}$$

The risk indicator $\rho(x_i)$ that formula (5.124) correlates with the indicators $\Pi(x_i)$ and $D(x_i)$ is called the *operative risk characteristic*. Figure 5.25 shows it graphically. Table 5.8 lists the theoretical data for plotting the operative risk characteristic.

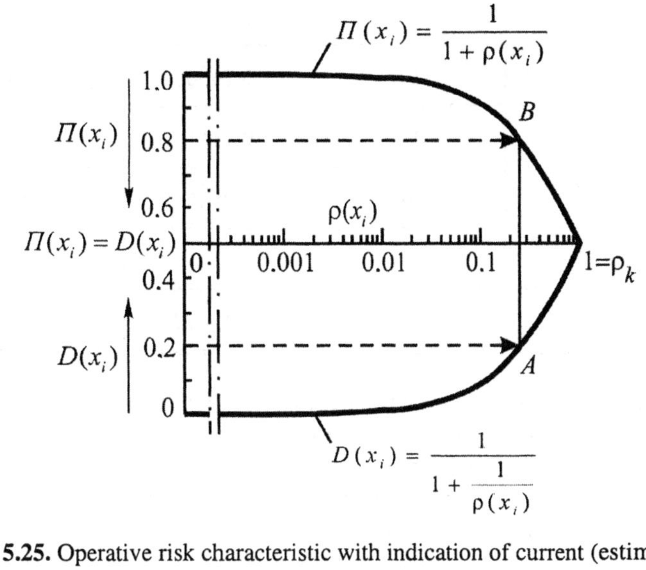

Fig. 5.25. Operative risk characteristic with indication of current (estimated) value of $\rho(x_i)$ (straight line *AB*)

The lower risk boundary $\rho(x_i) = 0$ corresponds to the case when $D(x_i) = 0$. The upper value of the analysed risks $\rho(x_i) = 1$ is limited by the condition that $\Pi(x_i) = D(x_i) = 0.5$. The value $\rho_k = 1$ is considered *critical*. Since the numerical value of the risk indicator may change by several orders, the logarithmic scale $\rho(x_i)$ is used for the operative characteristic (Fig. 5.25).

Table 5.8. Theoretical data for plotting operative risk characteristics

$D(x)$	$\Pi(x) = 1 - D(x)$	$\rho = D(x) / \Pi(x)$	$R_\rho = 1 - \rho(x)$
0.01	0.99	0.0101	0.9899
0.05	0.95	0.0526	0.9474
0.1	0.9	0.1111	0.9999
0.2	0.8	0.2500	0.7500
0.3	0.7	0.4286	0.5714
0.4	0.6	0.6666	0.3334
$D(x) = 0.5 = \Pi(x)$	0.5	$\rho_k = 1$	0
0.6	0.4	0.6666	0.3334
0.7	0.3	0.4286	0.5714
0.8	0.2	0.2500	0.7500
0.9	0.1	0.1111	0.9999
0.95	0.05	0.0526	0.9474
0.99	0.01	0.0101	0.9899
$\Pi(x)$	$D(x) = 1 - \Pi(x)$	$\rho(x) = D(x) / \Pi(x)$	$R_\rho = 1 - \rho(x)$

The current (estimated with formulas (5.124), (5.125) or (5.126)) risk value is shown by the full vertical line AB in the operative characteristic (cf. Fig. 5.25).

A *normative risk value* $[\rho]$ can be established in respect to any characteristic of WFD. It is validated by relevant technical and economic calculations with the account of the gravity of consequences in case some unfavorable events occur. Standard STB 1234–2000 establishes *three quality categories* and the normative risk corresponding to them (Table 5.9). The vertical dotted line in the operative risk characteristic shows the normative risk value ρ (Fig. 5.26).

Any event and situation for which

$$\rho(x_i) \leq [\rho], \tag{5.127}$$

is considered acceptable or *tolerable*. The values are considered unacceptable or *intolerable*.

Table 5.9. Categories of quality of active system

Categories	Normative values of indices		
	$\Pi(x)$ not less than	$D(x_i)$, %, not more than	$[\rho(x)]$
Top	0.995	0.5	0.0050
First	0.990	1.0	0.0101
Second	0.950	5.0	0.0526

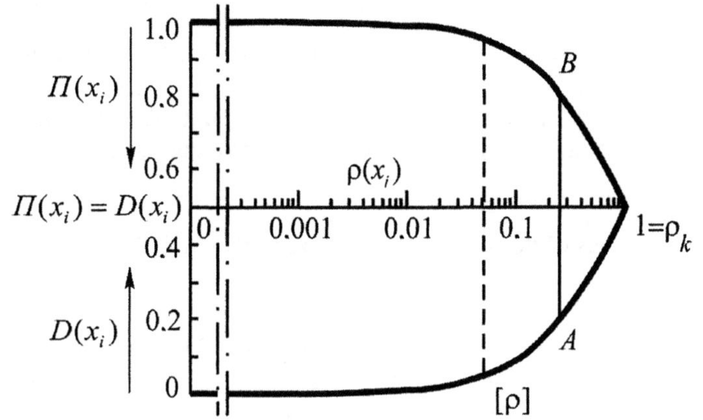

Fig. 5.26. Indication of normative risk [ρ] (dotted line) on its operative characteristic

$$[\rho] < \rho(x_i) < \rho_k \qquad (5.128)$$

When it is necessary to compare relatively the risk on the basis of numerous and different characteristics, a *risk diagram* is plotted (Fig. 5.27). Each sector on the diagram is an operative risk characteristic (cf. Fig. 5.25) plotted according to a certain attribute. The scale of numerical values of the risk indicator $\rho(x_i)$ is shown on the radius A of the circle. The pie diagram contains the following information:

- the critical risk $\rho_k = 1$ corresponds to the external circle;
- the normative risk value [ρ] reflects the internal circle (dots);
- the risk regions are shaded up to the value of the estimated risk $\rho(x_i)$ depicted by the arc of the circle with the radius $r = \rho(x_i)$ within each sector.

When normative risk is not established, the dotted circle is absent on the diagram. The external circle can correspond to another value of risk considered to be representative instead of the critical risk, this other value is indicated at the end ▯ of the circle radius.

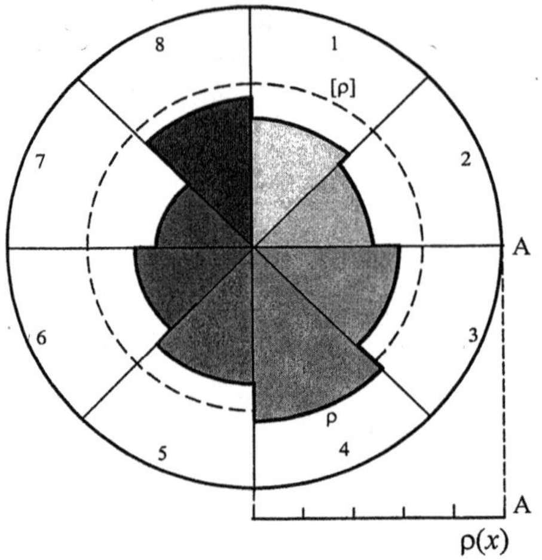

Fig. 5.27. Scheme of pie risk diagram based on 8 (1, 2, ..., 8) symptoms

Safety is treated as an antipode of risk. The *safety indicator* $R_\rho(x_i)$ is determined as a value that adds to risk (5.124) up to a unity, i.e.

$$R_\rho(x_i) = 1 - \rho(x_i). \tag{5.129}$$

Numerical values of the safety indicator are analyzed within the range

$$0 \le R_\rho(x_i) \le 1, \tag{5.130}$$

that corresponds to the interval (5.123) of variations of the risk indicator $\rho(x_i)$.

Zero safety ($R_\rho(x) = 0$) corresponds to critical risk ($\rho_k = 1$). Unconditional safety ($R_\rho(x) = 1$) corresponds to zero risk ($\rho(x_i) = 0$).

From Table 5.8 two general conclusions can be drawn:

– safety is always lower than estimated quality (i.e. $R_\rho(x) < \Pi(x)$), the difference between these indicators is more significant, the poorer the quality;

– risk is always higher than the estimated quality deteriation indicator (i.e. $R_\rho(x) > D(x)$), but the difference between these indicators smoothes out (reduces) as quality improves ($\Pi(x)$).

If a full-scale active system is considered, its *risk in time* $\rho(t)$ is determined from formula (5.121) as a ratio between failure probability $P(t)$ and failure free probability $Q(t)$ that are determined in accordance with Fig. 1.86:

$$P(t) = \int_0^{t_n} \varphi(t)dt\,; \tag{5.131}$$

$$Q(t) = \int_{t_n}^{\infty} \varphi(t)dt = 1 - P(t)\,. \tag{5.132}$$

Note that the risk $\rho(t)$ is set for a full-scale active system operating in predictable conditions, while the risk $\rho(x_i)$ relates only to some or other properties of the modeled active system undergoing laboratory tests. Hence, there is a *transitional function* λ_x between them (cf. formulas (5.120) and (5.121)) that is determined experimentally or from the corresponding theoretical considerations provided that the interval $0 \le \rho(t) \le 1$ is set.

Yet, in a number of cases, for example, through unforeseen changes in the conditions of operation or force majeure the risk $\rho(t)$ can change within a broad range of values, namely

$$0 \le \rho(t) = P(t)/Q(t) \le \infty. \tag{5.133}$$

On the other hand, numerous phenomena and factors can affect risk variations within interval (5.133), for example, lessening the risk (when the hardening processes evolve) or, on the opposite, increasing it (when the softening processes evolve). A *dialectic function* $\delta(y, z, t)$ is used to allow for such changes that acquires the values $\delta \gtrless 1$. The risk then is

$$\rho(\delta, t) = \rho(t)\delta. \tag{5.134}$$

The dialectic function is

$$\delta = \delta(y,z,t) = e^{-\gamma+\beta} = e^{\beta\left(1-\frac{\gamma}{\beta}\right)} = \frac{1}{e^{\gamma\left(1-\frac{\beta}{\gamma}\right)}}, \tag{5.135}$$

where the time function of damaging phenomena and factors

$$\beta = \beta(z_1, z_2, \ldots, z_n, t) \tag{5.136}$$

and the time function of hardening phenomena and factors

$$\gamma = \gamma(y_1, y_2, \ldots, y_n, t). \tag{5.137}$$

Dialectic function (5.135) in the simplest case decomposes into the *hardening function* (at $\beta = 0$)

$$\delta_\gamma = e^{-\gamma} \tag{5.137a}$$

and the *softening function* (at $\gamma = 0$)

$$\delta_\beta = e^{\beta}. \tag{5.137b}$$

Figure 5.28 shows their graphs. The general regularities of evolution of hardening-softening processes predicted by Eqs. (5.137a) and (5.137b) are the following, On the one hand, growth of the parameter β leads to the augmenting value of function (5.137b). On the other hand, no matter how much the parameter γ grows, hardening function (5.137a) remains attenuating: it is asymptotically approaches zero never reaching it. That is why when both parameters grow alike, the system always tends to soften and then to degrade fully, and that may occur in the explosive manner (Fig. 5.29). Therefore, *hardening is a limited phenomenon,* its extent is always finite. Meanwhile softening is an *unlimited phenomenon,* its extent may grow infinitely, i.e. until a *catastrophe* occurs.

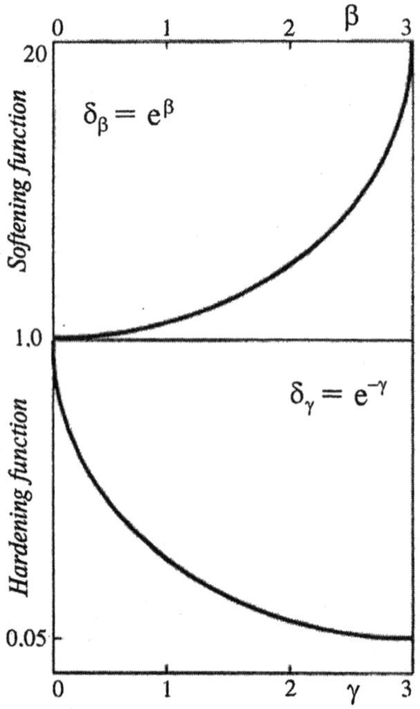

Fig. 5.28. Graphs of functions of hardening and softening

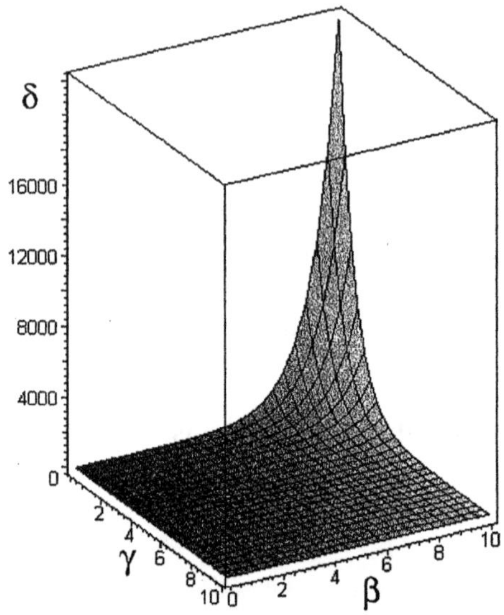

Fig. 5.29. Function (5.135)

Functions (5.136) and (5.137) are specified to apply to the analysed conditions of the events (situations) evolution, risk is determined by characteristic signs.

Thus, the QRR approach (quality–risk–reliability) described by general algorithm (5.120) is made specifically tailored for practical application using formulas (5.122)–(5.137).

5.7 Control over processes of wear-fatigue damage

The general sense of the *problem of control over* WFD processes is the following (Fig. 5.30). The active system (AS) is treated as an object of control. The problem of formulation of an optimal control program becomes a problem of optimization: $F(\sigma, p) \Rightarrow \min$, $C_0 \Rightarrow \min$, i.e. a parameter opt F of optimization is established on the basis of technical and economic estimates. A variety $\{a_j\}$ of parameters a_j of the state of AS in operation is measured. The results serve to formulate a current measure $\omega_\Sigma(t)$ of complex wear-fatigue damage that is the function of time t and particular measures of damage ω_σ, ω_p, ω_{ch}, ω_T due to cyclic stresses (the index σ), contact pressure (the index p), the processes of electrochemical corrosion (the index ch), temperature (the index T) in the zone of contact interactions between the components of the AS. The state of the AS at any moment of its operation is rated using the integral parameter $F_t(\sigma, p, \omega_\Sigma(t))$. Another task is to correlate (compare) the optimal (opt F) with the current (F_t) value of the integral parameter F. The obtained mismatch ΔF of the parameters opt F and F_t serves to solve the problem of synthesizing dynamic or optimal control $U = \varphi(\Delta F)$. The feasibility of control of the AS is implemented by the executive organ (EO).

Of course, the problem of control of WFD is highly intricate, yet it is very practical when it is necessary to control effectively the reliability of most essential active systems of a machine in operation based on the most important criteria of performance and in order to achieve simultaneously savings of costs of labor, means and materials in operation and production.

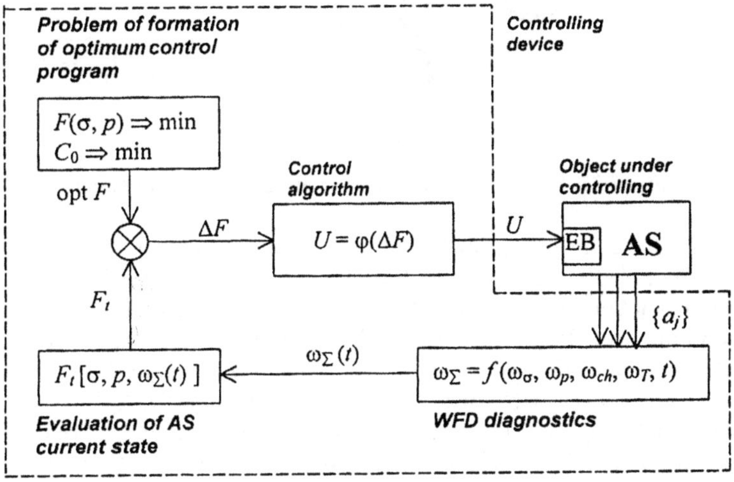

Fig. 5.30. Content of task of optimization by active system

Figure 5.31 shows the main *tf*-channels (*tribo-fatigue*) of WFD *control* to which the following belong: *a*) design and technological parameters ($V_{P\gamma}/V_0, S_{P\gamma}/S_k$); *b*) parameters m_j, $j = 1, 2, \ldots, n$, of composition and structure (mechano-physico-chemical properties) of materials; *c*) loading parameters (σ, T, τ_W, D); *d*) parameters of time (t); *e*) parameters of state (damage) of components (ω_σ, ω_τ, $\omega_{\square}$, ω_{ch}); *f*) parameters of interaction between damage ($R_{\sigma/\tau}$, $R_{T/M}$, $D_{\sigma,T,\tau}$).

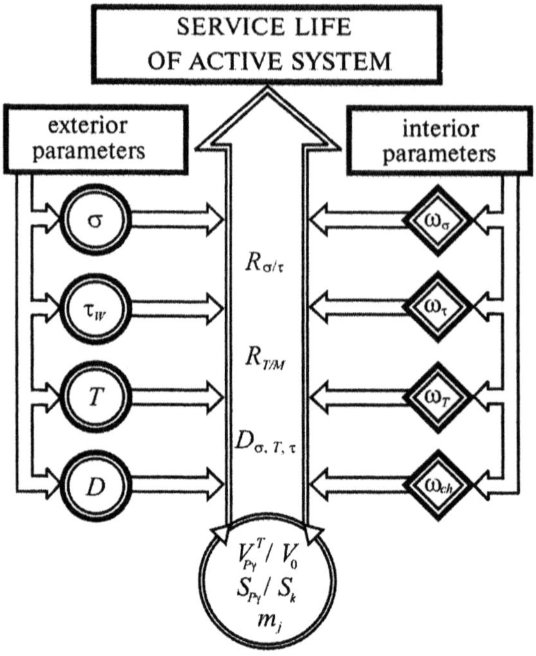

Fig. 5.31. Main *tf*-channels of controlling life of active system

While the problem of optimal control over the processes of WFD has been solved (cf. Figs. 5.30, 5.31) from technical and economical standpoint, it does not mean that a machine designed satisfies the modern concept of safety, economics and ergonomics (cf. Fig. 5.23). According to the concept, though the *required durability is ensured, the safety guarantee of the personnel (and the environment) can be violated.* Hence it follows the need to implement the QRS (quality–risk–safety) approach (Fig. 5.32). On the other hand, note that quality, risk and safety may turn out not only insufficient but excessive that operation of a machine definitely reveals; hence the controlling effect is written with the signs $\pm U$.

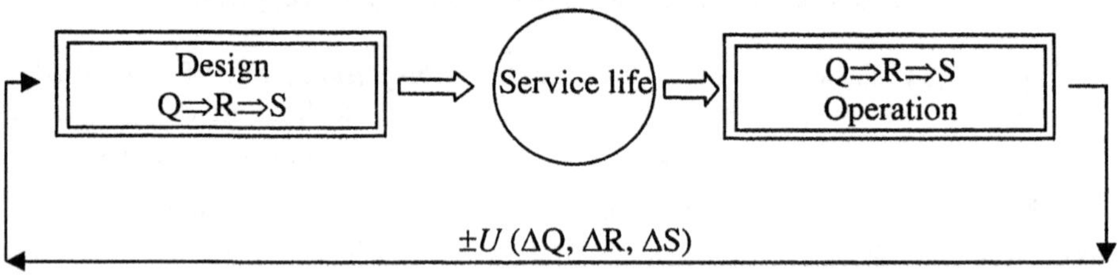

Fig. 5.32. Problem of wear-fatigue damage control with allowance for QRS

Specific (though simplest) examples will show the benefit and effectiveness of the QRR (quality–risk–reliability) approach.

Table 5.10. Indicators of quality and risk of use (accurate to three digits after point) based on yield point of three grades of steel

Parameters	40X		40XH		18XГТ	
	40 melts	1 melt	40 melts	1 melt	40 melts	1 melt
$\bar{\sigma}_y$	580	568	730	706	480	465
$S_{\bar{\sigma}_y}$	48.9	24.4	40.3	27.2	34.5	20.5
$D(\sigma_y)$	0.047	0	0	0	0.079	0.044
$\Pi(\sigma_y)$	0.953	1	1	1	0.921	0.956
$\rho(\sigma_y)$	0.049	0	0	0	0.073	0.046
$R_p(\sigma_y)$	0.951	1	1	1	0.927	0.954

An experimental study of the quality and risk of three grades of structural steel used for producing components of essential active systems was performed based on the yield limit: 40X and 40XH (after martempering, the diameters of blanks were 80 and 100 mm, respectively) and 18XГТ (after normalizing, the diameter of blanks was 80 mm). 200 specimens of each grade of steel were tested for tension, with first 100 specimens taken from blanks of one and the same melt, the following 100 specimens were cut out from blanks belonging to forty different melting. The parameters of distribution of yield points (the mean value $\bar{\sigma}_y$ and root-mean-square deviation $S_{\bar{\sigma}_y}$) are listed in Table 5.10; it also contains the indicators of quality, risk and safety estimated with formulas (5.122), (5.123) and (5.129). It is apparent that the quality of steel 40XH under study causes no doubt. Steel 40X behaves somewhat differently: while the quality of a single melt was undoubtedly ensure, many melts turned out to show that $D(\sigma_y) = 4.7\%$, i.e. the quality satisfies only the requirements of the second category (according to Table 5.7). Regarding steel 18XГТ the risk of its application is too high: while one melt has $\rho(\sigma_y) = 0.046 < [\rho] = 0.0526$, other 40 melts show $\rho(\sigma_y) = 0.073 > [\rho]$ (Fig. 5.33).

Thus, the steel 18XГТ is inadmissible to fabricate essential pieces because it does not ensure the required safety of operation ($R_\rho(\sigma_y) = 0.927 < 0.947$). Therefore, the problem is to establish the causes of quality loss of larger pieces from this steel using one of most important characteristics of mechanical properties, viz. the yield point and to undertake corresponding actions to eliminate them.

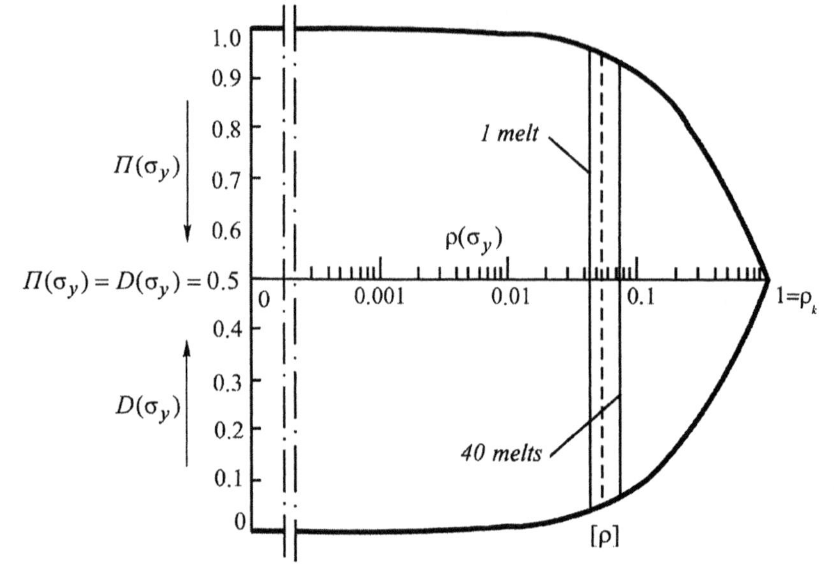

Fig. 5.33. Analysis of risk of using steel 18 ХГТ based on yield point

Figure 5.34 shows pie risk diagrams (cf. Fig. 5.27) based on six characteristics of mechanical properties of steel 45 for crankshafts (*a*) and steel 38XA for connecting rod bolts (*b*) of heavy-duty compressors. These characteristics describe the resistance of metal to static (σ_y, σ_b, δ, ψ), cyclic (σ_{-1}) and impact (KCV) loads with the allowance for both strength and plasticity. Each sector of the diagram contains the normative (dotted line) and actual (a continuous thick line) risk for a given characteristic of mechanical properties. It is apparent that the risk indicator exceeds significantly the normative boundary in a number of cases.

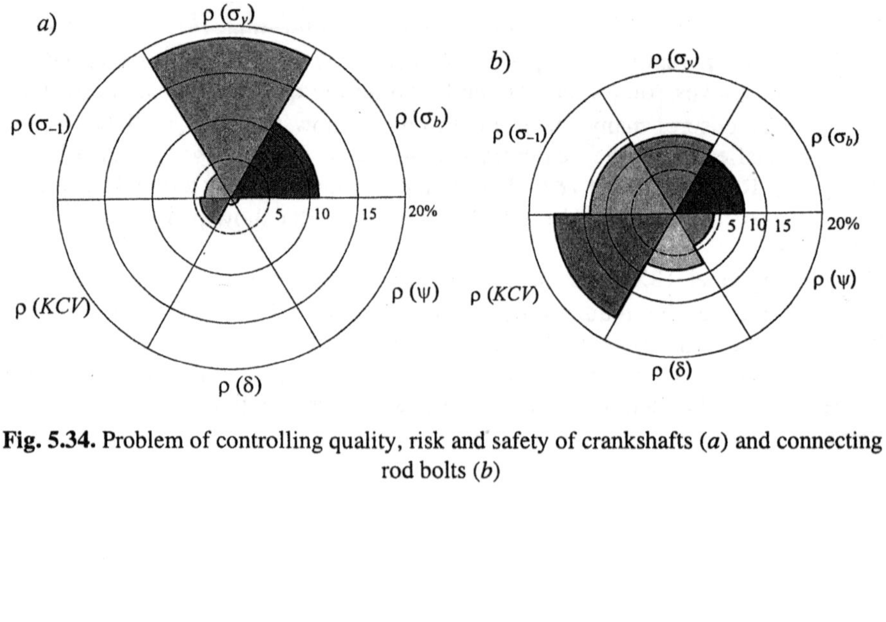

Fig. 5.34. Problem of controlling quality, risk and safety of crankshafts (*a*) and connecting rod bolts (*b*)

In this connection two types of problems can be considered.

The *problem of the first type* appears, for example, when risk is analyzed with the use of steel 45 for crankshafts. It follows from the risk pie diagram (cf. Fig. 5.34, *a*) that $\rho(x) > [\rho]$ based on two characteristics: σ_y and σ_b. However, the experience of operation shows that there are practically no premature fatigue damage in the shafts from this steel. This contradiction can be obviated on the assumption that the requirements of the GOST to the values $\sigma_y^{\text{GOST}} = 300$ MPa and $\sigma_b^{\text{GOST}} = 560$ MPa of steel 45 for crankshafts are too overstated. These requirements should be modified, i.e. they should be correlated with the properties of strength obtained in reality. If $\sigma_y^{\text{GOST}} = 270$ MPa and $\sigma_b^{\text{GOST}} = 550$ MPa are assumed, it does not change in any way the properties of the metal, hence the strength reliability of the shafts, yet the risk indicators calculated with modified requirements will be $\rho(x) < [\rho]$, i.e. their values become tolerable.

The *problem of the second type* appears when analyzing risk indicators of steel 38XA for connecting rod bolts. The risk pie diagram (cf. Fig. 5.34, *b*) shows that $\rho(x) < [\rho]$ for five indicators. The experience of operations corroborates this analysis: premature fatigue failures of a given component take place. Therefore the problem is to improve the quality of steel 38XA for connecting rod bolts so that the metal of better quality favors lower risk indicators. In the case in question it is tolerated to reduce quality based on the indicator $\Pi(\psi)$. It is known that reduction of plastic properties enables to boost the characteristics σ_y and σ_b, by applying, for example, suitable heat treatment, in its turn, it leads to a higher fatigue limit σ_{-1}, hence, to improved reliability of connecting rod bolts.

Calculations show that in the case in question it is enough to reduce the value ψ 11% with a corresponding rise of other characteristics by 5...6% to bring the quality of steel 38□□ into the tolerable limits, i.e. to exclude $\rho(x) \leq [\rho]$.

Statistical data of reliability of machine parts in operation were used in the above conclusions from the risk analysis. If such data are unavailable (for example, at the stage of designing), risk predictions are made when studying performance characteristics of a metal for essential parts. In this way the problem is successfully solved of making a proper choice of a material (or its state) for each active system.

A practical example.

Steel 18XΓT and 25XΓT (both after cementation) are widely used to produce toothed wheels, with the fatigue limit of both grades based on deterministic tests for fatigue (Fig. 5.35) is practically the same. Models shown in Fig. 2.2 were tested. The tests yield the characteristics of bending and contact strength at once by plotting the relevant fatigue curves (Fig. 5.36) under changes of only the values of contact load F_N, like it happens in real conditions of operation of gears. Note that this test method enables to identify the most dangerous region of changes of F_N, in which both contact and bending fatigue of gears can lead to the limiting state ($F_N \approx 1600\ldots2400$ N); the limiting state above and below this region occurs, as a rule, following *only one* of these criteria.

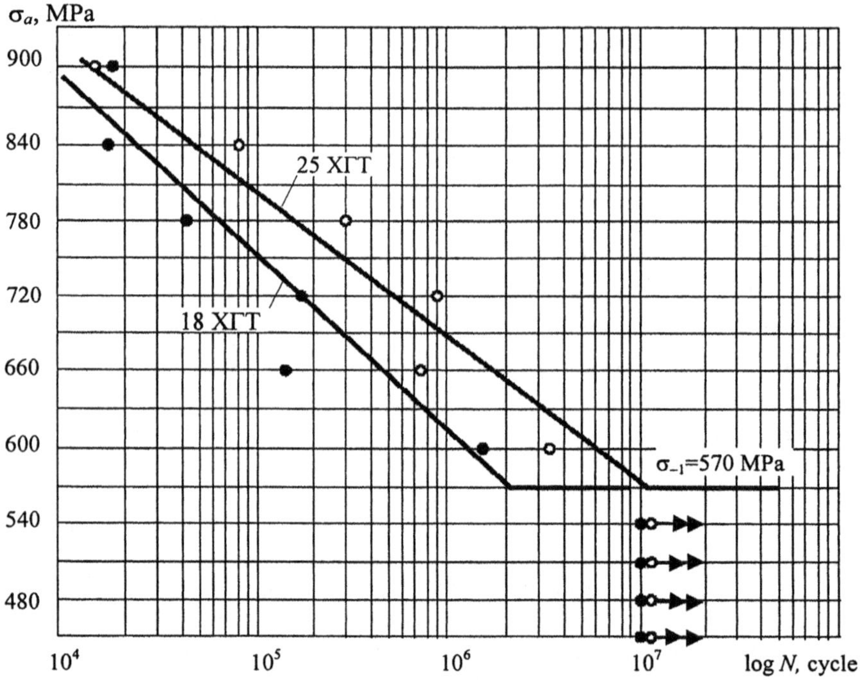

Fig. 5.35. Curves of mechanical fatigue (in bending) of steels 18XΓT and 25XΓT

It has been established from the results of the statistical tests of gearing (Table 5.11) that steel 25XΓT demonstrates substantially higher quality indicators, hence much lower risk of its use than steel 18XΓT.

In accordance with Fig. 5.32 the results of the statistical tests have served to elaborate a method of controlling WFD of gearing from steel 25XΓT for a gearbox with the set range of speeds (cf. Fig. 2.13 and Table 2.7). The method is based on statistically interrelated limits of contact fatigue p_f, fatigue limits to bending σ_{-1} and hardness *HRC* (Fig. 5.37) proceeding from the condition that wheels' quality should satisfy the first category according to GOST 1234–2000 (see Table 5.7), hence, the risk of their use should be $\rho(x) < [\rho] = 0{,}0101$. Hardness is convenient to use from the practical viewpoint because it is easy to measure it very quickly.

According to Fig. 5.37 the following normative values of the parameters of *HRC* hardness distribution that satisfy the first category of quality of gears:

Mean value $\overline{HRC}^* = 58.25$;

Mean square deviation – $S_{\overline{HRC}} = 1.6$;

Minimum value $HRC^*_{\min} = 54.2$.

Table 5.11. Characteristics of performance of gearings

Mechanical fatigue								
Endurance limit in bending			Statistical indicators			Quality category		
normative σ^*_{-1}, MPa	mean $\bar{\sigma}_{-1}$, MPa	$S_{\sigma_{-1}}$, MPa	quality $\Pi(x)$	loss of quality D(x)	risk $\rho(x)$	top	first	second
25XГТ								
360			0.9832	0.0168	0.0171	–	–	X
350	462	48.0	0.9902	0.00983	0.00992	–	X	–
335			0.9959	0.00408	0.0041	X	–	–
18XГТ								
When normative requirements are the same as for steel 25XГТ								
360			0.9318	0.0682	0.0732	Intolerable		
350	430	47.0	0.9803	0.0197	0.0201	loss		
335			0.978	0.0216	0.0221	of quality		
Rolling fatigue								
Contact fatigue limit			Statistical indicators			Quality category		
normative p^*_f, MPa	mean $\bar{p}_f$, MPa	S_{p_f}, MPa	quality $\Pi(x)$	quality loss D(x)	risk $\rho(x)$	top	first	second
25XГТ								
1000			0.9743	0.0257	0.0263	–	–	X
925	1304	156	0.9944	0.00757	0.00763	–	X	–
895			0.9956	0.00438	0.0044	X	–	–
18XГТ								
When normative requirements are the same as for steel 25XГТ								
1000			0.9616	0.0384	0.03990	–	–	X
925	1269	152	0.9882	0.0118	0.01200	–	–	X
895			0.9931	0.00695	0.00700	–	X	–

The above normative parameters of hardness makes it possible to reach the normative values of the parameters of distribution of fatigue limits in bending σ_{-1} and rolling fatigue limits p_f listed in Table 5.12.

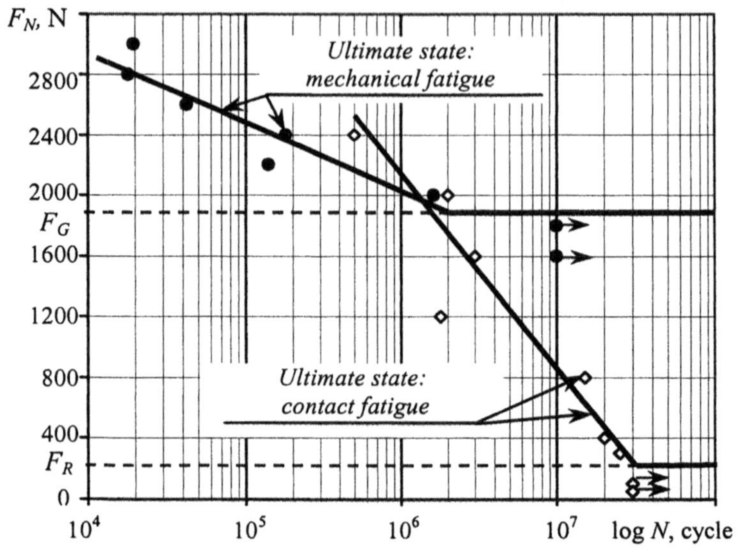

Fig. 5.36. Results of tests of models of gearings from steel 25XГТ: F_G – ultimate load based on bending fatigue criterion; F_R – ultimate load based on rolling fatigue criterion

All the strength properties are ensured if the quality of thermochemical treatment of toothed wheels complies with the following basic requirements:

a) the microstructure of the carburized layer of tempered and annealed wheels should be martensite (retained austenite is allowed, fine carbides; individual inclusions of ferrite are allowed in the microstructure of the core of teeth);

b) the number of troostite islands in the carburized layer should not exceed the score of five;

c) the quantity of retained austenite should not exceed the score of two;

d) the quantity of carbides in the carburized layer should not exceed the score of five;

e) the quantity of retained ferrite in the core of teeth should not exceed the score of four.

To control statistically the quality of toothed wheels in the process of their manufacture a random sample of at least 100 hardness values is prepared:

a) if the quality of a batch of 1000 toothed wheels of a given steel grade is checked, a random sample of 100 wheels is prepared. Each wheel is tested for hardness only once resulting in a sample of 100 values;

b) if the quality of a batch of 100 toothed wheels of a given steel grade is checked, a random sample of 10 wheels is prepared. Hardness of each wheel is measured 10 times resulting in a sample of 100 values.

HRC hardness of cylindrical toothed wheels is measured along the axis of teeth at the level of the diameter of tooth spaces. Hardness of teeth with rounded tooth faces is measured along the axis of teeth at the diameter (d_f = 3...5 mm). Hardness of conical toothed wheels is measured along the axis of teeth over the external face.

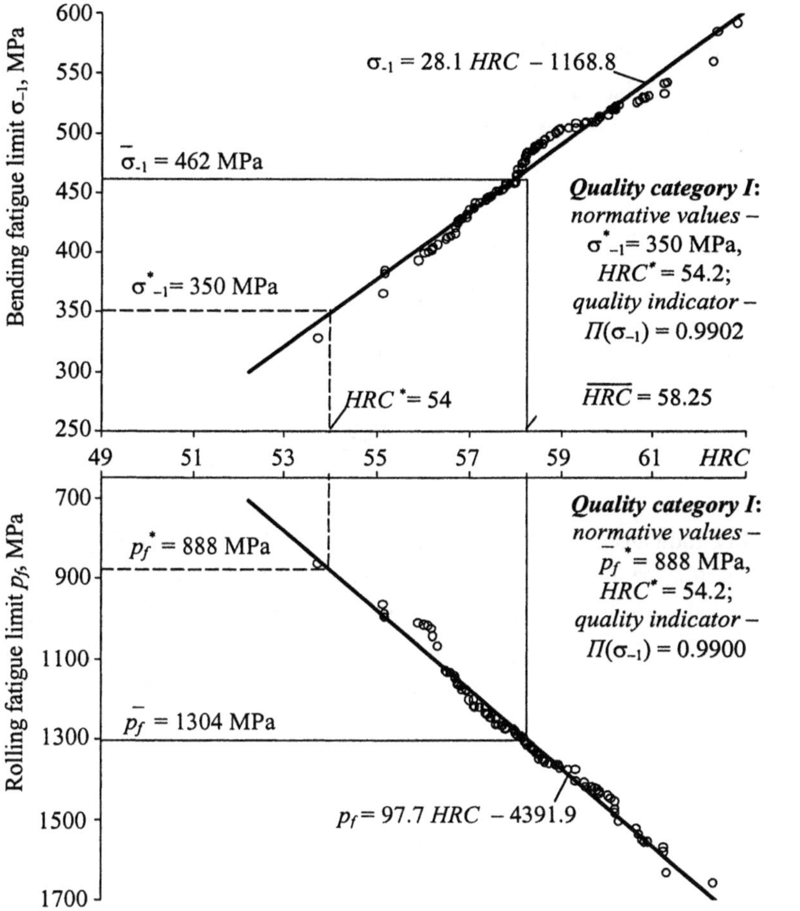

Fig. 5.37. Problem of wear-fatigue damage control of gearings from steel 25 XГТ

Table 5.12. Normative characteristics of resistance to wear-fatigue damage

Characteristic	Value, MPa	
Mean value	$\bar{\sigma}_{-1} = 462$	$\bar{p}_f = 1304$
Mean square deviation	$S_{\sigma_{-1}} = 48$	$S_{p_f} = 156$
Minimum value	$\sigma^*_{-1} = 350$	$p^*_f = 888$

The obtained sample is analyzed statistically to obtain measured (empirical) parameters of hardness distribution:

- mean arithmetic value $\overline{HRC}_{\text{meas}}$;
- mean square deviation S_{HRC}^{meas}.

Normative and measured values of the parameters of hardness distribution are compared:

a) if

$$\left.\begin{aligned} \overline{HRC}_{\text{meas}} &\geq \overline{HRC}^*, \\ S_{HRC}^{\text{meas}} &\leq S_{HRC}^*, \end{aligned}\right\}$$

the first category of quality of toothed wheels is considered achieved;

b) if

$$\left.\begin{aligned} \overline{HRC}_{\text{meas}} &< \overline{HRC}^*, \\ S_{HRC}^{\text{meas}} &< S_{HRC}^*, \end{aligned}\right\}$$

then quality indicators $\Pi(HRC)$, statistical quality violation $D(HRC)$, risk of use $\rho(HRC)$ should be calculated using the sample of measured hardness values HRC_{meas} and the minimum normative value $HRC_{\min}^* = 54.2$. The obtained values of the indicators $\Pi(HRC)$ and $\rho(HRC)$ are compared with the required standard and a conclusion is made whether the toothed wheels satisfy the first category of quality;

c) if

$$\left.\begin{aligned} \overline{HRC}_{\text{meas}} &< \overline{HRC}^*, \\ S_{HRC}^{\text{meas}} &> S_{HRC}^*, \end{aligned}\right\}$$

then the first category of quality of the material of toothed wheels is considered not achieved.

To achieve the first category of quality means that the risk of use of toothed wheels is $\rho(x) < [\rho] = 0.0101$.

From ratios (5.120) and (5.121) it is easy to obtain the transitional function $\lambda_{\sigma_{-1}} = 495.5$ for the bending fatigue and $\lambda_{p_f} = 505$ for the rolling fatigue (i.e. $\overline{\lambda} = 500.25$, on the average) if it is assumed that $P(t) = 2 \cdot 10^{-5}$. To conclude it can be said that the safety of the toothed wheels in question in operation is not less than $R_\rho(t) = 0.99998$, providing the requirement to toothed gears (cf. Fig. 5.37) is satisfied.

5.8 Designing

5.8.1 General

Instead of exemplifying design calculations of a specific active system (that belongs to the calculation and designing work), our task is to analyze in comparison various results of calculation of active systems (*a*) by traditional methods (following individual criteria of resistance to fatigue and/or wear resistance) and (*b*) by methods of tribo-fatigue (using the complex criteria of WFD). Consequently, we will consider just the simplest active systems schematically shown in Fig. 2.1. These systems should be considered *basic* because they are affected both by contact and off-contact (alternating) loads. They can also be considered *typical* because three basic types of friction occur here, viz. sliding, rolling and slippage.

It is apparently impossible to describe the whole variety of cases that may happen in calculations within a brief paragraph, hence, below basic methods of calculations at the stage of designing are illustrated.

5.8.2 Determination of cross sectional dimension

Determine the diameter of the shaft of the active system that operates with mechano-sliding fatigue (cf. Fig. 2.1, *a*). The condition of strength with the safety factor $n_{\sigma\tau}$ should be recorded with the allowance for the direct effect

$$\sigma \leq [\sigma] = \sigma_{-1\tau} / n_{\sigma\tau}, \tag{5.138}$$

where $[\sigma]$ – admissable normal stress. The value $\sigma_{-1\tau}$ (the fatigue limit of the shaft with the allowance for the effect of friction and wear processes) can be determined either experimentally (see Chapter 3) or from one of formulas (4.1), (4.4), (4.9), (5.23), (5.29), (5.29a) etc. depending on the conditions of operation of the system.

Assume the shaft is bent with the moment M. The maximum normal stress in it is

$$\sigma = \frac{M}{W} = \frac{M}{\pi d^3 / 32}. \tag{5.139}$$

Taking into account (5.139) in (5.138), we find the required diameter of the shaft using, for example, formula (5.29a) for estimating the values $\sigma_{-1\tau}$:

$$d_{TF} \geq \sqrt[3]{\frac{32 M n_{\sigma\tau}}{\pi \sigma_{-1\tau}}} = \sqrt[3]{\frac{32 M n_{\sigma\tau}}{\pi \sigma_{-1} \sqrt{\dfrac{1}{R_{\sigma/\tau}} - \dfrac{\tau_W^2}{\tau_f^2}}}}. \tag{5.140}$$

Figure 5.38 shows a graphic analysis of formula (5.140) where the ordinate axis is the ratio $d_{TF}/\,d_F$ of the diameter of the shaft determined with the WFD criterion (d_{TF}) or with the mechanical fatigue criterion (d_F); it is assumed here that the safety factors coincide in both cases ($n_{\sigma\tau} = n_\sigma$). A horizontal dotted line in Fig. 5.38 corresponds to the case $d_{TF} = d_F$. The allowance for friction and wear processes at $R_{\sigma/\tau} = 1$ is illustrated by a curvilinear dotted line, full lines characterize cases when $R_{\sigma/\tau} > 1$ or $R_{\sigma/\tau} < 1$.

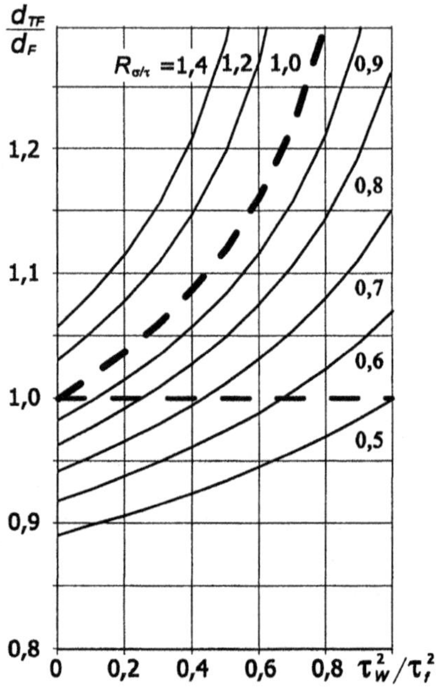

Fig. 5.38. Determination of shaft diameter

For example, assume $\tau_W^2/\tau_f^2 = 0.7$. Then $d_{TF}/\,d_F \approx 1.22$ when $R_{\sigma/\tau} = 1$, and $d_{TF}/\,d_F \approx 1.4$ when $R_{\sigma/\tau} \approx 1.2$. In cases that $R_{\sigma/\tau} \approx 0.5$ and 0.3, we obtain $d_{TF} \approx 0.96 d_F$ and $d_{TF}/\,d_F \approx 1.85$, respectively. Specific values of the parameter $R_{\sigma/\tau}$ of damage interaction are determined as a function of the properties of hardening–softening of the materials assumed for components of the active system and conditions of its operation (cf., for example, Figs. 5.5, 5.15–5.17).

The general conclusion is the following: the shaft's diameter determined with the criterion of tribo-fatigue can be substantially smaller or larger than the diameter determined with the criterion of mechanical fatigue provided the bending moment coincides in both cases. It is clear therefore that the traditional method of calculating components of the active system cannot be considered sufficient.

Now determine the radius of the roller for the active system operating with mechano-rolling fatigue (cf. Fig. 2.2, *b*). In this case the condition of strength with the safety factor $n_{p\sigma}$ should allow for the back effect:

$$p_0 \leq [p] = p_{f\sigma} / n_{p\sigma}, \tag{5.141}$$

where $[p]$ – admissable contact pressure. The value $p_{f\sigma}$ (the contact fatigue limit with the allowance for cyclic stresses) is determined either experimentally (see Chapter 3) or with one of formulas (4.10) of type (5.24), (5.31), etc.

Pressure p_0 for the cylindrical roller with the axis parallel to the axis of the shaft is estimated from formula (1.104a) if the contact load F_N is specified. Taking into account (1.104a) in (5.141), a formula is obtained to calculate the required (nominal) contact area A_{TF} using, for example, formulas (5.31) for estimating the value $p_{f\sigma}$:

$$A_{TF} \geq \frac{2F_N n_{p\sigma}}{\pi p_{f\sigma}} = \frac{2F_N n_{p\sigma}}{\pi p_{f\sigma}\sqrt{\dfrac{1}{R_{\sigma/p}} - \dfrac{\sigma^2}{\sigma_{-1}^2}}}. \tag{5.142}$$

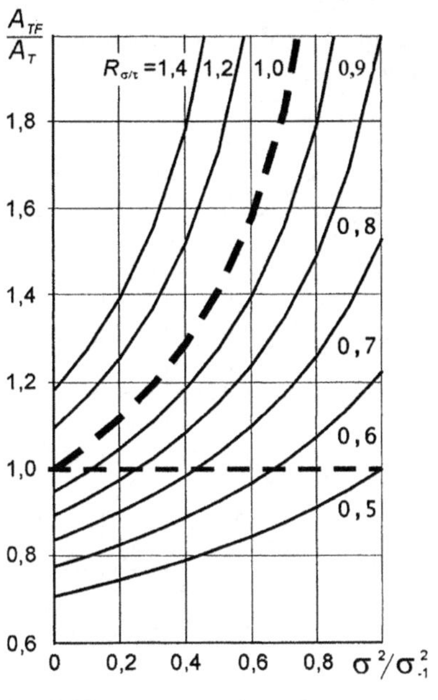

Fig. 5.39. Determination of contact area

If the value of the contact area (5.142) and the shaft's radius are known, it is easy to calculate (for example, with a corresponding Hertzian solution) the roller's radius. The analysis of condition (5.142) is similar to that of condition (5.140) and it is shown in Fig. 5.39. From the latter it follows that depending on the conditions of the system operation and selected materials for fabricating its components, $A_{TF} \gtreqless A_T = A_a$ is possible.

5.8.3 Selection of material

As an example let us consider the problem of selecting a material for the shaft using condition (5.138). Rewrite it in the following form:

$$\sigma_{-1\tau} / n_{\sigma\tau} = [\sigma] \geq \sigma. \tag{5.143}$$

Formula (5.29) for determination of $\sigma_{-1\tau}$, in accordance with condition (5.143) yields

$$[\sigma]_{TF} = \frac{\sigma_{-1}\sqrt{\dfrac{1}{R_{\sigma/\tau}} - \dfrac{\tau_W^2}{\tau_f^2}}}{n_{\sigma\tau}} \geq \sigma. \tag{5.144}$$

From (5.144) the admissable stress is determined using the WFD criterion $[\sigma]_{TF}$, and then the latter serves to determine a specific brand of a material and its condition that maintain the required safety factor.

Figure 5.40 shows a graphic analysis of condition (5.144) providing that the strength safety factors are $n_{\sigma\tau} = n_{\sigma}$.

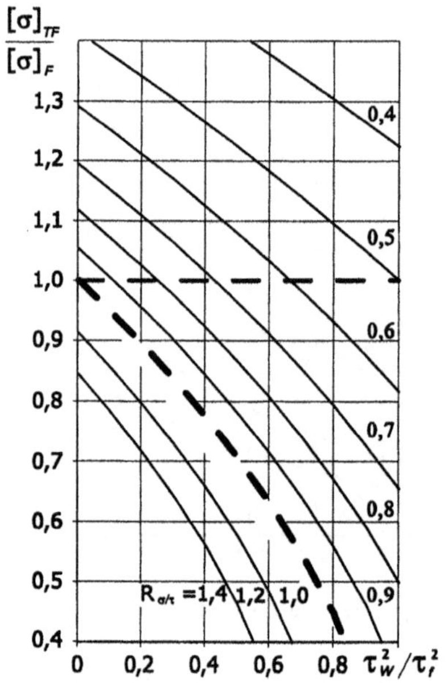

Fig. 5.40. Selection of material

The *normative values* $[n_\sigma]$ of the *safety factor* for calculating structural components for mechanical fatigue are listed in Table 5.13; it also provides the recommended factors $n_{\sigma\tau}$ as applied to calculating components of the active systems using the WFD criterion. It is apparent that $n_{\sigma\tau} < n$; it is one of the effective ways of reducing the design mass of a piece while ensuring the required reliability of most essential active systems in operation.

Table 5.13. Strength safety factors

Safety factors	Required reliability	
	average	higher
$[n]$	1.4...1.6	1.5...1.8
$n_{\sigma\tau}$	1.2...1.3	1.3...1.5

Below we consider a similar problem of selecting a material for the sliding bearing. We have

$$\tau_{f\sigma} / n_{\tau\sigma} = [\tau] \geq \tau_W. \tag{5.145}$$

From (5.145) formula (5.37) for determination of $\tau_{f\sigma}$, yields

$$[\tau]_{TF} = \frac{\tau_f \sqrt{\dfrac{1}{R_{\sigma/\tau}} - \dfrac{\sigma^2}{\sigma_{-1}^2}}}{n_{\tau\sigma}} \geq \tau_W. \tag{5.146}$$

If the admissable stress with the WFD criterion $[\tau]_{TF}$ is known, we choose a specific brand of the material and its condition that satisfy the accepted safety factor. If no recommendations are available which $n_{\tau\sigma}$ is to choose, it can be assumed, as a first approximation, that $n_{\tau\sigma} = n_{\sigma\tau}$ according to Table 5.13.

5.8.4 Requirements to friction coefficient

The friction coefficient plays a significant role in the dynamic analysis of mechanisms of motion. We will briefly consider a different problem: what the friction coefficient should be to ensure the required safety factor of the active system. It is thus admitted that the friction coefficient affects wear processes in the active system, at least indirectly.

This problem regarding, for example, mechano-sliding fatigue is solved with the help of the condition of strength validity:

$$\tau_W = fp_a \leq [\tau], \tag{5.147}$$

whence it follows that

$$f_{TF} \leq [\tau]_{TF} / p_a, \tag{5.148}$$

or, with the account of (5.146),

$$f_{TF} \leq \frac{\tau_f \sqrt{\dfrac{1}{R_{\sigma/\tau}} - \dfrac{\sigma^2}{\sigma_{-1}^2}}}{p_a n_{\tau\sigma}}. \qquad (5.148a)$$

It is to emphasize that formulas (5.148) or (5.148a) answer the question what friction coefficient the active system should have to ensure the required reliability in operation. The problem as to how to determine the friction coefficient in a specific active system is solved with different methods (see Sect. 5.4). Based on (5.105) it is established that the active system should have the following friction index:

$$f_F = \frac{\sigma_a}{\tau_W} \leq [f_{TF}]. \qquad (5.149)$$

In fact, expression (5.149) is *the condition of strength according to the friction parameter*, while the value $[f_{TF}]$ should be considered *the admissable friction parameter.*

5.8.5 Assessment of reliability indicators

Calculation of the complex indicator of reliability of the metal-to-polymer active system is carried out with formula (5.70) and of the metal-to-metal active system with formula (5.81).

For example, it is established in Sect. 5.2.1 that the probability of failure of the active system $P(\sigma, \tau_W)$ is always higher than particular probabilities of failure during friction $P(\tau_W)$ or mechanical fatigue $P(\sigma)$ (see Table 5.4) under comparable conditions. The value of the parameter of interaction between damage $R_{\sigma/\tau}$ can on its part either increase or reduce the value $P(\sigma, \tau_W)$ in response to the conditions of operation of the active system (cf., for example, Figs. 5.15–5.17).

Note that the indicator $P(\sigma, \tau_W)$ is the probability of failure of the active system as a whole; if the latter is known it is impossible to indicate what criterion exactly determines failure. In other words, it is impossible to analyze generally the direct or back effect using formulas (5.70) or (5.81).

In order to make it possible we use the concept of the complex dangerous volume (see Sect. 2.4.3). Then formula (5.73) is recorded as

$$P_\tau(\sigma) = 1 - \exp\left\{ - C_W \frac{W_{P\gamma}}{V_0} \left[\left(\frac{T_M}{T_0} \right)^{m_T} \frac{\sigma - \sigma_{-1\min}}{\sigma_W} \right]^{m_V} \right\}, \qquad (5.150)$$

where $P_\tau(\sigma)$ – the probability of failure with the criterion of resistance to fatigue with the allowance for the friction and wear processes (during direct effect).

Formula (5.74) is similarly written as

$$P_\sigma(\tau_W) = 1 - \exp\left\{-C_W \frac{W_{P\gamma}}{S_k}\left(\frac{\tau_{tf}^{(1)}\Delta T}{\tau_d - \tau_W}\right)^{m_S}\right\}, \tag{5.151}$$

where $P_\sigma(\tau_W)$ – the probability of failure with the criterion of wear resistance with the account of the effect of cyclic stresses (during back effect).

Finally, for the analysis of reliability of active systems it is possible to use the *information indicator of failure free operation*

$$J_{TF} = -k(J_F + J_T) = -k[Q(\sigma)\ln Q(\sigma) + Q(\tau_W)\ln Q(\tau_W)], \tag{5.152}$$

where $Q(\sigma) = 1 - P(\sigma)$ and $Q(\tau_W) = 1 - P(\tau_W)$ – the probability of failure free operation. Based on (5.152) the following relations are obtained:

$$\frac{J_{TF}}{J_F} = k\left[1 + \frac{\varphi_\tau}{\varphi_\sigma}\exp(-\varphi_\tau + \varphi_\sigma)\right]; \tag{5.153}$$

$$\frac{J_{TF}}{J_T} = k\left[1 + \frac{\varphi_\sigma}{\varphi_\tau}\exp(+\varphi_\tau - \varphi_\sigma)\right], \tag{5.154}$$

where the strength function φ_σ is determined from expression (5.53) with substitution of σ_{-1} with σ, and the tribological function is determined from Eq. (5.54) with substitution of τ_f with τ_W.

Note that formulas (5.153) and (5.154) contain dialectic function (5.135), namely

$$\delta_1 = \exp(-\varphi_\tau + \varphi_\sigma), \quad \delta_2 = \exp(+\varphi_\tau - \varphi_\sigma), \tag{5.155}$$

that is analyzed in Sect. 5.6. It is just to indicate φ_σ in formula (5.153) plays the role of the softening function, while φ_τ in formula (5.154), on the contrary, plays this role.

5.8.6 Calculation of durability

Service life designing is based on the models of durability assessment under given conditions of operation of objects.

Calculation of the durability of active systems based on the different criteria of the limiting state during regular loading is performed with formulas (5.82), (5.91), during block loading with formulas (5.97), (5.98) and during random loading with formulas (5.99), (5.100). Figure 5.20 shows the analysis of relations between the durabilities of active systems and the corresponding structural components or the friction pair (under comparable conditions).

If necessary, the durability at stage *I* (before the main crack appears) and at stage *II* (survivability with the crack) is estimated with respective formulas (2.67) and (2.71) and the like.

The condition of strength when designing service life is

$$N \geq N_R,$$

i.e. the calculated durability N should exceed the normative service life N_R.

5.8.7 Assessment of damage intensity

Formulas for estimating the linear wear intensity of friction pairs, the intensity of fatigue damage of structural components and the intensity of damage of the active systems are given in Sect. 5.5 and analyzed there.

An estimate of the volume intensity of damage of the above objects has been performed with the account of ideas about corresponding dangerous volumes (see Sects. 2.4 and 2.5):

$$I_S = \frac{S_{P\gamma}}{\pi d N_\tau}, \quad I_V = \frac{V_{P\gamma}}{\pi d N_\sigma}, \quad I_W = \frac{W_{P\gamma}}{\pi d N_W}. \tag{5.156}$$

The advantage of formula (5.156) is that simple and physically clear considerations lead to uniform calculations of *comparable characteristics of damage of any objects*. The general sense of the characteristics is that they define the volume of the material damaged (under given conditions of operation or tests) per loading cycle per unit of length of the periphery of the shaft (roller) with the diameter d. For a sliding bearing's insert d is its linear dimension in the direction of motion.

Basic relations between the characteristics (5.156) are the following:

$$\frac{I_W}{I_S} = \frac{W_{P\gamma}}{S_{P\gamma}} \frac{N_\tau}{N_W}, \quad \frac{I_W}{I_V} = \frac{W_{P\gamma}}{V_{P\gamma}} \frac{N_\sigma}{N_W}. \tag{5.157}$$

It follows from (5.156) and (5.157) that it is enough to obtain experimentally (or theoretically) only the values N_σ and N_τ, in order to calculate all the characteristics of intensity of damage of any objects (I_S, I_V, I_W). It is the easiest (and most accurate) to determine the durability N_σ and N_τ under given conditions experimentally; the methods of relevant tests are disclosed in Sects. 1.3.2, 1.4.5–1.4.7, 3.2.3.

Note that formulae (5.156) yield *mean* damage intensity throughout the time of operation (or tests) of the active system. If it is necessary to estimate the damage intensity at a specified moment of time, it can be done with kinetic function (5.92) of WFD accumulation. By expressing the complex measure of damage through the complex dangerous volume

$$\omega_{\Sigma n} = \frac{W_{P\gamma}(n)}{V},$$

we obtain

$$W_{P\gamma} = V\left[1-\left(1-\frac{n(\sigma,\tau_W)}{N_W}\right)^h\right]^q. \tag{5.158}$$

Formula (5.158) is similarly written for a structural component (by substituting $W_{P\gamma}$, $n(\sigma, \tau_W)$ and N_W with $V_{P\gamma}$, $n(\sigma)$ and N_σ) and for the friction pair (for the parameters $S_{P\gamma}$, $n(\tau_W)$ and N_τ). In the latter case the linear wear i_h is used to derive the formula for calculations as a function of duration of operation $n(\tau_W)$

$$i_h = \Delta h\left[1-\left(1-\frac{n(\tau_W)}{N_\tau}\right)^h\right]^q, \tag{5.159}$$

where Δh – ultimate admissable wear.

5.8.8 Analysis of states

The analysis of damaging conditions, limiting and translimiting states of the active system can be performed on the basis of the theory disclosed in Sect. 5.1. As has been shown that value growth of loading parameters (σ, τ_W, T, D) is followed initially by a stronger damage of the active system ω_Σ determined from formula (5.12 ($0 < \omega_\Sigma < 1$) that later reaches the limiting state ($\omega_\Sigma = 1$) and then the onset of the translimiting states ($\omega_\Sigma > 1$) is predicted. If the value is introduced into consideration inverse to damage,

$$k_\omega = \frac{1}{\omega_\Sigma}, \tag{5.160}$$

it is clear that it is the *coefficient of safety*: its growth means stronger reliability of the active system in operation.

Assume that corresponding *degradation* of materials and objects evolves in various (damaged and translimiting) states. The *degradation indicator* D_e should be determined without ambiguity both from the viewpoint of damage (ω_Σ) of the object and from the viewpoint of its safety (k_ω):

$$0 \le \frac{1}{e^{k_\omega}} = D_e = \frac{1}{e^{1/\omega_\Sigma}} \le 1. \tag{5.161}$$

Table 5.14 correlates various numerical values of the indicators ω_Σ, k_ω, D_e and intervals of their possible variations with predictable states and/or situations they may be in.

A-state is the intact state: stresses, though effective, are insufficient to cause any damage.

Table 5.14. Characteristics of objects states

	Damage state	Damage functions ω_Σ	Coefficient of safety k_ω	Degradation indicator D_e	Situations
A	Undamaged	0	∞	0	Safety
B	Damaged	$0 < \omega_\Sigma < 1$	$1 < k_\omega < \infty$	$0 < D_e < 1/e$	Incidents
C	Limiting (critical)	$1 = \omega_c$	$1 = k_c$	$1/e \approx 0.37 = D_c$	Break-down
D	Translimiting	$1 < \omega_\Sigma < \infty$	$0 < k_\omega < 1$	$1/e < D_e < 1$	Catastrophe
E	Decomposition	∞	0	1	Cataclysms

B-state is the stage of damage. *C-state* is the limiting (or critical) state. *D-state* is the stage of degradation (translimiting or transcritical state). *E-state* is decomposition of the object, i.e., for example, its transformation into a different substance.

So, three states of an object out of five (*A*-, *B*- and *D*-states) can be considered as peculiar stages of *evolution* (quantitative changes occur) and two (*C*- and *E*-states) as peculiar *qualitative leaps* (transformations occur).

Basic conclusions from the data in Table 5.14 are the following.

(1) The *safety state* is due to the intactness of the object.

(2) *Incidents as violations of guaranteed safety* occur when damage appears.

(3) The moment when the object reaches the limiting state is equivalent to an emergency.

(4) Transition of objects into some translimiting state can be due to the appearance of catastrophic situations.

(5) *decomposition* is interpreted as an infinitely extensive damage of the object; it may result from the phenomena called cataclysms.

So:

(*a*) *absolute safety* is characterized by the following governing parameters: $\omega_\Sigma = 0$, $k_\omega = \infty$, $D_e = 0$;

(*b*) decomposition is characterized by the following governing parameters: $\omega_\Sigma = \infty$, $k_\omega = 0$, $D_e = 1$;

(*c*) the limiting state is a critical boundary between the object as an operable entity and its "fate" of becoming incapacitated parts, pieces, fragments, etc.; this boundary is described by the following critical governing parameters: $\omega_c = k_c = 1$, $D_c = 1/e$.

It is worthwhile to note that the value $D_c \approx 0.37$ corresponds to the so-called *golden section*. In fact, decomposition means that such a state has set in which is insurmountable: it is unimaginable that $\omega_\Sigma > \infty$, $k_\omega < 0$, $D_e > 1$. Moreover, pay attention to the relation between the criterion of damage concentration (see Sect. 1.3.2) and the parameter of degradation D_e. Indeed, if it is assumed that $\omega_\Sigma = 1/C_*^{-1/3}$, then

$$0 \le D_e = \frac{1}{e^{C_*^{-1/3}}} = \frac{1}{e^{1/d_{Te}}} \le 1 . \qquad (5.162)$$

In accordance with (5.162), if the mean spacing between cracks is $C_*^{-1/3} = \infty$ (or the mean statistical size of a crack is $d_T = 0$), it is an undamaged object for which $\omega_\Sigma = 0$ and $D_e = 0$. If the spacing between cracks is $C_*^{-1/3} = 0$, so that they coalesce into a single continuous crack of any large size $d_T = \infty$, then it is an infinitely damaged object for which $\omega_\Sigma = \infty$, $D_e = 1$.

In fact, the extent of damage (ω_Σ) that can be identified just with defects creates unambiguously a limit of resistance of the object to damage (σ_{lim}): the stronger the damage the weaker the strength (Fig. 5.41, curve *AB*). The parameter $R > 1$ of interaction between damage (of any origin) increases too when the concentration of damage (defects) grows. Hence, the object reaches the limiting state under different loads depending on its *B*-state (see Table 5.14). Application of special techniques and processes of hardening enables to obtain a *splash of strength* from σ_{C1} to σ_{C2} on the *degradation curve AB* (curve C_1C_2, $R < 1$), but, as it is noted above, any hardening is finite and limited, a catastrophic drop of resistance to damage inevitably follows (curve C_2B, $R >> 1$). An ultimately damaged object ($\omega_\Sigma = 1$), for example, a destroyed shaft has *zero strength* (point *B*).

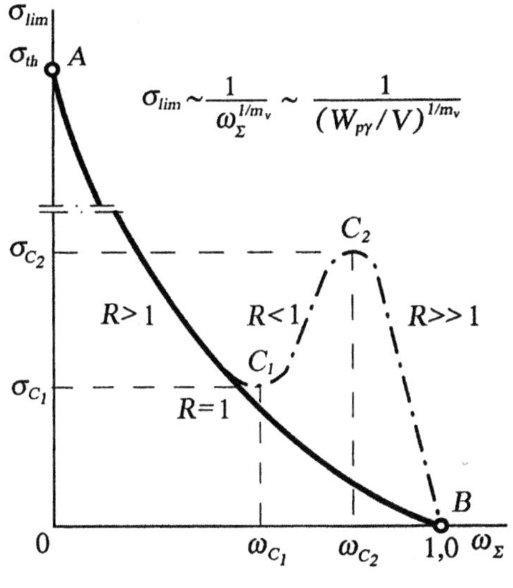

Fig. 5.41. Relation between ultimate stresses and damage concentration (density)

The analysis of the conditions of damage and translimiting states performed here using the data in Table 5.14 is applicable not only to individual active systems but sooner to essential technical objects in which *emergency situations* are possible in operation (such as severe breakdowns, technogenic disasters, etc.); such objects, of course and as a rule, incorporate a great number of active systems.

5.8.9 Prediction of risk and safety

The method of analyzing risk as an expectation of quality loss is disclosed in Sect. 5.6 (see formula (5.124)). Risk in accordance with (5.133) is determined as a relation between the probability of failure of the system and the probability of its failure free operation. Summarizing, *risk is expectation of any unfavorable events (situations)* and numerically this expectation is characterized figuratively by the *share of the bad in the good.* Then, if $P(A_1, A_2, ..., A_n) = P(A_i)$ – the probability of unfavorable events (situations), and $Q(B_1, B_2, ..., B_n) = Q(B_i))$ – the probability of opposite, or favorable events (situations, $i = 1, 2, ..., n$, then the risk

$$0 \leq \rho(A, B) = \rho = P(A_i) \,/\, Q(B_i) \leq \infty \qquad (5.163)$$

is determined providing there is a *peculiar law of preservation* in the form of

$$R_{A/B}\,[P(A_i) + Q(B_i)] = \text{Const} = 1, \qquad (5.164)$$

so that neither $P(A_i)$ nor $Q(B_i)$ vanish, but they interact and pass one into the other:

$$P(A_i) \Leftrightarrow Q(B_i).$$

It means, for example, that in case the probability $P(A_i)$ increases, the probability $Q(B_i)$ cannot grow or stay unchanged, it reduces in accordance with law (5.164). Graphically it is reflected on the *full operative risk characteristic* (Fig. 5.42). This characteristic has the abscissa axis according to interval (5.163) of possible variations of the risk indicator that is why it differs from that in Fig. 5.25, since the interval of risk variations there is limited on the right with the critical value $\rho_k = 1$. Point C corresponds to this value in Fig. 5.42. The current risk indicator in Fig. 5.42 is represented by intercepts $A_\rho B_\rho$, two probability values correspond to each intercept: $Q(B)$ along the line $B_1 B_\rho C B_\rho B_0$ and $P(A)$ along the line $A_0 A_\rho C A_\rho A_1$. Note, that while the segments $A_\rho B_\rho$ are equally long to the right and to the left of the critical point C (they are shown by a full and dotted lines in Fig. 5.42), it does not mean that the numerical values of the risk parameter characterize identical states of damage of the system. For example, if $\rho = 0.1/0.9 < 1$ (the full line $A_\rho B_\rho$), then the B-state occurs according to the classification in Table 5.14, it is vice versa if $\rho = 0.9/0.1 > 1$ (the dotted line $A_\rho B_\rho$), the D-state occurs.

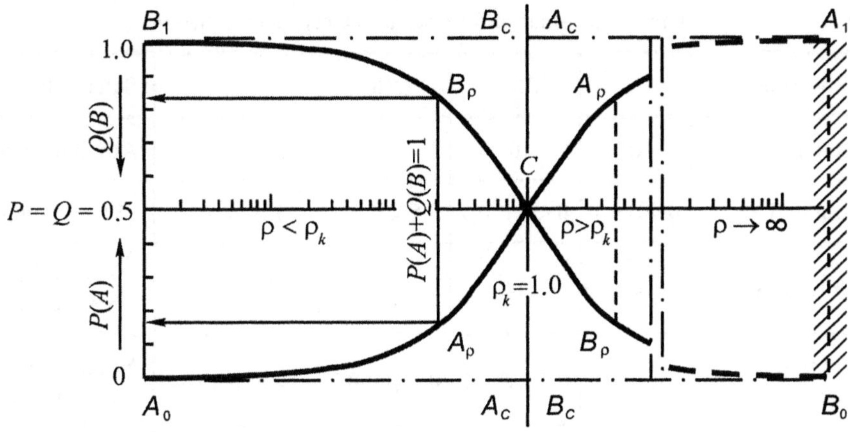

Fig. 5.42. Full operative risk characteristic

Taking into account interval (5.163), Table 5.15 lists quantitative parameters for *risk prediction,* hence, analysis of *safety conditions* because the risk and safety indicators are combined by the relation

$$S_\rho + \rho = 1. \tag{5.165}$$

According to (5.165), if risk reduces, safety grows correspondingly and vice versa. No matter how much the values S_ρ and ρ may vary, their sum remains unchanged and equal to a unity. Assuming for analysis the interval in (5.163), (5.165) establishes that the safety indicator may contain not only positive (at $\rho < 1$) but negative (at $\rho > 1$) values (Fig. 5.43).

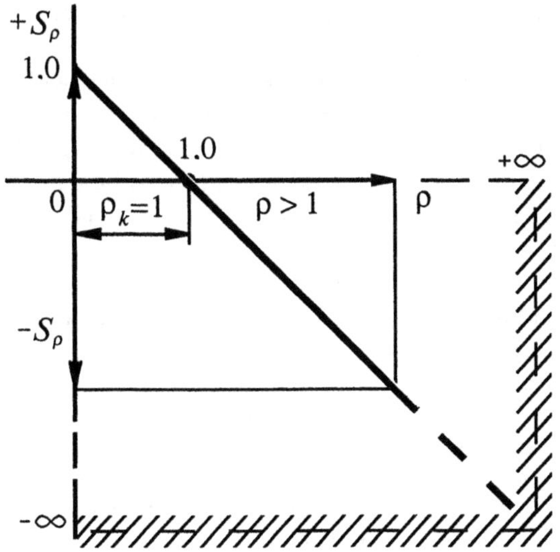

Fig. 5.43. *Risk–safety* diagram

Table 5.15. Analysis of risks and safety: main situations

Risk condition	ρ	$P(A_i)$	$Q(B_i)$	S_ρ	Safety situation
Zero risk	0	0	1	1	Absolute safety
Scanty risk	$0 < \rho < 1$	$0 < P < 0.5$	$0.5 < Q < 1$	$1 > S_\rho > 0$	Scanty safety
Critical risk	$1 = \rho_k$	0.5	0.5	0	Zero safety (break-down)
Transcritical risk	$1 < \rho < \infty$	$0.5 < P < 1$	$0 < Q < 0.5$	$0 > S_\rho > 1 - \infty$	Negative safety (catastrophe)
Infinite risk	∞	1	0	$1 - \infty \approx -\infty$	Absolute danger (cataclysms)

The basic conclusions that can be made from the data in Table 5.15 and Fig. 5.42 together with Table 5.14 are the following.

(1) Only intact objects ($\omega_\Sigma = 0$) are absolutely safe ($S_\rho = 1$), naturally there is no risk to operate them ($\rho = 0$).

(2) *Scanty safety is typical for objects with damage* ($\omega_\Sigma < 1$), the risk of operating them is within the interval $0 < \rho < 1$; any higher risk entails corresponding safety reduction

(3) When risk becomes critical ($\rho_k = 1$), the system reaches its limiting state characterized by the ultimate level of damage ($\omega_c = 1$); in this case the parameter of degradation is $D_c = 1/e$, it is an emergency situation characterized by *zero safety* ($S_\rho = 0$).

(4) *Negative safety* is a menacing risk (a catastrophe) without any guarantee of safety. A catastrophe is possible when the system reaches a corresponding translimiting state ($\omega_\Sigma > 1$) for which risk is transcritical ($\rho > 1$).

(5) Absolute danger ($S_\rho = -\infty$) is identical to infinite damage ($\omega_\Sigma = \infty$) and risk ($\rho = \infty$). Cataclysms apparently happen in such conditions when the system decomposes ($D_e = 1$).

Thus, prediction of risk and safety using the parameters in Tables 5.14 and 5.15 is possible providing the models for calculating damage ω_Σ (see Sect. 5.1) and estimating reliability indicators (see Sect. 5.2) are available.

It is mentioned above (see Sect. 5.1) that the theory of translimiting states of technical objects has not been developed yet. It seems it cannot be developed using exceptionally the ideas of mechanics of a deformable solid. Yet, the *mechanics of damage (including cracking) and degradation, including the concept of risk and safety*, can probably serve the foundation for constructing the theory. The last two paragraphs just outline a possible variant of the theory of translimiting states of active systems.

Self-test questions

1. Describe the basic methods of calculations of active systems. Say about the purposes and tasks of the methods you know.
2. Formulate the basic notions serving the basis of the theory of limiting states of active systems.
3. What is the effective energy – thermal, force, frictional?
4. Formulate and write down the energy criterion of the limiting state of active systems. Indicate its main features and advantages. May be you have any critical comments.
5. Write down the particular energy criteria of isothermal mechanical and isothermal sliding fatigue. In what way are they different? What do they have in common?
6. Is a thermochemical explosion in the active system possible? Say what you know about it.
7. What is the fundamental feature of the parameter U_0 as the energy of breaking atomic bonds? What does not it depend on? What determines it?
8. Write down the boundary conditions from which the coefficients a are determined and which are used to release the effective portions from full energies.
9. How can one determine experimentally the parameters of WFD interaction ($R_{T/M}$, $R_{\sigma/\tau}$) ? What do these parameters signify? What is their role in estimating damage of the active system?
10. How can one determine a specific contribution of each type of corrosive damage (friction corrosion, thermal corrosion, stress-induced corrosion) and how do they affect the resistance to WFD?
11. What are the symptoms of asymmetry of WFD processes? What parameter serves to rate them? What numerical values can this parameter have? Illustrate your answer with a suitable energy diagram.
12. Give a graphic interpretation of the energy criterion of limiting states of active systems in case of the direct effect. Based on what criterion is the limiting state reached? What damages are attendant?
13. Give a graphic interpretation of the energy criterion of limiting states of active systems in case of the back effect. Based on what criterion is the limiting state reached? What damages are attendant?
14. What is the role of the friction and wear processes in determining the durability (the bearing capacity) of an active system following the criterion of resistance to fatigue? Can wear reduce or promote the reliability of a system?
15. What is the role of cyclic stresses in determining the wear resistance of an active system? Can these stresses boost wear resistance rather than impair it?
16. What is the governing parameter of WFD? Within what interval does it vary? What value is critical and what is the critical state of an active system?
17. Depict schematically the graphs of five possible diagrams of the limiting state of active systems during the direct and back effects. Describe features of each graph.
18. What is the relation between the parameter of interaction of damage $R_{\sigma/\tau}$ and the relative value of the parameter of asymmetry of WFD processes?

19. Describe the role of hardening-softening processes in determining the damage and the limiting state of active systems. With what parameters can the ratio between these processes be controlled?
20. Linearize the relation between the parameter of interaction between damages ($R_{\sigma/\tau}$) and the relative parameter ($\overline{\rho}$) of asymmetry of WFD processes. Evaluate the typical graphs.
21. Analyze the basic regularities of interactions between damages during isothermal fatigue. How are they described graphically and analytically?
22. What are the grounds to assert that ultimate stresses σ_{-1} and τ_f are random values? Can they be treated independent?
23. Write down the three-parameter distribution of Waybull and describe the sense of the parameters of distribution.
24. How are the strength parameters $\sigma_{-1\min}$, σ_W, m_V determined that the distribution of Waybull contains?
25. How are the tribological parameters τ_d, $\tau_{tf}^{(1)}$, m_S determined that the distribution of Freshet contains?
26. What do you know about the distribution of Freshet? How many parameters does this distribution contain?
27. How is the parameter of thermal activation of fatigue damage (m_T) determined? What is its sense?
28. Write down the function of failure probability of the metal-to-polymer active system and analyze how the failure probability is affected.
    a) by effective friction and cyclic stresses;
    b) by the strength parameters;
    c) by the tribological parameters;
    d) by the temperature of the metal and the polymer.
29. How does the function of probability of failure of the metal-to-polymer active system allow for the processes of interaction between fatigue and friction damage? What can this interaction lead to?
30. Is it true to assert that the numerical values of the effective stresses (σ and τ_W) fully determine the contribution of the latter into the probability of failure of the metal-to-polymer active system? What does the overloading coefficient imply? And the coefficient of removal?
31. Can the inequality occur in the metal-to-polymer active system: a) $\sigma_{-1\tau} < \sigma_{-1}$? b) $\sigma_{-1\tau} < \sigma_{-1\min}$?
32. Can the inequality occur in the metal-to-polymer active system: a) $\tau_{f\sigma} > \tau_d$? b) $\tau_{f\sigma} < \tau_f$?
33. How are the conditions of reliability recorded when the direct and back effects are studied? Formulate these conditions. Write down the generalized conditions of failure and the probabilities of failure-free operation using the statistical model of the deformable solid with a damaged volume.

34. Write down the target functions for solving the problem of optimization of the metal-to-polymer active system. How do they change for the metal-to-metal system?
35. How to calculate the intensity of wear of a polymer such as to satisfy the required condition of reliability of the metal-to-polymer active system?
36. Can the wear intensity in the metal-to-polymer active system be smaller than in the corresponding friction pair? Why? Indicate the equation that can provide an answer.
37. If the necessary functions to calculate the reliability of the metal-to-polymer active system are available, can they be used to calculate the reliability of the metal-to-metal active system? Should any principal changes be made? What exactly?
38. Find the Table that summarizes the conditions of reliability for three objects: a structural component, a friction pair, and an active system. Analyze these conditions in comparison.
39. What is technical service life? What units serve to measure it?
40. If the mechanical fatigue curve is known (and its parameters, naturally), can you plot the curve of mechano-sliding fatigue (when the direct effect occurs)?
41. If the sliding fatigue curve is known (and its parameters, naturally), can you plot the curve of mechano-sliding fatigue (when the back effect occurs)?
42. In what way are the corresponding (what?) fatigue curves correlated in order to obtain some (what?) formula for calculating the service life of an active system during regular stress?
43. What typical conditions of block loading of active systems do you know? Depict them graphically and compare.
44. What principles (hypotheses) serve as the basis of the method of calculating service life in case of block loading?
45. How does random loading differ from block loading? What is the fundamental difference between the methods of calculating service life in these two situations?
46. What individual characteristics can describe the process of friction in the active system? How many characteristics do you know?
47. How is the cyclic force of friction determined ?
48. How is the cyclic friction coefficient calculated?
49. What is the index of friction in the active system? How can it be calculated? What is its relation with the common friction coefficient?
50. Can the friction index be related to the relative parameter of asymmetry of WFD and its governing parameter?
51. What value is called the parameter of friction in the active system?
52. What combination of indicators determines the state-of-the-art and competitiveness of a modern machine?
53. Make a general analysis of the approach "quality – risk – safety". Can you write the algorithms of its implementation?
54. What indicators serve to estimate the quality of an active system? What WFD characteristics serve to determine these indicators?
55. What is risk? Can it be calculated with the statistical quality indicators? Is a risk indicator a random or deterministic value?

56. How is the operative risk characteristic constructed? Does it relate to the quality indicators?
57. What is the normative (or admissible) risk? How can it be established?
58. What is the critical risk? What is the condition of its occurrence?
59. How is a risk pie diagram plotted? What purpose does it serve?
60. How is the safety notion interpreted when risk appears? How is the safety indicator calculated?
61. What is the dialectic function? What does it describe? What purposes does it serve?
62. Formulate the problem of optimal control over the WFD processes. How is the current state of an active system diagnosed and rated? How is the algorithm of control constructed?
63. List the main *tf*-channels of WFD control in the active system that you know. Describe the role (significance) of each.
64. What methods of calculation of the active systems do you believe the most (the least) essential? Corroborate your view. What methods of calculation should be additionally disclosed (or developed)?
65. Describe the procedure of determination of the cross section size of a component of the active system? In what way does it differ from the calculation of strength of a structural component?
66. How can the material for a component of the active system be selected in order to achieve a normal safety factor?
67. How can you justify the requirements to the friction coefficient needed to maintain the reliability of the active system in operation?
68. How can you assess the intensity of damage of a structural component, a friction pair, and an active system using the idea about corresponding damaged volumes? If it is possible, in what way can you do it?
69. What states of damage of the active system do you know? What situations correspond to them?
70. What is the indicator of degradation of the active system? Within what interval do its numerical values vary? Which are typical values?
71. Is there any relation between the indicator of degradation and the coefficient of safety of the active system? Compare their possible numerical values. Show corresponding situations of risk and safety.
72. Give a definition of risk as expectation of unfavorable events (situations). What is the interval of changes of the risk indicator?
73. Can you depict the full operative risk characteristic and analyze it?
74. What is the relation between the indicators of risk and safety? Can safety be negative?
75. What is limited risk? In what situations is its appearance predictable?
76. Is it possible to characterize numerically absolute safety and absolute danger? How? What do these notions mean?
77. How do you interpret zero safety? Can you describe it numerically? With what indicators?

78. Can you say about the significance of tribo-fatigue in solving general problems of ensuring guaranteed operation reliability and durability of modern machines and equipment?

Tasks for research

1. Using reference manuals analyze the scale dependencies of the fatigue limits of several brands of materials. Depict them graphically and calculate the parameter of mechanical non-homogeneity m_V of each brand. Now determine the strength limits (yield limits) of the same brands of materials and plot their relations with the parameter m_V. Is there any stable correlation between them? If it is there, describe it with the empirical equation. If not, try to understand why it is missing in this case. You realize how essential it is to establish this relation: an opportunity appears to assess the parameter m_V from the results of simplest tests for tension.
2. Write a PC program to calculate the probability of failure of the metal-to-polymer active system (Eq. (5.70)). Perform a graphic analysis how the probability of failure changes: a) if strength parameters $\sigma_{-1\min}$, σ_W, m_V change; b) if tribological parameters τ_d, $\tau_{tf}^{(1)}$, m_S change: c) if the thermodynamic parameter m_T changes. Which of these parameters affect the probability of failure change of the system more strongly? Can it be said that in principle the tribological or strength parameters are more effective to control the failure probability?
3. Under the guidance of the instructor try to formulate, specify and solve the problem of optimization (5.78). Definitely it is the most challenging (and appealing) among all the tasks you have been proposed.
4. Prepare a PC program to analyze variations of the intensity of damage of the metal-to-polymer active system due to possible variations of basic parameters (formulas (5.118) and (5.119)). Indicate the effective ways of controlling the intensity of damage that ensue from your analysis.
5. Investigate the function of probability of failure for the metal-to-metal system (formula (5.81)) similarly to what is done in the manual in relation to the metal-to-polymer active system.
6. Try to propose (proceeding from any consideration, for example, mathematical) two or three functions possessing a common (obligatory) property: their numerical values in case of changes of variables should acquire the values $R \gtreqless 1$. If you manage it, check whether they can be used (specified) as a function of interaction between damage (at least under limited conditions).
7. Study the methods of estimating the coefficient of friction of the *wheel/ rail* pair in special literature. Proceeding from the fact that the wheel/rail pair is indeed an active system, try to calculate the cyclic friction coefficient like (5.104) and the complex friction indicator like (5.108). Analyze how the cyclic (bending) load affects the characteristics of friction in the *wheel/rail* system?
8. Formulate the problem of determination of the required moment of resistance of the rail at some (specified by the instructor) axial loads. Make two alternative calculations: (1) with the allowance for the effect of friction processes (using formula (5.40)); and

(2) when these processes are ignored, i.e. using the traditional method. Compare the results.

9. Make a sample (for example, during one year) of violations of traffic safety on railways, including train wreck, bad quality, and poor workmanship. The data of types of accidents and their gravity are usually compiled in statistical reports (prepared by each section). Analyze the obtained data and plot the operative risk characteristics (cf. Fig. 5.25) for different service-wagon, locomotive, track services, etc. Summarize and plot a risk pie diagram (cf. Fig. 5.27) for these services. What will you recommend to improve traffic safety?

BIBLIOGRAPHY

To hear others' views is almost the same as to visit other countries and other epochs.

Johann Wolfgang von Goethe

Chapter 1

Standards

1. GOST 23207–78. Fatigue resistance. Basic terms, definitions and designations. (in Russian).
2. GOST 25.502–79. Calculations and tests for strength in machine building. (in Russian). Methods of mechanical tests of metals. Methods of fatigue tests. (in Russian).
3. GOST 27.002–78. Reliability in engineering. Basic terms and definitions. (in Russian).
4. GOST 27674–88. Friction, wear and lubrication. Terms and definitions. (in Russian).

General problems of machine building

1. Frolov K V, Methods of improving machines and modern problems of machine science. Mashinostroenie, Moscow, 1984. (in Russian).
2. Mechanics of machines on the eve of the third millennium. Proc. of Int. Sci. Conf. Ed. by M Vysotsky, L Krasnevsky, V Algin, Belavtotraktorostroenie, Minsk, 2001. (in Russian).
3. Orlov P I, Fundamentals of designing in 3 vol., Mashinostroenie, Moscow, 1977. (in Russian).
4. Problems of reliability and service life in machine building. Ed. by K Frolov, A Gusenkov, Nauka, Moscow, 1986. (in Russian).

Strength of materials

The theory of elasticity and plasticity

1. Bartenev Y M, Strength and mechanics of polymer fracture. Khimiya, Moscow, 1984. (in Russian).
2. Bezukhov N I, Fundamentals of the theory of elasticity, plasticity and creep. Textbook, Vysshaya Shkola, Moscow, 1968. (in Russian).
3. Feodosyev B I, Strength of materials. Textbook, Nauka, Moscow, 1986. (in Russian).
4. Panin V E, Egorushkin V E, Makarov P V et al., Physical mezomechanics and computer-aided designing of materials. Nauka, Novosibirsk, v. 1, v. 2, 1995. (in Russian).
5. Pisarenko G S, Agaryov V A, Kvitka A L et al., Strength of materials. Textbook, Ed. by G Pisarenko, Vysshaya Shkola, Kiev, 1986. (in Russian).
6. Rabotnov Yu N, Mechanics of deformable solid. Nauka, Moscow, 1979. (in Russian).
7. Regel V R, Slutsker A I, Tomashevsky E E, Kinetic nature of strength of solids. Nauka, Moscow, 1974. (in Russian).
8. Sosnovskiy L A, Basic types of deformation: individual, particular, general. Byel. Inst. of Railway Engineers, Comel, 1987. (in Russian).
9. Sosnovskiy L A, Theories of limiting states of stress. Byel. Inst. of Railway Engineers, Comel, 1991. (in Russian).
10. Starovoytov E I, Fundamentals of the theory of elasticity, plasticity and viscoelasticity. Textbook, Bel. State University of Transport, Gomel, 2001. (in Russian).
11. Starovoytov E I, Strength of materials. Textbook, Bel. State University of Transport, Gomel, 1999. (in Russian).
12. Timoshenko S P, Gere G, Mechanics of materials. Ed. by E Grigolyuk, Mir Publishers, Moscow, 1976. (in Russian).
13. Troshchenko V T, Krasovsky A Ya, Pokrovsky V V, Sosnovskiy L A, Strizhalo V A, Resistance of materials to deformation and fracture. Ed. by V Troshchenko, Naukova Dumka, Kiev, 1993, in two parts. (in Russian).

Cracking resistance

1. Broek D, Fundamentals of mechanics of fracture. Vysshaya Shkola, Moscow, 1980. (in Russian).
2. Kolesnikov Yu V, Morozov E M, Mechanics of contact fracture. Nauka, Moscow, 1989. (in Russian).
3. Kuzmenko I M, Mechanics of fracture. Textbook, Mogilyov State Tech. Uni., Mogilyov, 2001. (in Russian).
4. Makhutov N A, Deformation criteria of fracture and calculation of strength of structural components. Mashinostroenie, Moscow, 1981. (in Russian).
5. Parton V Z, Morozov E M, Mechanics of elastoplastic fracture. Nauka, Moscow, 1985. (in Russian).

6. Troshchenko V T, Pokrovsky V V, Prokopenko A V, Resistance of metals to cracking under cyclic loading. Naukova Dumka, Kiev, 1986. (in Russian).

Mechanical fatigue

1. Bogdanovich A V, Sosnovskiy L A, Starovoytov E I, Laboratory practical work in mechanical fatigue. Byel. Inst. of Railway Engineers, Comel, 1990. (in Russian).
2. Filimonov G N, Balatsky L T, Fretting in joints of ship parts. Sudostroenie, Leningrad, 1973. (in Russian).
3. Heywood R B, Designing with allowance for fatigue. Mashinostroenie, Moscow, 1969. (in Russian).
4. Ivanova V S, Terentyev V F, Nature of fatigue of metals. Metallurgiya, Moscow, 1975. (in Russian).
5. Karpenko V G, Effect of the environment on strength and durability of metals. Naukova Dumka, Kiev, 1976. (in Russian).
6. Kogaev V P, Calculation of strength under stresses variable in time. Mashinostroenie, Moscow, 1977. (in Russian).
7. Kogaev V P, Makhutov N A, Gusenkov A P, Calculation of strength and durability of machines and components. Reference book. Mashinostroenie, Moscow, 1985. (in Russian).
8. Kudryavtsev I V, Naumchenkov N E, Savvina N M, Fatigue of large machine parts. Mashinostroenie, Moscow, 1981. (in Russian).
9. Oleynik N V, Carrying capacity of structural components under cyclic loading. Naukova Dumka, Kiev, 1985. (in Russian).
10.Sosnovskiy L A, Statistical machanics of fatigue fracture. Navuka i tekhnika, Minsk, 1987. (in Russian).
11.Troshchenko V T, Fatigue and non-elasticity of metals. Naukova Dumka, Kiev, 1971. (in Russian).
12.Troshchenko V T, Sosnovskiy L A, Resistance of metals and alloys to fatigue. Reference book in 2 volumes. Naukova Dumka, Kiev, 1987. (in Russian).

Friction and wear

1. Bogdanovich P N, Prushak V Ya, Friction and wear in machines. Vysshaya Shkola, Minsk, 1999. (in Russian).
2. Buckley D, Surface phenomena during adhesion and contact interaction. Ed. by A Sviridyonok, Mashinostroenie, Moscow, 1986. (in Russian).
3. Chichinadze A V, Braun E D, Bushe N A et al., Fundamentals of tribology. Textbook. Mashinostroenie, Moscow, 2001. (in Russian).
4. Drozdov Yu N, Pavlov V G, Puchkov V N, Friction in extreme conditions. Mashinostroenie, Moscow, 1986. (in Russian).
5. Garkunov D N, Triboengineering. Mashinostroenie, Moscow, 1985. (in Russian).

6. Goryacheva I G, Dobychin N M, Contact problems in tribology. Mashinostroenie, Moscow, 1988. (in Russian).
7. Kragelsky I V, Dobychin M N, Kombalov V S, Fundamentals of calculation of friction and wear. Mashinostroenie, Moscow, 1977. (in Russian).
8. Surface strength of materials in friction. Ed. by B Kostetsky, Tekhnika, Kiev, 1976. (in Russian).
9. Sysoev P V, Bogdanovich P N, Lizarev A D, Deformation and wear of polymers in friction. Navuka i Tekhnika, Minsk, 1985. (in Russian).
10. Waterhouse R B, Fretting corrosion. Ed. by G Filimonov, Mashinostroenie, Leningrad, 1976. (in Russian).

Reliability

1. Algin V B, Dynamics, reliability and life designing of durable transmissions of mobile machinery. Navuka i Tekhnika, Minsk, 1995. (in Russian).
2. Biryukov M P, Biryukov P M, Machine parts. Calculation of reliability. VUZYUNITY, Minsk, 1999. (in Russian).
3. Bolotin V V, Statistical methods in civil mechanics. Stroyizdat, Moscow, 1965. (in Russian).
4. Kapur L, Lamberson L, Reliability and designing of systems. Mir Publishers, Moscow, 1978. (in Russian).
5. Pronikov A S, Reliability of machines. Mashinostroenie, Moscow, 1978. (in Russian).
6. Reshetov D N, Operability and reliability of machine parts. Textbook. Vysshaya Shkola, Moscow, 1974. (in Russian).
7. Sosnovskiy L A, Elements of theory of probabilities, mathematical statistics and theory of reliability. Textbook. Bel. States Uni. of Transport, Gomel, 1994. (in Russian).
8. Sosnovskiy L A, Probabilistic methods of designing of strength in linear and complex states of stress. Two parts, Byel. Inst. of Railway Engineers, 1984. (in Russian).
9. Tsitovich I S, Berestnev O V, Ways of improving reliability of machinery. Navuka i Tekhnika, Minsk, 1979. (in Russian).

Machine parts

1. Course designing of machine parts. Ed. by V Kudryavtsev, Mashinostroenie, Leningrad, 1984. (in Russian).
2. Ivanov M N, Machine parts. Textbook. Vysshaya Shkola, Moscow, 2000. (in Russian).
3. Kurmaz L V, Skoybeda A T, Machine parts. Designing. Tekhnoprint, Minsk, 2002. (in Russian).
4. Skoybeda A T, Kuzmin A V, Makeychik N N, Machine parts and basics of designing. Textbook. Vysshaya Shkola, Minsk, 2000. (in Russian).

Chapters 2–5

Standards

1. GOST 30638–99. Tribo-fatigue. Terms and Definitions. (in Russian).
2. GOST 30754–2001. Tribo-fatigue. Methods of tribo-fatigue tests. Tests for mechano-rolling fatigue. (in Russian).
3. GOST 30755–2001. Tribo-fatigue. Machines for tribo-fatigue tests. General technical requirements. (in Russian).
4. STB 1233–2000. Tribo-fatigue. Methods of tribo-fatigue tests. Accelerated tests for mechano-rolling fatigue. (in Russian).
5. STB 1234–2000. Tribo-fatigue. Active systems. Statistical quality indicators. (in Russian).
6. Tribo-fatigue – 95. Standardization. Chronicle. Gomel, 1996. (in Russian).

Books

1. Frolov K V, Sosnovskiy L A, Makhutov N A, Drozdov Yu N, Tribo-fatigue: new ideas in promising direction. S&P Group TRIBOFATIGUE, Gomel, 1990. (in Russian).
2. SI-series machines for wear-fatigue tests. Ed. by M S Vysotsky, S&P Group TRIBOFATIGUE, Gomel, 1995. (in Russian).
3. Sosnovskiy L A, Troshchenko V T, Makhutov N A, Gao Vanzhen, Shcherbakov S S, Wear-fatigue damage and its prediction (tribo-fatigue). S&P Group TRIBOFATIGUE, Gomel, 2001. (in Russian).
4. Sosnovskiy L A, Makhutov N A, Tribo-fatigue: Wear-fatigue damage in problems of life and safety of machinery. FCNTP 'Safety', Moscow, 2000. (in Russian).
5. Frolov K V, Makhutov N A, Troshchenko V T et al., Some words about tribo-fatigue. Remika, Minsk, 1996. (in Russian).
6. Sosnovskiy L A, Bogdanovich A V, Theory of accumulation of wear-fatigue damage. S&P Group TRIBOFATIGUE, Gomel, 2000. (in Russian).
7. Koreshkov V N, Ozhigar Y P, Sosnovskiy L A, Modeling active systems. S&P Group TRIBOFATIGUE, Gomel, 1999. (in Russian).
8. Zhmaylik V A, Podobedov V N, Sosnovskiy L A, Quality of materials and active systems. S&P Group TRIBOFATIGUE, Gomel, 1998. (in Russian).
9. Proc. of III Int. Sym. on tribo-fatigue. Ed by Gao Vanzhen and Li Jian, Hunan University Press, China, 2000.
10. Wear-fatigue tests, methods, some results. S&P Group TRIBOFATIGUE, Gomel, 1996. (in Russian).
11. Draygor D A, Valchuk G I, Effect of wear on fatigue strength of steel with allowance for scale effect. Academy of Sciences of the Ukrainian SSR Publishers, Kiev, 1962. (in Russian).
12. Sharay V T, Comprehensive study of wear and fatigue of steels. Tech. Literature of the Ukrainian SSR Publishers, Kiev, 1959. (in Russian).

Dictionaries

1. Sosnovskiy L A, Mechanics of fatigue fracture. Reference-dictionary in two volumes. S&P Group TRIBOFATIGUE, Gomel, 1994. (in Russian).
2. Трибофатика. Tribo-fatigue. Tribo-Ermiidung. Four-language terminological dictionary. Ed. by L Sosnovskiy, S&P Group TRIBOFATIGUE, Comel, 1995. (in Russian).

Bibliographic indicators

1. 100 first publications on tribo-fatigue. Literature for the period 1985–1996. Compiled by T Eseeva, S&P Group TRIBOFATIGUE, Gomel, 1997. (in Russian).
2. Bibliographic list of publications on tribo-fatigue for the period 1985–2000. Compiled by T Eseeva, S Tyurin, S&P Group TRIBOFATIGUE, Gomel, 2001. (in Russian).

Manuals

1. Bogdanovich A V, Elovoy O M, Marchenko A V, Fundamentals of tribo-fatigue. Laboratory practical work, part 1, ed. by L Sosnovskiy, Belarusian State University of Transport, Gomel, 1999. (in Russian).
2. Bogdanovich A V, Elovoy O M, Sosnovskiy L A, Assessment of basic indicators of reliability and risk of nonreconditioned parts. Byel. Inst. of Railway Engineers, Gomel, 1995. (in Russian).
3. Bogdanovich A V, Elovoy O M, Sosnovskiy L A, Assessment of reliability of simple crankshaft, part 2, Belarusian State University of Transport, Gomel, 2002. (in Russian).
4. Sosnovskiy L A, Complex estimate of reliability of active systems based on the criteria of fatigue and wear resistance. (Fundamentals of tribo-fatigue). Byel. Inst. of Railway Engineers, Gomel, 1988. (in Russian).
5. Sosnovskiy L A, Indicator of safety and operative risk characteristic. Byel. Inst. of Railway Engineers, Gomel, 1991. (in Russian).
6. Sosnovskiy L A, Skrezhendevsky VV, Elovoy O M, Assessment of reliability of simple crankshaft, part 1, Byel. Inst. of Railway Engineers, Gomel, 1994. (in Russian).

SUBJECT INDEX

APPENDIX

Appendix I
SCIENTISTS ABOUT TRIBO-FATIGUE

One should obey the helmsman so as to sail over the seas ...

Pythagoras

... It is in Belarus that a new science – tribo-fatigue – was born and has been developing successfully.

Professor **Konstantin V. Frolov,**
Vice-president of the Russian Academy of Sciences,
Academician of the Russian Academy of Sciences and the Belarusian National Academy of Sciences,
Hero of Labor,
Doctor of Technical Sciences,
Head of A. A. Blagonravov Institute of Machine Studies of the Russian Academy of Sciences

(*September, 1993*)

... From the engineering point of view, tribo-fatigue is based on the study of the interaction of elements of mechanical systems. It is this interaction that governs finally the reliability both of parts and a machine as a whole. It means that we seize making calculation for each specific element but start making calculation and design of mechanical systems.

Professor **Michael S. Vysotsky,**
Vice-President of the Academy of Sciences of Belarus,
Academician of the Academy of Sciences of Belarus,
Doctor of Technical Sciences,
General Designer of the Industrial Group "BelautoMAZ",
Laureate of the State Prize of Belarus and the USSR

(*September, 1993*)

... A new science, tribo-fatigue... will allow critical units of machines and structures to be properly designed.

Professor **Valery T. Troshchenko,**
Academician of the National Academy of Sciences of Ukraine,
Doctor of Technical Sciences,
Head of the Institute of Problems of Strength of the National Academy of Sciences of Ukraine

(*July, 1995*)

Tribo-fatigue as a science most likely will hold its place in one of the new trends, in which the Russian Academy of Sciences and the Academies of Sciences of other countries... are engaged. This trend includes: ... physics, chemistry, the mechanics of catastrophes, safety and protection methods. Its aim may be formulated briefly as follows: the elaboration of methods, means, criteria and ways of protection against technical accidents and catastrophes.

Professor **Nickolai A. Makhutov,**
Corresponding Member of the Russian Academy of Sciences, Doctor of Technical Sciences,
Head of the Department of A. A. Blagonravov Institute of Machine Studies of the Russian Academy of Sciences,
Chairman of Scientific Council of The Ministry of Extraordinary Situations of Russia

(*September, 1993*)

...The problems of the development of aircraft-cosmic equipment, atomic energetics, engineering facilities for conquering the Ocean, deep entrails of the Earth, the creation of artificial joints, heart-valves, different organs of a man and an animal, the industrial development of high density magnetic recording, the improvement of reliability of a new generation of machines and mechanisms, raising the efficiency, accuracy and safety are closely linked with the level of knowledge in tribology, corrosion, fatigue, with an understanding of the phenomena of the above-mentioned processes which occur simultaneously. This direction may be called tribo-corro-fatigue.

Professor **Yuri N. Drosdov,**
Doctor of Technical Sciences,
Academician of the Academy of Cosmonautics of Russia,
Head of the Department of A.A. Blagonravov Institute of Machine Studies of the Russian Academy of Sciences

(*June, 1995*)

...The development of tribo-fatigue brings about the introduction and standardization of methods of complex wear-fatigue tests.

Valery N. Koreshkov,
Chairman of the State Committee on Standardization, Metrology and Certification of the Republic of Belarus

(*September, 1993*)

...The birth of a new science happens not so often nowadays, therefore it is undoubtedly a credit to any state... The Republic of Belarus is the cradle of tribo-fatigue. It was acknowledged by the great scientists of our time.

Professor **Vasily I. Strazhev,**
Minister of Education and Science of the Republic of Belarus,
Doctor of Physics and Mathematics

(*July, 1995*)

... Tribo-fatigue is a new and a rapidly developing branch of mechanics... Some years ago the course "Fundamentals of Tribo-fatigue" was introduced into the curriculum of the BelGUT... By now necessary text-books have been published dealing with the course. The presentation of the latest achievements of the science in the framework of the subject contributes undeniably to the improvement of training level of future mechanical engineers.

Professor **Veniamin I. Senko,**
Rector of the Belarusian State University of Transport,
Doctor of Technical Sciences,
Honored Scientist of Education of the Republic of Belarus

(*September, 2002*)

Appendix II
TRIBO-FATIGUE: TERMS AND DEFINITIONS

(According to GOST 30638–99 "Tribo-fatigue. Terms and Definitions")

ТРИБОФАТИКА: ТЕРМИНЫ И ОПРЕДЕЛЕНИЯ

(по ГОСТ 30638–99 "Трибофатика. Термины и определения")

An accurate logical definition of concepts is the main condition of true knowledge.

Socrates

1 General terms

1 Общие понятия

1.1 трибофатика: наука об износоусталостных повреждениях и разрушении силовых систем машин и оборудования

tribo-fatigue: science concerned with wear-fatigue damages and fractures of active systems of machines and equipment

1.2 силовая система: механическая система, в которой реализуется процесс трения в любых его проявлениях и которая одновременно воспринимает и транзитно передает повторно-переменную нагрузку

active system (AS): mechanical system in which any form of friction processes is realized and alternative load is taken up and transmitted through simultaneously

1.3 надежность по критериям трибофатики: надежность силовой системы, обусловленная ее способностью сохранять во времени и в установленных пределах значения характеристик сопротивления износоусталостным повреждениям

reliability on tribo-fatigue criteria: reliability of an active system determined by its ability to remain values of wear-fatigue resistance characteristics in the established intervals during specified time

1.4 износоусталостное повреждение: повреждение, обусловленное кинетическим взаимодействием явлений усталости, трения в любых его проявлениях, изнашивания и (или) эрозии (рисунок 1)

wear-fatigue damage (WFD): damage caused by the kinetic interaction of fatigue, friction, wear and / or erosion phenomena (figure 1)

1.5 сопротивление износоусталостным повреждениям: свойство материала противостоять износоусталостным повреждениям

wear-fatigue resistance: ability of a material to withstand wear-fatigue damage

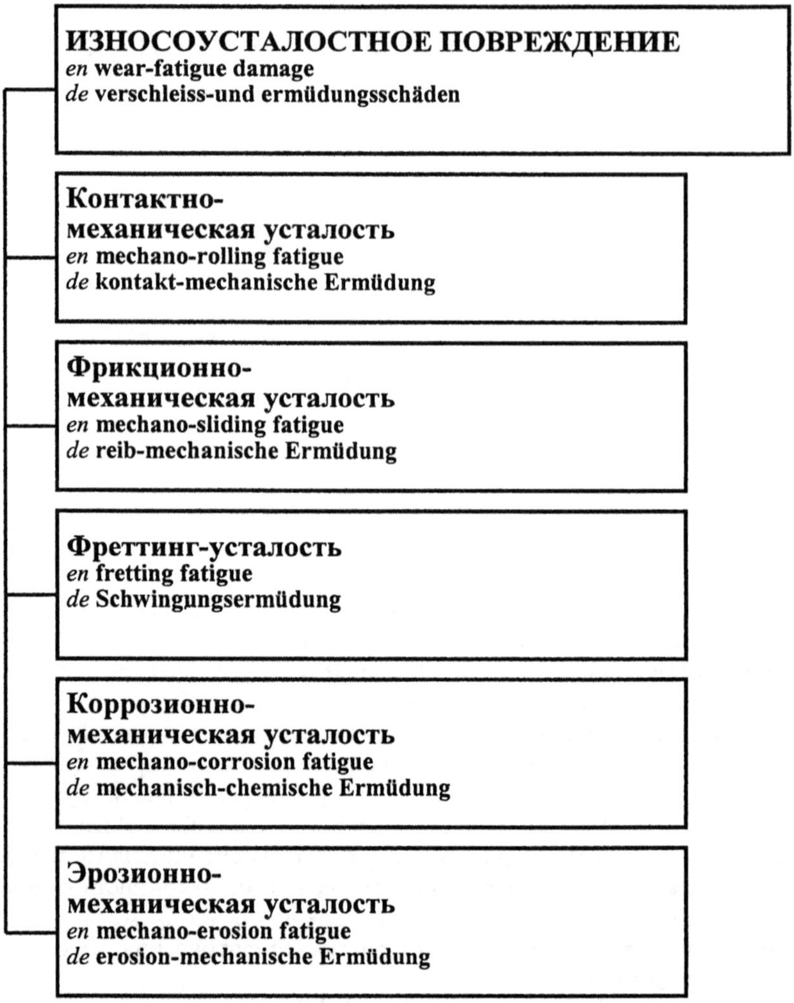

Рисунок 1. Основные виды износоусталостного повреждения

Figure 1. General types of wear-fatigue damage

1.6 прямой эффект: влияние процессов и условий трения и изнашивания на изменение характеристик сопротивления усталости силовой системы и/или ее элементов

direct effect (DE): changes of fatigue resistance characteristics of an active system and/or its elements produced by friction and wear processes

1.7 обратный эффект: влияние повторно-переменных напряжений (деформаций) на изменение характеристик трения и изнашивания силовой системы и/или ее элементов

back effect (BE): changes of friction and wear characteristics of an active system and/or its elements produced by alternative stresses (strains) on

1.8 износоусталостные испытания: испытания, при которых определяют количественные характеристики сопротивления износоусталостным повреждениям

wear-fatigue tests (WFT): tests used to determine quantitative characteristics of wear-fatigue resistance

1.9 машина для износоусталостных испытаний: машина, предназначенная для закрепления силовой системы или ее модели, реализации комплексного воздействия на нее повторно-переменных нагрузок и процессов трения при заданных условиях и режимах, обеспечения требуемой продолжительности испытаний, измерения и регистрации характеристик сопротивления износоусталостным повреждениям

wear-fatigue test machine: machine used for fastening an active system or its model, exposing the system to the complex effect of alternative loads and friction processes under prescribed conditions and modes, providing the required test longivity, measuring and recording its wear-fatigue resistance characteristics

1.10 контактно-механическая усталость: износоусталостное повреждение, обусловленное кинетическим взаимодействием явлений механической усталости и трения качения

mechano-rolling fatigue (MRF): wear-fatigue damage caused by the kinetic interaction of mechanical fatigue and rolling friction phenomena

1.11 фрикционно-механическая усталость: износоусталостное повреждение, обусловленное кинетическим взаимодействием явлений механической усталости и трения скольжения

mechano-sliding fatigue (MSF): wear-fatigue damage caused by the kinetic interaction of mechanical fatigue and sliding friction phenomena

1.12 коррозионно-механическая усталость: усталость материала при одновременном воздействии повторно-переменных напряжений и коррозионной среды

mechano-corrosion fatigue (MCF): wear-fatigue damage caused by the kinetic interaction of the mechanical fatigue and corrosion phenomena

1.13 фреттинг-усталость: износоусталостное повреждение, обусловленное кинетическим взаимодействием явлений механической усталости и фреттинга

fretting fatigue (FF): wear-fatigue damage caused by the kinetic interaction of mechanical fatigue and fretting phenomena

1.14 эрозионно-механическая усталость: износоусталостное повреждение, обусловленное кинетическим взаимодействием явлений механической усталости и эрозии

mechano-erosion fatigue (MEF): wear-fatigue damage caused by the kinetic interaction of the mechanical fatigue and erosion phenomena

1.15 поверхностное крошение: отделение с поверхностей взаимодействия элементов силовой системы мелкодисперсных частиц материала, образующихся в результате множественного микросдвига по пересекающимся плоскостям и дробления зерен при износоусталостном повреждении

surface chipping (SC): detaching fine-dispersed particles of the material formed as a result of multiple microshift in intersecting planes and fragmentation of grains from interacting surfaces of active system's elements under wear-fatigue damage

1.16 опасный объем, $V_{P\gamma}$**:** часть рабочего объема элемента силовой системы, в пределах которой с вероятностью P, соответствующей вероятности усталостного повреждения, установленной с доверительной вероятностью γ, действующие циклические напряжения превышают нижнюю границу рассеивания предельных напряжений

damaged volume, $V_{P\gamma}$**:** part of the working volume of an active system's element within which acting cyclic stresses exceed the lower boundary of limiting stress dispersion with fatigue damage probability P found with confidence γ

1.17 опасная поверхность, $S_{P\gamma}$**:** часть объема рабочего поверхностного слоя элемента силовой системы, в пределах которой с вероятностью P, соответствующей вероятности повреждения при трении, установленной с доверительной вероятностью γ, действующие контактные (фрикционные) напряжения превышают нижнюю границу рассеивания предельных напряжений

damaged surface, $S_{P\gamma}$**:** part of volume of the working surface layer of an active system's element within which acting contact stresses exceed the lower boundary of the limiting stress dispersion with friction damage probability P found with confidence γ

1.18 комплексный опасный объем, $W_{P\gamma}$**:** часть рабочего объема элемента силовой системы, в пределах которой возможно возникновение и развитие износоусталостных повреждений, определяемая как объединение опасного объема $V_{P\gamma}$ и опасной поверхности $S_{P\gamma}$, умноженное на функцию взаимодействия $\varphi_{VS} \gtreqless 1$

complex damaged volume, $W_{P\gamma}$**:** part of the working volume of an active system's element within which there is a possibility of initiating and developing of wear-fatigue damages, this volume being the sum of damaged volume $V_{P\gamma}$ and damaged surface $S_{P\gamma}$ multiplied by the interaction function $\varphi_{VS} \geq 1$

$$W_{P\gamma} = (V_{P\gamma} \cup S_{P\gamma})\ \varphi_{VS}\,.$$

1.19 предельное состояние силовой системы: неработоспособное состояние силовой системы по одному или одновременно нескольким признакам: достижению износом предельного значения, образованию рассеянных трещин или ямок выкрашивания (питтингов) критических размеров или критической концентрации, достижению остаточной деформацией недопустимой величины, образованию слоя коррозионного повреждения критической толщины, появлению поверхностных волн пластичности, образованию магистральной трещины определенной длины, разрушению элементов

limiting state of an active system: inoperative state of an active system manifested by one or several signs simultaneously: achievement limiting value of wear, formation of scattered cracks or spalling stains (pittings) of critical size or critical concentration, achievement intolerable value of residual deformation, generation a critically thick layer damaged by corrosion, apperance of the surface waves of plasticity, formation of a main crack of certain length, fracture of elements

1.20 долговечность силовой системы: продолжительность испытания или эксплуатации силовой системы в заданных условиях до наступления ее предельного состояния

active system life: duration of active system testing or operation under specified conditions until it reaches the limiting state

2 Friction characteristics in an active system

2 Характеристики трения в силовой системе

2.1 сила трения в силовой системе, $F_{f\sigma}$: сила трения, определяемая с учетом действия на пару трения или на один из ее элементов повторно-переменной нагрузки

friction force in an active system, $F_{f\sigma}$: friction force, defined upon alternating load activity in a friction joint or its elements

$$F_{f\sigma} = F_f\,\varphi(\sigma).$$

2.2 циклическая составляющая силы трения, F_c: составляющая силы трения в силовой системе, обусловленная циклическими напряжениями (деформациями) в одном или обоих ее элементах

cyclic component of a friction force, F_c: component of a friction force in an active system caused by cyclic stresses (strains) in one or both its elements

2.3 коэффициент трения в силовой системе, f_σ: отношение силы трения в силовой системе к контактной нагрузке, прижимающей ее элементы друг к другу в процессе трения движения

friction coefficient in an active system, f_σ: ratio of friction force in an active system and contact load pressing its elements to each other during friction process

$$f_\sigma = F_{f\sigma}/F_N.$$

2.4 циклическая составляющая коэффициента трения, f_c: отношение циклической составляющей силы трения к контактной нагрузке

cyclic component of a friction coefficient, f_c: ratio of cyclic component of a friction force and contact load

$$f_c = F_c/F_N.$$

3 Wear-fatigue damage characteristics

3 Характеристики сопротивления износоусталостным повреждениям

3.1 кривая контактно-механической усталости, $N(\sigma_a,\ p_0\ = \text{const})$, $N(p_0,\ \sigma_a = \text{const})$**:** график, характеризующий зависимость между амплитудой напряжений цикла σ_a и циклической долговечностью $N_{\sigma p}$ для одинаковых моделей силовой системы, построенный по параметру контактных напряжений при трении качения p_0 = const (рисунок 2, *а*), либо график, характеризующий зависимость между контактными напряжениями при трении качения p_0 и контактной долговечностью $N_{p\sigma}$ для одинаковых моделей силовой системы, построенный по параметру амплитуды напряжений цикла σ_a = const (рисунок 2, *б*)

mechano-rolling fatigue curve, $N(\sigma_a,\ p_0 = \text{const})$, $N(p_0,\ \sigma_a = \text{const})$**:** either a graph of cycle stress amplitude σ_a versus fatigue life, $N_{\sigma p}$, for the same models of an active system plotted with respect to contact stress under rolling friction p_0 = const (figure 2, *a*), or a graph of contact stress p_0 under rolling friction versus rolling fatigue life, $N_{p\sigma}$, for the same models of an active system plotted with respect to cycle stress amplitude σ_a = const (figure 2, *b*)

3.2 предел контактно-механической выносливости, σ_{-1p}, $p_{f\sigma}$ **:** предел выносливости по параметру контактных напряжений при трении качения, σ_{-1p} (рисунок 2, *а*), либо предел контактной выносливости по параметру амплитуды напряжений цикла, $p_{f\sigma}$ (рисунок 2, *б*)

mechano-rolling fatigue limit, σ_{-1p}, $p_{f\sigma}$ **:** fatigue limit on parameter of contact stress under rolling friction, σ_{-1p} (figure 2, *a*), or rolling fatigue limit on parameter of cycle stress amplitude, $p_{f\sigma}$ (figure 2, *b*)

3.3 предел ограниченной контактно-механической усталости, σ_{-1pN}, $p_{f\sigma N}$**:** предел ограниченной усталости по параметру контактных напряжений при трении качения, σ_{-1pN} (рисунок 2, *а*), либо предел ограниченной контактной усталости по параметру амплитуды напряжений цикла, $p_{f\sigma N}$ (рисунок 2, *б*)

mechano-rolling fatigue limit at *N* cycles: fatigue threshold on parameter of contact stresses under rolling friction, σ_{-1pN} (figure 2, *a*), or rolling fatigue threshold on parameter of cycle stresses amplitude, $p_{f\sigma N}$ (figure 2, *b*)

3.4 абсцисса точки перелома кривой контактно-механической усталости, $N_{\sigma pG}$, $N_{p\sigma G}$**:** абсцисса точки перелома кривой механической усталости, построенной по параметру контактных напряжений при трении качения, $N_{\sigma pG}$ (рисунок 2, *а*), либо абсцисса точки перелома кривой контактной усталости, построенной по параметру амплитуды напряжений цикла, $N_{p\sigma G}$ (рисунок 2, *б*)

turning point of mechano-rolling fatigue curve: turning point of mechanical fatigue curve, plotted on parameter of contact stresses under rolling friction, $N_{\sigma pG}$ (figure 2, *a*), or turning point of rolling fatigue curve, plotted on parameter of cycle stresses amplitude, $N_{p\sigma G}$ (figure 2, *b*)

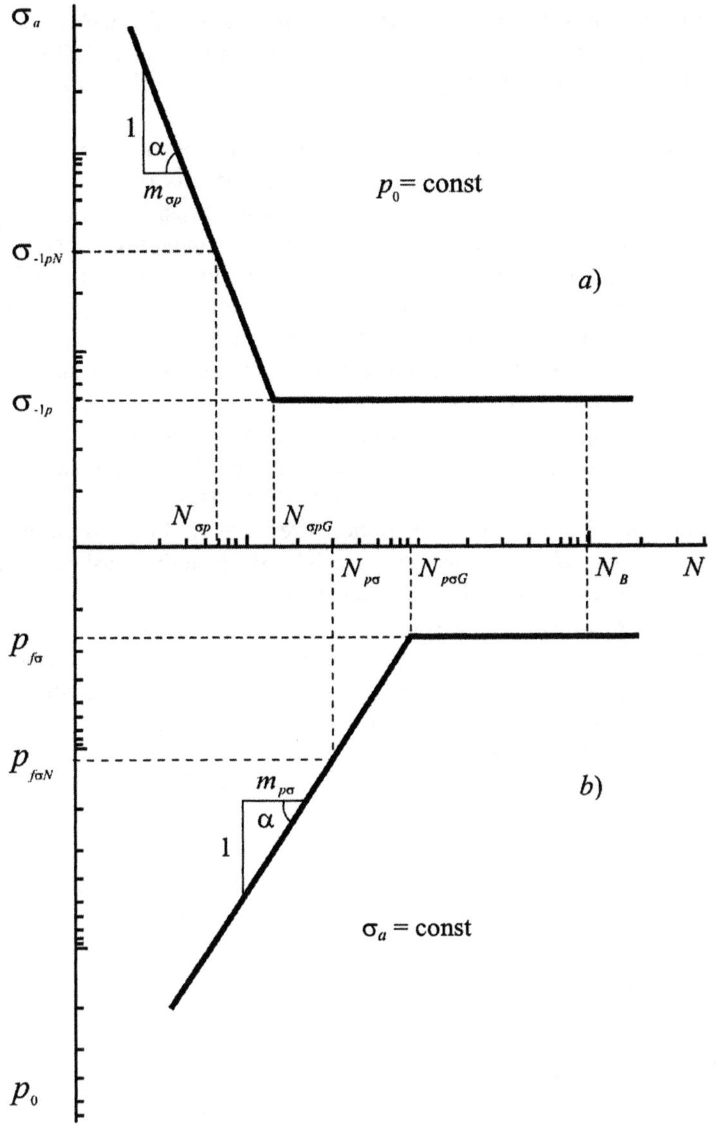

Рисунок 2. Схемы кривых контактно-механической усталости

Figure 2. Schemes of mechano-rolling fatigue curves

3.5 показатель наклона кривой контактно-механической усталости, $m_{\sigma p}$, $m_{p\sigma}$**:** показатель наклона кривой механической усталости, построенной по параметру контактных напряжений при трении качения, $m_{\sigma p}$ (рисунок 2, *а*), либо показатель наклона кривой контактной усталости, построенной по параметру амплитуды напряжений цикла, $m_{p\sigma}$ (рисунок 2, *б*)

mechano-rolling fatigue curve exponent: mechanical fatigue curve exponent, plotted on parameter of contact stresses under rolling friction, $m_{\sigma p}$ (figure 2, *a*), or rolling fatigue curve exponent, plotted on parameter of cycle stresses amplitude, $m_{p\sigma}$ (figure 2, *b*)

3.6 кривая фрикционно-механической усталости, $N(\sigma_a, \tau_W = \text{const})$, $N(\tau_W, \sigma_a = \text{const})$**:** график, характеризующий зависимость между амплитудой напряжений цикла σ_a и циклической долговечностью $N_{\sigma\tau}$ для одинаковых моделей силовой системы, построенный по параметру фрикционных напряжений при трении скольжения τ_W = const (рисунок 3, *а*), либо график, характеризующий зависимость между фрикционными напряжениями при трении скольжения τ_W и фрикционной долговечностью $N_{\tau\sigma}$ для одинаковых моделей силовой системы, построенный по параметру амплитуды напряжений цикла σ_a = const (рисунок 3, *б*)

mechano-sliding fatigue curve, $N(\sigma_a, \tau_W = \text{const})$, $N(\tau_W, \sigma_a = \text{const})$**:** graph of cycle stress amplitude, σ_a, versus cyclic fatigue life, $N_{\sigma\tau}$, for the same models of an active system plotted with respect to friction stress under sliding friction, τ_W = const (figure 3, *a*), or graph of friction stress under sliding friction, τ_W versus sliding fatigue life, $N_{\tau\sigma}$, for the same models of an active system plotted with respect to cycle stress amplitude σ_a = const (figure 3, *b*)

3.7 предел фрикционно-механической выносливости, $\sigma_{-1\tau}$, $\tau_{f\sigma}$**:** предел выносливости по параметру фрикционных напряжений при трении скольжения, $\sigma_{-1\tau}$ (рисунок 3, *а*), либо предел фрикционной выносливости по параметру амплитуды напряжений цикла, $\tau_{f\sigma}$ (рисунок 3, *б*)

mechano-sliding fatigue limit: fatigue limit on parameter of contact stresses under sliding friction, $\sigma_{-1\tau}$ (figure 3, *a*), or sliding fatigue limit on parameter of cycle stress amplitude, $\tau_{f\sigma}$ (figure 3, *b*)

3.8 предел ограниченной фрикционно-механической усталости, $\sigma_{-1\tau N}$, $\tau_{f\sigma N}$**:** предел ограниченной усталости по параметру фрикционных напряжений при трении скольжения, $\sigma_{-1\tau N}$ (рисунок 3, *а*), либо предел ограниченной фрикционной усталости по параметру амплитуды напряжений цикла, $\tau_{f\sigma N}$ (рисунок 3, *б*)

mechano-sliding fatigue limit at N cycles: fatigue threshold on parameter of contact stresses under sliding friction, $\sigma_{-1\tau N}$ (figure 3, *a*), or sliding fatigue threshold on parameter of cycle stresses amplitude, $\tau_{f\sigma N}$ (figure 3, *б*)

3.9 абсцисса точки перелома кривой фрикционно-механической усталости, $N_{\sigma\tau G}$, $N_{\tau\sigma G}$**:** абсцисса точки перелома кривой механической усталости, построенной по параметру фрикционных напряжений при трении скольжения, $N_{\sigma\tau G}$ (рисунок 3, *а*), либо абсцисса точки перелома кривой фрикционной усталости, построенной по параметру амплитуды напряжений цикла, $N_{\tau\sigma G}$ (рисунок 3, *б*)

turning point of mechano-sliding fatigue curve: turning point of mechanical fatigue curve, plotted on parameter of contact stresses under sliding friction, $N_{\sigma\tau G}$ (figure 3, *a*), or turning point of sliding fatigue curve, plotted on parameter of cycle stresses amplitude, $N_{\tau\sigma G}$ (figure 3, *b*)

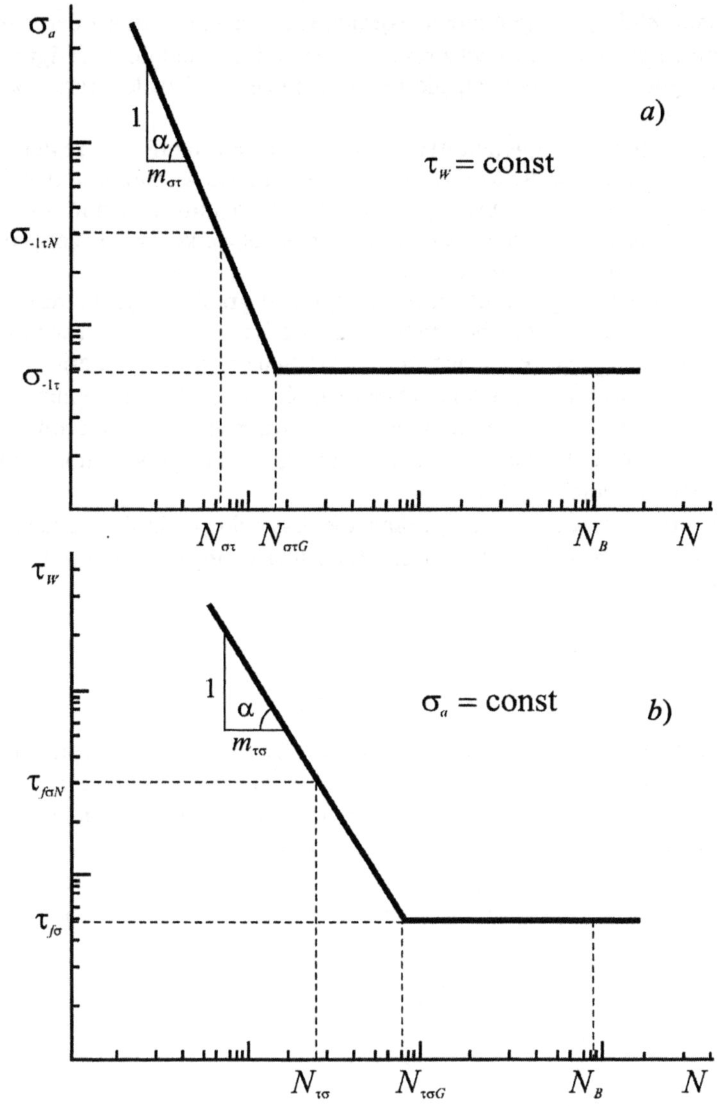

Рисунок 3. Схемы кривых фрикционно-механической усталости

Figure 3. Schemes of mechano-sliding fatigue curves

3.10 показатель наклона кривой фрикционно-механической усталости, $m_{\sigma\tau}$, $m_{\tau\sigma}$: показатель наклона кривой механической усталости, построенной по параметру фрикционных напряжений при трении скольжения, $m_{\sigma\tau}$ (рисунок 3, *а*), либо показатель наклона кривой фрикционной усталости, построенной по параметру амплитуды напряжений цикла, $m_{\tau\sigma}$ (рисунок 3, *б*)

mechano-sliding fatigue curve exponent: mechanical fatigue curve exponent, plotted on parameter of contact stresses under sliding friction, $m_{\sigma\tau}$ (figure 3, *a*), or sliding fatigue curve exponent, plotted on parameter of cycle stresses amplitude, $m_{\tau\sigma}$ (figure 3, *b*)

3.11 диаграмма предельных состояний силовой системы: график, характеризующий зависимость между амплитудой напряжений цикла и контактным давлением для предельных состояний силовой системы по критериям сопротивления усталости, износостойкости и сопротивления износоусталостным повреждениям

limiting states diagram of an active system: graph of cycle stress amplitude versus contact pressure for the limiting states of an active system with respect to fatigue resistance, wear resistance and wear-fatigue resistance criteria

3.12 коэффициент прямого эффекта, K_{Dp}, $K_{D\tau}$: безразмерная величина, определяемая как отношение предела контактно-механической σ_{-1p} либо фрикционно-механической $\sigma_{-1\tau}$ выносливости к пределу выносливости при механической усталости σ_{-1}

direct effect index, K_{Dp}, $K_{D\tau}$: dimensionless value defined as a ratio of either mechano-rolling fatigue limit, σ_{-1p}, or mechano-sliding fatigue limit, σ_{-1F}, to mechanical fatigue limit, σ_{-1}

$$K_{Dp} = \sigma_{-1p} / \sigma_{-1} ,$$

$$K_{D\tau} = \sigma_{-1\tau} / \sigma_{-1} .$$

3.13 коэффициент обратного эффекта, K_{Bp}, $K_{B\tau}$: безразмерная величина, определяемая как отношение предела контактно-механической $p_{f\sigma}$ либо фрикционно-механической $\tau_{f\sigma}$ выносливости к пределу выносливости при контактной p_f либо фрикционной τ_f усталости

back effect index, K_{Bp}, $K_{B\tau}$: dimensionless value defined as a ratio of mechano-rolling fatigue limit, $p_{f\sigma}$, or mechano-sliding fatigue limit, $\tau_{f\sigma}$, to either rolling fatigue limit, p_f, or sliding fatigue limit, τ_f

$$K_{Bp} = p_{f\sigma} / p_f ,$$

$$K_{BF} = F_{f\sigma} / F_f .$$

3.14 управляющий параметр износоусталостного повреждения, ψ_{-1p}, $\psi_{-1\tau}$ **:** безразмерная величина, определяемая либо как отношение предела выносливости по параметру контактных напряжений при трении качения σ_{-1p} к пределу контактной выносливости по параметру амплитуды напряжений цикла $p_{f\sigma}$, либо как отношение предела выносливости по параметру фрикционных напряжений при трении скольжения $\sigma_{-1\tau}$ к пределу фрикционной выносливости по параметру амплитуды напряжений цикла $\tau_{f\sigma}$

governing parameter of Wear-Fatigue damage: dimensionless value defined either ratio of fatigue limit on parameter of contact stresses under rolling friction σ_{-1p} to rolling fatigue limit on parameter of cycle stresses $p_{f\sigma}$, or ratio of fatigue limit on parameter of contact stresses under sliding friction $\sigma_{-1\tau}$ to sliding fatigue limit on parameter of cycle stresses $\tau_{f\sigma}$

$$\psi_{-1p} = \sigma_{-1p} / p_{f\sigma},$$

$$\psi_{-1\tau} = \sigma_{-1\tau} / \tau_{f\sigma}.$$

3.15 граничное значение управляющего параметра износоусталостного повреждения, $\psi^{\lim}_{-1p}, \psi^{\lim}_{-1\tau}$**:** значение управляющего параметра износоусталостного повреждения, определяемое как отношение предельных напряжений при механической σ_{-1} и контактной p_f, либо фрикционной τ_f усталости

governing parameter threshold of Wear-Fatigue damage: value of governing parameter of WFD defined as ratio of limiting stresses under mechanical fatigue σ_{-1} or rolling p_f or sliding τ_f fatigue

$$\psi^{\lim}_{-1p} = \sigma_{-1} / p_f,$$

$$\psi^{\lim}_{-1\tau} = \sigma_{-1} / \tau_f.$$

3.16 относительный опасный объем, ω_{VP}**:** мера усталостного повреждения элемента силовой системы, определяемая как отношение опасного объема $V_{P\gamma}$ к его рабочему объему V_0

relative damaged volume, ω_{VP}**:** extent of fatigue damage of an active system's element defined as a ratio of damaged volume, $V_{P\gamma}$, to its working volume V_0

$$\omega_{VP} = V_{P\gamma} / V_0,$$

$$0 \le \omega_{VP} \le 1$$

3.17 относительная опасная поверхность, ω_{SP}**:** мера поврежденности элемента силовой системы при трении, определяемая как отношение опасной поверхности $S_{P\gamma}$ к его рабочей поверхности S_0

relative damaged surface, ω_{SP}**:** extent of damage of an active system's element under friction (sliding or rolling) defined as a ratio of damaged surface, $S_{P\gamma}$, to its working surface, S_0

$$\omega_{SP} = S_{P\gamma} / S_0,$$

$$0 \le \omega_{SP} \le 1$$

3.18 относительная опасная зона, ω_{WP}**:** мера износоусталостного повреждения, определяемая как отношение комплексного опасного объема $W_{P\gamma}$ к рабочему объему элемента силовой системы V

relative damaged zone, ω_{WP}**:** extent of wear-fatigue damage defined as a ratio of complex damaged volume, $W_{P\gamma}$, to the working volume of an active system's element V

$$\omega_{WP} = W_{P\gamma} / V,$$

$$0 \le \omega_{WP} \le 1$$

3.19 критерий подобия износоусталостных повреждений, Π_{SK}**:** безразмерный параметр, определяемый как отношение комплексного опасного объема $W_{P\gamma}$ элемента любой силовой системы к рабочему объему аналогичного элемента модельной силовой системы V_{st}, которая принимается в качестве стандартной

similarity criterion of Wear-Fatigue damage, Π_{SK}**:** dimensionless parameter defines as ratio of complex damage volume $W_{P\gamma}$ of element of any active system to working volume of the same element of an active system model V_{st} which is standard one

$$\Pi_{SK} = W_{P\gamma} / V_{st}$$

3.20 износ силовой системы, i_σ**:** износ, обусловленный процессами трения (в любых его проявлениях) и циклического нагружения одновременно

wear of an active system, i_σ**:** wear caused by any form of friction processes and cyclic loading simultaneously

$$i_\sigma = i_h \varphi_1(\sigma)$$

3.21 интенсивность изнашивания силовой системы, I_σ**:** интенсивность изнашивания, обусловленная процессами трения (в любых его проявлениях) и циклического нагружения одновременно

wear intensity of an active system, I_σ**:** wear intensity caused by any form of friction processes and cyclic loading simultaneously

$$I_\sigma = I_h \varphi_2(\sigma)$$

3.22 группа износостойкости силовой системы: характеристика износоусталостного повреждения, определяемая интервалом изменения интенсивности изнашивания силовой системы при ее испытаниях или эксплуатации (таблица 1)

wear resistance group of an active system: characteristic of WFD defined by interval of changing of wear intensity of an active system under its testing or exploration (table 1)

Table 1

Группа Group	A	B	C
I_σ	$< 10^{-7}$	$10^{-7}...10^{-5}$	$> 10^{-5}$
$\log(1/I_\sigma)$	> 7	7…5	< 5

4 Alphabetical index

4 Алфавитные указатели терминов

4.1 Russian alphabetical index

4.1 Алфавитный указатель терминов на русском языке

4.2 English alphabetical index

5 Definitions and units of measurement

5 Обозначения и единицы измерения величин

Обозначение Symbol	Параметр, пункт по стандарту Name, No. as per standard	Единица измерения Unit of measurement
A, B, C	Условное обозначение группы износостойкости силовой системы; 3.22 Wear resistance group of an active system; 3.22	–
Π_{SK}	Критерий подобия износоусталостных повреждений; 3.19 Similarity criterion of wear-fatigue damage; 3.19	–
F_c	Циклическая составляющая силы трения; 2.2 Cyclic component of a friction force; 2.2	Н N
F_f	Сила трения в узле трения; 2.1 Friction force in assembly friction; 2.1	Н N
F_f	Сила трения в силовой системе; 2.1 Friction force in an active system; 2.1	Н N
F_N	Контактная нагрузка; 2.3 Contact load; 2.3	Н N
f_c	Циклическая составляющая коэффициента трения; 2.4 Cyclic component of a friction coefficient; 2.4	–
f_σ	Коэффициент трения в силовой системе; 2.3 Friction coefficient in an active system; 2.3	–
I_h	Интенсивность изнашивания узла трения; 3.21 Wear intensity of assembly friction; 3.21	–
I_σ	Интенсивность изнашивания силовой системы; 3.21 Wear intensity of an active system; 3.21	–
i_h	Износ узла трения; 3.20 Wear of assembly friction; 3.20	м; м^2; кг m; m^2; kg

i_σ	Износ силовой системы; 3.20 Wear of an active system; 3.20	м; м^2; кг m; m^2; kg
K_{Bp}, $K_{B\tau}$	Коэффициент обратного эффекта; 3.13 Back effect index; 3.13	–
K_{Dp}, $K_{D\tau}$	Коэффициент прямого эффекта; 3.12 Direct effect index; 3.12	–
$m_{\sigma p}$, $m_{p\sigma}$	Показатель наклона кривой контактно-механической усталости; 3.5 Mechano-rolling fatigue curve exponent; 3.5	–
$m_{\sigma\tau}$, $m_{\tau\sigma}$	Показатель наклона кривой фрикционно-механической усталости; 3.10 Mechano-sliding fatigue curve exponent; 3.10	–
$N(\sigma_a, p_0 = \text{const})$ $N(p_0, \sigma_a = \text{const})$	Кривая контактно-механической усталости; 3.1 Mechano-rolling fatigue curve; 3.1	–
$N_{\sigma pG}$, $N_{p\sigma G}$	Абсцисса точки перелома кривой контактно-механической усталости; 3.4 Turning point of mechano-rolling fatigue curve; 3.4	Цикл cycle
$N(\sigma_a, \tau_W = \text{const})$ $N(\tau_W, \sigma_a = \text{const})$	Кривая фрикционно-механической усталости; 3.6 Mechano-sliding fatigue curve; 3.6	–
$N_{\sigma\tau G}$, $N_{\tau\sigma G}$	Абсцисса точки перелома кривой фрикционно-механической усталости; 3.9 Turning point of mechano-sliding fatigue curve; 3.9	Цикл cycle
p_0	Контактное напряжение при трении качения; 3.1 Contact stress under rolling friction; 3.1	МПа MPa
p_f	Предел выносливости при контактной усталости; 3.13 Rolling fatigue limit; 3.13	МПа MPa
$S_{p\gamma}$	Опасная поверхность; 1.17 Damaged surface; 1.17	м^2 m^2
V_{st}	Рабочий объем элемента силовой системы, принимаемой в качестве стандартной; 3.19 Working volume of the same element of an active system which is standard one; 3.19	м^3 m^3

$V_{P\gamma}$	Опасный объем; 1.16 Damaged volume; 1.16	м^3 m^3
$W_{P\gamma}$	Комплексный опасный объем; 1.18 Complex damaged volume; 1.18	м^3 m^3
$\varphi(\sigma)$	Функция влияния циклических напряжений на силу трения в силовой системе; 2.1 Function of influence of cycle stresses on friction force in an active system; 2.1	–
$\varphi_1(\sigma)$	Функция влияния циклических напряжений на износ силовой системы; 3.20 Function of influence of cycle stresses on wear of an active system; 3.20	–
$\varphi_2(\sigma)$	Функция влияния циклических напряжений на интенсивность изнашивания силовой системы; 3.21 Function of influence of cycle stresses on wear intensity of an active system; 3.21	–
φ_{VS}	Функция взаимодействия опасного объема и опасной поверхности; 1.18 Interaction function of damaged volume and damaged surface; 1.18	–
σ_{-1}	Предел выносливости при механической усталости; 3.12 Mechanical fatigue limit; 3.12	МПа MPa
σ_a	Амплитуда напряжений цикла; 3.1 Cycle stress amplitude; 3.1	МПа MPa
σ_{-1p}, $p_{f\sigma}$	Предел контактно-механической выносливости; 3.2 Mechano-rolling fatigue limit; 3.2	МПа MPa
σ_{-1pN}, $p_{f\sigma N}$	Предел ограниченной контактно-механической усталости; 3.3 Mechano-rolling fatigue limit at N cycles; 3.3	МПа MPa
$\sigma_{-1\tau}$, $\tau_{f\sigma}$	Предел фрикционно-механической выносливости; 3.7 Mechano-sliding fatigue limit; 3.7	МПа MPa
$\sigma_{-1\tau N}$, $\tau_{f\sigma N}$	Предел ограниченной фрикционно-механической усталости; 3.8 Mechano-sliding fatigue limit at N cycles; 3.8	МПа MPa

τ_W	Фрикционное напряжение при трении скольжения; 3.6 Friction stress under sliding friction; 3.6	МПа MPa
τ_f	Предел выносливости при фрикционной усталости; 3.15 Sliding fatigue limit; 3.15	МПа MPa
ω_{SP}	Относительная опасная поверхность; 3.17 Relative damaged surface; 3.17	–
ω_{VP}	Относительный опасный объем; 3.16 Relative damaged volume; 3.16	–
ω_{WP}	Относительная опасная зона; 3.18 Relative damaged zone; 3.18	–
ψ_{-1p}, $\psi_{-1\tau}$	Управляющий параметр износоусталостного повреждения; 3.14 Governing parameter of wear-fatigue damage; 3.14	–
$\psi_{-1p}^{\lim}, \psi_{-1\tau}^{\lim}$	Граничное значение управляющего параметра износоусталостного повреждения; 3.15 Boundary value of the governing parameter of wear-fatigue damage; 3.15	–

Appendix III
ON METHODOLOGY OF TRIBO-FATIGUE*

L A SOSNOVSKIY, N A MAKHUTOV, GAO WANZHEN

Modern western civilization has reached exceptional heights in the art of partitioning the integrity... We have greatly progressed in this art, so much we have that we often overlook to re-assemble the separated parts into the integrity that they used to make up.

Alvin Toffler

Introduction

Many definitions have appeared in publications during the last 15 years of the previous century of what is commonly termed now as *tribo-fatigue*. Tribo-fatigue has been determined as a "new scientific trend" either in tribology or mechanics of fatigue fracture, or as a "new sphere of activity" in creating reliable mechanical systems, or as a "new branch of mechanics", and finally as a "science of wear-fatigue damage and fracture of active systems of machines and equipment" [1–5]. It is premature right now either to criticize or support any of these definitions. In our view, it should be stated outright that it is emergence and evolution of the *mechanics of wear-fatigue damage* [6–8]. This emergence and evolution occur at the borders between three disciplines: *tribology, mechanics of fatigue fracture and the theory of reliability of mechanical systems*. It is to emphasize that they occur at the borders rather than within any single of these disciplines. Hence, each of these disciplines is a source of tribo-fatigue, its particular, definitely powerful foundation.

Since the mechanics of wear-fatigue damage is dealt with, the term tribo-fatigue [3, 6, 9] quite sufficiently but in short reflects its notion: *tribos* (from Greek) – friction symbolizes modern tribology, including tribological reliability, *fatigue* (from French) symbolizes modern mechanics of fatigue fracture, including strength reliability. It is apparent that the term *tribo-fatigue* combines two words (notions), which are completely equal. If we deal with physico-mechanical processes based on shown above notions, it should be stressed on their dialectic interaction and interaffect correlation (friction (and wear) ⇔ fatigue = tribo-fatigue).

Each scientific discipline has its own specific attributes. Exactly these specific attributes allow to discriminate explicitly what is what.

---

* Presentation at III International symposium on tribo-fatigue (Beijing, China, 2000) and a paper in the journal "Zavodskaya Laboratoriya" (2002, No. 6).

Objects of studies

The *object of study* is the first and, presumably the most essential specific attribute of a scientific discipline. It is a structural element in the mechanics of fatigue fracture. It is a friction pair in tribology. Tribo-fatigue studies *active systems* [3, 6]. An active system implies any mechanical system in which the process of friction is realized in its any manifestation (like in sliding, rolling, slippage, impact, etc.) and which simultaneously bears and transmits cyclic workload. Figure 1 shows an example of two typical active systems: a *wheel / rail*, and *a connecting rod / a crank pin*. A space system of cyclic forces (stresses, deformations) appears in the zone of contact between the wheel and the rail accompanied by the process of rolling friction (including slippage). Hence, *mechano-rolling fatigue* should be considered as the dominating type of damage of this system in operation. A space system of cyclic stresses (deformations) may also appear in the zone of contact between the crank pin and the connecting rod due to the effect of both contact and non contact loads, but in this case in combination with the process of sliding friction. It can be termed as *mechano-sliding fatigue* being a dominating type of damage of this system in operation. Block shaft bearings experience pure friction, crank webs and the connecting rod experience pure fatigue. Hence, the block bearings are definitely the object of tribology (T), connecting rod or shaft webs are definitely the object of the mechanics of fatigue fracture (F), while the systems like *crank-up pin / connecting rod* or *wheel / rails* are objects of tribo-fatigue (TF). Yet, traditionally, the interests of tribology and mechanics of fatigue fracture cover also the systems of *crank pins / connecting rods* and *wheels / rails*. Later the implications will be discussed [5, 6, 9].

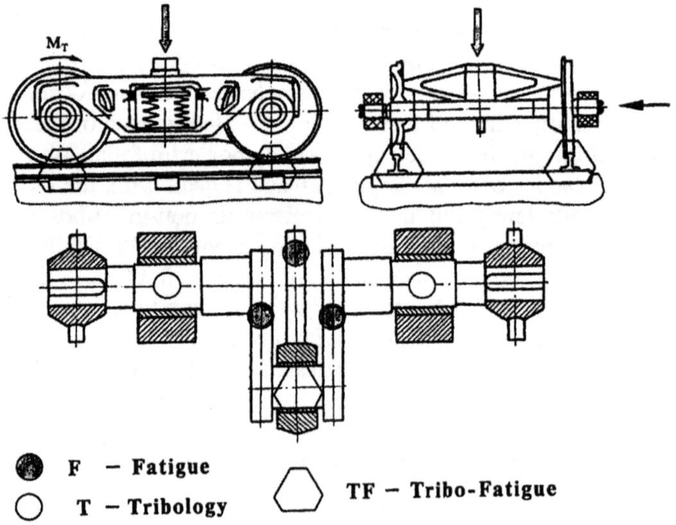

Fig. 1. Examples of typical active systems

Let's take a look on how the serviceability of one such system is assessed in practice.

Assume it is a mechanical system comprising *a connecting rod (with a polymeric sliding bearing) / a steel crank-pin*. It is a common friction pair for the tribologist. Since the steel shaft does not wear in contact with a polymer, the wear resistance of the polymeric bearing is to be assessed (by calculation or experimentally). The configuration in Fig. 2 can serve for this assessment. In case p is contact pressure, then the probability of failure $F(p)$ is determined in the general case using the criteria of wear resistance, which are usually assumed to be the wear rate $I(p)$ and / or the durability $N(p)$.

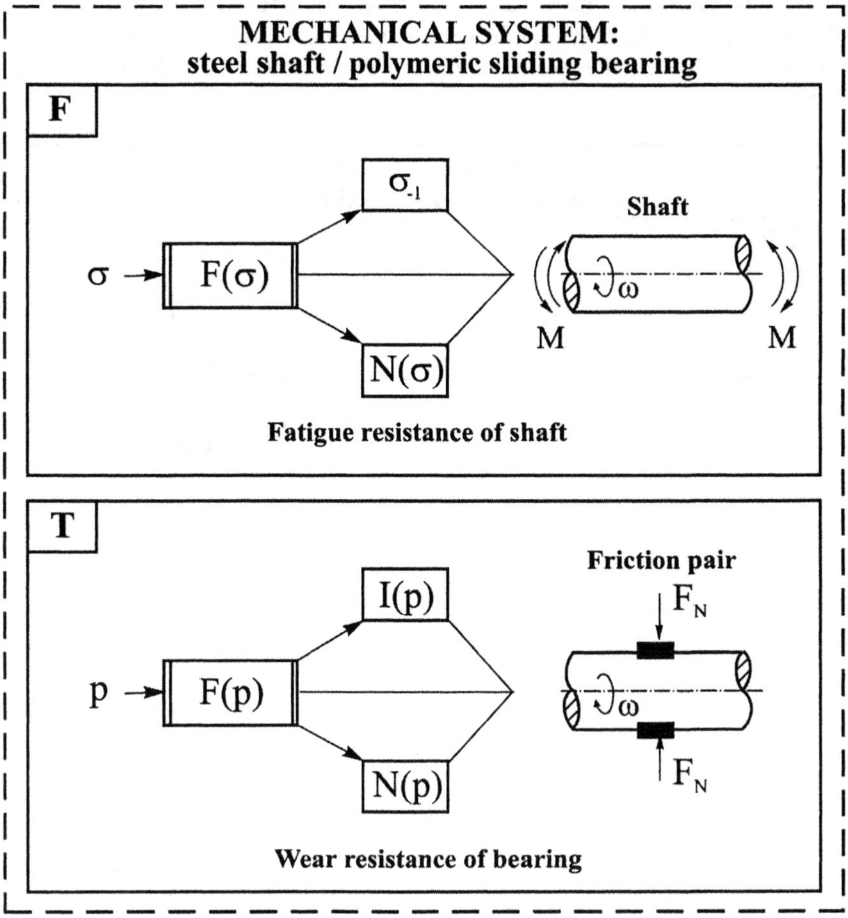

Fig. 2. Traditional approach to the calculation of a mechanical system

From the point of view of a specialist of strength, it is the structural element or the crank-pin. If σ is the effective cyclic stress, then in the general case the probability of failure $F(\sigma)$ is determined (by calculation or experimentally) using the criteria of fatigue resistance, which usually imply the fatigue limit σ_{-1} and/or fatigue life $N(\sigma)$.

Hence, the traditional assessment of reliability of a certain mechanical system using individual criteria (either resistance to fatigue or resistance to wear) implies that the relationship between its elements (the crank-pin ↔ the sliding bearing) is either weak (tribology: only the friction pair is studied) or totally absent (the mechanics of fatigue strength: only an individual structural element is studied). In reality in this case a specialist in tribo-fatigue considers it as a metal-to-polymer active system, which operates in the conditions of *mechano-sliding fatigue*. It means that there is a force interaction between the elements of the system governed by the simultaneous and joint effect of both contact pressure p and cyclic stresses σ, which are induced by non-contact loads. Then the scheme of assessment of the serviceability of the system should resemble that shown in Fig. 3.

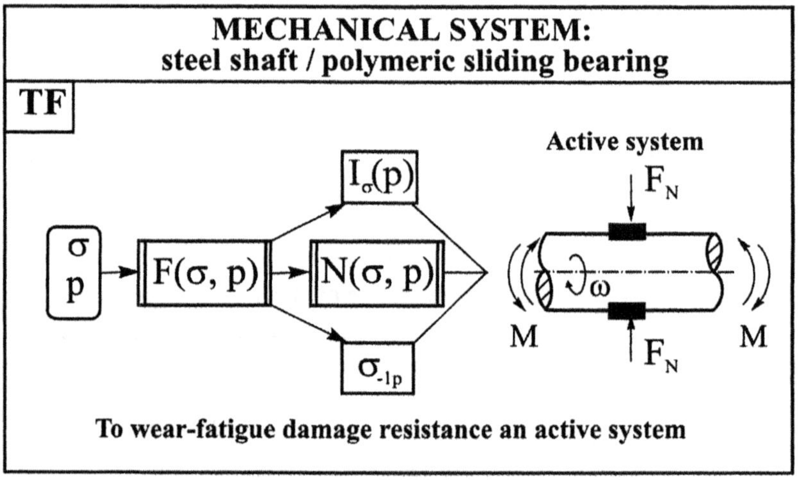

Fig. 3. To the calculation of an active system

The reliability of an active system can be described with the following *complex indicators*: $F(\sigma, p)$ – the probability of failure of the system determined by the probability of failure of either the shaft or the bearing, or the probability of failure of both these elements simultaneously; $N(\sigma, p)$ is the durability of the system determined by the durability of either the shaft or the bearing, or by the durability of these two elements simultaneously; σ_{-1p} is the fatigue limit of the shaft with the account of the effect of friction and wear processes under contact pressure p; $I_\sigma(p)$ is the wear rate of the bearing under the effect of cyclic stresses σ.

The effect of friction and wear processes upon the change in characteristics of fatigue resistance of the elements of the system is termed *direct effect*. It is clear then that characteristic σ_{-1p} describes the direct effect quantitatively. The effect of cyclic stresses upon the characteristics of friction and wear is termed *back effect*. Then it is clear that the characteristic $I_\sigma(p)$ describes the back effect quantitatively.

Figure 2 and 3 clearly show that only the active system is actually adequate to the real mechanical system to be studied, meanwhile the friction pair or the structural element are just its particular schematizations. The indicators of reliability of a studied object determined with the methods of tribo-fatigue sufficiently reflect the real conditions of its operation, meanwhile similar indicators determined with the methods of tribology or the mechanics of fatigue fracture describe the behavior of the object in idealized conditions. It also relates to such an essential characteristic as the friction coefficient.

The *friction force in the active system* F_σ can be considered [1, 3, 10] as a function of common friction force in sliding F_s which appears in the circumferential direction (the object – the friction pair), and the *cyclic component* F_c of the *friction force* which additionally appears in the zone of contact due to the excitation of cyclic stresses (deformations) in the axial direction (Fig. 4). Then *the cyclic component of the friction coefficient is* $f_c = F_c / F_N$ *and the friction coefficient in the active system*

$$f_y = f_s \varphi_1(\sigma) = \varphi_2(f_s, f_c), \qquad (1)$$

where $f_s = F_s / F_N$ is the usual friction coefficient in sliding.

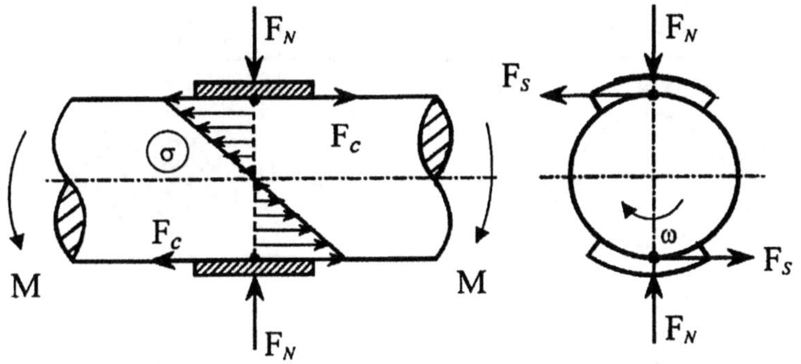

Fig. 4. Force and friction coefficient in an active system

Shown above ideas are summarized in Fig. 5. On the one hand, the theories of friction, wear, lubrication were integrated into tribology (T) as an integral scientific discipline. It is natural since the friction and wear processes, including those with lubrication, really combine and interact in friction pairs.

On the other hand, the mechanics of fatigue fracture (F) has emerged among general problems of dynamics, strength and stability, as a discipline of specific practical significance for modern machine building. Tribo-fatigue (TF) became essential to solve effectively complex problems of reliability (R) of the key systems of machines and equipment – active systems using the most significant criteria of serviceability.

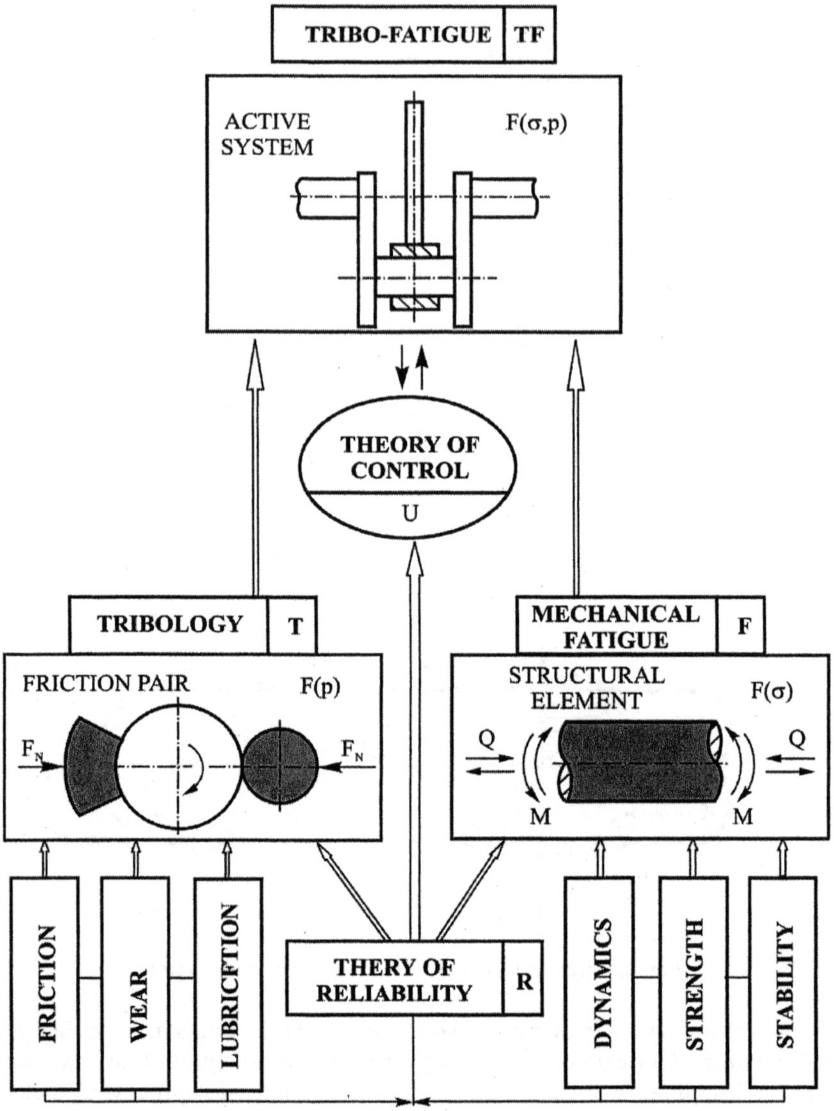

Fig. 5. – Tribo-fatigue as a complex scientific discipline

Method of studies

Another essential attribute of each scientific discipline is the *methods of studies of objects* (Table 1). Let's begin our analysis with *experimental methods* (Fig. 6) [11, 12].

Table 1

Discipline	Object of study	Basic methods of studies		Scale of damage
		experimental	theoretical	
T (tribology)	Friction pair	Friction tests	Mechanics of contact interactions	Surface damage (wear, pitting, etc.)
F (mechanics of fatigue fracture)	Structural element	Fatigue tests	Mechanics of deformation and fracture	Volume (fatigue) fracture
TF (tribo-fatigue)	Active system	Wear-fatigue tests	Mechanics of wear-fatigue damage	Complex surface damage and volume fracture

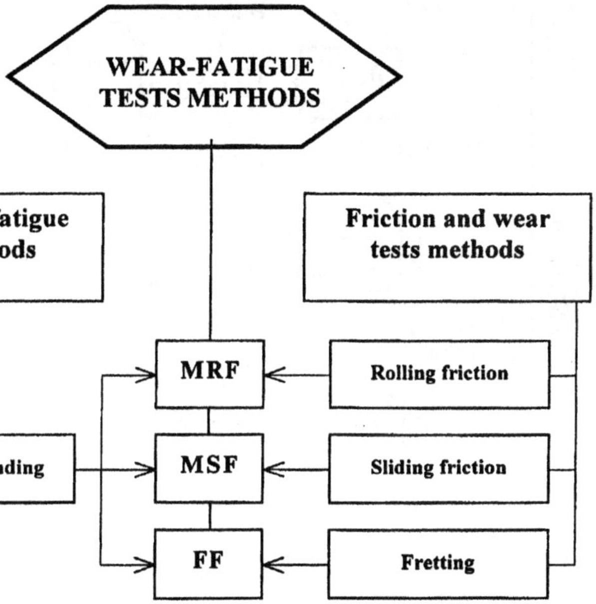

Fig. 6. Development of methods of wear-fatigue tests: MRF – mechano-rolling fatigue, MSF – mechano-sliding fatigue, FF – fretting fatigue

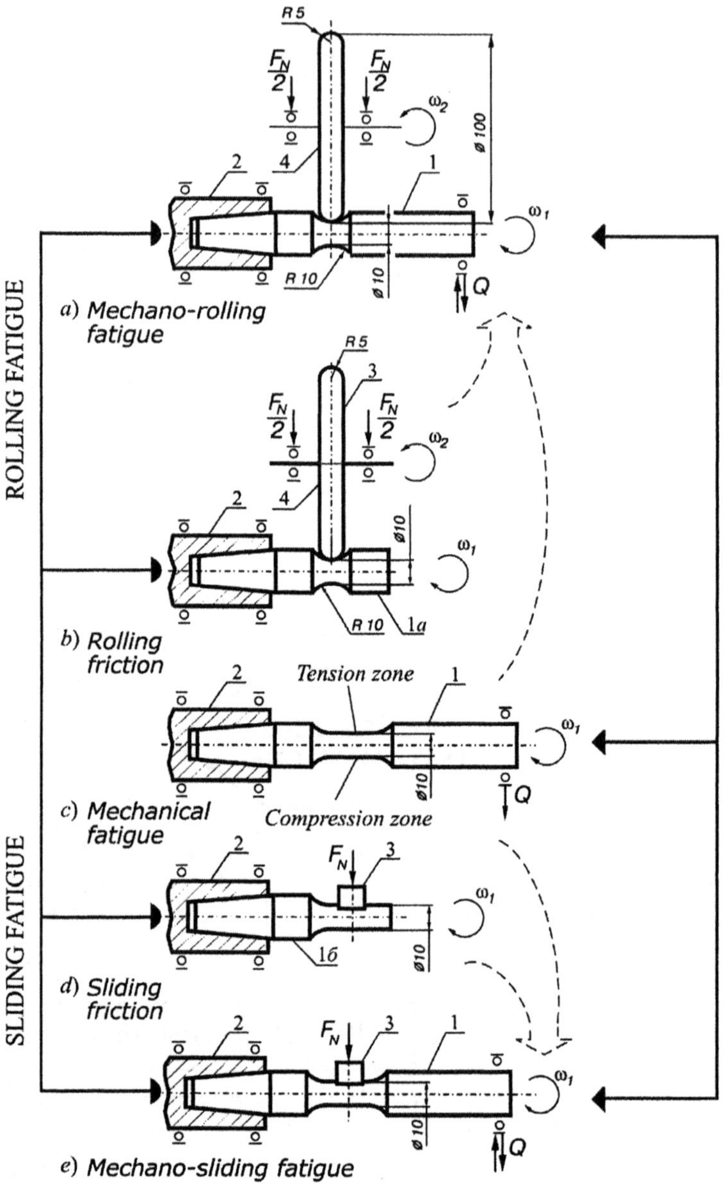

Fig. 7. Typical methods of wear-fatigue tests: *1, 1a, 1b* – specimen; *2* – test apparatus spindle; *3, 4* – counterspecimen; *Q* – bending load; F_N – contact load; ω_1, ω_2 – speed of rotation of specimen, counterspecimen

Specialists in the mechanics of fatigue fracture elaborate and apply the methods and machines for testing structural elements under various conditions of cyclic loading. Figure 6 shows one such method for rotation bending of a cylindrical specimen. Figure 7, *c* shows the tests scheme. Tribologists elaborate and apply the methods and machines for testing friction pairs under various conditions of contact interactions. Figure 6 shows all three methods, Figs. 7, *b* and 7, *d* show the schemes of tests under rolling and sliding friction. Specialists in tribo-fatigue elaborate the methods and machines for *complex wear-fatigue tests* of models of active systems. Figure 6 shows three methods of the tests, Figs. 7, *a* and 7, *e* show the schemes of tests for *mechano-sliding* and *mechano-rolling fatigue*. They are the combination of test schemes implemented by specialists in tribology and strength. The difference is the following. Machines for friction tests do not allow to investigate the resistance of structural elements to fatigue. Machines for fatigue tests do not allow to investigate the friction and wear processes. Meanwhile *SI series machines for wear-fatigue tests* allow to investigate both, as it should be, but it is more essential that they allow to carry out complex tests under any combination of cyclic and contact loads acting simultaneously. Naturally it becomes possible to obtain fundamentally new experimental results.

Figure 8 shows an example of the results of the tests of the active system, such as carbon steel 45 (the cylindrical specimen) / alloyed steel 25XГТ (the roller), for mechano-rolling fatigue [5, 13, 14].

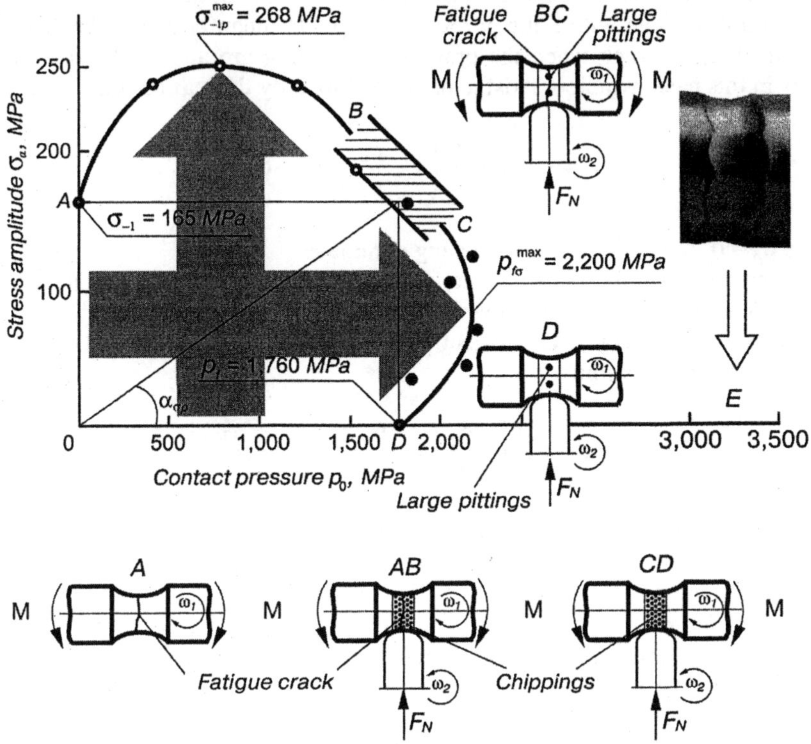

Fig. 8. Multicriterial diagram of limiting states of active system in mechano-rolling fatigue (SPW — surface plasticity waves)

The *ABCD* diagram is plotted in the coordinates of pressure p_0 in the center of the contact area (the x-axis) and amplitude σ_a of cyclic stresses in bending (the y-axis).

The point A is the fatigue limit σ_{-1} of steel 45 specimens, it is determined by common mechanical fatigue tests of the scheme shown in Fig. 7, *c*. The criterion of the limiting state is disintegration of the specimen into two parts due to the growth of the main fatigue crack in the dangerous section. Hence, this point implies the mechanics of fatigue fracture. Generally the y-axis σ_a is the scale of strength: the results of fatigue tests of any structural elements of any materials may and should lie within this scale.

The point D is the critical pressure p_f under rolling friction without slippage, it is determined by common friction tests. The criterion of the limiting state is the appearance of pittings of critical density along the rolling path. Hence, this point implies tribology. Generally the x-axis p_0 is the tribological scale: the results of tests of any friction pairs with the elements of any materials may and should lie within this scale.

The curves *ABCD* are a *diagram of limiting states of an active system under mechano-rolling fatigue*. The diagram is plotted using the results of wear-fatigue tests of the scheme shown in Fig. 7, *a*. Hence, it is tribo-fatigue.

The limiting state within the portion *AB* is predominantly due to the growth of the main fatigue crack when the processes of pitting are attendant. Direct effect occurs in this case, which is satisfactorily described by the expression

$$\sigma_{-1p} = \sigma_{-1}\left[1 - \frac{p_0}{p_f}\left(1 - \frac{p_0}{p_f}\right)\ln(1 - \sigma_p)\right], \tag{2}$$

where $\mu_p = 0.92$ is the contact hardening parameter.

On the contrary, the limiting state within the portion CD is determined by the critical density of pittings, meanwhile the evolution of mechanical fatigue cracks is an attendant damage. Back effect occurs in this case, which is satisfactorily described by the equation

$$p_{f\sigma} = p_f\left[1 - \frac{\sigma_a}{\sigma_{-1}}\left(1 - \frac{\sigma_a}{\sigma_{-1}}\right)\ln(1 - \mu_\sigma)\right], \tag{3}$$

where $\mu_\sigma = 0.65$ is the cyclic hardening parameter.

The portion *BC* is transient, hence it is of particular interest since the kinetic processes of interactions between friction phenomena (together with wear) and mechanical fatigue take place here at a very high level of loading parameters σ_a and p_0. In these test conditions it is stated that surface waves of plasticity emerge along the rolling path, though the profile radius of the counterspecimen (the roller) remains practically perfectly smooth. Appearance of the waves is another attribute of the limiting state since impermissible vibrations are generated in the system.

Analysis of the *ABCD* diagram allows to make the following basic conclusions.

1. Fatigue limit of a specimen increases 1.5—1.6 times providing the process of rolling friction comes into effect simultaneously (the direct effect — *AB* portion). The *direct effect coefficient* advanced in tribo-fatigue

$$K_D = \sigma_{-1p}/\sigma_{-1} \tag{4}$$

is in reality the characteristic of strength; in experiments its maximum is $K_{D\max}$ = 268/165 = 1.62. Coefficient (4) is naturally included into equation (2).

2. The critical (limiting) pressure under rolling friction increases 1.2—1.25 times providing cyclic stresses are induced in the specimen simultaneously (the back effect — *BC* portion). The *back effect coefficient* advanced in tribo-fatigue

$$K_B = p_{f\sigma}/p_f \tag{5}$$

is in reality a tribological characteristic; in experiments its maximum is $K_{B\max}$ = 2200/1760 = 1.25. Coefficient (5) is naturally included into equation (3).

3. Within the optimum range of contact pressures ($p \approx$ 400–1300 MPa) the process of wear under rolling results in significant improvement of the reliability of a system based on the criterion of fatigue resistance, hence any tendency towards wearless friction in this case is unjustifiable.

4. Tensile stresses under cyclic loading within the optimum conditions ($\sigma_a \approx$ 50–100 MPa) are positive significantly improving the reliability of a system based on the criterion of rolling friction resistance.

Better characteristics of the limiting state σ_{-1p} and $p_{f\sigma}$ in the process of wear-fatigue tests compared with the characteristics under rolling friction (p_f) and mechanical fatigue (σ_{-1}) can be explained from the standpoint of mechanics by the following major causes:

- addition of stresses with opposite signs (contact and bending), which leads to shifting the mean stress of the cycle towards negative values and therefore leads to the reduction of the maximum cycle stress;
- hardening of the working portion of the specimen by surface plastic deformation;
- appearance of favorable residual compressive stresses;
- healing primary fatigue cracks during elastoplastic deformation in the process of rolling friction.

The *governing parameter of wear-fatigue damage* (see Fig. 8)

$$m_{\sigma p} = \tan \alpha_{\sigma p} = \sigma_a/p_0 \tag{6}$$

has the critical value

$$m_k = \sigma_{-1}/p_f = 165/1760 = 0.094. \tag{6a}$$

This critical value separates the spheres of direct and back effects in the diagram of the limiting states of an active system. If $m_{\sigma p} < m_k$, the *CD* curve is obtained. If $m_{\sigma p} > m_k$, the *AB* curve is obtained. $m_{\sigma p} = \infty$ corresponds to the point *A* (pure mechanical fatigue), and $m_{\sigma p} = 0$ corresponds to the point *D* (pure rolling friction).

Hence, only methods of wear-fatigue tests allow to obtain a number of characteristics which truly reflect (describe) the serviceability of a real active system as the object studied by tribo-fatigue. Naturally the methods of tests for friction and the methods of fatigue tests reflect (describe) the performance of friction pairs (the object studied by tribology) and the structural elements (the objects studied by the mechanics of fatigue fracture), respectively. Yet, from Fig. 8 it follows that characteristics of serviceability determined experimentally within the frameworks of the reviewed disciplines should not be opposed. On the contrary, the point A (pure fatigue) and the point D (pure friction) naturally belong to the $ABCD$ diagram of limiting states (wear-fatigue damage) so that the characteristics σ_{-1} (in the point A) and p_f (in the point D) are basic for tribo-fatigue (see also equations (2), (3) and coefficients (4), (5), (6*a*)).

Now let's examine the *methods of theoretical studies* (see Table 1). The theory is known to rely upon experience. Hence, tribologists use their own experience to elaborate primarily the mechanics of contact interactions. Specialists in strength use their experience to elaborate the mechanics of deformation and fracture. Of course, specialists in tribo-fatigue use both as an inseparable entity. Yet, in order to study a more extensive object new approaches are to be sought for investigating complex phenomena.

Hence, a non-traditional approach towards the analysis of contact problems and the problems of the mechanics of deformation and fracture is being developed recently in tribo-fatigue. The approach is based on using a *statistical model of the deformable solid with a dangerous volume* (DSDV model) [15–17]. According to this model the strength of a specimen (including its surface strength) is determined by the region of finite dimensions containing the critical level of stresses. This region is termed as the *dangerous volume*.

The concept of the approach advanced in tribo-fatigue is the following [18, 19].

Assume the steel shaft is cyclically bended by moment M, so that in some region of the shaft the field of normal stresses σ is a damaging one. It means that a dangerous volume $V_{P_\gamma} > 0$ (cases F in Fig. 9), limited by the condition $\sigma \geq \sigma_{-1\min}$, where $\sigma_{-1\min}$ is the lowest value of the dispersion of fatigue limits, is formed on the surface of the shaft.

Assume the process of rolling friction or sliding friction is realized in the dangerous zone of the shaft. Assume that the field of contact pressures p in the (shaft – counterspecimen) active system is such that it produces a dangerous volume S_{P_γ} (cases T in Fig. 9) in the contact region. In case of sliding friction this dangerous volume $S_{P_\gamma} = S_{WP}$ is formed within a fine surface layer of the shaft (case *T-1*). In case of rolling friction the dangerous volume $S_{P_\gamma} = S_{WP}$ (case *T-1*) and / or $S_{P_\gamma} = S_{VP}$ (case *T-2*) is formed both on and under the surface. In all these cases it is limited by the damaging level of tangential stresses $\tau \geq \tau_{-1\min}$, where $\tau_{-1\min}$ is the lowest value of dispersion of fatigue limit in shear (or torsion of thin-walled tubes).

Assume wear-fatigue tests are a combination of cyclic bending and friction (sliding or rolling). Then two situations are possible when dangerous volumes appear (the right column in Fig. 9). First, volumes V_{P_γ} and S_{P_γ} combine on the surface (case *FT-1*). Second, they combine under the surface (case *FT-2*).

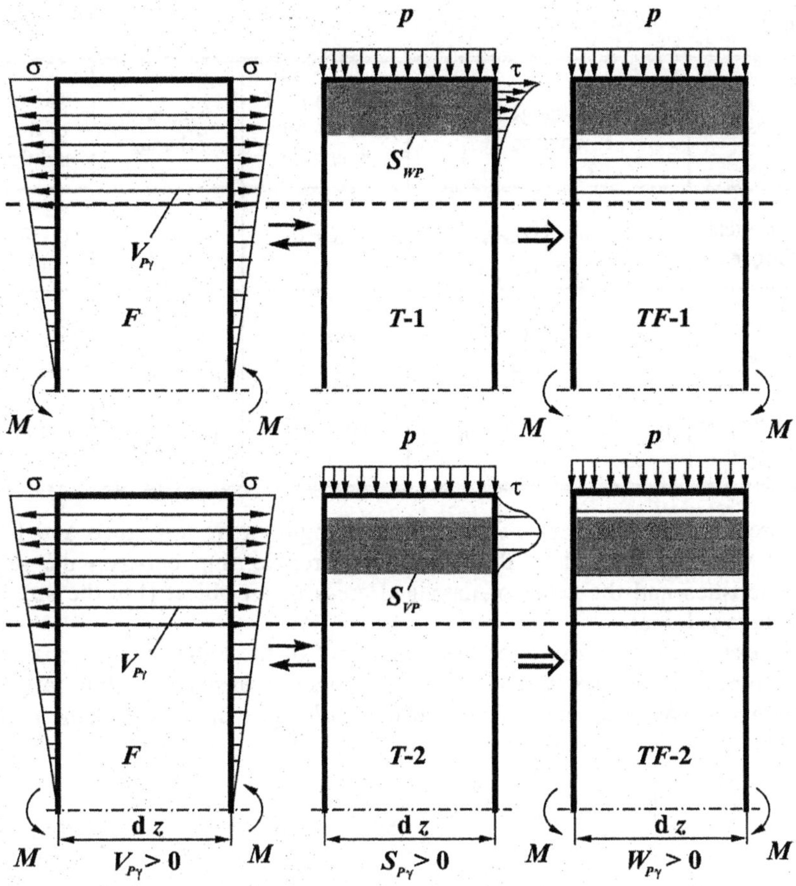

Fig. 9. Scheme of emergence of dangerous volumes during friction tests ($S_{P\gamma} > 0$), fatigue tests ($V_{P\gamma} > 0$) and wear-fatigue tests ($W_{P\gamma} > 0$)

In both these cases a combined dangerous volume $W_{P\gamma} > 0$ appears as a function of volumes $V_{P\gamma}$ and $S_{P\gamma}$, i.e.

$$W_{P\gamma} = (V_{P\gamma} + S_{P\gamma})\varphi_{SV.} \tag{7}$$

where φ_{SV} is the function of interaction; it can be assumed that in some cases it is enough to treat this function as the parameter of interaction.

The formulas for determination of dangerous volumes using any component of normal σ and tangent τ stresses are given in Table 2. Applying the DSDV model to mechanical fatigue and friction it is easy to derive the conditions of failure-free operation and / or the conditions of damage and fracture if suitable measures of damage ω are introduced (S_0 and V_0 are the working volumes in friction and fatigue, respectively). All the solutions are easily expandable to cover the case of complex wear-fatigue damage.

Table 2

Damage	Mean dangerous volume	Condition of failure-free operation	Measure of damage	Condition of damage and fracture
Mechanical fatigue	$V_{0.5\gamma} = \iiint\limits_{\sigma \geq \sigma_{-1\min}} dxdydz$	$V_{P\gamma} = 0$	$\omega_V = \frac{V_{P\gamma}}{V_0}$	$0 < \omega_V \leq 1$
Friction and wear	$S_{0.5\gamma} = \iiint\limits_{\tau \geq \tau_{-1\min}} dxdydz$	$S_{P\gamma} = 0$	$\omega_S = \frac{S_{P\gamma}}{S_0}$	$0 < \omega_S \leq 1$
Wear-fatigue damage	$W_{0.5\gamma} = (V_{0.5\gamma} + S_{0.5\gamma})\varphi_{SV}$	$W_{P\gamma} = 0$	$\omega_\Sigma = \frac{W_{P\gamma}}{V_0}$	$0 < \omega_\Sigma \leq 1$

Moreover, the problems of reliability, strength, wear resistance and life are solved with the account of the fundamental relationship between the relevant characteristics and the scale of damaged areas of an object (i.e. the size of its dangerous volume).

For example basic results of the assessment of the indicators of reliability of metal-to-polymer active system (see Fig. 7, *e*) under mechano-sliding fatigue can be shown. It has been established that the *probability of failure* of the metal-to-polymer system is determined by the function:

$$F(\sigma, p) = \left(1 - \exp\left\{-C_V \frac{V_{P\gamma}}{V_0}\left[\left(\frac{T_M}{T_0}\right)^{m_T}\left(\frac{\sigma - \sigma_{-1\min}}{\sigma_W}\right)\right]^{m_V} - C_S \frac{S_{P\gamma}}{S_0}\left(\frac{p_{tf}^{(1)} \Delta T_P}{p_d - p}\right)^{m_S}\right\}\right)\Psi(\sigma, p) \quad (8)$$

where $\Psi(\sigma, p)$ is the function of interactions between the phenomena of friction, wear (under the effect of contact pressure p) and mechanical fatigue (under the effect of cyclic stresses σ).

According to (8) the probability of failure of the metal-to-polymer system under mechano-sliding fatigue increases with growth of the number of defects in the metal and the polymer (the parameters of isotropy m_V and m_S), the sizes of dangerous areas of deformation under cyclic loading and friction (the dangerous volumes $V_{P\gamma}$ and $S_{P\gamma}$), effective cyclic ($\sigma > \sigma_{-1\min}$) and contact $p < p_d$ stresses, temperatures of the metal and polymer (T_M and ΔT_P). It also depends on the shape of the bodies, the configuration of their contact interaction, the method of cyclic deformation of the body (the coefficients C_V and C_S), geometrical dimensions of the contacting solids (the working volumes V_0 and S_0), the mechano-physical properties of the polymer (the destruction limit p_d and singular thermofluctuating stress $p_{tf}^{(1)}$) and the metal (the parameters $\sigma_{-1\min}$ and σ_w of the function of distribution of durability limits), thermal activation of the chemico-physical processes of fatigue damage (the parameter m_T).

After performing relevant limiting transitions function (8) yields the formulas for predicting two more indicators of the wear-fatigue damage (see Fig. 3). First, a formula to calculate the average value $\bar{\sigma}_{-1p}$ of the fatigue limit of the metallic shaft with the account of the effect of the friction and wear processes under contact pressure p (the direct effect):

$$\bar{\sigma}_{-1p} = \bar{\sigma}_{-1T}\left[1 - b_S \frac{S_{0.5\gamma}}{S_0}\left(\frac{p_{if}^{(1)} \Delta T_p}{p_d - p}\right)^{m_S}\right]^{1/m_V}, \tag{9}$$

where $\bar{\sigma}_{-1T}$ is the average value of the fatigue limit of the shaft with the account of temperature T_M, b_S – the coefficient. According to formula (9) usually $\bar{\sigma}_{-1p} < \bar{\sigma}_{-1T}$ in the metal-to-polymer system during mechano-sliding fatigue, the bigger is the contact pressure p the smaller is the value $\bar{\sigma}_{-1p}$.

Second, a formula for calculating the average wear rate $I_\sigma(p)$ of the polymeric counterbody with the account of the effect of cyclic stresses σ in the conjugated metallic specimen under the effect of contact pressure p (the back effect):

$$\bar{I}_\sigma(p) = \frac{\bar{I}(p)}{1 - b_V \frac{V_{0.5\gamma}}{V_0}\left[\left(\frac{T_M}{T_0}\right)^{m_T}\left(\frac{\sigma - \sigma_{-1\min}}{\sigma_w}\right)\right]^{m_V}}, \tag{10}$$

where $\bar{I}(p)$ is the average wear rate in a similar friction unit (in which $\sigma = 0$), b_V is the coefficient. It follows from formula (10) that $\bar{I}_\sigma(p) > \bar{I}(p)$ is always observed in the metal-to-polymer system under mechano-sliding fatigue under equal contact pressures in the friction unit and the active system.

According to (10), *wear rate can be controlled in a non-traditional manner* by exciting cyclic stresses in one of the elements of the friction unit. Experimental studies have confirmed that such control is highly effective. To exemplify it Fig. 10 shows the results of MSF (see Fig. 7, *e*) tests of the alloyed steel 40X (the specimen) / formaldehyde copolymer (the counterspecimen) system under constant contact pressure p_a = 5.7 MPa. If the amplitude of cyclic stresses increases from 160 to 300 MPa, an incriment in the wear rate due to these stresses grows from 110 to 180%.

It should be noted that $\sigma_{-1p} > \sigma_{-1}$ in case of the mechano-rolling fatigue of the metal-to-metal system (see Fig. 8 and Eq. (2)), on the contrary, in case of the mechano-sliding fatigue of the metal-to-polymer system it turns out that $\sigma_{-1p} < \sigma_{-1}$ in accordance with Eq. (9). The latter regularity is well corroborated by the experimental data. Figure 11 exemplifies it by showing the back dependence of the limiting stresses on the contact pressure in the alloyed steel / polymer system. It means that the regularities of the direct effect can be highly variable.

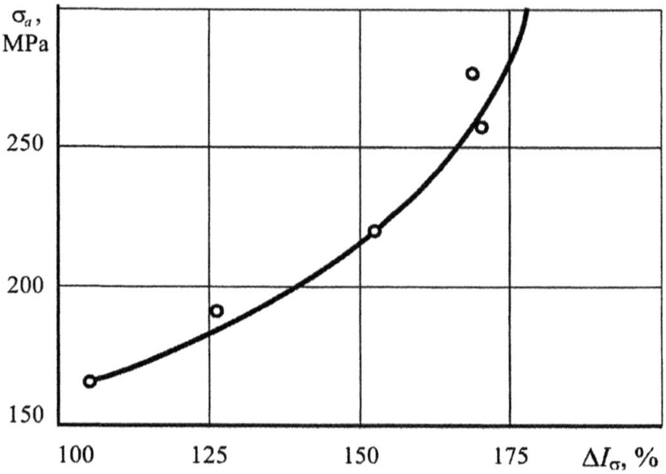

Fig. 10. Dependence of the increment of polymer wear rate intensification on amplitude of cyclic stresses (active system is alloyed steel 40X / formaldehyde copolymer)

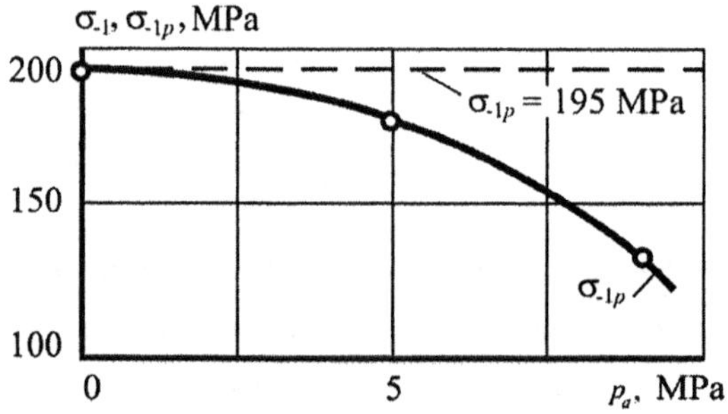

Fig. 11. Dependence of limiting stresses on nominal contact pressure in the active system of chromium steel 40X (specimen) / glass-filled (~25%) polyamide Duretan BKV-30H (counterspecimen)

Equations (9) and (10) approximate sufficiently well the experimental data (the points in Figs. 10 and 11). One of the causes is that both these equations are constructed with the account of the DSDV model. Equation (9) contains the value $S_{P\gamma}/S_0$, meanwhile equation (10) contains the value $V_{P\gamma}/V_0$.

Processes and phenomena

Now let's consider the third most essential attribute of any scientific discipline, viz. the *processes and phenomena* it studies.

Surface damage is the basic process of degradation of the specimen and the counterspecimen under the effect of contact loading in the friction pair; it is studied by tribology (see Table 1). Volume (fatigue) fracture is the basic process of degradation of the structural element under the effect of alternating loading; it is studied by the mechanics of fatigue fracture. In case of friction processes (under the effect of contact loading) and the processes of cyclic deformation (under the effect of both contact and non-contact alternating loading) combined in the element of an active system, complex surface damage (due to any mechanism) is just an initial process of its degradation which evolves in time and inevitably ends in volume fracture; this type of *complex surface damage and fracture* is studied by tribo-fatigue (see Table 1) [21–25].

It is easy to discriminate between the surface damage and the volume fracture. It is harder to solve the following problem: what makes the surface damage of the specimen in a friction pair different from the complex surface damage of the element of an active system termed as *wear-fatigue damage* in tribo-fatigue. This difference should be established with one essential and obligatory provision: the contact loading should be the same whether it applies to an active system or to a similar friction pair.

Application of accurate experimental methods of studies allows to investigate and understand the features of complex – wear-fatigue damage [26, 27]. Figure 12 exemplifies it with the results of studies (method of the atom force microscopy) of the processes of cracking of steel 45 specimens in rolling friction (the left column of figures) and under wear-fatigue tests (the remaining figures) as a function of the level of contact pressure p_0 and the amplitude of cyclic stresses σ_a. Figures (~35×35 μm^2 in size) show the morphology of cracks typical for the relevant conditions of tests. The histogram shows the dependence of the critical depth h of the damaged layer on the level of cyclic stresses (under constant contact pressure $p_0 = 2130$ MPa). These experimental data allow to conclude the following.

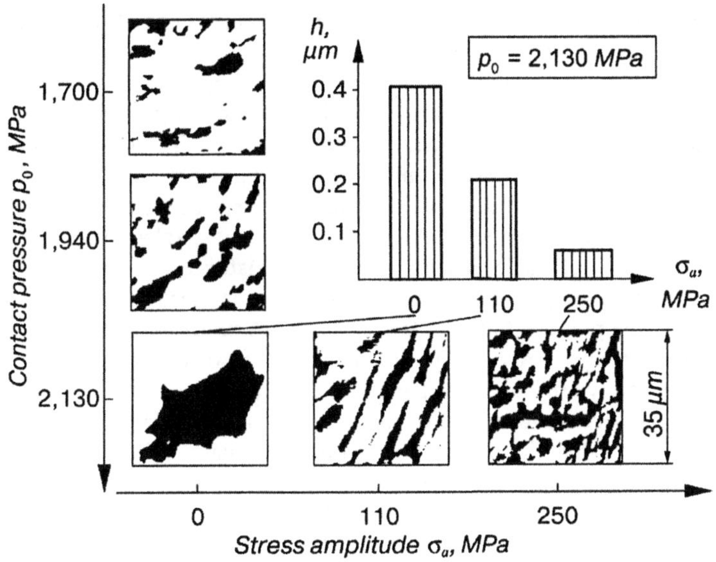

Fig. 12. Microtopography of surface damage in rolling friction (vertical column of images) and during wear-fatigue tests (remaining images)

In case of pure rolling friction any higher contact pressure intensifies plastic deformation and hence deformed fragmentation of grains, it initiates the appearance of discrete pores and cracks followed by their coalescence into chains. The system of deformed grains, chains of pores and cracks is unidirectional and it is oriented in the direction of rolling. This process produces relatively large discrete pitting. Two types of wear, viz. delaminating and pitting, come to dominate. The critical depth of the damaged layer is estimated to be ~ 0.4–0.5 μm.

Similar deformed fragmentation of grains and origination of pores and cracks are observed during wear-fatigue tests. But the process of damage changes significantly as the amplitude of cyclic stresses increases, the processes of origination of the second system of cracks accelerate and these cracks arrange transverse to the rolling direction. Hence, damage becomes dissipated and an almost regular grid of intersecting cracks and pores appears around finely dispersed particles (fragments of grains) of the material. The stronger the cyclic stresses are the denser is the grid of cracks and pores, the finer and thinner the separating particles become, the critical depth of the damaged layer reduces to 0.05 μm. It prevents the appearance of larger pittings, they are not observed under these conditions. *Surface chipping* is the dominating wear process in this case. It is characterized by the separation of fine dispersed particles from the working surface, these particles result from multiple microshearing of intersecting planes and fine shattering of grains. This mechanism of complex surface damage is termed the *multiple microshearing dissipated effect* (MMSD phenomenon) [5] in tribo-fatigue.

The above results allow to establish additionally the following causes why wear-fatigue damage under definite conditions becomes less dangerous than damage by friction (under similar contact pressures), as it follows from the experimental data in Fig. 8.

1 Superposition of the fields of contact and bending stresses causes dissipation of a larger share of applied energy in a finer surface layer of the material and localization of the processes of cracking and wear in it. The energy of deformation is expended more on finer shattering of grain fragments and their separation in multitudes than on penetration of damage into the depth of the specimen material.

2 Wear of the surface layer damaged by a grid of cracks and pores exposes a new, relatively intact surface with a higher resistance to fracture. Hence, the appearance of relatively large pittings, which have dangerous microconcentrations of stresses at their bottom and a dangerous main crack is postponed in time or even entirely prevented (depending on the conditions of loading).

3 Approximately tenfold rejuvenation of the working surface by fragmentation, chipping and separation of metallic particles during wear-fatigue tests is required before damage reaches the depth comparable with that reached in rolling friction, providing contact pressure is the same in both these cases.

Hence, wear-fatigue damage is *a specific, peculiar type of surface damage* of the basic element of the active system. Its feature under these conditions is surface chipping due to the MMSD phenomenon occurring on the intersecting planes of sliding. Its characteristic property is that, notwithstanding the fact that it is a damaging process, it is useful since it improves significantly the reliability and durability of the active system. The optimal combination of loading parameters σ_a and p_a apparently yields such a state of the active system when its load carrying capacity is maintained spontaneously and during a long period of time (or it is automatically controled) by fine wear of the surface layer and its removal from the friction zone. To summarize it is worthwhile to note that the active system is a peculiar dynamic system and its behavior can and should be controlled (U) by adjusting the wear rate in a non-traditional manner (see the example in Eq. (10) and in Fig. 10). *Optimal control of the process of wear-fatigue damage* is a fundamentally new and practically essential trend of studies evolving within the framework of tribo-fatigue and involving the methods of the *theory of control.* It is shown in Fig. 5 by the arrows indicating the relationship between these two scientific disciplines.

In relation to the studies of the direct and back effects two circumstances are to be highlighted.

First, the experimental results obtained in tribo-fatigue and presented in Fig. 11 are unexpected from the traditional point of view: a significant reduction of the fatigue limit (32% at contact pressure 8.5 MPa) is observed when no physical wear of specimens occurs at all. In other words, wear produces no damaging effect (the diameter of the specimen does not reduce, no marks as concentrators of stresses appear on the friction surface, no seizure occurs over the real contact spots, etc.), meanwhile fatigue resistance drops drastically. It is explained by the effect of little studied mechano-physical phenomena evolving in the steel-to-polymer friction zone. In particular, products of tribo-destruction of polymer are known to act as surfactants. They accumulate in the contact zone and stimulate the migration and proliferation of dislocations on the metallic friction surface (the Rebinder effect). This leads to the acceleration of its fatigue damage. Also, as the contact pressure goes up and the time of tests extends, the mean temperature in the friction zone increases (up to 70 °C in the experimental conditions). It induces thermal activation of damaging chemi-physical processes, hence the fatigue resistance of specimens reduces even more. A theoretical analysis (from the standpoint of tribo-fatigue) corroborates these regularities: the wear characteristic in formula (9) is absent, still a significant variation of the magnitude σ_{-1p} is predicted due to the effect of the parameters of the mechano-physical state of the polymer $(p_{tf}^{(1)}, m_s, p_d)$, and due to the changes of its temperature (ΔT_p).

Second, the back effect is not known in tribology or in the mechanics of fatigue fracture; tribo-fatigue was the first to establish it [20]. According to the experimental data presented in Fig. 10, the durability of a MPS assessed basing on the wear criterion is in many respects determined by the back effect.

Let us analyze Fig. 13 in order to answer the question why the wear of the polymeric counterspecimen is strongly intensified when cyclic stresses are induced in the contacting metallic counterspecimen. The specimen is shown as a rotating disk *1* with a smooth working surface, the counterspecimen is shown as a stationary single indentor *2*. In common friction tests (Fig. 13, *a*) only contact load q_r is effective, indentor *2* is statically bent (in the direction opposite to the direction of rotation ω_1) and the area of deformation on the working surface of a disk is a strip (the friction path). In case of wear-fatigue tests (Fig. 13, *b*) additional cyclic stresses (deformations) $\sigma = \pm \sigma_Z$ are excited in the disk. Oscillations of the working surface of the disk in the direction z make the friction path on the surface zigzag shaped, the indentor experiences cyclic bending (in the direction *z*). The wear processes of the two elements naturally intensify in accordance with the magnitude of the cyclic stresses $\sigma = \pm\sigma_Z$. If the indentor is polymeric and the disk is steel, only the wear of the polymer as a weaker material is intensified.

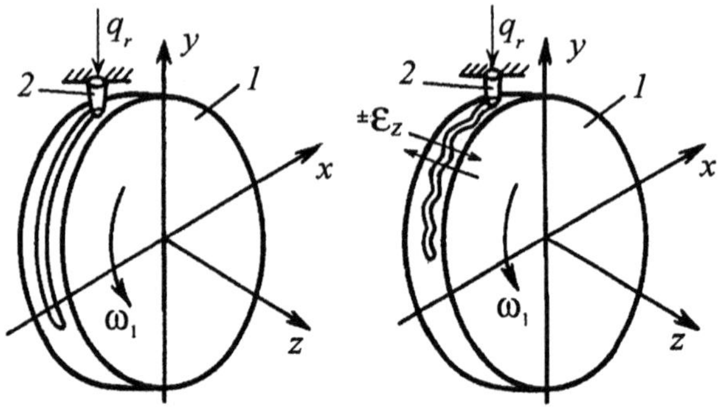

Fig. 13. To the analysis of the back effect

Hence, in the general case of wear-fatigue tests, the back effect has two manifestations: wear accelerates in both elements under the effect of cyclic stresses σ excited in one of the elements of the active system. Analytically this effect for a MPS is described by Eq. (10).

Objectives and tasks

Finally the fourth attribute of any scientific discipline is its *objectives and tasks* [28].

The task challenging the specialists in the mechanics of fatigue fracture is to combat fatigue breakdowns. Prediction and prevention of breakdowns is the apex of their achievements. Though fatigue fracture of a structural element relates, as a rule, to hazards for personnel, in practice its prevention is not always absolute. Economic considerations come into play and specialists of strength set the problem of strength optimization formulated as

$$\text{(I)} \qquad \begin{cases} F(\sigma) \Rightarrow \min, \\ C_0(\overline{\sigma}_{-1}, S_{\sigma_{-1}}, \overline{\sigma}_a, S_{\sigma_a}) \Rightarrow \min \end{cases}$$

according to which the probability of fatigue fracture $F(\sigma)$ should be minimal providing the total cost C_0 in the spheres of production and maintenance is also minimized. This cost definitely depends on the parameters of distribution of effective and limiting stresses to be achieved ($\overline{\sigma}_{-1}$, $\overline{\sigma}_a$ — mean values of limiting and effective stresses; $S_{\sigma_{-1}}, S_{\sigma_a}$ – their root-mean-square deviations.

The major problem of tribologists is to combat wear. The consummation is the achievement of nearly wearless friction. They similarly formulate the problem of optimization as

(II)
$$\begin{cases} I(p) \Rightarrow \min, \\ F(p) \Rightarrow \min, \\ C_0(\overline{p}, S_p, \overline{p}_f, S_{p_f}) \Rightarrow \min \end{cases}$$

since they have to bear in mind the powerful economic factor; $\overline{p}_f$ and $\overline{p}$ — mean values of limiting and effective pressure, S_{p_f} and S_p — their root–mean–square deviations.

The main problem of the specialists in tribo-fatigue is to control the processes of complex (wear-fatigue) damage in order to achieve optimum (and feasible) service life of a specific active system. Efforts are made to use wear and fatigue damage in the process of operation useful to extend durability of a unit. The essence is simple. Tribologists consider additional cyclic stresses in a friction pair as a damaging factor. Specialists in strength, in their turn, consider wear as the damaging factor of a structural element. Specialists in tribo-fatigue consider that friction, wear, fatigue are the phenomena which interact kinetically and may either intensify strongly the degradation of a material, or, on the contrary, to produce spontaneous and extended preservation of the load carrying capacity depending on conditions. To understand the conditions and mechanisms of evolution of these processes is to gain a simple key to control them. Thus, the specialists in tribo-fatigue should formulate and solve the *problem of optimum control of a dynamic (active) system (III)*:

(III)

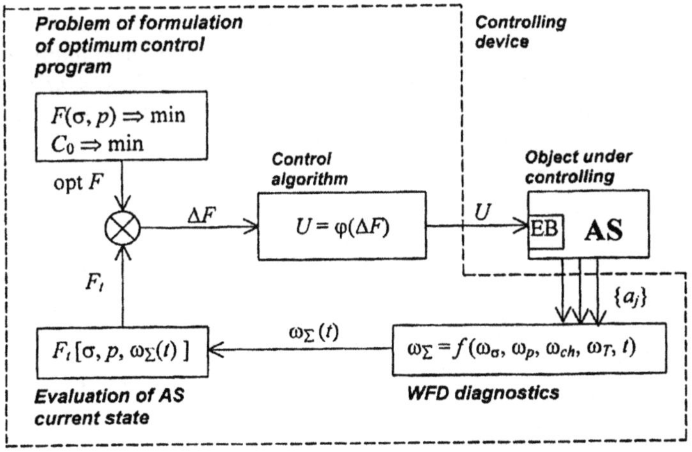

The content of the problem is the following. An active system (AS) is considered as an object under controlling. The problem of formulation of an optimum control program is set as the problem of optimization: $F(\sigma, p) \Rightarrow \min$, $C_0 \Rightarrow \min$, i.e. feasibility of calculations yield the parameter of optimization opt F. In operation a set $\{a_j\}$ of parameters a_j of the state of the AS is measured. The results yield a current measure $\omega_\Sigma(t)$ of the complex wear-fatigue damage (WFD) which is a function of time t and particular measures of damage ω_σ, ω_p, ω_{ch}, ω_T induced by cyclic stresses (index σ), contact pressure (index p), the processes of electrochemical corrosion (index ch), temperature (index T) in the zone of contact interactions between elements of the AS. The condition of the AS at any moment of its operation is estimated using the integral parameter $F_t(\sigma, p, \omega_\Sigma(t))$. Another objective is to correlate (compare) the optimum (opt F) and the current (F_t) values of the integral parameter F. The obtained disagreement ΔF of the parameters opt F and F_t serves to solve the problem of synthesizing dynamic or optimal control $U = \varphi\ (\Delta F)$. The executive body (EB) makes management of the AS feasible.

Comparison of problems I, II and III indicates that problem III is definitely more intricate, yet more practical: it implies effective management of reliability of the most essential active systems of a machine in operation based on the most significant criteria of performance and simultaneous reduction of the cost of labor, means and materials in production and maintenance.

Interactions between scientific disciplines

It has already been discussed and shown in Fig. 5 that tribology, the mechanics of fatigue fracture and the theory of reliability of mechanical systems are the foundation of tribo-fatigue. Another aspect of the problem (Fig. 14) is that the results yielded by tribo-fatigue, on their hand, strongly contribute to tribology, the mechanics of fatigue fracture and the theory of reliability of mechanical systems. Tribology (T) due to tribo-fatigue (TF) formulates a new class of essential problems, such as studies of external (in relation to contact) loading and its effect upon the characteristics of friction and wear of units [29–34 and others]. Tribo-fatigue briefly determines this class of problems as the back effect (BE in Fig. 14, for example, see formulas (3), (5), (10)). Then, it is easy to see that, thanks to tribo-fatigue (TF) again, the mechanics of fatigue fracture (F) deals with a new class of essential problems, investigation of friction and wear processes and their effect upon the characteristics of fatigue resistance of materials and structural elements [35–39 and others]. Tribo-fatigue determines this class of problems as the direct effect (DE in Fig. 14, for examples see formulas (2), (4), (9)). Finally, when the theory of reliability (R) is concerned, again due to tribo-fatigue (TF), an individual class of practically essential problems is elaborated: studies of failures of active systems using complex criteria of wear-fatigue damage and fracture [40–42 and others] (see function (8) for example). It is a *new S-function (8) of two-dimensional distribution* of random values x and y:

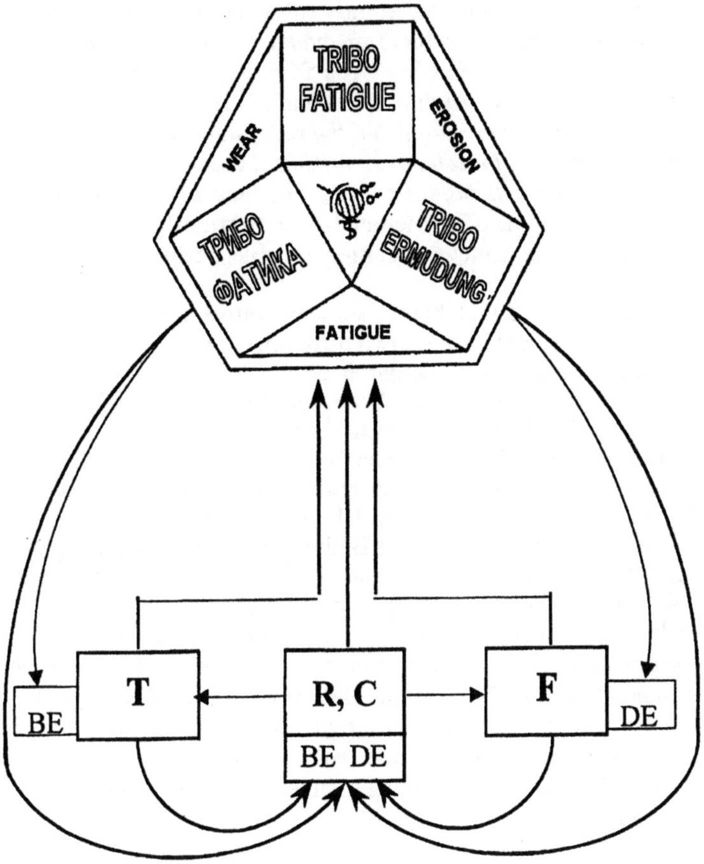

Fig. 14. Relationships and interactions between scientific disciplines

$$F(x,y) = 1 - \exp\left[-a\left(\frac{x - x_{\min}}{x_*}\right)^{m_x} - b\left(\frac{y_*}{y_{\max} - y}\right)^{m_y}\right], \tag{11}$$

where $x_{\min}$, $y_{\max}$, a, b, m_x, m_y, x_*, y_* are the parameters of distribution. An essential feature of the function is that one of the random values has the lowermost limit ($x > x_{\min}$) and the other has the uppermost limit ($y < y_{\max}$). Function (11) can be apparently useful in the theory of reliability and in mathematical statistics to analyze a variety of random events.

Nevertheless, once again it should be explicitly emphasized that tribo-fatigue not only has yielded new results useful "for tribology", "for the mechanics of fatigue fracture", "for the theory of reliability of mechanical systems", but the most essential is that tribo-fatigue has yielded fundamentally new scientific concepts useful to formulate and successfully solve the problems of great practical significance. For example, the problem of dynamic (optimal) control (U) of the process of WFD of active systems has been formulated. It implies that the time has come to *proceed from traditional calculations and designing of basic (individual) parts of machines based on their strength and wear resistance to calculations and designing of mechanical systems, i.e. the same parts, yet with the account of actual interactions between them.* The new principle of designing the most essential active systems of a machine based on tribo-fatigue criteria requires to assess precisely and to achieve specified reliable performance with the least cost.

Still it should be admitted that there is some cognitive and psychological barrier some specialists confront and have known difficulties to overcome when digesting new ideas of tribo-fatigue. In particular it is due to the fact that the theory of fatigue wear of friction surface has been longly and fruitfully developed in tribology. Following this deep-rooted tradition some tribologists treat tribo-fatigue as fatigue in the process of friction. The II International Symposium on Tribo-Fatigue (Moscow, 1996) much contributed to overcoming this erroneous approach. It can be presumed that the III International Symposium on Tribo-Fatigue to be held in October, 2000, in Beijing [8] will further integrate various scientific schools and stimulate mutual understanding for the purpose of successful solution of theoretical and practical problems of promoting reliability of modern technology using the complex criteria of performance.

Hence, it seems natural that some engineering universities, like the Gomel State University named after P. O. Sukhoy, the Belarusian State University of Transport have included the fundamentals of tribo-fatigue as a university discipline into their curricula for mechanical engineers. Experience shows that universities are fast to respond to practical demands for training engineers as soon as new trends emerge in science and technology.

Interests of tribo-fatigue

So far Byelorussian standard СТБ 994—95 [1] and Interstate (for CIS states) standard GOST 30638–99 “Tribo-Fatigue. Terms and Definitions” [3] have been officially approved and put into effect. Figure 15 outlines the “range of interests of tribo-fatigue” determined by these standards. Table 3 shows typical examples of active systems and types of wear-fatigue damage (their definitions).

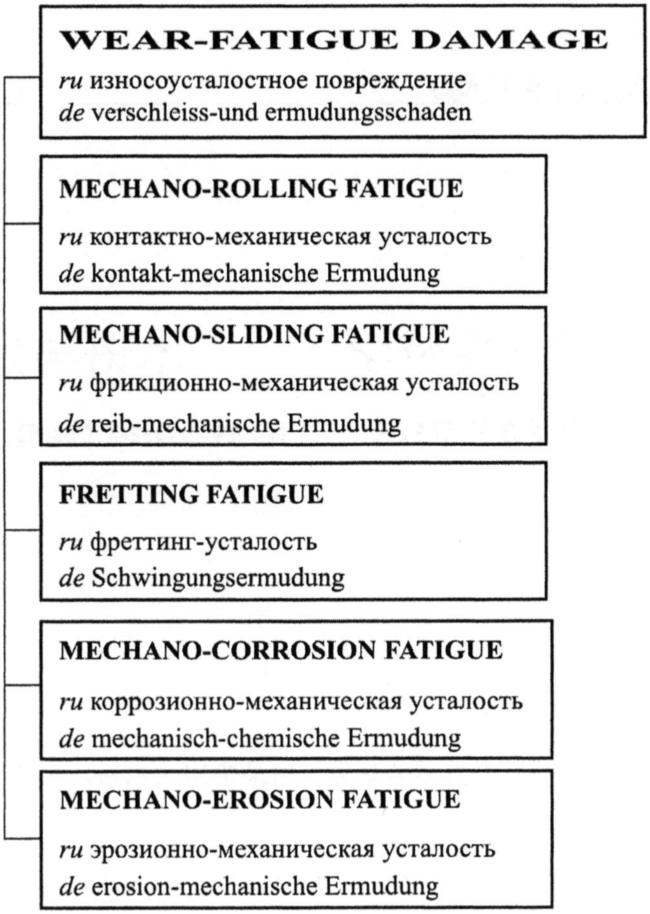

Fig. 15. Basic types of wear-fatigue damage (GOST 30638–99)

GOST 30638 has been translated (Fig. 16) into Chinese and its approval in the Chinese People's Republic is pending.

国际标准（申报稿） ГОСТ 30638

摩擦疲劳学 术语与定义

TRIBO-FATIGUE Terms and Definitions

范围及应用领域

本标准给出摩擦疲劳学领域中最常用的术语，包括它们的定义及分类。

对于某些术语和词组，本标准给出它们的缩写形式；当它们不发生歧意时，推荐使用这些缩写形式。

注意一本标准同时给出相对应的俄文的术语。然而，只有以法定语言给出的术语和定义才可能被视为国际标准化组织（ISO）的术语和定义。

本标准还给出一份特征量清单，推荐将它们用于计量和示性。

Fig. 16. Title-page of the Interstate Standard GOST 30638–99 in the Chinese language

Table 3

Typical active system	Complex damage and fracture	Definition
Crank-pin / connecting rod end with sliding bearing	Mechano-sliding fatigue	Wear-fatigue damage due to the effect of kinetic interaction between the phenomena of mechanical fatigue and sliding friction
Wheel / rail	Mechano-rolling fatigue	Wear-fatigue damage due to the effect of kinetic interaction between the phenomena of mechanical fatigue and rolling friction (rolling friction with slippage)
Spline shaft / bushing	Fretting fatigue	Wear-fatigue damage due to the effect of kinetic interaction between the phenomena of mechanical fatigue and fretting
(Screw) propeller shaft / sea water	Mechano-corrosion fatigue	Fatigue of the material under the simultaneous effect of alternating stresses and corrosive environment
Turbine blades / fluid or gas stream carrying solid particles	Mechano-erosion fatigue	Wear-fatigue damage due to the effect of kinetic interaction between the phenomena of mechanical fatigue and erosion

Thus, tribo-fatigue is a new, dynamically evolving part of mechanics. The integrity of studies, implying that an entity is viewed as multiple, is its basic methodological principle.

Bibliography

1. STB 994–95. Tribo-fatigue. Terms and definitions. (Standard of Belarus). BelStandard, Minsk, 1995. (in Russian).
2. Sosnovskiy L A, Tribo-fatigue: main terms and definitions. Friction and Wear, 1992, No. 4. (in Russian).
3. GOST 30638–99. Tribo-fatigue. Terms and definitions. (Interstate Standard). Interstate Standardization, metrology and certification board, Minsk, 1999. (in Russian).
4. Some words about tribo-fatigue. (Essays). Authors: Vysotsky M C, Frolov K V, Troshchenko V T, and others. Ed. by A Bogdanovich, Gomel, Minsk, Moscow, Kiev, "Remika", 1996. (in Russian).
5. Sosnovskiy L A, Makhutov N A, Methodological problems of complex assessment of damage and limiting states of active systems. Zavodskaya Laboratoriya, Moscow, 1991, No. 5. (in Russian).
6. Sosnovskiy L A, Gao Wanzhen, On methodology of tribo-fatigue. Proc. int. sci. tech conference, Coll. Modern problems of machine studies, State tech university named after P Sukhoy, Gomel, 2000. (in Russian).
7. Sosnovskiy L A, Mechanics of wear-fatigue. Proc. of the III int. symp. on tribo-fatigue, Ed. by Gao Wanzhen and Li Jian, Hunan University Press, Beijing, 2000.
8. Frolov K V, Foreword. Proc. of the III int. symp. on tribo-fatigue, Ed. by Gao Wanzhen and Li Jian, Hunan University Press, Beijing, 2000.
9. Sosnovskiy L A, Tribo-fatigue: problems and prospects. Paper at the USSR Academy of Sciences Exhibition "Mathematics and mechanics to national economy", Gomel, 1989. (in Russian).
10. Shcherbakov S S, The force and the coefficient of friction in active systems. Proc. of the III int. symp. on tribo-fatigue, Ed. by Gao Wanzhen and Li Jian, Hunan University Press, Beijing, 2000.
11. Sosnovskiy L A, Methods of wear-fatigue tests of materials. Zavodskaya Laboratoriya, Moscow, 1990, No. 6. (in Russian).
12. Sosnovskiy L A, Methods of wear-fatigue tests of active systems and their models. Friction and Wear, 1993, No. 5. (in Russian).
13. Sosnovskiy L A, Makhutov N A, Bogdanovich A V, Tyurin S A, Diagram of steel 45 limiting states during mechano-rolling fatigue. Zavodskaya Laboratoriya, 1996, No. 2. (in Russian).
14. STB 1066–97. Tribo-fatigue. Methods of wear-fatigue tests. Tests for mechano-rolling fatigue. (State Standard of Belarus). Minsk, 1997. (in Russian).
15. Sosnovskiy L A, The statistical model of a deformed solid body and some its application. Strength of Materials, 1990, No. 5, part 1, 1992, No. 11, part 2. (in Russian).
16. Sosnovskiy L A, Statistical mechanics of fatigue fracture. Nauka i Tekhnika, Minsk, 1987. (in Russian).
17. Sosnovskiy L A, Calculations of fatigue of machine parts with dangerous volume in the probabilistic definition. Applied Mechanics, 1979, No. 4. (in Russian).
18. Sosnovskiy L A, Koreshkov V N, The statistical model of a deformed solid body and its application. Strength of Materials, 1999, No. 6, part 3. (in Russian).
19. Tribo-fatigue–96/97: Modelling of active systems. Annual edition. Ed. by V Shurinov, Sci. Prod. Group "Tribofatigue", Gomel, 1990. (in Russian).

20. Sosnovskiy L A, Reliability and durability of elements during wear-fatigue tests. Reliability and Durability of Machines and Equipment, 1986, No. 9. (in Russian).
21. Sosnovskiy L A, Mechano-sliding fatigue of active systems. Vestnik Machinostroeniya, 1992, No. 8–9. (in Russian).
22. Sosnovskiy L A, Makhutov N A, Shurinov V A, Fretting-fatigue: main regularities. Zavodskaya Laboratoriya, 1992, No. 8. (in Russian).
23. Sosnovskiy L A, Makhutov N A, Shurinov V A, Mechano-sliding fatigue: main regularities. Zavodskaya Laboratoriya, 1992, No. 9. (in Russian).
24. Sosnovskiy L A, Makhutov N A, Shurinov V A, Mechano-rolling fatigue: main regularities. Zavodskaya Laboratoriya, 1992, No. 11. (in Russian).
25. Sosnovskiy L A, Makhutov N A, Mechano-corrosion fatigue: main regularities. Zavodskaya Laboratoriya, 1993, No. 7. (in Russian).
26. Chizhik C A, Sosnovskiy L A, Gorbunov V V, Lisitsyn S. D, Senkova E L, Peculiarities of emergence and growth of small surface cracks in carbon steel during mechano-rolling fatigue. Zavodskaya Laboratoriya, 1996, No. 3. (in Russian).
27. Sosnovskiy L A, Experimental grounds of tribo-fatigue. Strength of Materials, 1997, Nos. 3–4. (in Russian).
28. Sosnovskiy L A, Gao Wanzhen, Tribo-fatigue. Proc. of the III int. symp. on tribo-fatigue, Hunan University Press, Beijing, 2000.
29. Yahata N, Hirata T, Kato T, Watanabe M, Wear, 1988, v. 121.
30. Marchenko E A, Petrova I M, Concurent effect of friction and volume loading on fatigue damage accumulation in surface layers of metals. Paper thesis at II int. symp. on tribo-fatigue, "SPAS"– Sci. Prod. Group "Tribofatigue", Moscow, Gomel, 1996. (in Russian).
31. Kudryavtsev I A, Yakovlev V F, Effect of alternate (variable) mode of loading on contact durability of rail steel. Paper thesis at II int. symp. on tribo-fatigue, "SPAS"– Sci. Prod. Group "Tribofatigue", Moscow, Gomel, 1996. (in Russian).
32. Volkov V M, Petukhov A N, The fracture simulation of metals with cavitation fluid and cyclic mechanical loads. Paper thesis at II int. symp. on tribo-fatigue, "SPAS"– Sci. Prod. Group "Tribofatigue", Moscow, Gomel, 1996. (in Russian).
33. Licyuk V S, Zhelnin G G, Sharapov S N, Rail damage and wheel wear. Permanent Railway and Track Facilities, 1997, No. 6. (in Russian).
34. Bufeev V A, Effect of spatial system of active forces on the friction process (supercoulomb and coulomb friction phenomenon). Friction and Wear, 1996, No. 1. (in Russian).
35. Troshchenko V T, Tsybanev G V, Khotyanovsky A O, Strength of Materials, 1988, No. 6. (in Russian).
36. Nosovsky I G, Tsybanev G V, Belas O N, Strength of Materials, 1990, No. 4. (in Russian).
37. Blagodarny V M, The strength rating for gears in terms of their wearing. Paper thesis at II int. symp. on tribo-fatigue, "SPAS"– Sci. Prod. Group "Tribofatigue", Moscow, Gomel, 1996. (in Russian).
38. Bugov Kh I, Egozhev A M, Mechanism of wear-fatigue fracture of a group bolted joint in the conditions of deformable slip. Paper thesis at II int. symp. on tribo-fatigue, "SPAS"– Sci. Prod. Group "Tribofatigue", Moscow, Gomel, 1996. (in Russian).
39. Makhutov N A, Firsov V T, Yastrebov S S, Grechushkin G M, Fretting-fatigue of press joints. Vestnik Mashinostroenya, 1991, No. 1. (in Russian).

40. Algin V B, Tribo-fatigue objects as a system in accord with their element behaviour life. Paper thesis at II int. symp. on tribo-fatigue, "SPAS"– Sci. Prod. Group "Tribofatigue", Moscow, Gomel, 1996. (in Russian).

41 Morozov L V, Statistical and physical prerequisites of using a new class of distribution in tribo-fatigue. Paper thesis at II int. symp. on tribo-fatigue, "SPAS"– Sci. Prod. Group "Tribofatigue", Moscow, Gomel, 1996. (in Russian).

42 Konovalov L V, Lunkova S M, Complex reliability of equipment due to different criteria of failure. Paper thesis at II int. symp. on tribo-fatigue, "SPAS"– Sci. Prod. Group "Tribofatigue", Moscow, Gomel, 1996. (in Russian).

Appendix IV
SOME STAGES OF PROGRESS AND PROSPECTS OF TRIBO-FATIGUE*

A V KUKHAREV

In any thing it is most significant to establish the natural origin.

Plato

1 Introduction

The term *tribo-fatigue* is defined in Interstate Standard GOST 30638–99 as a science of wear-fatigue damage and fracture of active systems of machines and equipment. Tribo-fatigue made an arduous way during 15 years after this term was first published to become acknowledged. I would like to tell briefly about the stages of its progress and about some events I was involved in.

2 Tribo-fatigue: 1995

In 1995 I was invited to a meeting to talk about tribo-fatigue. A condensed text was published in 1996 under the heading "Tribo-fatigue is and will be at the service of the people" [1] that is reproduced below

When the term *tribo-fatigue* became public in 1986 (it happened in Minsk) initially, it did not provoke any emotions. A stormy discussion took place in 1990 at the All-Union scientific conference in Gomel. Specialists from famous research centers of the Soviet Union were anxious to verify what kind of science tribo-fatigue was. Is it just a new scientific trend?

Most of the participants of the International Tribo-fatigue Symposium in 1993 (Gomel, Belarus) had already some first-hand knowledge of tribo-fatigue and for many it had become their domain of research. However, for some scientists and engineers the word "tribo-fatigue" sounded quite new. Thus, the discussion whether it is a science or not is still going on.

* Presintation at the 4th International Tribo-fatigue Symposium (2002, Ternopil, Ukraine).

I would like to say that it is an earnest problem and its solution is not a toy in the hands of theoreticians. In my view, it is a matter of primary importance for machine building, in the first place. To prove this statement I would like to refer to recent events and... battles.

Studies of the processes of friction and wear continue uninterrupted globally, including studies of the role of lubrication between contacting solid bodies. The leading role in these studies belongs to the so-called complex studies, i.e. tribological studies. All three processes are studied in their integrity, or as a complex. What do they yield?

Professor Jost (UK) noted specifically what science had to pay for the battle against the new tendencies. He writes that initially tribology was scorned because of its versatility. This scornful attitude inhibited directly the progress of research and development in machine building; heavy expenses to reduce friction and wear together with their consequences cost hugely... Only in Great Britain 5,5 million pounds (the assessment of 1965) could have been saved had tribology been regarded with more respect. Now it is believed that due respect of tribology could have saved from 1.3 to 1.6% of the General National Product. It is the scale of losses only within one country!

The fundamentals of tribo-fatigue have been developed in our country. It studies the regularities of wear-fatigue damage. It has been reported repeatedly that this damage causes 70, 80 and even 90% of premature failures of modern machinery and equipment. It is tribo-fatigue that discovers new effective ways of controlling and preventing damage. Just think! Engineers in the whole world (including us) have been trying to devise methods of combating wear aiming at the so-called wearless friction; specialists in tribo-fatigue have established that wear, on the contrary, is needed in many essential cases to increase the durability of an active system. This approach leads both to creating more reliable machines and to huge economic savings.

Instead of combating wear there is a different tendency of battling against tribo-fatigue. There are appeals to stop financing tribo-fatigue research, expel it from universities, exclude it from major subjects, to stop designing machines for wear-fatigue tests...

In this connection I would like to recall how Francis Bacon cherished science dearly. He wrote that if science by itself had not yielded any profits, it could not be called useless just because it would sharpen our mind and put it in order. It is the purpose of science in our time to serve people. That is the view of Leo Tolstoy. We cannot but share this view. And we have no doubt that tribo-fatigue is already and will be at the service of people...

...What is tribo-fatigue essentially?

Some tribologists claim that tribo-fatigue is a new scientific trend in tribology, and they are right in part. In part because only one side of the medal is taken into account, namely, the effect of cyclic loading on variations of friction and wear characteristics. Exactly this side of the medal corresponds to the long-lasting ideas and traditional scientific views of an "inveterate tribologist".

On the contrary, when some specialists in strength assert that tribo-fatigue is a new trend in the mechanics of fatigue fracture, I believe that they are also right in part. In part because they take into account only one side (naturally the reverse versus tribologists) of the medal, namely, the effect of friction and wear processes on variations of fatigue resistance characteristics. Exactly this reverse side corresponds to the long-lasting ideas and traditional scientific views of a "inveterate strength specialist".

If the cognitive psychological barrier separating narrow specialists is crossed, and both sides of the medal are viewed in their inseparable entity, then two new scientific trends have appeared at the junction between tribology and the mechanics of fatigue fracture, that have been combined dialectically and generated grounds for tribo-fatigue as a new science. As any other science, tribo-fatigue has its own object of study (active systems), its own methods of studies (wear-fatigue tests), its own models and criteria (complex indicators of wear-fatigue damage). While supported by tribology, the mechanics of fatigue fracture, reliability of mechanical systems, it does not minimize their significance as parents of tribo-fatigue and just proves the vitality and power of the new science. Remember that Isaac Newton said that we would see farther because we stand on the shoulders of the giants. Specialists in tribo-fatigue have already demonstrated that they see farther than others.

3 Essential stages in the progress of tribo-fatigue

Below is a brief enumeration of events that, in my view, have been essential for the progress of tribo-fatigue [2, 3].

September 29, 1984	The term tribo-fatigue was proposed (in a letter of L A Sosnovskiy to K V Frolov)
1986	The term tribo-fatigue was first mentioned in a publication (Minsk) [4]
1986/87	The first course of lectures on tribo-fatigue was delivered at the Byelorussian State Institute of Railway Engineers [5]
November 28, 1989	The first award for tribo-fatigue (L A Sosnovskiy was awarded a silver medal of the USSR Exhibition of Economic Achievements. "For development of methodological and theoretical principles of tribo-fatigue") [6]
September 5, 1990	The first All-Union round-table sitting devoted to "Problems of tribo-fatigue" of scientists and specialists from the CIS countries took place (Gomel, co-chaired by N A Makhutov and L A Sosnovskiy) [7]
1990	The first tribo-fatigue research program is published [8]
1992	A separate "Tribo-fatigue" research program is approved in the Republic of Belarus for the first time
August, 1992	Tribofatigue Ltd. is set up which was transformed into Tribofatigue Production & Research Group in 1993

September 14–17, 1993	The first International tribo-fatigue symposium is held in Gomel, Belarus [9]. The first press-conference (K V Frolov and L A Sosnovsky) on tribo-fatigue (September 14)
1994	The Tribofatigue Production Group creates a pilot multipurpose SI machine for wear-fatigue tests of materials and models of active systems, the first prototype of a new generation of testing equipment for tribo-fatigue research [10–12]
June, 1995	The Academies of Sciences of Russia, Belarus and Ukraine approve "Plan of International Tribo-fatigue Research" (see in [1])
September 30, 1995	The BelStandard Committee approves the first tribo-fatigue standard STB 994–95 "Tribo-fatigue. Terms and Definitions" [13]
December, 1995	The research Center of the Academy of Sciences of Belarus and the Gomel Fodder Harvester Production Group set up a tribo-fatigue laboratory
1996	The first annual publication "Tribo-fatigue" appears [14]; The first four-language dictionary of terms "Tribofatika, Trybofatyka, Tribo-fatigue, Triboermüdung" [15] appears; a collection of scientific essays "Some words about tribo-fatigue" written by 17 eminent researchers and scientists [1] appears
October 15–17, 1996	The second International Tribo-fatigue Symposium (Moscow, Russia) [16] is held
December 20, 1996	The International Coordinating Tribo-fatigue Board is set up (co-chairmen N A Makhutov, L A Sosnovskiy, V T Troshchenko and Gao Wanzhen since 1999)
May 14, 1998	The Extraordinary and Plenipotentiary Ambassador of the People's Republic of China in Belarus Madam U Siaco visits the Tribofatigue Production Group starting scientific cooperation in tribo-fatigue between Belarus and the Chinese People's Republic
1998	The course "Fundamentals of tribo-fatigue" is included into the curriculum of the Gomel State Polytechnic University
1999	The first Interstate Standard GOST 30638 "Tribo-fatigue. Terms and Definitions" is approved [17]. The monograph is published, the methodology of tribo-fatigue being used to analyse life as a special method of damage accumulation.
October 22–26, 2000	The III International Tribo-fatigue Symposium (Beijing, China) [19] is held
2001	Scientists of Russia, Belarus, Ukraine and China publish the first international monograph devoted to tribo-fatigue [20]
September 23–27, 2002	The IV International Tribo-fatigue Symposium (Ternopil, Ukraine) is held

4 Tribo-fatigue: 2000

At present the results have been obtained in the sphere of tribo-fatigue that the II International Symposium rated as most important. Below some achievements are listed:

- experimentally validated new methods and processes of wear-fatigue tests have been advanced;
- basic regularities of wear-fatigue damage (direct and back effects) have been experimentally studied);
- several theoretical problems have been formulated and solved, their summarization has led to formulation of the principles of the mechanics of wear-fatigue damage and fracture;
- a problem is being formulated and solved how to control wear-fatigue damage in newly designed active systems of machines and equipment;
- first books reviewing tribo-fatigue have been written and published;
- a number of standards of tribo-fatigue have been approved and introduced;
- several modification of SI machines for wear-fatigue tests of materials and models of active systems have been developed.

In October of 2000 scientists from many countries at the III International Tribo-fatigue Symposium (China, Beijing) evaluated the state of its progress during 15 years [19]. A summary monograph [20] written by specialists from Belarus, Russia, Ukraine and China is published. I will make a brief quotation from the foreword.

"Five of us participated at the III International Tribo-fatigue Symposium in Beijing (October 2000) and made our presentations there. The other five both made presentations and had been busy organizing two preceding symposia. Though our presentations seemed often as individual, our task was common: to contribute with our research to the progress of tribo-fatigue. We have integrated our results and we believe that we have obtained an entity that is now called tribo-fatigue.

It is rather hard to write a monograph when there are six authors and they are separated by huge distances, still it was a relatively enjoyable task: we were inspired by the problem that, in our view, has paramount significance for modern machine building ".

The tribo-fatigue bibliography during 1995–2000 [21] includes only the publications that basically relate to the research in Gomel and tribo-fatigue R&D accomplished in Belarus. The list contains over 200 scientific works authored by almost 70 scientists and engineers from over 50 institutions.

Thus, it can be asserted that tribo-fatigue is a new vigorously developing part of mechanics.

5 Some results and prospects

The prospects were outlined at the II International Tribo-fatigue Symposium (Moscow, 1996) [22].

(A) The progress of tribo-fatigue will be determined theoretically by new more profound insight into the basic regularities of wear-fatigue damage, the conditions governing the limiting state of active systems and the search for new principles and methods of predicting durability and preventing emergencies in operation of essential and intricate technical systems.

(B) The progress of tribo-fatigue practically leads to transition from designing individual units of machines and equipment to designing service life of active systems, so a complex of methods of controlling wear-fatigue damage of specific active systems is to be developed and introduced to reduce labour cost, to save means and materials in production and operation of modern machinery while strengthening its durability.

(C) Concerning improvement of testing equipment, the progress of tribo-fatigue leads to development and introduction of new methods and processes of wear-fatigue tests, including shortcut tests, and, therefore, to the development of a new class of testing equipment.

(D) Concerning development of standard and engineering base, the progress of tribo-fatigue leads to development and introduction of a complex of standards of methods of wear-fatigue tests in order to formulate and solve in future the problem of certification of active systems using the most essential criteria of their serviceability.

(E) Concerning training specialists, the progress of tribo-fatigue leads to the need of teaching the course of the "Fundamentals of tribo-fatigue" to students majoring in machine building (dynamics, strength and wear resistance of machines, instruments and equipment); it is time to train students and researchers in this sphere as well.

(F) Concerning research, the progress of tribo-fatigue leads to development and export of hitech products (new and hitech methods of tests, fundamentally new testing equipment, new standards).

This list of prospects in the sphere of tribo-fatigue needs just one explanation and addition.

The explanation is that in 2002 five Ph. D. theses and one doctorship thesis dealing with tribo-fatigue were prepared. The Belarusian State University of Transport is publishing the first manual "Fundamentals of Tribo-fatigue" [23] for students of higher technical schools; it is supplemented with a laboratory practical course [24] and an assignment for designing [25]. Hence the problem of scientific procedures and training qualified specialists in the new and promising domain of knowledge is being successfully solved.

Furthermore, the promising methods of tribo-fatigue have been recently applied to the analysis of biological objects and humans specifically [18]. Life is considered as a specific way of damage accumulation; principles of its quantitative assessment in dialectics have been developed. Tribo-fatigue thus becomes useful in the humanitarian sphere.

6 Conclusion

We have been eye witnessing the integration of individual sciences into a new, more common integral discipline. It is another example of the current day evolution of science characterized by the tendencies towards integration: from particular to general.

The IV International Tribo-fatigue Symposium in Ternopil (September, 2002) will probably decide when and where the V International Tribo-fatigue Symposium will take place. I would like to wish its participants a fruitful work at the next symposium.

Acknowledgments

I would like to express my sincere appreciation of discussions of the problems of tribo-fatigue research I had to many scientists, engineers and organizers of science, particularly the Russian Academician K V Frolov, the Corresponding Member of the Russian Academy of Science N A Makhutov, the Ukrainian Academician V T Troshchenko, the Belarusian Academician M S Vysotsky, Professor L A Sosnovskiy.

Bibliography

1 Vysotsky M C, Frolov K V, Troshchenko V T, and others. Ed. by A V Bogdanovich, Some words about tribo-fatigue. (Essays). Gomel, Minsk, Moscow, Kiev, "Remika", 1996. (in Russian).
2 Elovoy O M, First steps of tribo-fatigue. Gomel, 1996. (in Russian).
3 Kukharev A V, Some stages of progress of tribo-fatigue. Proc. of the III int. symp. on tribo-fatigue, Hunan University Press, Beijing, 2000.
4 Sosnovskiy L A, Complex assessment of the reliability of active systems. Minsk, 1986. (in Russian).
5 Sosnovskiy L A, Complex assessment of the reliability of active systems using the criteria of fatigue and wear resistance. ByelIIZhT, Gomel, 1988. (in Russian).
6 Sosnovskiy L A, Tribo-fatigue: problems and prospects. ByelIIZhT, Gomel, 1989. (in Russian).
7 Sosnovskiy L A, Problems of complex assessment of damage and limiting state of active systems. Basic terms. ByelIIZhT, Gomel, 1990. (in Russian).
8 Frolov K V, Sosnovskiy L A, Makhutov N A, Drozdov Yu N, Tribo-fatigue: new ideas in the promising direction. Gomel, 1990. (in Russian).

9 Friction. Wear. Fatigue. Ed. by L A Sosnovskiy, Proc. int. symp. on tribo-fatigue, Gomel, 1993. (in Russian).

10 Makhutov N A, Sosnovskiy L A, Bogdanovich A V, Andropov P V, Marchenko A V, Tyurin S A, Methods of wear-fatigue tests and their implementation with a SI testing machine. Zavodskaya Laboratoriya, 1995, No. 6. (in Russian).

11 Indman N L, Ozhigar G P, Sosnovskiy L A, Constructional features of SI machine. Zavodskaya Laboratoriya, 1995, No. 6. (in Russian).

12 Rozhdestvensky A Yu, Kovalev V V, Elovoy O M, Belits F Yu, Control and measuring system of SI machine. Zavodskaya Laboratoriya, 1995, No. 6. (in Russian).

13 СТБ 994–95. Tribo-fatigue. Terms and definitions. BelStandard, Minsk, 1995.

14 Tribo-fatigue-95. Annual edition. Ed. by M Vysotsky, Sci. Prod. Group "Tribofatigue", Gomel, 1996. (in Russian).

15 Sosnovskiy L A, Four-language dictionary of terms. Ed. by L A Sosnovskiy, Sci. Prod. Group "Tribofatigue", Minsk, Gomel, 1996. (in Russian).

16 Theses of papers of II Intern. Symp. on Tribo-Fatigue. Ed. by V A Andriyashin and others, "SPAS"– Sci. Prod. Group "Tribofatigue", Moscow, Gomel, 1996. (in Russian).

17 GOST 30638–99. Tribo-fatigue. Terms and definitions. (Interstate Standard). Interstate Standardization, metrology and certification board, Minsk, 1999.

18 Sosnovskiy L A, Tribo-fatigue: on dialectics of life. Sci. Prod. Group "Tribofatigue", Gomel, 1999. (in Russian).

19 Proc. of III int. symp. on tribo-fatigue. Ed. by Gao Wanzhen and Li Jian, Hunan University Press, Beijing, 2000.

20 Sosnovskiy L A, Troshchenko V T, Makhutov N A, Gao Wanzhen, Bogdanovich A V, Shcherbakov S S, Wear-fatigue damage and its prediction (tribo-fatigue). Gomel, Kiev, Moscow, Uhan, 2001. (in Russian).

21 Bibliography of tribo-fatigue publications. Ed. by T Eseva, S Tyurin, Int. tribo-fatigue coordinating board, Sc. Prod. Group "Tribofatigue", Gomel, 2001. (in Russian).

22 Resolutions of II int. symp. on tribo-fatigue. Sci. Prod. Group "Tribofatigue", Gomel, 1996. (in Russian).

23 Sosnovskiy L A, Fundamentals of tribo-fatigue. (Manual). Bel. University of Transport, (in press), Gomel, 2002. (in Russian).

24 Bogdanovich A V, and others, A laboratory practical course, part 1. Bel. University of Transport, (in press), Gomel, 1999. (in Russian).

25 Bogdanovich A V, Elovoy O M, Sosnovskiy L A, Assessment of reliability of simple crakshaft: textbook. Bel. University of Transport, Gomel, 2002. (in Russian).

Printing: Strauss GmbH, Mörlenbach
Binding: Schäffer, Grünstadt

GPSR Compliance

The European Union's (EU) General Product Safety Regulation (GPSR) is a set of rules that requires consumer products to be safe and our obligations to ensure this.

If you have any concerns about our products, you can contact us on ProductSafety@springernature.com

In case Publisher is established outside the EU, the EU authorized representative is:

Springer Nature Customer Service Center GmbH
Europaplatz 3
69115 Heidelberg, Germany

Batch number: 10584518

Printed by Printforce, the Netherlands